Computational Statistics Handbook with MATLAB®

Third Edition

Chapman & Hall/CRC
Computer Science and Data Analysis Series

The interface between the computer and statistical sciences is increasing, as each discipline seeks to harness the power and resources of the other. This series aims to foster the integration between the computer sciences and statistical, numerical, and probabilistic methods by publishing a broad range of reference works, textbooks, and handbooks.

SERIES EDITORS
David Blei, Princeton University
David Madigan, Rutgers University
Marina Meila, University of Washington
Fionn Murtagh, Royal Holloway, University of London

Proposals for the series should be sent directly to one of the series editors above, or submitted to:

Chapman & Hall/CRC
Taylor and Francis Group
3 Park Square, Milton Park
Abingdon, OX14 4RN, UK

Published Titles

Semisupervised Learning for Computational Linguistics
Steven Abney

Visualization and Verbalization of Data
Jörg Blasius and Michael Greenacre

Design and Modeling for Computer Experiments
Kai-Tai Fang, Runze Li, and Agus Sudjianto

Microarray Image Analysis: An Algorithmic Approach
Karl Fraser, Zidong Wang, and Xiaohui Liu

R Programming for Bioinformatics
Robert Gentleman

Exploratory Multivariate Analysis by Example Using R
François Husson, Sébastien Lê, and Jérôme Pagès

Bayesian Artificial Intelligence, Second Edition
Kevin B. Korb and Ann E. Nicholson

Computer Science and Data Analysis Series

Computational Statistics Handbook with MATLAB®
Third Edition

Wendy L. Martinez

Angel R. Martinez

CRC Press
Taylor & Francis Group
Boca Raton London New York

CRC Press is an imprint of the
Taylor & Francis Group, an **informa** business

A CHAPMAN & HALL BOOK

CRC Press
Taylor & Francis Group
6000 Broken Sound Parkway NW, Suite 300
Boca Raton, FL 33487-2742

First issued in paperback 2021

© 2016 by Taylor & Francis Group, LLC
CRC Press is an imprint of Taylor & Francis Group, an Informa business

No claim to original U.S. Government works

ISBN-13: 978-1-4665-9273-5 (hbk)
ISBN-13: 978-1-03-217958-2 (pbk)
DOI: 10.1201/b19035

Library of Congress Cataloging-in-Publication Data

Martinez, Wendy L.
   Computational statistics handbook with MATLAB / Wendy L. Martinez and Angel R. Martinez. -- Third edition.
      pages cm. -- (Chapman & Hall/CRC computer science and data analysis series)
   "A CRC title."
   Includes bibliographical references and indexes.
   ISBN 978-1-4665-9273-5 (alk. paper)
   1. Mathematical statistics--Data processing. 2. MATLAB. I. Martinez, Angel R. II. Title.

QA276.4.M272 2016
519.50285'53--dc22   2015033851

Visit the Taylor & Francis Web site at
http://www.taylorandfrancis.com

and the CRC Press Web site at
http://www.crcpress.com

To

Edward J. Wegman

Teacher, Mentor, and Friend

Table of Contents

Chapter 3
Sampling Concepts

Chapter 4
Generating Random Variables

Chapter 5
Exploratory Data Analysis

Chapter 6
Finding Structure

Chapter 7
Monte Carlo Methods for Inferential Statistics

Chapter 8
Data Partitioning

Chapter 9
Probability Density Estimation

Chapter 10
Supervised Learning

Chapter 11
Unsupervised Learning

Chapter 12
Parametric Models

Appendix D
Notation

Preface to the Third Edition

It has been almost ten years since the second edition of the *Computational Statistics Handbook with MATLAB®* was published, and MATLAB has evolved greatly in that time. There were also some relevant topics in computational statistics that we always wanted to include, such as support vector machines and multivariate adaptive regression splines. So, we felt it was time for a new edition.

To use all of the functions and code described in this book, one needs to have the MATLAB Statistics and Machine Learning Toolbox®, which is a product of The MathWorks, Inc. The name of this toolbox was changed to the longer title to reflect the connection with machine learning approaches, like supervised and unsupervised learning (Chapters 10 and 11). We will keep to the shorter name of "MATLAB Statistics Toolbox" in this text for readability.

We list below some of the major changes in the third edition.

- Chapter 10 has additional sections on support vector machines and nearest neighbor classifiers. We also added a brief description of naive Bayes classifiers.

- Chapter 12 was updated to include sections on stepwise regression, least absolute shrinkage and selection operator (lasso), ridge regression, elastic net, and partial least squares regression.

- Chapter 13 now has a section on multivariate adaptive regression splines.

- Spatial statistics is an area that uses many of the techniques covered in the text, but it is not considered part of computational statistics. So, we removed Chapter 15.

- The introduction to MATLAB given in Appendix A has been expanded and updated to reflect the new desktop environment, object-oriented programming, and more.

- The text has been updated for MATLAB R2015a and the corresponding version of the Statistics and Machine Learning Toolbox.

We retained the same philosophy and writing style used in the previous editions of the book. The theory is kept to a minimum and is included where it offers some insights to the data analyst. All MATLAB code, example files, and data sets are available for download at the CRC website for the book:

http://www.crcpress.com/product/ISBN/9781466592735

The latest version of the Computational Statistics Toolbox can be found at the CRC website and above at the following link:

http://www.pi-sigma.info/

We would like to acknowledge the invaluable help of the reviewers for the previous editions. Reviewers for this edition include Tom Lane, Terrance Savitsky, and Gints Jekabsons. We thank them for their insightful comments. We are especially indebted to Tom Lane (from The MathWorks, Inc.) for his vision and leadership, which we think were instrumental in making MATLAB a leading computing environment for data analysis and statistics. Finally, we are grateful for our editors at CRC Press (David Grubbs and Michele Dimont) and the MATLAB book program at The MathWorks, Inc.

Disclaimers

1.   Any MATLAB programs and data sets that are included with the book are provided in good faith. The authors, publishers, or distributors do not guarantee their accuracy and are not responsible for the consequences of their use.

2.   Some of the MATLAB functions provided with the Computational Statistics Toolbox were written by other researchers, and they retain the copyright. References are given in the **help** section of each function. Unless otherwise specified, the Computational Statistics Toolbox is provided under the GNU license specifications:

http://www.gnu.org/copyleft/gpl.html

3.   The views expressed in this book are those of the authors and do not necessarily represent the views of the United States government.

MATLAB® and SIMULINK® are registered trademarks of The MathWorks, Inc. For product information, please contact:

The MathWorks, Inc.
3 Apple Hill Drive
Natick, MA 01760-2098 USA
Tel: 508-647-7000
Fax: 508-647-7001
E-mail: info@mathworks.com
Web: www.mathworks.com

Wendy L. and Angel R. Martinez
November 2015

Preface to the Second Edition

We wrote a second edition of this book for two reasons. First, the Statistics Toolbox for MATLAB® has been significantly expanded since the first edition, so the text needed to be updated to reflect these changes. Second, we wanted to incorporate some of the suggested improvements that were made by several of the reviewers of the first edition. In our view, one of the most important issues that needed to be addressed is that several topics that should be included in a text on computational statistics were missing, so we added them to this edition.

We list below some of the major changes and additions of the second edition.

- Chapters 2 and 4 have been updated to include new functions for the multivariate normal and multivariate t distributions, which are now available in the Statistics Toolbox.

- Chapter 5 of the first edition was split into two chapters and updated with new material. Chapter 5 is still on exploratory data analysis, but it now has updated information on new MATLAB functionality for univariate and bivariate histograms, glyphs, parallel coordinate plots, and more.

- Topics that pertain to touring the data and finding structure can be found in Chapter 6. This includes new content on independent component analysis, nonlinear dimensionality reduction, and multidimensional scaling.

- Chapter 9 of the first edition was divided into two chapters (now Chapters 10 and 11), and new content was added. Chapter 10 of the new edition includes methods pertaining to supervised learning with new topics on linear classifiers, quadratic classifiers, and voting methods (e.g., bagging, boosting, and random forests).

- Methods for unsupervised learning or clustering have been moved to Chapter 11. This content has been expanded and updated to include model-based clustering and techniques for assessing the results of clustering.

- Chapter 12 on parametric models has been added. This has descriptions of spline regression models, logistic regression, and generalized linear models.

- Chapter 13 on nonparametric regression has been expanded. It now includes information on more smoothers, such as bin smoothing, running mean and line smoothers, and smoothing splines. We also describe additive models for nonparametric modeling when one has many predictors.

- The text has been updated for MATLAB R2007a and the Statistics Toolbox, V6.0.

We tried to keep to the same philosophy and style of writing that we had in the first book. The theory is kept to a minimum, and we provide references at the end of the text for those who want a more in-depth treatment. All MATLAB code, example files, and data sets are available for download at the CRC website and StatLib:

```
http://lib.stat.cmu.edu
http://www.crcpress.com/e_products/downloads/
```

We also have a website for the text, where up-to-date information will be posted. This includes the latest version of the Computational Statistics Toolbox, code fixes, links to useful websites, and more. The website can be found at

```
http://www.pi-sigma.info/
```

The first edition of the book was written using an older version of MATLAB. Most of the code in this text should work with earlier versions, but we have updated the text to include new functionality from the Statistics Toolbox, Version 6.0.

We would like to acknowledge the invaluable help of the reviewers for the second edition: Tom Lane, David Marchette, Carey Priebe, Jeffrey Solka, Barry Sherlock, Myron Katzoff, Pang Du, and Yuejiao Ma. Their many helpful comments made this book a much better product. Any shortcomings are the sole responsibility of the authors. We greatly appreciate the help and patience of those at CRC Press: Bob Stern, Jessica Vakili, Russ Heap, and James Yanchak. We are grateful to Hsuan-Tien Lin for letting us use his boosting code and Jonas Lundgren for his spline regression function. Finally, we are especially indebted to Tom Lane and those members of the MATLAB book program at The MathWorks, Inc. for their special assistance with MATLAB.

Disclaimers

1.   Any MATLAB programs and data sets that are included with the book are provided in good faith. The authors, publishers, or distributors do not guarantee their accuracy and are not responsible for the consequences of their use.

2. Some of the MATLAB functions provided with the Computational Statistics Toolbox were written by other researchers, and they retain the copyright. References are given in the **help** section of each function. Unless otherwise specified, the Computational Statistics Toolbox is provided under the GNU license specifications:

http://www.gnu.org/copyleft/gpl.html

3. The views expressed in this book are those of the authors and do not necessarily represent the views of the United States Department of Defense or its components.

We hope that readers will find this book and the accompanying code useful in their educational and professional endeavors.

Wendy L. and Angel R. Martinez
November 2007

Preface to the First Edition

Computational statistics is a fascinating and relatively new field within statistics. While much of classical statistics relies on parameterized functions and related assumptions, the computational statistics approach is to let the data tell the story. The advent of computers with their number-crunching capability, as well as their power to show on the screen two- and three-dimensional structures, has made computational statistics available for any data analyst to use.

Computational statistics has a lot to offer the researcher faced with a file full of numbers. The methods of computational statistics can provide assistance ranging from preliminary exploratory data analysis to sophisticated probability density estimation techniques, Monte Carlo methods, and powerful multi-dimensional visualization. All of this power and novel ways of looking at data are accessible to researchers in their daily data analysis tasks. One purpose of this book is to facilitate the exploration of these methods and approaches and to provide the tools to make of this, not just a theoretical exploration, but a practical one. The two main goals of this book are

- To make computational statistics techniques available to a wide range of users, including engineers and scientists, and
- To promote the use of MATLAB® by statisticians and other data analysts.

We note that MATLAB and Handle Graphics® are registered trademarks of The MathWorks, Inc.

There are wonderful books that cover many of the techniques in computational statistics and, in the course of this book, references will be made to many of them. However, there are very few books that have endeavored to forgo the theoretical underpinnings to present the methods and techniques in a manner immediately usable to the practitioner. The approach we take in this book is to make computational statistics accessible to a wide range of users and to provide an understanding of statistics from a computational point of view via methods applied to real applications.

This book is intended for researchers in engineering, statistics, psychology, biostatistics, data mining, and any other discipline that must deal with the analysis of raw data. Students at the senior undergraduate level or beginning graduate level in statistics or engineering can use the book to supplement course material. Exercises are included with each chapter, making it suitable as a textbook for a course in computational statistics and data analysis. Scientists who would like to know more about programming methods for analyzing data in MATLAB should also find it useful.

We assume that the reader has the following background:

- Calculus: Since this book is computational in nature, the reader needs only a rudimentary knowledge of calculus. Knowing the definition of a derivative and an integral is all that is required.

- Linear Algebra: Since MATLAB is an array-based computing language, we cast several of the algorithms in terms of matrix algebra. The reader should have a familiarity with the notation of linear algebra, array multiplication, inverses, determinants, an array transpose, etc.

- Probability and Statistics: We assume that the reader has had introductory probability and statistics courses. However, we provide a brief overview of the relevant topics for those who might need a refresher.

We list below some of the major features of the book.

- The focus is on implementation rather than theory, helping the reader understand the concepts without being burdened by the theory.

- References that explain the theory are provided at the end of each chapter. Thus, those readers who need the theoretical underpinnings will know where to find the information.

- Detailed step-by-step algorithms are provided to facilitate implementation in any computer programming language or appropriate software. This makes the book appropriate for computer users who do not know MATLAB.

- MATLAB code in the form of a Computational Statistics Toolbox is provided. These functions are available for download.

- Exercises are given at the end of each chapter. The reader is encouraged to go through these because concepts are sometimes explored further in them. Exercises are computational in nature, which is in keeping with the philosophy of the book.

- Many data sets are included with the book, so the reader can apply the methods to real problems and verify the results shown in the book. The data are provided in MATLAB binary files (`.mat`) as well as text, for those who want to use them with other software.

- Typing in all of the commands in the examples can be frustrating. So, MATLAB scripts containing the commands used in the examples are also available for download.

- A brief introduction to MATLAB is provided in Appendix A. Most of the constructs and syntax that are needed to understand the programming contained in the book are explained.
- Where appropriate, we provide references to Internet resources for computer code implementing the algorithms described in the chapter. These include code for MATLAB, S-plus, Fortran, etc.

We would like to acknowledge the invaluable help of the reviewers: Noel Cressie, James Gentle, Thomas Holland, Tom Lane, David Marchette, Christian Posse, Carey Priebe, Adrian Raftery, David Scott, Jeffrey Solka, and Clifton Sutton. Their many helpful comments made this book a much better product. Any shortcomings are the sole responsibility of the authors. We owe a special thanks to Jeffrey Solka for some programming assistance with finite mixtures. We greatly appreciate the help and patience of those at CRC Press: Bob Stern, Joanne Blake, and Evelyn Meany. We also thank Harris Quesnell and James Yanchak for their help with resolving font problems. Finally, we are indebted to Naomi Fernandes and Tom Lane at The MathWorks, Inc. for their special assistance with MATLAB.

Disclaimers

1. Any MATLAB programs and data sets that are included with the book are provided in good faith. The authors, publishers, or distributors do not guarantee their accuracy and are not responsible for the consequences of their use.
2. The views expressed in this book are those of the authors and do not necessarily represent the views of the Department of Defense or its components.

Wendy L. and Angel R. Martinez
August 2001

Chapter 1

Introduction

1.1 What Is Computational Statistics?

Obviously, computational statistics relates to the traditional discipline of statistics. So, before we define computational statistics proper, we need to get a handle on what we mean by the field of statistics. At a most basic level, statistics is concerned with the transformation of raw data into knowledge [Wegman, 1988].

When faced with an application requiring the analysis of raw data, any scientist must address questions such as:

- What data should be collected to answer the questions in the analysis?
- How many data points should we obtain?
- What conclusions can be drawn from the data?
- How far can those conclusions be trusted?

Statistics is concerned with the science of uncertainty and can help the scientist deal with these questions. Many classical methods (regression, hypothesis testing, parameter estimation, confidence intervals, etc.) of statistics developed over the last century are familiar to scientists and are widely used in many disciplines [Efron and Tibshirani, 1991].

Now, what do we mean by computational statistics? Here we again follow the definition given in Wegman [1988]. Wegman defines *computational statistics* as a collection of techniques that have a strong "focus on the exploitation of computing in the creation of new statistical methodology."

Many of these methodologies became feasible after the development of inexpensive computing hardware since the 1980s. This computing revolution has enabled scientists and engineers to store and process massive amounts of data. However, these data are typically collected without a clear idea of what they will be used for in a study. For instance, in the practice of data analysis today, we often collect the data and then we design a study to gain some useful information from them. In contrast, the traditional approach has been to first design the study based on research questions and then collect the required data.

Because the storage and collection is so cheap, the data sets that analysts must deal with today tend to be very large and high-dimensional. It is in situations like these where many of the classical methods in statistics are inadequate. As examples of computational statistics methods, Wegman [1988] includes parallel coordinates for visualizing high dimensional data, nonparametric functional inference, and data set mapping, where the analysis techniques are considered fixed.

Efron and Tibshirani [1991] refer to what we call computational statistics as *computer-intensive statistical methods*. They give the following as examples for these types of techniques: bootstrap methods, nonparametric regression, generalized additive models, and classification and regression trees. They note that these methods differ from the classical methods in statistics because they substitute computer algorithms for the more traditional mathematical method of obtaining an answer. An important aspect of computational statistics is that the methods free the analyst from choosing methods mainly because of their mathematical tractability.

Volume 9 of the *Handbook of Statistics: Computational Statistics* [Rao, 1993] covers topics that illustrate the "... trend in modern statistics of basic methodology supported by the state-of-the-art computational and graphical facilities...." It includes chapters on computing, density estimation, Gibbs sampling, the bootstrap, the jackknife, nonparametric function estimation, statistical visualization, and others.

Gentle [2005] also follows the definition of Wegman [1988] where he states that computational statistics is a discipline that includes a "... class of statistical methods characterized by computational intensity...". His book includes Monte Carlo methods for inference, cross-validation and jackknife methods, data transformations to find structure, visualization, probability density estimation, and pattern recognition.

We mention the topics that can be considered part of computational statistics to help the reader understand the difference between these and the more traditional methods of statistics. Table 1.1 [Wegman, 1988] gives an excellent comparison of the two areas.

TABLE 1.1

Comparison Between Traditional Statistics and Computational Statistics [Wegman, 1988]

Traditional Statistics	Computational Statistics
Small to moderate sample size	Large to very large sample size
Independent, identically distributed data sets	Nonhomogeneous data sets
One or low dimensional	High dimensional
Manually computational	Computationally intensive
Mathematically tractable	Numerically tractable
Well focused questions	Imprecise questions
Strong unverifiable assumptions: Relationships (linearity, additivity) Error structures (normality)	Weak or no assumptions: Relationships (nonlinearity) Error structures (distribution free)
Statistical inference	Structural inference
Predominantly closed form algorithms	Iterative algorithms possible
Statistical optimality	Statistical robustness

Reprinted with permission from the *Journal of the Washington Academy of Sciences*

1.2 An Overview of the Book

Philosophy

The focus of this book is on methods of computational statistics and how to implement them. We leave out much of the theory, so the reader can concentrate on how the techniques may be applied. In many texts and journal articles, the theory obscures implementation issues, contributing to a loss of interest on the part of those needing to apply the theory. The reader should not misunderstand, though; the methods presented in this book are built on solid mathematical foundations. Therefore, at the end of each chapter, we include a section containing references that explain the theoretical concepts associated with the methods covered in that chapter.

What Is Covered

In this book, we cover some of the most commonly used techniques in computational statistics. While we cannot include all methods that might be a part of computational statistics, we try to present those that have been in use for several years.

Since the focus of this book is on the implementation of the methods, we include step-by-step descriptions of the procedures. We also provide examples that illustrate the use of the methods in data analysis. It is our hope that seeing how the techniques are implemented will help the reader understand the concepts and facilitate their use in data analysis.

Some background information is given in Chapters 2, 3, and 4 for those who might need a refresher in probability and statistics. In Chapter 2, we discuss some of the general concepts of probability theory, focusing on how they will be used in later chapters of the book. Chapter 3 covers some of the basic ideas of statistics and sampling distributions. Since many of the approaches in computational statistics are concerned with estimating distributions via simulation, this chapter is fundamental to the rest of the book. For the same reason, we present some techniques for generating random variables in Chapter 4.

Some of the methods in computational statistics enable the researcher to explore the data before other analyses are performed. These techniques are especially important with high dimensional data sets or when the questions to be answered using the data are not well focused. In Chapters 5 and 6, we present some graphical exploratory data analysis techniques that could fall into the category of traditional statistics (e.g., box plots, scatterplots). We include them in this text so statisticians can see how to implement them in MATLAB® and to educate scientists and engineers as to their usage in exploratory data analysis. Other graphical methods in this book *do* fall into the category of computational statistics. Among these are isosurfaces, parallel coordinates, the grand tour, and projection pursuit.

In Chapters 7 and 8, we present methods that come under the general heading of resampling. We first cover some of the main concepts in hypothesis testing and confidence intervals to help the reader better understand what follows. We then provide procedures for hypothesis testing using simulation, including a discussion on evaluating the performance of hypothesis tests. This is followed by the bootstrap method, where the data set is used as an estimate of the population and subsequent sampling is done from the sample. We show how to get bootstrap estimates of standard error, bias, and confidence intervals. Chapter 8 continues with two closely related methods called the jackknife and cross-validation.

One of the important applications of computational statistics is the estimation of probability density functions. Chapter 9 covers this topic, with an emphasis on the nonparametric approach. We show how to obtain estimates using probability density histograms, frequency polygons, averaged shifted histograms, kernel density estimates, finite mixtures, and adaptive mixtures.

Chapters 10 and 11 describe statistical pattern recognition methods for supervised and unsupervised learning. For supervised learning, we discuss Bayes decision theory, classification trees, and ensemble classifier methods. We present several unsupervised learning methods, such as hierarchical clustering, k-means clustering, and model-based clustering. In addition, we cover the issue of assessing the results of our clustering, including how one can estimate the number of groups represented by the data.

In Chapters 12 and 13, we describe methods for estimating the relationship between a set of predictors and a response variable. We cover parametric methods, such as linear regression, spline regression, and logistic regression. This is followed by generalized linear models and model selection methods. Chapter 13 includes several nonparametric methods for understanding the relationship between variables. First, we present several smoothing methods that are building blocks for additive models. For example, we discuss local polynomial regression, kernel methods, and smoothing splines. What we have just listed are methods for one predictor variable. Of course, this is rather restrictive, so we conclude the chapter with a description of regression trees, additive models, and multivariate adaptive regression splines.

An approach for simulating a distribution that has become widely used over the last several years is called Markov chain Monte Carlo. Chapter 14 covers this important topic and shows how it can be used to simulate a posterior distribution. Once we have the posterior distribution, we can use it to estimate statistics of interest (means, variances, etc.).

We also provide several appendices to aid the reader. Appendix A contains a brief introduction to MATLAB, which should help readers understand the code in the examples and exercises. Appendix B has some information on indexes for projection pursuit. In Appendix C, we include a brief description of the data sets that are mentioned in the book. Finally, we present a brief overview of notation that we use in Appendix D.

A Word About Notation

The explanation of the methods in computational statistics (and the understanding of them!) depends a lot on notation. In most instances, we follow the notation that is used in the literature for the corresponding method. Rather than try to have unique symbols throughout the book, we think it is more important to be faithful to the convention to facilitate understanding of the theory and to make it easier for readers to make the connection between the theory and the text. Because of this, the same symbols might be used in several places to denote different entities or different symbols could be used for the same thing depending on the topic. However, the meaning of the notation should be clear from the context.

In general, we *try* to stay with the convention that random variables are capital letters, whereas small letters refer to realizations of random variables. For example, X is a random variable, and x is an observed value of that random variable. When we use the term *log*, we are referring to the natural logarithm.

A symbol that is in bold refers to an array. Arrays can be row vectors, column vectors, or matrices. Typically, a matrix is represented by a bold capital letter such as **B**, while a vector is denoted by a bold lowercase letter such as **b**. Sometimes, arrays are shown with Greek symbols. For the most part, these will be shown in bold font, but we do not *always* follow this convention. Again, it should be clear from the context that the notation denotes an array.

When we are using explicit matrix notation, then we specify the dimensions of the arrays. Otherwise, we do not hold to the convention that a vector always has to be in a column format. For example, we might represent a vector of observed random variables as (x_1, x_2, x_3) or a vector of parameters as (μ, σ).

Our observed data sets will always be arranged in a matrix of dimension $n \times d$, which is denoted as X. Here n represents the number of observations we have in our sample, and d is the number of variables or dimensions. Thus, each row corresponds to a d-dimensional observation or data point. The ij-th element of X will be represented by x_{ij}. Usually, the subscript i refers to a row in a matrix or an observation, and a subscript j references a column in a matrix or a variable.

For the most part, examples are included after we explain the procedures, which include MATLAB code as we describe next. We indicate the end of an example by using a small box (❑), so the reader knows when the narrative resumes.

1.3 MATLAB® Code

Along with the explanation of the procedures, we include MATLAB commands to show how they are implemented. To make the book more readable, we will indent MATLAB code when we have several lines of code, and this can always be typed in as you see it in the book. Since all examples are available for download, you could also copy and paste the code into the MATLAB command window and execute them.

Any MATLAB commands, functions, or data sets are in courier bold font. For example, **plot** denotes the MATLAB plotting function. We note that due to typesetting considerations, we often have to continue a MATLAB command using the continuation punctuation (**. . .**). See Appendix A for more information on how this punctuation is used in MATLAB.

Since this is a book about computational statistics, we assume the reader has the MATLAB Statistics and Machine Learning Toolbox. In the rest of the book, we will refer to this toolbox with its shortened name—the Statistics Toolbox. We note in the text what functions are part of the main MATLAB software package and which functions are available only in the Statistics Toolbox.

We try to include information on MATLAB functions that are relevant to the topics covered in this text. However, this book is about the methods of computational statistics and is not meant to be a user's guide for MATLAB. Therefore, we do not claim to include all relevant MATLAB capabilities. Please see the documentation for more information on statistics in MATLAB and current functionality in the Statistics Toolbox. We also recommend the text by Martinez and Cho [2014] for those who would like a short user's guide with a focus on statistics.

The choice of MATLAB for implementation of the methods in this text is due to the following reasons:

- The commands, functions, and arguments in MATLAB are not cryptic. It is important to have a programming language that is easy to understand and intuitive, since we include the programs to help teach the concepts.
- It is used extensively by scientists and engineers.
- Student versions are available.
- It is easy to write programs in MATLAB.
- The source code or M-files can be viewed, so users can learn about the algorithms and their implementation.
- User-written MATLAB programs are freely available.
- The graphics capabilities are excellent.

It is important to note that the MATLAB code given in the body of the book is for *learning purposes*. In many cases, it is not the most efficient way to program the algorithm. One of the purposes of including the MATLAB code is to help the reader understand the algorithms, especially how to implement them. So, we try to have the code match the procedures and to stay away from cryptic programming constructs. For example, we use **for** loops at times (when unnecessary!) to match the procedure. We make no claims that our code is the best way or the only way to program the algorithms.

When presenting the syntax for a MATLAB function we usually just give the basic command and usage. Most MATLAB functions have a lot more capabilities, so we urge the reader to look at the documentation. Another very useful and quick way to find out more is to type **help *function_name*** at the command line. This will return information such as alternative syntax, definitions of the input and output variables, and examples.

In some situations, we do not include all of the code in the text. These are cases where the MATLAB program does not provide insights about the algorithms. Including these in the body of the text would distract the reader from the important concepts being presented. However, the reader can always consult the M-files for the functions, if more information is needed.

Computational Statistics Toolbox

Some of the methods covered in this book are not available in MATLAB. So, we provide functions that implement most of the procedures that are given in the text. Note that these functions are a little different from the MATLAB code provided in the examples. In most cases, the functions allow the user to implement the algorithms for the general case. A list of the functions and their purpose is given at the end of each chapter.

The MATLAB functions for the book are in the Computational Statistics Toolbox. To make it easier to recognize these functions, we put the letters **cs** in front of *most* of the functions. We included several new functions with the second edition, some of which were written by others. We did not change these functions to make them consistent with the naming convention of the toolbox. The latest toolbox can be downloaded from

    http://www.pi-sigma.info/

    http://www.crcpress.com/product/ISBN/9781466592735

Information on installing the toolbox is given in the **readme** file.

Internet Resources

One of the many strong points about MATLAB is the availability of functions written by users, most of which are freely available on the Internet. With each chapter, we provide information about Internet resources for MATLAB programs (and other languages) that pertain to the techniques covered in the chapter.

The following are some Internet sources for MATLAB code. Note that these are not necessarily specific to statistics, but are for all areas of science and engineering.

- The main website at The MathWorks, Inc. has downloadable code written by MATLAB users. The website for contributed M-files and other useful information is called MATLAB Central. The link below will take you to the website where you can find MATLAB code, links to websites, news groups, webinar information, blogs, and a lot more.

http://www.mathworks.com/matlabcentral/

- A good website for user-contributed statistics programs is StatLib at Carnegie Mellon University. They have a section containing MATLAB code. The home page for StatLib is

http://lib.stat.cmu.edu

1.4 Further Reading

To gain more insight on what is computational statistics, we refer the reader to the seminal paper by Wegman [1988]. Wegman discusses many of the differences between traditional and computational statistics. He also includes a discussion on what a graduate curriculum in computational statistics should consist of and contrasts this with the more traditional course work. A later paper by Efron and Tibshirani [1991] presents a summary of the new focus in statistical data analysis that came about with the advent of the computer age. Other papers in this area include Hoaglin and Andrews [1975] and Efron [1979]. Hoaglin and Andrews discuss the connection between computing and statistical theory and the importance of properly reporting the results from simulation experiments. Efron's article presents a survey of computational statistics techniques (the jackknife, the bootstrap, error estimation in discriminant analysis, nonparametric methods, and more) for an audience with a mathematics background, but little knowledge of statistics. Chambers [1999] looks at the concepts underlying computing with data, including the challenges this presents and new directions for the future.

There are very few general books in the area of computational statistics. One is a compendium of articles edited by C. R. Rao [1993]. This is a fairly comprehensive summary of many topics pertaining to computational statistics. The texts by Gentle [2005; 2009] are excellent resources for the student or researcher. The edited volume by Gentle, Härdle, and Mori [2004] is a wonderful resource with up-to-date articles on statistical computing, statistical methodology, and applications. The book edited by Raftery, Tanner, and Wells [2002] is another source for articles on many of the topics covered in this text, such as nonparametric regression, the bootstrap, Gibbs sampling, dimensionality reduction, and many others.

For those who need a resource for learning MATLAB, we recommend a book by Hanselman and Littlefield [2011]. This gives a comprehensive overview of MATLAB, and it has information about the many capabilities of MATLAB, including how to write programs, graphics and GUIs, and much more. Martinez and Cho [2014] published a primer or short user's guide on using MATLAB for statistics.

The documentation for the Statistics Toolbox and base MATLAB is also a very good resource for learning about many of the approaches discussed in this book. See Appendix A for information about accessing these documents.

Chapter 2

Probability Concepts

2.1 Introduction

A review of probability is covered here at the outset because it provides the foundation for what is to follow: computational statistics. Readers who understand probability concepts may safely skip over this chapter.

Probability is the mechanism by which we can manage the uncertainty underlying all real world data and phenomena. It enables us to gauge our degree of belief and to quantify the lack of certitude that is inherent in the process that generates the data we are analyzing. For example:

- To understand and use statistical hypothesis testing, one needs knowledge of the sampling distribution of the test statistic.
- To evaluate the performance (e.g., standard error, bias, etc.) of an estimate, we must know its sampling distribution.
- To adequately simulate a real system, one needs to understand the probability distributions that correctly model the underlying processes.
- To build classifiers to predict what group an object belongs to based on a set of features, one can estimate the probability density function that describes the individual classes.

In this chapter, we provide a brief overview of probability concepts and distributions as they pertain to computational statistics. In Section 2.2, we define probability and discuss some of its properties. In Section 2.3, we cover conditional probability, independence, and Bayes' theorem. Expectations are defined in Section 2.4, and common distributions and their uses in modeling physical phenomena are discussed in Section 2.5. In Section 2.6, we summarize some MATLAB® functions that implement the ideas from Chapter 2. Finally, in Section 2.7 we provide additional resources for the reader who requires a more theoretical treatment of probability.

2.2 Probability

Background

A *random experiment* is defined as a process or action whose outcome cannot be predicted with certainty and would likely change when the experiment is repeated. The variability in the outcomes might arise from many sources: slight errors in measurements, choosing different objects for testing, etc. The ability to model and analyze the outcomes from experiments is at the heart of statistics. Some examples of random experiments that arise in different disciplines are given next.

- Engineering: Data are collected on the number of failures of piston rings in the legs of steam-driven compressors. Engineers would be interested in determining the probability of piston failure in each leg and whether the failure varies among the compressors [Hand, et al.; Davies and Goldsmith, 1972].

- Medicine: The oral glucose tolerance test is a diagnostic tool for early diabetes mellitus. The results of the test are subject to variation because of different rates at which people absorb the glucose, and the variation is particularly noticeable in pregnant women. Scientists would be interested in analyzing and modeling the variation of glucose before and after pregnancy [Andrews and Herzberg, 1985].

- Manufacturing: Manufacturers of cement are interested in the tensile strength of their product. The strength depends on many factors, one of which is the length of time the cement is dried. An experiment is conducted where different batches of cement are tested for tensile strength after different drying times. Engineers would like to determine the relationship between drying time and tensile strength of the cement [Hand, et al., 1994; Hald, 1952].

- Software Engineering: Engineers measure the failure times in CPU seconds of a command and control software system. These data are used to obtain models to predict the reliability of the software system [Hand, et al., 1994; Musa, et al., 1987].

The *sample space* is the set of all outcomes from an experiment. It is possible sometimes to list all outcomes in the sample space. This is especially true in the case of some discrete random variables. Examples of these sample spaces are listed next.

- When observing piston ring failures, the sample space is $\{1, 0\}$, where 1 represents a failure and 0 represents a non-failure.
- If we roll a six-sided die and count the number of dots on the face, then the sample space is $\{1, 2, 3, 4, 5, 6\}$.

The outcomes from random experiments are often represented by an uppercase variable such as X. This is called a *random variable*, and its value is subject to the uncertainty intrinsic to the experiment. Formally, a random variable is a real-valued function defined on the sample space. As we see in the remainder of the text, a random variable can take on different values according to a probability distribution. Using our examples of experiments from above, a random variable X might represent the failure time of a software system or the glucose level of a patient. The observed value of a random variable X is denoted by a lowercase x. For instance, a random variable X might represent the number of failures of piston rings in a compressor, and $x = 5$ would indicate we observed 5 piston ring failures.

Random variables can be discrete or continuous. A *discrete random variable* can take on values from a finite or countably infinite set of numbers. Examples of discrete random variables are the number of defective parts or the number of typographical errors on a page. A *continuous random variable* is one that can take on values from an interval of real numbers. Examples of continuous random variables are the inter-arrival times of planes at a runway, the average weight of tablets in a pharmaceutical production line, or the average voltage of a power plant at different times.

We cannot list all outcomes from an experiment when we observe a continuous random variable because there are an infinite number of possibilities. However, we could specify the interval of values that X can take on. For example, if the random variable X represents the tensile strength of cement, then the sample space might be $(0, \infty)$ kg/cm^2.

An *event* is a subset of outcomes in the sample space. An event might be that a piston ring is defective or that the tensile strength of cement is in the range 40 to 50 kg/cm^2. The probability of an event is usually expressed using the random variable notation illustrated next.

- Discrete Random Variables: Letting 1 represent a defective piston ring and letting 0 represent a good piston ring, then the probability of the event that a piston ring is defective would be written as

$$P(X = 1).$$

- Continuous Random Variables: Let X denote the tensile strength of cement. The probability that an observed tensile strength is in the range 40 to 50 kg/cm^2 is expressed as

$$P(40 \text{ kg/cm}^2 \le X \le 50 \text{ kg/cm}^2).$$

Some events have a special property when they are considered together. Two events that cannot occur simultaneously or jointly are called *mutually exclusive events*. This means that the intersection of the two events is the empty set and the probability of the events occurring together is zero. For example, a piston ring cannot be both defective and good at the same time. So, the event of getting a defective part and the event of getting a good part are mutually exclusive events. The definition of mutually exclusive events can be extended to any number of events by considering all pairs of events. Every pair of events must be mutually exclusive for all of them to be mutually exclusive.

Probability

Probability is a measure of the likelihood that some event will occur. It is also a way to quantify or to gauge the likelihood that an observed measurement or random variable will take on values within some set or range of values. Probabilities always range between 0 and 1. A *probability distribution* of a random variable describes the probabilities associated with each possible value for the random variable.

We first briefly describe two somewhat classical methods for assigning probabilities: the *equal likelihood model* and the *relative frequency method*. When we have an experiment where each of n outcomes is equally likely, then we assign a probability mass of $1/n$ to each outcome. This is the equal likelihood model. Some experiments where this model can be used are flipping a fair coin, tossing an unloaded die, or randomly selecting a card from a deck of cards.

With the relative frequency method, we conduct the experiment n times and record the outcome. The probability of event E is then assigned by $P(E) = f/n$, where f denotes the number of experimental outcomes that satisfy event E.

Another way to find the desired probability that an event occurs is to use a *probability density function* when we have continuous random variables or a *probability mass function* in the case of discrete random variables. Section 2.5 contains several examples of probability density (mass) functions. In this text, $f(x)$ is typically used to represent the probability mass or density function for either discrete or continuous random variables, respectively. We now discuss how to find probabilities using these functions, first for the continuous case and then for discrete random variables.

To find the probability that a continuous random variable falls in a particular interval of real numbers, we have to calculate the appropriate area under the curve of $f(x)$. Thus, we have to evaluate the integral of $f(x)$ over the interval of random variables corresponding to the event of interest. This is represented by

$$P(a \le X \le b) = \int_a^b f(x)dx. \tag{2.1}$$

The area under the curve of $f(x)$ between a and b represents the probability that an observed value of the random variable X will assume a value between a and b. This concept is illustrated in Figure 2.1 where the shaded area represents the desired probability.

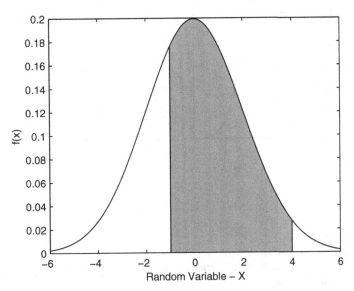

FIGURE 2.1
The area under the curve of f(x) between -1 and 4 is the same as the probability that an observed value of the random variable will assume a value in the same interval.

It should be noted that a valid probability density function should be non-negative, and the total area under the curve must equal 1. If this is not the case, then the probabilities will not be properly restricted to the interval [0, 1]. This will be an important consideration in Chapter 9 when we discuss probability density estimation techniques.

The *cumulative distribution function* $F(x)$ is defined as the probability that the random variable X assumes a value less than or equal to a given x. This is calculated from the probability density function, as follows:

$$F(x) = P(X \le x) = \int_{-\infty}^x f(t)dt. \tag{2.2}$$

It is obvious from Equation 2.2 that the cumulative distribution function takes on values between 0 and 1, so $0 \le F(x) \le 1$. A probability density function, along with its associated cumulative distribution function, are illustrated in Figure 2.2.

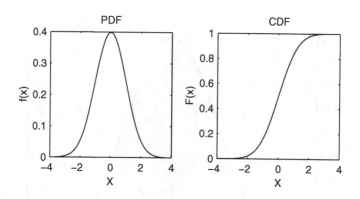

FIGURE 2.2
This shows the probability density function on the left with the associated cumulative distribution function on the right. Notice that the cumulative distribution function takes on values between 0 and 1.

For a discrete random variable X, that can take on values $x_1, x_2, \ldots$, the probability mass function is given by

$$f(x_i) = P(X = x_i); \qquad i = 1, 2, \ldots, \tag{2.3}$$

and the cumulative distribution function is

$$F(a) = \sum_{x_i \le a} f(x_i); \qquad i = 1, 2, \ldots. \tag{2.4}$$

Axioms of Probability

Probabilities follow certain axioms that can be useful in computational statistics. We let S represent the sample space of an experiment and E represent some event that is a subset of S.

AXIOM 1
The probability of event E must be between 0 and 1:

$$0 \leq P(E) \leq 1.$$

AXIOM 2

$$P(S) = 1.$$

AXIOM 3
For mutually exclusive events, $E_1, E_2, ..., E_k$,

$$P(E_1 \cup E_2 \cup ... \cup E_k) = \sum_{i=1}^{k} P(E_i).$$

Axiom 1 has been discussed before and simply states that a probability must be between 0 and 1. Axiom 2 says that an outcome from our experiment must occur. Axiom 3 enables us to calculate the probability that at least one of the mutually exclusive events $E_1, E_2, ..., E_k$ occurs by summing the individual probabilities.

2.3 Conditional Probability and Independence

Conditional Probability

Conditional probability is an important concept. It is used to define independent events and enables us to revise our degree of belief given that another event has occurred. Conditional probability arises in situations where we need to calculate a probability based on some partial information concerning the experiment, and we will see that it plays a vital role in supervised learning applications.

The *conditional probability* of event E given event F is defined as follows:

CONDITIONAL PROBABILITY

$$P(E|F) = \frac{P(E \cap F)}{P(F)}; \quad P(F) > 0. \tag{2.5}$$

Here $P(E \cap F)$ represents the **joint probability** that both E and F occur together, and $P(F)$ is the probability that event F occurs. We can rearrange Equation 2.5 to get the following rule:

MULTIPLICATION RULE

$$P(E \cap F) = P(F)P(E|F). \tag{2.6}$$

Independence

Often we can assume that the occurrence of one event does not affect whether or not some other event happens. For example, say a couple would like to have two children, and their first child is a boy. The gender of their second child does not depend on the gender of the first child. The fact that we know they have a boy already does not change the probability that the second child is a boy. Similarly, we can sometimes assume that the value we observe for a random variable is not affected by the observed value of other random variables.

These types of events and random variables are called **independent**. If *events* are independent, then knowing that one event has occurred does not change our degree of belief or the likelihood that the other event occurs. If *random variables* are independent, then the observed value of one random variable does not affect the observed value of another.

In general, the conditional probability $P(E|F)$ is not equal to $P(E)$. In these cases, the events are called **dependent**. Sometimes, we can assume independence based on the situation or the experiment, which was the case with our example above. However, to show independence mathematically, we must use the following definition.

INDEPENDENT EVENTS
Two events E and F are said to be independent if and only if any of the following are true:

$$P(E \cap F) = P(E)P(F),$$
$$P(E) = P(E|F). \tag{2.7}$$

Note that if events E and F are independent, then the Multiplication Rule in Equation 2.6 becomes

$$P(E \cap F) = P(F)P(E),$$

which means that we simply multiply the individual probabilities for each event together. This can be extended to k events to give

$$P(E_1 \cap E_2 \cap \ldots \cap E_k) = \prod_{i=1}^{k} P(E_i), \qquad (2.8)$$

where events E_i and E_j (for all i and j, $i \neq j$) are independent.

Bayes' Theorem

Sometimes we start an analysis with an initial degree of belief that an event will occur. Later on, we might obtain some additional information about the event that would change our belief about the probability that the event will occur. The initial probability is called a *prior probability*. Using the new information, we can update the prior probability using Bayes' theorem to obtain the *posterior probability*.

The experiment of recording piston ring failure in compressors mentioned at the beginning of the chapter is an example of where Bayes' theorem might be used, and we derive Bayes' theorem using this example. Suppose our piston rings are purchased from two manufacturers: 60% from *manufacturer A* and 40% from *manufacturer B*.

Let M_A denote the event that a part comes from manufacturer A and M_B represent the event that a piston ring comes from manufacturer B. If we select a part at random from our supply of piston rings, we would assign probabilities to these events as follows:

$$P(M_A) = 0.6,$$
$$P(M_B) = 0.4.$$

These are our prior probabilities that the piston rings are from the individual manufacturers.

Say we are interested in knowing the probability that a piston ring that subsequently failed came from *manufacturer A*. This would be the posterior probability that it came from *manufacturer A*, given that the piston ring failed. The additional information we have about the piston ring is that it failed, and we use this to update our degree of belief that it came from *manufacturer A*.

Bayes' theorem can be derived from the definition of conditional probability (Equation 2.5). Writing this in terms of our events, we are interested in the following probability:

$$P(M_A|F) = \frac{P(M_A \cap F)}{P(F)}, \qquad (2.9)$$

where $P(M_A|F)$ represents the posterior probability that the part came from *manufacturer A*, and F is the event that the piston ring failed. Using the Multiplication Rule (Equation 2.6), we can write the numerator of Equation

2.9 in terms of event F and our prior probability that the part came from *manufacturer A*, as follows:

$$P(M_A|F) = \frac{P(M_A \cap F)}{P(F)} = \frac{P(M_A)P(F|M_A)}{P(F)}. \tag{2.10}$$

The next step is to find $P(F)$. The only way that a piston ring will fail is if: (1) it failed and it came from *manufacturer A*, or (2) it failed and it came from *manufacturer B*. Thus, using the third axiom of probability, we can write

$$P(F) = P(M_A \cap F) + P(M_B \cap F).$$

Applying the Multiplication Rule as before, we have

$$P(F) = P(M_A)P(F|M_A) + P(M_B)P(F|M_B). \tag{2.11}$$

Substituting this for $P(F)$ in Equation 2.10, we write the posterior probability as

$$P(M_A|F) = \frac{P(M_A)P(F|M_A)}{P(M_A)P(F|M_A) + P(M_B)P(F|M_B)}. \tag{2.12}$$

Note that we need to find the probabilities $P(F|M_A)$ and $P(F|M_B)$. These are the probabilities that a piston ring will fail given it came from the corresponding manufacturer. These must be estimated in some way using available information (e.g., past failures). When we revisit Bayes' theorem in the context of statistical pattern recognition (Chapter 10), these are the probabilities that are estimated to construct a certain type of classifier.

Equation 2.12 is Bayes' theorem for a situation where only two outcomes are possible. In general, Bayes' theorem can be written for any number of mutually exclusive events, $E_1, ..., E_k$, whose union makes up the entire sample space. This is given next.

BAYES' THEOREM

$$P(E_i|F) = \frac{P(E_i)P(F|E_i)}{P(E_1)P(F|E_1) + ... + P(E_k)P(F|E_k)}. \tag{2.13}$$

2.4 Expectation

Expected values and variances are important concepts in statistics. They are used to describe distributions, to evaluate the performance of estimators, to obtain test statistics in hypothesis testing, and many other applications.

Mean and Variance

The *mean* or *expected value* of a random variable is defined using the probability density or mass function. It provides a measure of central tendency of the distribution. If we observe many values of the random variable and take the average of them, we would expect that value to be close to the mean. The expected value is defined below for the discrete case.

EXPECTED VALUE - DISCRETE RANDOM VARIABLES

$$\mu = E[X] = \sum_{i=1}^{\infty} x_i f(x_i).$$ \hfill (2.14)

We see from the definition that the expected value is a sum of all possible values of the random variable where each one is weighted by the probability that X will take on that value.

The *variance* of a discrete random variable is given by the following definition.

VARIANCE - DISCRETE RANDOM VARIABLES

For $\mu < \infty$,

$$\sigma^2 = V(X) = E[(X-\mu)^2] = \sum_{i=1}^{\infty} (x_i - \mu)^2 f(x_i).$$ \hfill (2.15)

From Equation 2.15, we see that the variance is the sum of the squared distances from the mean, each one weighted by the probability that $X = x_i$. Variance is a measure of dispersion in the distribution. If a random variable has a large variance, then an observed value of the random variable is more likely to be far from the mean μ. The standard deviation σ is the square root of the variance.

The mean and variance for continuous random variables are defined similarly, with the summation replaced by an integral. The mean and variance of a continuous random variable are given next.

EXPECTED VALUE - CONTINUOUS RANDOM VARIABLES

$$\mu = E[X] = \int_{-\infty}^{\infty} xf(x)dx. \tag{2.16}$$

VARIANCE - CONTINUOUS RANDOM VARIABLES

For $\mu < \infty$,

$$\sigma^2 = V(X) = E[(X-\mu)^2] = \int_{-\infty}^{\infty} (x-\mu)^2 f(x)dx. \tag{2.17}$$

We note that Equation 2.17 can also be written as

$$V(X) = E[X^2] - \mu^2 = E[X^2] - (E[X])^2.$$

Other expected values that are of interest in statistics are the **moments** of a random variable. These are the expectation of powers of the random variable. In general, we define the **r-th moment** as

$$\mu'_r = E[X^r], \tag{2.18}$$

and the **r-th central moment** as

$$\mu_r = E[(X-\mu)^r]. \tag{2.19}$$

The mean corresponds to μ'_1, and the variance is given by μ_2.

Skewness

The third central moment μ_3 is often called a measure of asymmetry or skewness in the distribution. The uniform and the normal distribution are examples of symmetric distributions. The gamma and the exponential are examples of skewed or asymmetric distributions. The following ratio is called the **coefficient of skewness,** which is often used to measure this characteristic:

$$\gamma_1 = \frac{\mu_3}{\mu_2^{3/2}}. \tag{2.20}$$

Distributions that are skewed to the left will have a negative coefficient of skewness, and distributions that are skewed to the right will have a positive value [Hogg and Craig, 1978]. The coefficient of skewness is zero for symmetric distributions. However, a coefficient of skewness equal to zero does not imply that the distribution must be symmetric.

Kurtosis

Skewness is one way to measure a type of departure from normality. *Kurtosis* measures a different type of departure from normality by indicating the extent of the peak (or the degree of flatness near its center) in a distribution. The *coefficient of kurtosis* is given by the following ratio:

$$\gamma_2 = \frac{\mu_4}{\mu_2^2}. \tag{2.21}$$

We see that this is the ratio of the fourth central moment divided by the square of the variance. If the distribution is normal, then this ratio is equal to 3. A ratio greater than 3 indicates more values in the neighborhood of the mean (is more peaked than the normal distribution). If the ratio is less than 3, then it is an indication that the curve is flatter than the normal.

Sometimes the *coefficient of excess kurtosis* is used as a measure of kurtosis. This is given by

$$\gamma_2' = \frac{\mu_4}{\mu_2^2} - 3. \tag{2.22}$$

In this case, distributions that are more peaked than the normal correspond to a positive value of γ_2', and those with a flatter top have a negative coefficient of excess kurtosis.

2.5 Common Distributions

In this section, we provide a review of some useful probability distributions and briefly describe some applications to modeling data. Most of these distributions are used in later chapters, so we take this opportunity to define them and to fix our notation. We first cover two important discrete distributions: the binomial and the Poisson. These are followed by several continuous distributions: the uniform, the normal, the exponential, the gamma, the chi-square, the Weibull, the beta, the Student's *t* distribution, the multivariate normal, and the multivariate *t* distribution.

Binomial

Let's say that we have an experiment, whose outcome can be labeled as a "success" or a "failure." If we let $X = 1$ denote a successful outcome and $X = 0$ represent a failure, then we can write the probability mass function as

$$
\begin{aligned}
f(0) &= P(X = 0) = 1 - p, \\
f(1) &= P(X = 1) = p,
\end{aligned}
\tag{2.23}
$$

where p represents the probability of a successful outcome. A random variable that follows the probability mass function in Equation 2.23 for $0 < p < 1$ is called a ***Bernoulli random variable***.

Now suppose we repeat this experiment for n trials, where each trial is independent (the outcome from one trial does not influence the outcome of another) and results in a success with probability p. If X denotes the number of successes in these n trials, then X follows the binomial distribution with parameters n and p. Examples of binomial distributions with different parameters are shown in Figure 2.3.

To calculate a binomial probability, we use the following formula:

$$
f(x;n, p) = P(X = x) = \binom{n}{x} p^x (1 - p)^{n-x}; \qquad x = 0, 1, ..., n. \tag{2.24}
$$

The mean and variance of a binomial distribution are given by

$$
E[X] = np,
$$

and

$$
V(X) = np(1 - p).
$$

Some examples where the results of an experiment can be modeled by a binomial random variable are

- A drug has probability 0.90 of curing a disease. It is administered to 100 patients, where the outcome for each patient is either cured or not cured. If X is the number of patients cured, then X is a binomial random variable with parameters (100, 0.90).

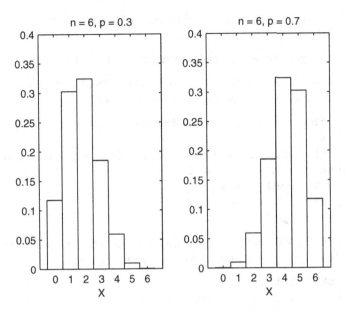

FIGURE 2.3
Examples of the binomial distribution for different success probabilities.

- The National Institute of Mental Health estimates that there is a 20% chance that an adult American suffers from a psychiatric disorder. Fifty adult Americans are randomly selected. If we let X represent the number who have a psychiatric disorder, then X takes on values according to the binomial distribution with parameters $(50, 0.20)$.

- A manufacturer of computer chips finds that on the average 5% are defective. To monitor the manufacturing process, they take a random sample of size 75. If the sample contains more than five defective chips, then the process is stopped. The binomial distribution with parameters $(75, 0.05)$ can be used to model the random variable X, where X represents the number of defective chips.

Example 2.1

Suppose there is a 20% chance that an adult American suffers from a psychiatric disorder. We randomly sample 25 adult Americans. If we let X represent the number of people who have a psychiatric disorder, then X is a binomial random variable with parameters $(25, 0.20)$. We are interested in the probability that at most 3 of the selected people have such a disorder. We can use the MATLAB Statistics Toolbox function **binocdf** to determine $P(X \leq 3)$, as follows:

```
prob = binocdf(3,25,0.2);
```

We could also sum up the individual values of the probability mass function from $X = 0$ to $X = 3$:

```
prob2 = sum(binopdf(0:3,25,0.2));
```

Both of these commands return a probability of 0.234. We now show how to generate the binomial distributions shown in Figure 2.3.

```
% Get the values for the domain, x.
x = 0:6;
% Get the values of the probability mass function.
% First for n = 6, p = 0.3:
pdf1 = binopdf(x,6,0.3);
% Now for n = 6, p = 0.7:
pdf2 = binopdf(x,6,0.7);
```

Now we have the values for the probability mass function (or the heights of the bars). The plots are obtained using the following code:

```
% Do the plots.
subplot(1,2,1),bar(x,pdf1,1,'w')
title(' n = 6, p = 0.3')
xlabel('X'),ylabel('f(X)')
axis square
subplot(1,2,2),bar(x,pdf2,1,'w')
title(' n = 6, p = 0.7')
xlabel('X'),ylabel('f(X)')
axis square
```

❑

Poisson

If a random variable X is a ***Poisson random variable*** with parameter λ, $\lambda > 0$, then it has the probability mass function given by

$$f(x;\lambda) = P(X = x) = e^{-\lambda}\frac{\lambda^x}{x!}; \quad x = 0, 1, \ldots , \tag{2.25}$$

where $x!$ denotes the factorial of x. The factorial of a non-negative integer x is the product of all positive integers less than or equal to x.

The expected value and variance of a Poisson random variable are both λ, thus,

$$E[X] = \lambda,$$

and

$$V(X) = \lambda.$$

The Poisson distribution can be used in many applications. Examples where a discrete random variable might follow a Poisson distribution are

- the number of typographical errors on a page,
- the number of vacancies in a company during a month, or
- the number of defects in a length of wire.

The Poisson distribution is often used to approximate the binomial. When n is large and p is small (so np is moderate), then the number of successes occurring can be approximated by the Poisson random variable with parameter $\lambda = np$.

The Poisson distribution is also appropriate for some applications where events occur at points in time or space. Examples include the arrival of jobs at a business, the arrival of aircraft on a runway, and the breakdown of machines at a manufacturing plant. The number of events in these applications can be described by a *Poisson process*.

Let $N(t)$, $t \geq 0$, represent the number of events that occur in the time interval $[0, t]$. For each interval $[0, t]$, $N(t)$ is a random variable that can take on values $0, 1, 2, \ldots$. If the following conditions are satisfied, then the counting process $\{N(t), t \geq 0\}$ is said to be a Poisson process with mean rate λ [Ross, 2000]:

1. $N(0) = 0$.

2. The process has independent increments.

3. The number $N(t)$ of events in an interval of length t follows a Poisson distribution with mean λt. Thus, for $s \geq 0$ and $t \geq 0$,

$$P(N(t+s) - N(s) = k) = e^{-\lambda t}\frac{(\lambda t)^k}{k!}; \qquad k = 0, 1, \ldots. \qquad (2.26)$$

From the third condition, we know that the process has stationary increments. This means that the distribution of the number of events in an interval depends only on the length of the interval and not on the starting point. The second condition specifies that the number of events in one interval does not affect the number of events in other intervals. The first condition states that the counting starts at time $t = 0$. The expected value of $N(t)$ is given by

$$E[N(t)] = \lambda t.$$

Example 2.2

In preparing this text, we executed the spell check command, and the editor reviewed the manuscript for typographical errors. In spite of this, some mistakes might be present. Assume that the number of typographical errors per page follows the Poisson distribution with parameter $\lambda = 0.25$. We calculate the probability that a page will have at least two errors as follows:

$$P(X \geq 2) = 1 - \{P(X = 0) + P(X = 1)\} = 1 - e^{-0.25} - e^{-0.25}0.25 \approx 0.0265.$$

We can get this probability using the MATLAB Statistics Toolbox function **poisscdf**. Note that $P(X = 0) + P(X = 1)$ is the Poisson cumulative distribution function for $a = 1$ (see Equation 2.4), which is why we use **1** as the argument to **poisscdf**.

```
prob = 1-poisscdf(1,0.25);
```

❑

Example 2.3

Suppose that accidents at a certain intersection occur in a manner that satisfies the conditions for a Poisson process with a rate of 2 per week ($\lambda = 2$). What is the probability that at most 3 accidents will occur during the next 2 weeks? Using Equation 2.26, we have

$$P(N(2) \leq 3) = \sum_{k=0}^{3} P(N(2) = k).$$

Expanding this out yields

$$P(N(2) \leq 3) = e^{-4} + 4e^{-4} + \frac{4^2}{2!}e^{-4} + \frac{4^3}{3!}e^{-4} \approx 0.4335.$$

As before, we can use the **poisscdf** function with parameter given by $\lambda t = 2 \cdot 2$.

```
prob = poisscdf(3,2*2);
```

❑

Uniform

Perhaps one of the most important distributions is the *uniform distribution* for continuous random variables. One reason is that the uniform (0, 1) distribution is used as the basis for simulating most random variables as we discuss in Chapter 4.

A random variable that is uniformly distributed over the interval (a, b) follows the probability density function given by

$$f(x;a, b) = \frac{1}{b-a}; \quad a < x < b. \tag{2.27}$$

The parameters for the uniform are the interval endpoints, a and b. The mean and variance of a uniform random variable are given by

$$E[X] = \frac{a+b}{2}$$

and

$$V(X) = \frac{(b-a)^2}{12}.$$

The cumulative distribution function for a uniform random variable is

$$F(x) = \begin{cases} 0; & x \leq a \\ \dfrac{x-a}{b-a}; & a < x < b \\ 1; & x \geq b. \end{cases} \tag{2.28}$$

Example 2.4

In this example, we illustrate the uniform probability density function over the interval $(0, 10)$, along with the corresponding cumulative distribution function. The MATLAB Statistics Toolbox functions **unifpdf** and **unifcdf** are used to get the desired functions over the interval.

```
% First get the domain over which we will
% evaluate the functions.
x = -1:.1:11;
% Now get the probability density function
% values at x.
pdf = unifpdf(x,0,10);
% Now get the cdf.
cdf = unifcdf(x,0,10);
```

Plots of the functions are provided in Figure 2.4, where the probability density function is shown in the left plot and the cumulative distribution on the right. These plots are constructed using the following MATLAB commands.

```
% Do the plots.
subplot(1,2,1),plot(x,pdf)
title('PDF')
xlabel('X'),ylabel('f(X)')
axis([-1 11 0 0.2])
axis square
subplot(1,2,2),plot(x,cdf)
title('CDF')
xlabel('X'),ylabel('F(X)')
axis([-1 11 0 1.1])
axis square
```

❑

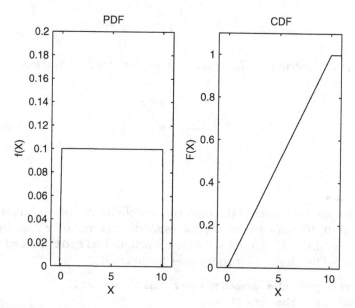

FIGURE 2.4
On the left is a plot of the probability density function for the uniform (0, 10). Note that the height of the curve is given by $1/(b-a) = 1/10 = 0.10$. The corresponding cumulative distribution function is shown on the right.

Normal

A well-known distribution in statistics and engineering is the ***normal distribution***. Also called the ***Gaussian distribution***, it has a continuous probability density function given by

$$f(x;\mu, \sigma^2) = \frac{1}{\sigma\sqrt{2\pi}} \exp\left\{-\frac{(x-\mu)^2}{2\sigma^2}\right\}, \tag{2.29}$$

where

$$-\infty < x < \infty; \quad -\infty < \mu < \infty; \quad \sigma^2 > 0 \ .$$

The normal distribution is completely determined by its parameters (μ and σ^2), which are also the expected value and variance for a normal random variable. The notation $X \sim N(\mu, \sigma^2)$ is used to indicate that a random variable X is normally distributed with mean μ and variance σ^2. Several normal distributions with different parameters are shown in Figure 2.5.

Some special properties of the normal distribution are given here.

- The value of the probability density function approaches zero as x approaches positive and negative infinity.
- The probability density function is centered at the mean μ, and the maximum value of the function occurs at $x = \mu$.
- The probability density function for the normal distribution is symmetric about the mean μ.

The special case of a ***standard normal*** random variable is one whose mean is zero ($\mu = 0$) and whose standard deviation is one ($\sigma = 1$). If X is normally distributed, then

$$Z = \frac{X - \mu}{\sigma} \tag{2.30}$$

is a standard normal random variable.

Traditionally, the cumulative distribution function of a standard normal random variable is denoted by

$$\Phi(z) = \frac{1}{\sqrt{2\pi}} \int_{-\infty}^{z} \exp\left\{-\frac{y^2}{2}\right\} dy \ . \tag{2.31}$$

The cumulative distribution function for a standard normal random variable can be calculated using the error function, denoted by *erf*. The relationship between these functions is given by

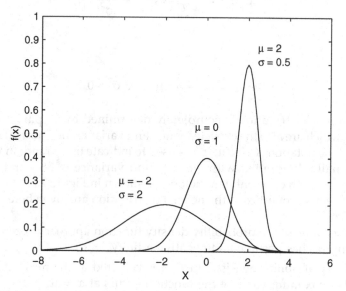

FIGURE 2.5
Examples of probability density functions for normally distributed random variables. Note that as the variance increases, the height of the probability density function at the mean decreases.

$$\Phi(z) = \frac{1}{2}erf\left(\frac{z}{\sqrt{2}}\right) + \frac{1}{2}. \tag{2.32}$$

The error function can be calculated in MATLAB using **erf(x)**. The MATLAB Statistics Toolbox has a function called **normcdf(x,mu,sigma)** that will calculate the cumulative distribution function for values in *x*. Its use is illustrated in the example given next.

Example 2.5

Similar to the uniform distribution, the functions **normpdf** and **normcdf** are available in the MATLAB Statistics Toolbox for calculating the probability density function and cumulative distribution function for the Gaussian. There is another special function called **normspec** that determines the probability that a random variable *X* assumes a value between two limits, where *X* is normally distributed with mean μ and standard deviation σ. This function also plots the normal density, where the area between the specified limits is shaded. The syntax is shown next.

```
% Set up the parameters for the normal distribution.
mu = 5;
```

```
sigma = 2;
% Set up the upper and lower limits. These are in
% the two element vector 'specs'.
specs = [2, 8];
prob = normspec(specs, mu, sigma);
```

The resulting plot is shown in Figure 2.6. By default, MATLAB will put the probability between the limits in the title, which in this case is 0.87. Note that the default title and labels can be changed easily using the **title, xlabel,** and **ylabel** functions. You can also obtain tail probabilities by using **-Inf** as the first element of **specs** to designate no lower limit or **Inf** as the second element to indicate no upper limit.
❑

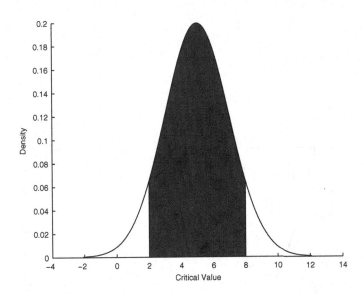

FIGURE 2.6
This shows the output from the function **normspec**. *Note that it shades the area between the lower and upper limits that are specified as input arguments. The probability between the limits is approximately 0.87.*

Exponential

The *exponential distribution* can be used to model the amount of time until a specific event occurs or to model the time between independent events. Some examples where an exponential distribution is appropriate are

- The time until the computer locks up,

- The time between arrivals of telephone calls, or
- The time until a part fails.

The exponential probability density function with parameter λ is

$$f(x;\lambda) = \lambda e^{-\lambda x}; \quad x \geq 0; \quad \lambda > 0. \tag{2.33}$$

The mean and variance of an exponential random variable are given by the following:

$$E[X] = \frac{1}{\lambda}$$

and

$$V(X) = \frac{1}{\lambda^2}.$$

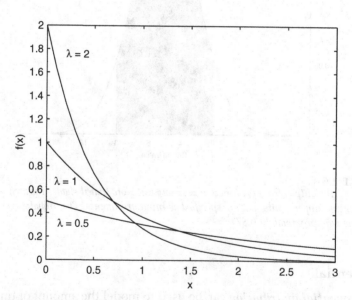

FIGURE 2.7
Exponential probability density functions for various values of λ.

The cumulative distribution function of an exponential random variable is given by

$$F(x) = \begin{cases} 0; & x < 0 \\ 1 - e^{-\lambda x}; & x \geq 0. \end{cases} \quad (2.34)$$

The exponential distribution is the only continuous distribution that has the *memoryless property*. This property describes the fact that the remaining lifetime of an object (whose lifetime follows an exponential distribution) does not depend on the amount of time it has already lived. This property is represented by the following equality, where $s \geq 0$ and $t \geq 0$:

$$P(X > s + t | X > s) = P(X > t).$$

In words, this means that the probability that the object will operate for time $s + t$, given it has already operated for time s, is simply the probability that it operates for time t.

When the exponential distribution is used to represent interarrival times, then the parameter λ is a rate with units of arrivals per time period. When the exponential is used to model the time until a failure occurs, then λ is the failure rate. Several examples of the exponential distribution are shown in Figure 2.7.

Example 2.6

The time between arrivals of vehicles at an intersection follows an exponential distribution with a mean of 12 seconds. What is the probability that the time between arrivals is 10 seconds or less? We are given the average interarrival time, so $\lambda = 1/12$. The required probability is obtained from Equation 2.34 as follows

$$P(X \leq 10) = 1 - e^{-(1/12)10} \approx 0.57.$$

You can calculate this using the MATLAB Statistics Toolbox function **expcdf(x, 1/λ)**. Note that this MATLAB function is based on a different definition of the exponential probability density function, which is given by

$$f(x;\mu) = \frac{1}{\mu} e^{-\frac{x}{\mu}}; \quad x \geq 0; \quad \mu > 0. \quad (2.35)$$

In the Computational Statistics Toolbox, we include a function called **csexpoc(x, λ)** that calculates the exponential cumulative distribution function using Equation 2.34.
❑

Gamma

The *gamma probability density function* with parameters $\lambda > 0$ and $t > 0$ is given by

$$f(x;\lambda, t) = \frac{\lambda e^{-\lambda x}(\lambda x)^{t-1}}{\Gamma(t)}; \quad x \geq 0, \tag{2.36}$$

where t is a shape parameter, and λ is the scale parameter. The gamma function $\Gamma(t)$ is defined as

$$\Gamma(t) = \int_0^\infty e^{-y} y^{t-1} dy. \tag{2.37}$$

For integer values of t, Equation 2.37 becomes

$$\Gamma(t) = (t-1)!. \tag{2.38}$$

Note that for $t = 1$, the gamma density is the same as the exponential. When t is a positive integer, the gamma distribution can be used to model the amount of time one has to wait until t events have occurred, if the inter-arrival times are exponentially distributed.

The mean and variance of a gamma random variable are

$$E[X] = \frac{t}{\lambda}$$

and

$$V(X) = \frac{t}{\lambda^2}.$$

The cumulative distribution function for a gamma random variable is calculated using [Meeker and Escobar, 1998; Banks, et al., 2001]

$$F(x;\lambda, t) = \begin{cases} 0; & x \leq 0 \\ \dfrac{1}{\Gamma(t)} \displaystyle\int_0^{\lambda x} y^{t-1} e^{-y} dy; & x > 0 . \end{cases} \tag{2.39}$$

Equation 2.39 can be evaluated in MATLAB using the **gammainc(λ*x,t)** function, where the above notation is used for the arguments.

Example 2.7

We plot the gamma probability density function for $\lambda = t = 1$ (this should look like the exponential), $\lambda = t = 2$, and $\lambda = t = 3$. You can use the MATLAB Statistics Toolbox function **gampdf(x,t,1/λ)** or the function **csgammp(x,t,λ)**. The resulting curves are shown in Figure 2.8.

```
% First get the domain over which to
% evaluate the functions.
x = 0:.1:3;
% Now get the functions values for
% different values of lambda.
y1 = gampdf(x,1,1/1);
y2 = gampdf(x,2,1/2);
y3 = gampdf(x,3,1/3);
% Plot the functions.
plot(x,y1,'r',x,y2,'g',x,y3,'b')
title('Gamma Distribution')
xlabel('X')
ylabel('f(x)')
```

❑

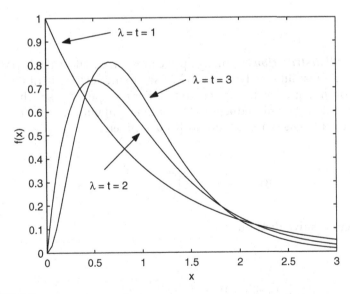

FIGURE 2.8
We show three examples of the gamma probability density function. We see that when $\lambda = t = 1$, we have the same probability density function as the exponential with parameter $\lambda = 1$.

Chi-Square

A gamma distribution where $\lambda = 0.5$ and $t = v/2$, with v a positive integer, is called a *chi-square distribution* (denoted as χ_v^2) with v degrees of freedom. The chi-square distribution is used to derive the distribution of the sample variance and is important for goodness-of-fit tests in statistical analysis [Mood, Graybill, and Boes, 1974].

The probability density function for a chi-square random variable with v degrees of freedom is

$$f(x;v) = \frac{1}{\Gamma(v/2)}\left(\frac{1}{2}\right)^{v/2} x^{v/2-1}e^{-\frac{1}{2}x}; \qquad x \geq 0. \tag{2.40}$$

The mean and variance of a chi-square random variable can be obtained from the gamma distribution. These are given by

$$E[X] = v$$

and

$$V(X) = 2v.$$

Weibull

The *Weibull distribution* has many applications in engineering. In particular, it is used in reliability analysis. It can be used to model the distribution of the amount of time it takes for objects to fail. For the special case where $v = 0$ and $\beta = 1$, the Weibull reduces to the exponential with $\lambda = (1/\alpha)$.

The Weibull density for $\alpha > 0$ and $\beta > 0$ is given by

$$f(x;v, \alpha, \beta) = \left(\frac{\beta}{\alpha}\right)\left(\frac{x-v}{\alpha}\right)^{\beta-1} e^{-\left(\frac{x-v}{\alpha}\right)^{\beta}}; \qquad x > v, \tag{2.41}$$

and the cumulative distribution is

$$F(x;v, \alpha, \beta) = \begin{cases} 0; & x \leq v \\ 1 - e^{-\left(\frac{x-v}{\alpha}\right)^{\beta}}; & x > v. \end{cases} \tag{2.42}$$

The location parameter is denoted by v, and the scale parameter is given by α. The shape of the Weibull distribution is governed by the parameter β.

The mean and variance [Banks, et al., 2001] of a random variable from a Weibull distribution are given by

$$E[X] = v + \alpha\Gamma(1/\beta + 1)$$

and

$$V(X) = \alpha^2\left\{\Gamma(2/\beta + 1) - [\Gamma(1/\beta + 1)]^2\right\}.$$

Example 2.8

Suppose the time to failure of piston rings for stream-driven compressors can be modeled by the Weibull distribution with a location parameter of zero, $\beta = 1/3$, and $\alpha = 500$. We can find the mean time to failure using the expected value of a Weibull random variable, as follows

$$E[X] = v + \alpha\Gamma(1/\beta + 1) = 500 \times \Gamma(3 + 1) = 3000 \text{ hours.}$$

Let's say we want to know the probability that a piston ring will fail before 2000 hours. We can calculate this probability using

$$F(2000; 0, 500, 1/3) = 1 - \exp\left\{-\left(\frac{2000}{500}\right)^{1/3}\right\} \approx 0.796.$$

❑

You can use the MATLAB Statistics Toolbox function for applications where the location parameter is zero ($v = 0$). This function is called **weibcdf** (for the cumulative distribution function), and the input arguments are $(x, \alpha^{-\beta}, \beta)$. The reason for the different parameters is that MATLAB uses an alternate definition for the Weibull probability density function given by

$$f(x; a, b) = abx^{b-1}e^{-ax^b}; \qquad x > 0. \tag{2.43}$$

Comparing this with Equation 2.41, we can see that $v = 0$, $a = \alpha^{-\beta}$, and $b = \beta$. You can also use the function **csweibc**(x, v, α, β) to evaluate the cumulative distribution function for a Weibull.

Beta

The *beta distribution* is very flexible because it covers a range of different shapes depending on the values of the parameters. It can be used to model a random variable that takes on values over a bounded interval and assumes one of the shapes governed by the parameters. A random variable has a beta distribution with parameters $\alpha > 0$ and $\beta > 0$ if its probability density function is given by

$$f(x;\alpha, \beta) = \frac{1}{B(\alpha, \beta)} x^{\alpha-1}(1-x)^{\beta-1}; \qquad 0 < x < 1, \qquad (2.44)$$

where

$$B(\alpha, \beta) = \int_0^1 x^{\alpha-1}(1-x)^{\beta-1} dx = \frac{\Gamma(\alpha)\Gamma(\beta)}{\Gamma(\alpha+\beta)}. \qquad (2.45)$$

The function $B(\alpha, \beta)$ can be calculated in MATLAB using the **beta(α,β)** function. The mean and variance of a beta random variable are

$$E[X] = \frac{\alpha}{\alpha + \beta}$$

and

$$V(X) = \frac{\alpha\beta}{(\alpha + \beta)^2(\alpha + \beta + 1)}.$$

The cumulative distribution function for a beta random variable is given by integrating the beta probability density function as follows

$$F(x;\alpha, \beta) = \int_0^x \frac{1}{B(\alpha, \beta)} y^{\alpha-1}(1-y)^{\beta-1} dy. \qquad (2.46)$$

The integral in Equation 2.46 is called the *incomplete beta function*. This can be calculated in MATLAB using the function **betainc(x,alpha,beta)**.

Example 2.9

We use the following MATLAB code to plot the beta density over the interval (0,1). We let $\alpha = \beta = 0.5$ and $\alpha = \beta = 3$.

```
% First get the domain over which to evaluate
```

```
% the density function.
x = 0.01:.01:.99;
% Now get the values for the density function.
y1 = betapdf(x,0.5,0.5);
y2 = betapdf(x,3,3);
% Plot the results.
plot(x,y1,'r',x,y2,'g')
title('Beta Distribution')
xlabel('x')
ylabel('f(x)')
```

The resulting curves are shown in Figure 2.9. You can use the MATLAB Statistics Toolbox function **betapdf(x,** α, β **)**, as we did in the example, or the function **csbetap(x,** α, β **)**.
❑

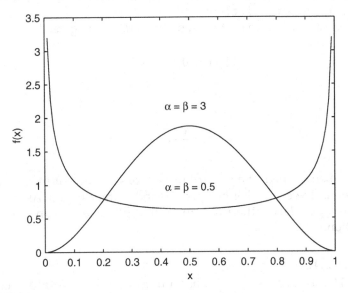

FIGURE 2.9.
Beta probability density functions for various parameters.

Student's *t* Distribution

An important distribution often used in inferential statistics is the *t* distribution. This distribution was first developed by William Gossett in 1908. He published his results under the pseudonym "Student," hence the distribution is sometimes known as the *Student's t distribution*.

The *t* distribution comes from the ratio of a standard normal random variable *Z* to the square root of an independently distributed chi-square random variable *U* divided by its degrees of freedom ν:

$$X = \frac{Z}{\sqrt{U/\nu}}. \tag{2.47}$$

It can be shown that the density of the random variable in Equation 2.47 is given by

$$f(x;\nu) = \frac{1}{\sqrt{\pi\nu}} \frac{\Gamma\left(\frac{\nu+1}{2}\right)}{\Gamma\left(\frac{\nu}{2}\right)} \left(1 + \frac{x^2}{\nu}\right)^{-(\nu+1)/2}. \tag{2.48}$$

The probability density function for the *t* distribution is symmetric and bell-shaped, and it is centered at zero.

The mean and variance for the *t* random variable are given by

$$E[X] = 0, \qquad \nu \geq 2,$$

and

$$V(X) = \frac{\nu}{\nu-2}, \qquad \nu \geq 3.$$

Since it is bell-shaped, the *t* distribution looks somewhat like the normal distribution. However, it has heavier tails and a larger spread. As the degrees of freedom gets large, the *t* distribution approaches a standard normal distribution.

Example 2.10

The MATLAB Statistics Toolbox has a function called **tpdf** that creates a probability density function for the Student's *t* distribution with ν degrees of freedom. The following steps will evaluate the density function for ν = 5.

```
% First we get the domain for the function.
x = -6:.01:6;
% Now get the values for the density function.
y = tpdf(x,5);
% Plot the results.
plot(x,y)
xlabel('x')
ylabel('f(x)')
```

The resulting curve is shown in Figure 2.10. Compare this with the probability density function for the standard normal shown in Figure 2.5, and note the fatter tails with the *t* distribution.
❑

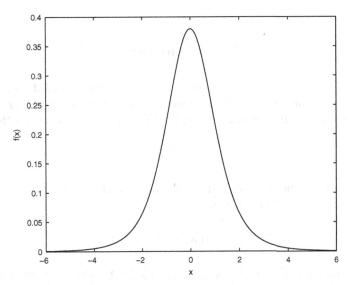

FIGURE 2.10
This illustrates the probability density function for a t random variable with 5 degrees of freedom.

Multivariate Normal

So far, we have discussed several univariate distributions for discrete and continuous random variables. In this section, we describe the *multivariate normal distribution* for continuous variables. This important distribution is used throughout the rest of the text. Some examples of where we use it are in exploratory data analysis, probability density estimation, and statistical pattern recognition.

The probability density function for a general multivariate normal density for *d* dimensions is given by

$$f(\mathbf{x}; \boldsymbol{\mu}, \boldsymbol{\Sigma}) = \frac{1}{(2\pi)^{d/2} |\boldsymbol{\Sigma}|^{1/2}} \exp\left\{ -\frac{1}{2}(\mathbf{x} - \boldsymbol{\mu})^T \boldsymbol{\Sigma}^{-1}(\mathbf{x} - \boldsymbol{\mu}) \right\}, \qquad (2.49)$$

where $\mathbf{x}$ is a d-component column vector, μ is the $d \times 1$ column vector of means, and Σ is the $d \times d$ covariance matrix. The superscript T represents the transpose of an array, and the notation $||$ denotes the determinant of a matrix.

The mean and covariance are calculated using the following formulas:

$$\mu = E[\mathbf{x}] \qquad (2.50)$$

and

$$\Sigma = E[(\mathbf{x} - \mu)(\mathbf{x} - \mu)^{T}], \qquad (2.51)$$

where the expected value of an array is given by the expected values of its components. Thus, if we let X_i represent the i-th component of $\mathbf{x}$ and μ_i the i-th component of μ, then the elements of Equation 2.50 can be written as

$$\mu_i = E[X_i].$$

If σ_{ij} represents the ij-th element of Σ, then the elements of the *covariance matrix* (Equation 2.51) are given by

$$\sigma_{ij} = E[(X_i - \mu_i)(X_j - \mu_j)].$$

The covariance matrix is symmetric ($\Sigma^{T} = \Sigma$) positive definite (all eigenvalues of Σ are greater than zero) for most applications of interest to statisticians and engineers.

We illustrate some properties of the multivariate normal by looking at the bivariate ($d = 2$) case. The probability density function for a bivariate normal is represented by a bell-shaped surface. The center of the surface is determined by the mean μ, and the shape of the surface is determined by the covariance Σ. If the covariance matrix is diagonal (all of the off-diagonal elements are zero), and the diagonal elements are equal, then the shape is circular. If the diagonal elements are not equal, then we get an ellipse with the major axis vertical or horizontal. If the covariance matrix is not diagonal, then the shape is elliptical with the axes at an angle. Some of these possibilities are illustrated in the next example.

Example 2.11

We first provide the following MATLAB function to calculate the multivariate normal probability density function and illustrate its use in the bivariate case. The function is called **csevalnorm**, and it takes input arguments **x, mu, cov_mat**. The input argument **x** is a matrix containing the points in the domain where the function is to be evaluated, **mu** is a *d*-dimensional row vector, and **cov_mat** is the $d \times d$ covariance matrix.

```
function prob = csevalnorm(x,mu,cov_mat);
[n,d] = size(x);
% center the data points
x = x-ones(n,1)*mu;
a = (2*pi)^(d/2)*sqrt(det(cov_mat));
arg = diag(x*inv(cov_mat)*x');
prob = exp((-.5)*arg);
prob = prob/a;
```

We now call this function for a bivariate normal centered at zero and covariance matrix equal to the identity matrix. The density surface for this case is shown in Figure 2.11.

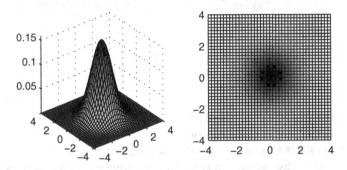

FIGURE 2.11
This figure shows a standard bivariate normal probability density function that is centered at the origin. The covariance matrix is given by the identity matrix. Notice that the shape of the surface looks circular. The plot on the right is for a viewpoint looking down on the surface.

```
% Get the mean and covariance.
mu = zeros(1,2);
cov_mat = eye(2);% Identity matrix
% Get the domain.
% Should range (-4,4) in both directions.
[x,y] = meshgrid(-4:.2:4,-4:.2:4);
% Reshape into the proper format for the function.
X = [x(:),y(:)];
Z = csevalnorm(X,mu,cov_mat);
% Now reshape the matrix for plotting.
z = reshape(Z,size(x));
subplot(1,2,1) % plot the surface
surf(x,y,z),axis square, axis tight
```

```
title('BIVARIATE STANDARD NORMAL')
```

Next, we plot the surface for a bivariate normal centered at the origin with non-zero off-diagonal elements in the covariance matrix. Note the elliptical shape of the surface shown in Figure 2.12.

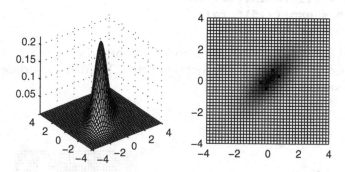

FIGURE 2.12
This shows a bivariate normal density where the covariance matrix has non-zero off-diagonal elements. Note that the surface has an elliptical shape. The plot on the right is for a viewpoint looking down on the surface.

```
subplot(1,2,2) % look down on the surface
pcolor(x,y,z),axis square
title('BIVARIATE STANDARD NORMAL')
% Now do the same thing for a covariance matrix
% with non-zero off-diagonal elements.
cov_mat = [1 0.7 ; 0.7 1];
Z = csevalnorm(X,mu,cov_mat);
z = reshape(Z,size(x));
subplot(1,2,1)
surf(x,y,z),axis square, axis tight
title('BIVARIATE NORMAL')
subplot(1,2,2)
pcolor(x,y,z),axis square
title('BIVARIATE NORMAL')
```

The Statistics Toolbox has a function called **mvnpdf** that will evaluate the multivariate normal density.
❑

The probability that a point $\mathbf{x} = (x_1, x_2)^T$ will assume a value in a region R can be found by integrating the bivariate probability density function over the region. Any plane that cuts the surface parallel to the x_1, x_2 plane intersects in an elliptic (or circular) curve, yielding a curve of constant density. Any plane perpendicular to the x_1, x_2 plane cuts the surface in a normal curve. This property indicates that in each dimension, the multivariate normal is a univariate normal distribution.

Multivariate *t* Distribution

The univariate Student's *t* distribution can also be generalized to the multivariate case. The *multivariate t distribution* is for *d*-dimensional random vectors where each variable has a univariate *t* distribution. Not surprisingly, it is used in applications where the distributions of the variables have fat tails, thus providing an alternative to the multivariate normal when working with real-world data.

A *d*-dimensional random vector

$$\mathbf{x}^T = (X_1, ..., X_d)$$

has the *t* distribution if its joint probability density function is given by

$$f(\mathbf{x};\mathbf{R},\nu,\mu) = \frac{\Gamma\left(\frac{\nu+d}{2}\right)}{(\pi\nu)^{d/2}\Gamma\left(\frac{\nu}{2}\right)|\mathbf{R}|^{1/2}}\left[1 + \frac{(\mathbf{x}-\mu)^T\mathbf{R}^{-1}(\mathbf{x}-\mu)}{\nu}\right]^{-(\nu+d)/2} \quad (2.52)$$

The multivariate *t* distribution has three parameters: the correlation matrix **R**, the degrees of freedom ν, and the *d*-dimensional mean μ. If the mean is zero ($\mu = \mathbf{0}$), then the distribution is said to be central. Similar to the univariate case, when the degrees of freedom gets large, then the joint probability density function approaches the *d*-variate normal.

The correlation matrix is related to the covariance matrix via the following relationship:

$$\rho_{ij} = \frac{\sigma_{ij}}{\sqrt{\sigma_{ii}\sigma_{jj}}},$$

where ρ_{ij} is the *ij*-th element of **R**. Note that the diagonal elements of **R** are equal to one.

Example 2.12

The MATLAB Statistics Toolbox has several functions for the central multivariate *t* distribution. These include the **mvtpdf** (evaluates the multivariate *t* probability density function) and the **mvtcdf** (computes the cumulative probabilities). These functions use the *d*-variate *t* distribution where $\mu = 0$. We now show how to plot a bivariate t probability density function with high negative correlation.

```
% First set up a correlation matrix with high
% negative correlation.
Rneg = [1, -0.8; -0.8, 1];
nu = 5;
% Get the domain for the pdf.
x = -4:.1:4;
y = -4:.1:4;
[X,Y] = meshgrid(x,y);
% Evaluate the pdf.
z = mvtpdf([X(:) Y(:)],Rneg,nu);
% Reshape to a matrix and plot.
z = reshape(z,length(x),length(y));
surf(x,y,z);
xlabel('x');
ylabel('y');
zlabel('f(x,y)');
axis tight
```

The surface plot is shown in Figure 2.13.
❏

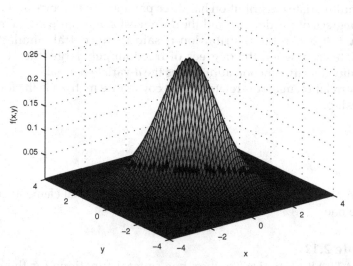

FIGURE 2.13
This is the surface plot for a bivariate t distribution with 5 degrees of freedom and centered at the origin. The variables have negative correlation.

2.6 MATLAB® Code

The MATLAB Statistics Toolbox has many functions for the more common distributions. It has functions for finding the value of the probability density (mass) function and the value of the cumulative distribution function. The reader is cautioned to remember that the definitions of the distributions (exponential, gamma, and Weibull) differ from what we describe in the text. For instance, the exponential and the gamma distributions are parameterized differently in the MATLAB Statistics Toolbox. See the toolbox documentation for a complete list of what is available for calculating probability density (mass) functions or cumulative distribution functions.

The Computational Statistics Toolbox contains functions for several of the distributions, as defined in this chapter. In general, those functions that end in **p** correspond to the probability density (mass) function, and those ending with a **c** calculate the cumulative distribution function. Table 2.1 provides a summary of the functions.

2.7 Further Reading

There are many excellent books on probability theory at the undergraduate and graduate levels. Ross [1994; 1997; 2000] is the author of several books on probability theory and simulation. These texts contain many examples and are appropriate for advanced undergraduate students in statistics, engineering, and science. Rohatgi [1976] gives a solid theoretical introduction to probability theory. This text can be used by advanced undergraduate and beginning graduate students. It has been updated with many new examples and special topics [Rohatgi and Saleh, 2000]. For those who want to learn about probability, but do not want to be overwhelmed with the theory, then we recommend Durrett [1994].

At the graduate level, there is a book by Billingsley [1995] on probability and measure theory. He uses probability to motivate measure theory and then uses measure theory to generate more probability concepts. Another good reference is a text on probability and real analysis by Ash [1972]. This is suitable for graduate students in mathematics and statistics. For a book that can be used by graduate students in mathematics, statistics, and engineering, see Port [1994]. This text provides a comprehensive treatment of the subject and can also be used as a reference by professional data analysts. Finally, Breiman [1992] provides an overview of probability theory that is accessible to statisticians and engineers.

TABLE 2.1

List of Functions from Chapter 2 Included in the
Computational Statistics Toolbox

Distribution	MATLAB Function
Beta	`csbetap`
	`csbetac`
Binomial	`csbinop`
	`csbinoc`
Chi-square	`cschip`
	`cschic`
Exponential	`csexpop`
	`csexpoc`
Gamma	`csgammp`
	`csgammc`
Normal – univariate	`csnormp`
	`csnormc`
Normal – multivariate	`csevalnorm`
Poisson	`cspoisp`
	`cspoisc`
Continuous uniform	`csunifp`
	`csunifc`
Weibull	`csweibp`
	`csweibc`

An excellent review of the multivariate t distribution has been written by Nadarajah and Kotz [2005]. This paper gives a readable introduction to the multivariate t distribution and its uses. Another resource on multivariate distributions has been written by Kotz, Balakrishnan, and Johnson [2000]. This has descriptions of several other distributions, such as the multivariate exponential, multivariate gamma, the Dirichlet distributions, and multivariate logistic distributions.

Exercises

2.1.   Write a function using MATLAB's functions for numerical integration such as **quad** or **quadl** that will find $P(X \leq x)$ when the random variable is exponentially distributed with parameter λ. See **help** for information on how to use these functions.

2.2.   Verify that the exponential probability density function with parameter λ integrates to 1. Use the MATLAB functions **quad** or **quadl**. See **help** for information on how to use these functions.

2.3.   Radar and missile detection systems warn of enemy attacks. Suppose that a radar detection system has a probability 0.95 of detecting a missile attack.

   a. What is the probability that one detection system will detect an attack? What distribution did you use?

   b. Suppose three detection systems are located together in the same area and the operation of each system is independent of the others. What is the probability that at least one of the systems will detect the attack? What distribution did you use in this case?

2.4.   When a random variable is equally likely to be either positive or negative, then the Laplacian or the double exponential distribution can be used to model it. The Laplacian probability density function for $\lambda > 0$ is given by

$$f(x) = \frac{1}{2}\lambda e^{-\lambda|x|}; \quad -\infty < x < \infty.$$

   a. Derive the cumulative distribution function for the Laplacian.

   b. Write a MATLAB function that will evaluate the Laplacian probability density function for given values in the domain.

   c. Write a MATLAB function that will evaluate the Laplacian cumulative distribution function.

   d. Plot the probability density function when $\lambda = 1$.

2.5.   Suppose X follows the exponential distribution with parameter λ. Show that for $s \geq 0$ and $t \geq 0$,

$$P(X > s + t \,|\, X > s) = P(X > t).$$

2.6.   The lifetime in years of a flat panel display is a random variable with the exponential probability density function given by

$$f(x;0.1) = 0.1e^{-0.1x}.$$

a. What is the mean lifetime of the flat panel display?

b. What is the probability that the display fails within the first two years?

c. Given that the display has been operating for one year, what is the probability that it will fail within the next year?

2.7.   The time to failure for a widget follows a Weibull distribution, with $v = 0$, $\beta = 1/2$, and $\alpha = 750$ hours.

a. What is the mean time to failure of the widget?

b. What percentage of the widgets will fail by 2500 hours of operation? That is, what is the probability that a widget will fail within 2500 hours?

2.8.   Let's say the probability of having a boy is 0.52. Using the Multiplication Rule, find the probability that a family's first and second children are boys. What is the probability that the first child is a boy and the second child is a girl?

2.9.   Repeat Example 2.1 for $n = 6$ and $p = 0.5$. What is the shape of the distribution?

2.10. Recall that in our piston ring example, $P(M_A) = 0.6$ and $P(M_B) = 0.4$. From prior experience with the two manufacturers, we know that 2% of the parts supplied by *manufacturer A are likely to* fail and 6% of the parts supplied by *manufacturer B are likely to* fail. Thus, $P(F|M_A) = 0.02$ and $P(F|M_B) = 0.06$. If we observe a piston ring failure, what is the probability that it came from manufacturer A?

2.11. Using the functions **fminbnd** (available in the standard MATLAB package), find the value for x where the maximum of the $N(3, 1)$ probability density occurs. Note that you have to find the minimum of $-f(x)$ to find the maximum of $f(x)$ using these functions. Refer to the **help** files on these functions for more information on how to use them.

2.12. Using **normpdf** or **csnormp**, find the value of the probability density for $N(0, 1)$ at $\pm\infty$. Use a small (large) value of x for $-\infty$ (∞).

2.13. Verify Equation 2.38 using the MATLAB functions **factorial** and **gamma**.

2.14. Find the height of the curve for a normal probability density function at $x = \mu$, where $\sigma = 0.5, 1, 2$. What happens to the height of the curve as σ gets larger? Does the height change for different values of μ?

2.15. Write a function that calculates the Bayes' posterior probability given a vector of conditional probabilities and a vector of prior probabilities.

2.16. Compare the Poisson approximation to the actual binomial probability $P(X = 4)$, using $n = 9$ and $p = 0.1, 0.2, ..., 0.9$.

2.17. Using the function **normspec**, find the probability that the random variable defined in Example 2.5 assumes a value that is less than 3. What is the probability that the same random variable assumes a value that is greater than 5? Find these probabilities again using the function **normcdf**.

2.18. Find the probability for the Weibull random variable of Example 2.8 using the MATLAB Statistics Toolbox function **weibcdf** or the Computational Statistics Toolbox function **csweibc**.

2.19. The MATLAB Statistics Toolbox has a GUI demo called **disttool**. First view the **help** file on **disttool**. Then run the demo. Examine the probability density (mass) and cumulative distribution functions for the distributions discussed in the chapter.

2.20. Create a sequence of t density functions for $v = 5, 15, 25, 35$ using the function **tpdf** in the Statistics Toolbox. Compare them with the standard normal distribution.

2.21. Evaluate the t probability density function at 0 for $v = 5, 15, 25, 35$ using the function **tpdf** in the Statistics Toolbox. Comment on how the value at the mode changes with v. Compare them with the value of the probability density function at zero for the standard normal.

2.22. Show that Equation 2.52 reduces to the univariate Student's t distribution when $d = 1$, $\mu = 0$, and $R = 1$.

2.23. Repeat Example 2.12 using other values for the matrix $\mathbf{R}$ and the degrees of freedom v.

2.24. Using the same procedure for Figures 2.11 and 2.12, construct a **pcolor** plot of the probability density functions from Example 2.12 and Exercise 2.23.

Chapter 3

Sampling Concepts

3.1 Introduction

In this chapter, we cover the concepts associated with random sampling and the sampling distribution of statistics. These notions are fundamental to computational statistics and are needed to understand the topics covered in the rest of the book. As with Chapter 2, those readers who have a basic understanding of these ideas may safely move on to more advanced topics.

In Section 3.2, we discuss the terminology and concepts associated with random sampling and sampling distributions. Section 3.3 contains a brief discussion of the Central Limit Theorem. In Section 3.4, we describe some methods for deriving estimators (maximum likelihood and the method of moments) and introduce criteria for evaluating their performance. Section 3.5 covers the empirical distribution function and how it is used to estimate quantiles. Finally, we conclude with a section on the MATLAB® functions available for calculating the statistics described in this chapter and a section on further readings.

3.2 Sampling Terminology and Concepts

In Chapter 2, we introduced the idea of a random experiment. We typically perform an experiment where we collect data that will provide information on the phenomena of interest. Using these data, we draw conclusions that are usually beyond the scope of our particular experiment. The researcher generalizes from that experiment to the class of all similar experiments. This is the heart of inferential statistics. The problem with this sort of generalization is that we cannot be absolutely certain about our conclusions.

However, by using statistical techniques, we can measure and manage the degree of uncertainty in our results.

Inferential statistics is a collection of techniques and methods that enable researchers to observe a subset of the objects of interest and using the information obtained from these observations make statements or inferences about the entire population of objects. Some of these methods include the estimation of population parameters, statistical hypothesis testing, and probability density estimation.

The *target population* is defined as the entire collection of objects or individuals about which we need some information. The target population must be well defined in terms of what constitutes membership in the population (e.g., income level, geographic area, etc.) and what characteristics of the population we are measuring (e.g., height, IQ, number of failures, etc.).

The following are some examples of populations, where we refer back to the situations described at the beginning of Chapter 2.

- For the piston ring example, our population is all piston rings contained in the legs of steam-driven compressors. We would be observing the time to failure for each piston ring.

- In the glucose example, our population might be all pregnant women, and we would be measuring the glucose levels.

- For cement manufacturing, our population would be batches of cement, where we measure the tensile strength and the number of days the cement is cured.

- In the software engineering example, our population consists of all executions of a particular command and control software system, and we observe the failure time of the system in seconds.

In most cases, it is impossible or unrealistic to observe the entire population. For example, some populations have members that do not exist yet (e.g., future batches of cement) or the population is too large (e.g., all pregnant women). So researchers measure only a part of the target population, called a *sample.* If we are going to make inferences about the population using the information obtained from a sample, then it is important that the sample be representative of the population. This can usually be accomplished by selecting a *simple random sample*, where all possible samples of size n are equally likely to be selected.

A random sample of size n is said to be *independent and identically distributed* (iid) when the random variables $X_1, X_2, ..., X_n$ each have a common probability density (mass) function given by $f(x)$. Additionally, when they are both independent and identically distributed (iid), the joint probability density (mass) function is given by

$$f(x_1, ..., x_n) = f(x_1) \times ... \times f(x_n),$$

which is simply the product of the individual densities (or mass functions) evaluated at each sample point.

There are two types of simple random sampling: *sampling with replacement* and *sampling without replacement*. When we sample with replacement, we select an object, observe the characteristic we are interested in, and return the object to the population. In this case, an object can be selected for the sample more than once. When the sampling is done without replacement, objects can be selected at most one time. These concepts will be used in Chapters 7 and 8 where the bootstrap and other resampling methods are discussed.

Alternative sampling methods exist. In some situations, these methods are more practical and offer better random samples than simple random sampling. One such method, called *stratified random sampling*, divides the population into levels, and then a simple random sample is taken from each level. Usually, the sampling is done in such a way that the number sampled from each level is proportional to the number of objects of that level that are in the population. Other sampling methods include *cluster sampling* and *systematic random sampling*. For more information on these and others, see the book by Levy and Lemeshow [1999].

The goal of inferential statistics is to use the sample to estimate or make some statement about a population parameter. Recall from Chapter 2 that a *parameter* is a descriptive measure for a population or a distribution of random variables. For example, population parameters that might be of interest include the mean (μ), the standard deviation (σ), quantiles, proportions, correlation coefficients, etc.

A *statistic* is a function of the observed random variables obtained in a random sample and does not contain any unknown population parameters. Often the statistic is used for the following purposes:

- As a point estimate for a population parameter,
- To obtain a confidence interval estimate for a parameter, or
- As a test statistic in hypothesis testing.

Before we discuss some of the common methods for deriving statistics, we present some of the statistics that will be encountered in the remainder of the text. In most cases, we assume that we have a random sample, $X_1, ..., X_n$, of independent, identically distributed (iid) random variables.

Sample Mean and Sample Variance

A familiar statistic is the *sample mean* given by

$$\overline{X} = \frac{1}{n}\sum_{i=1}^{n}X_i. \tag{3.1}$$

To calculate this in MATLAB, one can use the function called **mean**. If the argument to this function is a matrix, then it provides a vector of means, each one corresponding to the mean of a column. One can find the mean along any dimension (**dim**) of an array using the syntax: **mean(x,dim)**.

We note that Equation 3.1 is for 1D, but this is easily extended to more than one dimension. The d-dimensional sample mean is given by

$$\overline{\mathbf{x}} = \frac{1}{n}\sum_{i=1}^{n}\mathbf{x}_i,$$

where $\mathbf{x}_i$ is the i-th d-dimensional observation. The sample mean for the j-th variable is given by

$$\overline{x}_j = \frac{1}{n}\sum_{i=1}^{n}X_{ij}.$$

Another statistic that we will see again is the ***sample variance***, calculated from

$$S^2 = \frac{1}{n-1}\sum_{i=1}^{n}(X_i - \overline{X})^2 = \frac{1}{n(n-1)}\left(n\sum_{i=1}^{n}X_i^2 - \left(\sum_{i=1}^{n}X_i\right)^2\right). \tag{3.2}$$

The ***sample standard deviation*** is given by the square root of the variance (Equation 3.2) and is denoted by S. These statistics can be calculated in MATLAB using the functions **std(x)** and **var(x)**, where **x** is an array containing the sample values. As with the function **mean**, these can have matrices or multi-dimensional arrays as input arguments. For example, if the input is a matrix, then the functions work on the columns of the matrix.

Sample Moments

The sample moments can be used to estimate the population moments described in Chapter 2. The ***r-th sample moment*** about zero is given by

$$M'_r = \frac{1}{n}\sum_{i=1}^{n} X_i^r. \tag{3.3}$$

Note that the sample mean is obtained when $r = 1$. The r-th sample moments about the sample mean are statistics that estimate the population central moments and can be found using the following:

$$M_r = \frac{1}{n}\sum_{i=1}^{n}(X_i - \bar{X})^r. \tag{3.4}$$

We can use Equation 3.4 to obtain estimates for the coefficient of skewness γ_1 and the coefficient of kurtosis γ_2. Recall that these are given by

$$\gamma_1 = \frac{\mu_3}{\mu_2^{3/2}} \tag{3.5}$$

and

$$\gamma_2 = \frac{\mu_4}{\mu_2^2}. \tag{3.6}$$

Substituting the sample moments for the population moments in Equations 3.5 and 3.6, we have

$$\hat{\gamma}_1 = \frac{\dfrac{1}{n}\sum_{i=1}^{n}(X_i - \bar{X})^3}{\left(\dfrac{1}{n}\sum_{i=1}^{n}(X_i - \bar{X})^2\right)^{3/2}}, \tag{3.7}$$

and

$$\hat{\gamma}_2 = \frac{\dfrac{1}{n}\sum_{i=1}^{n}(X_i - \bar{X})^4}{\left(\dfrac{1}{n}\sum_{i=1}^{n}(X_i - \bar{X})^2\right)^2}. \tag{3.8}$$

We are using the hat (^) notation to denote an estimate. Thus, $\hat{\gamma}_1$ is an estimate for γ_1. The following example shows how to use MATLAB to obtain the sample coefficient of skewness and sample coefficient of kurtosis.

Example 3.1

In this example, we will generate a random sample that is uniformly distributed over the interval (0, 1). We would expect this sample to have a coefficient of skewness close to zero because it is a symmetric distribution. We would expect the kurtosis to be different from 3 because the random sample is not generated from a normal distribution.

```
% Generate a random sample from the uniform
% distribution.
n = 2000;
x = rand(1,n);
% Find the mean of the sample.
mu = mean(x);
% Find the numerator and denominator for gamma_1.
num = (1/n)*sum((x-mu).^3);
den = (1/n)*sum((x-mu).^2);
gam1 = num/den^(3/2);
```

This results in a coefficient of skewness of **gam1 = -0.0542**, which is not too far from zero. Now we find the kurtosis using the following MATLAB commands:

```
% Find the kurtosis.
num = (1/n)*sum((x-mu).^4);
den = (1/n)*sum((x-mu).^2);
gam2 = num/den^2;
```

This gives a kurtosis of **gam2 = 1.8766**, which is not close to 3, as expected. ❑

We note that these statistics might not be the best to use in terms of bias (see Section 3.4). However, they will prove to be useful as examples in Chapters 7 and 8, where we look at bootstrap methods for estimating the bias in a statistic. The MATLAB Statistics Toolbox function called **skewness** returns the coefficient of skewness for a random sample. The function **kurtosis** calculates the sample coefficient of kurtosis (*not* the coefficient of excess kurtosis).

Covariance

In the definitions given below (Equations 3.9 and 3.10), we assume that all expectations exist. The *covariance* of two random variables X and Y, with joint probability density function $f(x, y)$, is defined as

$$Cov(X, Y) = \sigma_{X,Y} = E[(X - \mu_X)(Y - \mu_Y)]. \tag{3.9}$$

The *correlation coefficient* of X and Y is given by

$$Corr(X, Y) = \rho_{X,Y} = \frac{Cov(X, Y)}{\sigma_X \sigma_Y} = \frac{\sigma_{X,Y}}{\sigma_X \sigma_Y}, \tag{3.10}$$

where $\sigma_X > 0$ and $\sigma_Y > 0$.

The correlation is a measure of the linear relationship between two random variables. The correlation coefficient has the following range: $-1 \leq \rho_{X,Y} \leq 1$. When $\rho_{X,Y} = 1$, then X and Y are perfectly positively correlated. This means that the possible values for X and Y lie on a line with positive slope. On the other hand, when $\rho_{X,Y} = -1$, then the situation is the opposite: X and Y are perfectly negatively correlated. If X and Y are independent, then $\rho_{X,Y} = 0$. Note that the converse of this statement does not necessarily hold.

There are statistics that can be used to estimate these quantities. Let's say we have a random sample of size n denoted as $(X_1, Y_1), ..., (X_n, Y_n)$. The sample covariance is typically calculated using the following statistic:

$$\hat{\sigma}_{X,Y} = \frac{1}{n-1} \sum_{i=1}^{n} (X_i - \bar{X})(Y_i - \bar{Y}). \tag{3.11}$$

This is the definition used in the MATLAB function **cov**. In some instances, the empirical covariance is used [Efron and Tibshirani, 1993]. This is similar to Equation 3.11, except that we divide by n instead of $n - 1$. We note that the function **cov** in the basic MATLAB package has this option; see the **help** documentation for more information.

We just showed how to determine the covariance between two variables, but this is easily extended to any number of variables. We first transform the data to have zero mean, which means that we subtract the d-dimensional sample mean from each row of the $n \times d$ data matrix **X**. We will denote this as $\mathbf{X}_c$. We then compute the sample covariance matrix **S** as

$$\mathbf{S} = \frac{1}{n-1} \mathbf{X}_c^T \mathbf{X}_c,$$

where the superscript T denotes a matrix transpose. The jk-th element of **S** is given by

$$s_{jk} = \frac{1}{n-1} \sum_{i=1}^{n} (x_{ij} - \bar{x}_j)(x_{ik} - \bar{x}_k); \qquad j, k = 1, ..., d.$$

Referring to Equation 3.10, we see that the sample correlation coefficient for two variables is given by

$$\hat{\rho}_{X,Y} = \frac{\sum_{i=1}^{n}(X_i - \bar{X})(Y_i - \bar{Y})}{\left(\sum_{i=1}^{n}(X_i - \bar{X})^2\right)^{1/2}\left(\sum_{i=1}^{n}(Y_i - \bar{Y})^2\right)^{1/2}}. \tag{3.12}$$

In the next example, we investigate the commands available in MATLAB that return the statistics given in Equations 3.11 and 3.12.

Example 3.2

In this example, we show how to use the MATLAB **cov** function to find the covariance between two variables and the **corrcoef** function to find the correlation coefficient. Both of these functions are available in the standard MATLAB language. We use the **cement** data [Hand, et al., 1994], which were analyzed by Hald [1952], to illustrate the basic syntax of these functions. The relationship between the two variables is nonlinear, so Hald looked at the log of the tensile strength as a function of the reciprocal of the drying time. When the **cement** data are loaded, we get a vector **x** representing the drying times and a vector **y** that contains the tensile strength. A scatterplot of the transformed data is shown in Figure 3.1.

```
% First load the data.
load cement
% Now get the transformations.
xr = 1./x;
logy = log(y);
% Now get a scatterplot of the data to see if
% the relationship is linear.
plot(xr,logy,'x')
axis([0 1.1 2.4 4])
xlabel('Reciprocal of Drying Time')
ylabel('Log of Tensile Strength')
```

We now show how to get the covariance matrix and the correlation coefficient for these two variables.

```
% Now get the covariance and
% the correlation coefficient.
cmat = cov(xr,logy);
cormat = corrcoef(xr,logy);
```

The results are

```
cmat =
      0.1020     -0.1169
     -0.1169      0.1393
cormat =
      1.0000     -0.9803
     -0.9803      1.0000
```

Note that the sample correlation coefficient (Equation 3.12) is given by the off-diagonal element of **cormat**, $\hat{\rho} = -0.9803$. We see that the variables are negatively correlated, which is what we expect from Figure 3.1 (the log of the tensile strength decreases with increasing reciprocal of drying time).
❑

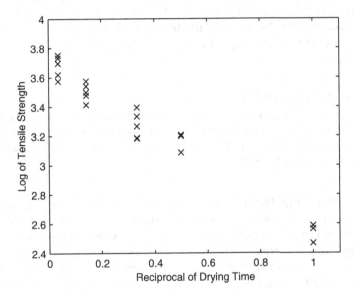

FIGURE 3.1
This scatterplot shows the observed drying times and corresponding tensile strength of the cement. Since the relationship is nonlinear, the variables are transformed as shown here. A linear relationship seems to be a reasonable model for these data.

3.3 Sampling Distributions

It was stated in the previous section that we sometimes use a statistic calculated from a random sample as a point estimate of a population parameter. For example, we might use $\bar{X}$ to estimate μ or use S to estimate σ.

Since we are using a sample and not observing the entire population, there will be some error in our estimate. In other words, it is unlikely that the statistic will equal the parameter. To manage the uncertainty and error in our estimate, we must know the sampling distribution for the statistic. The *sampling distribution* is the probability distribution for a statistic. To understand the remainder of the text, it is important to remember that *a statistic is a random variable.*

The sampling distributions for many common statistics are known. For example, if our random variable is from the normal distribution, then we know how the sample mean is distributed. Once we know the sampling distribution of our statistic, we can perform statistical hypothesis tests and calculate confidence intervals. If we do not know the distribution of our statistic, then we must use Monte Carlo simulation techniques or bootstrap methods to estimate the sampling distribution (see Chapter 7).

To illustrate the concept of a sampling distribution, we discuss the sampling distribution for $\overline{X}$, where the random variable X follows a distribution given by the probability density function $f(x)$. It turns out that the distribution for the sample mean can be found using the Central Limit Theorem.

CENTRAL LIMIT THEOREM

Let $f(x)$ represent a probability density with finite variance σ^2 and mean μ. Also, let $\overline{X}$ be the sample mean for a random sample of size n drawn from this distribution. For large n, the distribution of $\overline{X}$ is approximately normally distributed with mean μ and variance given by σ^2/n.

The Central Limit Theorem states that as the sample size gets large, the distribution of the sample mean approaches the normal distribution regardless of how the random variable X is distributed. However, if we are sampling from a normal population, then the distribution of the sample mean is exactly normally distributed with mean μ and variance σ^2/n.

This information is important because we can use it to understand the error in using $\overline{X}$ as an estimate of the population mean μ. We can also perform statistical hypothesis tests using $\overline{X}$ as a test statistic and can calculate confidence intervals for μ. In this book, we are mainly concerned with computational rather than theoretical methods for finding sampling distributions of statistics (e.g., Monte Carlo simulation or resampling). The sampling distribution of $\overline{X}$ is used to illustrate the concepts covered in the remaining chapters.

3.4 Parameter Estimation

One of the first tasks a statistician or an engineer undertakes when faced with data is to try to summarize or describe the data in some manner. Some of the statistics (sample mean, sample variance, coefficient of skewness, etc.) we covered in Section 3.2 can be used as descriptive measures for our sample. In this section, we look at methods to derive and to evaluate estimates of population parameters.

There are several methods available for obtaining parameter estimates. These include the method of moments, maximum likelihood estimation, Bayes estimators, minimax estimation, Pitman estimators, interval estimates, robust estimation, and many others. In this book, we discuss the maximum likelihood method and the method of moments for deriving estimates for population parameters. These somewhat classical techniques are included as illustrative examples only and are not meant to reflect the state of the art in this area. Many useful (and computationally intensive!) methods are not covered here, but references are provided in Section 3.7. Additionally, we do present some alternative methods for calculating interval estimates using Monte Carlo simulation and resampling methods (see Chapters 7 and 8).

Recall that a sample is drawn from a population that is distributed according to some function whose characteristics are governed by certain parameters. For example, our sample might come from a population that is normally distributed with parameters μ and σ^2. Or, it might be from a population that is exponentially distributed with parameter λ. The goal is to use the sample to estimate the corresponding population parameters. If the sample is representative of the population, then a function of the sample should provide a useful estimate of the parameters.

Before we undertake our discussion of maximum likelihood, we need to define what an estimator is. Typically, population parameters can take on values from a subset of the real line. For example, the population mean can be any real number, $-\infty < \mu < \infty$, and the population standard deviation can be any positive real number, $\sigma > 0$. The set of all possible values for a parameter θ is called the *parameter space*. The *data space* is defined as the set of all possible values of the random sample of size n. The estimate is calculated from the sample data as a function of the random sample. An *estimator* is a function or mapping from the data space to the parameter space and is denoted as

$$T = t(X_1, ..., X_n). \tag{3.13}$$

Since an estimator is calculated using the sample alone, it is a statistic. Furthermore, if we have a random sample, then an estimator is also a random variable. This means that the value of the estimator varies from one sample to another based on its sampling distribution. In order to assess the usefulness of our estimator, we need to have some criteria to measure the performance. We discuss four criteria used to assess estimators: bias, mean squared error, efficiency, and standard error.

Bias

The *bias* in an estimator gives a measure of how much error we have, on average, in our estimate when we use T to estimate our parameter θ. The *bias* is defined as

$$\text{bias}(T) = E[T] - \theta. \tag{3.14}$$

If the estimator is unbiased, then the expected value of our estimator equals the true parameter value, so $E[T] = \theta$.

To determine the expected value in Equation 3.14, we must know the distribution of the statistic T. In these situations, the bias can be determined analytically. When the distribution of the statistic is not known, then we can use methods such as the jackknife and the bootstrap (see Chapters 7 and 8) to estimate the bias of T.

Mean Squared Error

Let θ denote the parameter we are estimating and T denote our estimate, then the *mean squared error* (MSE) of the estimator is defined as

$$\text{MSE}(T) = E[(T - \theta)^2]. \tag{3.15}$$

Thus, the MSE is the expected value of the squared error. We can write this in more useful quantities such as the bias and variance of T. (The reader will see this again in Chapter 9 in the context of probability density estimation.) If we expand the expected value on the right-hand side of Equation 3.15, then we have

$$\text{MSE}(T) = E[(T^2 - 2T\theta + \theta^2)] = E[T^2] - 2\theta E[T] + \theta^2. \tag{3.16}$$

By adding and subtracting $(E[T])^2$ to the right-hand side of Equation 3.16, we have the following:

$$\text{MSE}(T) = E[T^2] - (E[T])^2 + (E[T])^2 - 2\theta E[T] + \theta^2. \tag{3.17}$$

The first two terms of Equation 3.17 are the variance of T, and the last three terms equal the squared bias of our estimator. Thus, we can write the mean squared error as

$$
\begin{aligned}
\text{MSE}(T) &= E[T^2] - (E[T])^2 + (E[T] - \theta)^2 \\
&= V(T) + [\text{bias}(T)]^2.
\end{aligned}
$$
(3.18)

Since the mean squared error is based on the variance and the squared bias, the error will be small when the variance and the bias are both small. When T is unbiased, then the mean squared error is equal to the variance only. The concepts of bias and variance are important for assessing the performance of any estimator.

Relative Efficiency

Another measure we can use to compare estimators is called *efficiency*, which is defined using the MSE. For example, suppose we have two estimators $T_1 = t_1(X_1, ..., X_n)$ and $T_2 = t_2(X_1, ..., X_n)$. If the MSE of one estimator is less than the other, e.g., $\text{MSE}(T_1) < \text{MSE}(T_2)$, then T_1 is said to be more efficient than T_2.

The *relative efficiency* of T_1 to T_2 is given by

$$
\textit{eff}(T_1, T_2) = \frac{\text{MSE}(T_2)}{\text{MSE}(T_1)}.
$$
(3.19)

If this ratio is greater than one, then T_1 is a more efficient estimator of the parameter.

Standard Error

We can get a measure of the precision of our estimator by calculating the standard error. The *standard error* of an estimator (or a statistic) is defined as the standard deviation of its sampling distribution:

$$
SE(T) = \sqrt{V(T)} = \sigma_T.
$$

To illustrate this concept, let's use the sample mean as an example. We know that the variance of the estimator is

$$
V(\overline{X}) = \frac{1}{n}\sigma^2
$$

for large n. So, the standard error is given by

$$SE(\overline{X}) = \sigma_{\overline{x}} = \frac{\sigma}{\sqrt{n}}. \tag{3.20}$$

If the standard deviation σ for the underlying population is unknown, then we can substitute an estimate for the parameter. In this case, we call it the estimated standard error:

$$\hat{SE}(\overline{X}) = \hat{\sigma}_{\overline{x}} = \frac{S}{\sqrt{n}}. \tag{3.21}$$

Note that the estimate in Equation 3.21 is also a random variable and has a probability distribution associated with it.

If the bias in an estimator is small, then the variance of the estimator is approximately equal to the MSE, $V(T) \approx \mathrm{MSE}(T)$. Thus, we can also use the square root of the MSE as an estimate of the standard error.

Maximum Likelihood Estimation

A *maximum likelihood estimator* is the value of the parameter (or parameters) that maximizes the likelihood function of the sample. The *likelihood function* of a random sample of size n from density (mass) function $f(x;\theta)$ is the joint probability density (mass) function, denoted by

$$L(\theta;x_1, ..., x_n) = f(x_1, ..., x_n;\theta). \tag{3.22}$$

Equation 3.22 provides the likelihood that the random variables take on a particular value $x_1, ..., x_n$. Note that the likelihood function L is a function of the parameter θ, and that we allow θ to represent a vector of parameters.

If we have a random sample (independent, identically distributed random variables), then we can write the likelihood function as

$$L(\theta) = L(\theta;x_1, ..., x_n) = f(x_1;\theta) \times ... \times f(x_n;\theta), \tag{3.23}$$

which is the product of the individual density functions evaluated at each x_i or sample point.

In most cases, to find the value $\hat{\theta}$ that maximizes the likelihood function, we take the derivative of L, set it equal to 0, and solve for θ. Thus, we solve the following likelihood equation:

$$\frac{d}{d\theta}L(\theta) = 0. \tag{3.24}$$

It can be shown that the likelihood function, $L(\theta)$, and logarithm of the likelihood function, $\ln[L(\theta)]$, have their maxima at the same value of θ. It is sometimes easier to find the maximum of $\ln[L(\theta)]$, especially when working with an exponential function. However, keep in mind that a solution to the above equation does not imply that it is a maximum; it could be a minimum. It is important to ensure this is the case before using the result as a maximum likelihood estimator.

When a distribution has more than one parameter, then the likelihood function is a function of all parameters that pertain to the distribution. In these situations, the maximum likelihood estimates are obtained by taking the partial derivatives of the likelihood function (or the logarithm of the likelihood), setting them all equal to zero, and solving the system of equations. The resulting estimators are called the joint maximum likelihood estimators. We see an example of this below, where we derive the maximum likelihood estimators for μ and σ^2 for the normal distribution.

Example 3.3

In this example, we derive the maximum likelihood estimators for the parameters of the normal distribution. We start off with the likelihood function for a random sample of size n given by

$$L(\theta) = \prod_{i=1}^{n} \frac{1}{\sigma\sqrt{2\pi}} \exp\left\{-\frac{(x_i - \mu)^2}{2\sigma^2}\right\} = \left(\frac{1}{2\pi\sigma^2}\right)^{n/2} \exp\left(-\frac{1}{2\sigma^2}\sum_{i=1}^{n}(x_i - \mu)^2\right).$$

Since this has the exponential function in it, we will take the logarithm to obtain

$$\ln[L(\theta)] = \ln\left[\left(\frac{1}{2\pi\sigma^2}\right)^{\frac{n}{2}}\right] + \ln\left[\exp\left(-\frac{1}{2\sigma^2}\sum_{i=1}^{n}(x_i - \mu)^2\right)\right].$$

This simplifies to

$$\ln[L(\theta)] = -\frac{n}{2}\ln[2\pi] - \frac{n}{2}\ln[\sigma^2] - \frac{1}{2\sigma^2}\sum_{i=1}^{n}(x_i - \mu)^2, \tag{3.25}$$

with $\sigma > 0$ and $-\infty < \mu < \infty$. The next step is to take the partial derivative of Equation 3.25 with respect to μ and σ^2. These derivatives are

$$\frac{\partial}{\partial \mu}\ln L = \frac{1}{\sigma^2}\sum_{i=1}^{n}(x_i - \mu) \tag{3.26}$$

and

$$\frac{\partial}{\partial \sigma^2}\ln L = -\frac{n}{2\sigma^2} + \frac{1}{2\sigma^4}\sum_{i=1}^{n}(x_i - \mu)^2. \tag{3.27}$$

We then set Equations 3.26 and 3.27 equal to zero and solve for μ and σ^2. Solving the first equation for μ, we get the familiar sample mean for the estimator.

$$\frac{1}{\sigma^2}\sum_{i=1}^{n}(x_i - \mu) = 0,$$

$$\sum_{i=1}^{n}x_i = n\mu,$$

$$\hat{\mu} = \bar{x} = \frac{1}{n}\sum_{i=1}^{n}x_i.$$

Substituting $\hat{\mu} = \bar{x}$ into Equation 3.27, setting it equal to zero, and solving for the variance, we get

$$-\frac{n}{2\sigma^2} + \frac{1}{2\sigma^4}\sum_{i=1}^{n}(x_i - \bar{x})^2 = 0$$

$$\hat{\sigma}^2 = \frac{1}{n}\sum_{i=1}^{n}(x_i - \bar{x})^2. \tag{3.28}$$

These are the sample moments about the sample mean, and it can be verified that these solutions jointly maximize the likelihood function [Lindgren, 1993].
❏

We know that the $E[\bar{X}] = \mu$ [Mood, Graybill, and Boes, 1974], so the sample mean is an unbiased estimator for the population mean. However, that is not the case for the maximum likelihood estimate for the variance. It can be shown [Hogg and Craig, 1978] that

$$E[\hat{\sigma}^2] = \frac{(n-1)\sigma^2}{n}$$

so we know (from Equation 3.14) that the maximum likelihood estimate, $\hat{\sigma}^2$, for the variance is biased. If we want to obtain an unbiased estimator for the variance, we simply multiply our maximum likelihood estimator by $(n/(n-1))$. This yields the familiar statistic for the sample variance given by

$$s^2 = \frac{1}{n-1}\sum_{i=1}^{n}(x_i - \bar{x})^2.$$

Method of Moments

In some cases, it is difficult finding the maximum of the likelihood function. For example, the gamma distribution has the unknown parameter t that is used in the gamma function, $\Gamma(t)$. This makes it hard to take derivatives and solve the equations for the unknown parameters. The method of moments is one way to approach this problem.

In general, we write the unknown population parameters in terms of the population moments. We then replace the population moments with the corresponding sample moments. We illustrate these concepts in the next example, where we find estimates for the parameters of the gamma distribution.

Example 3.4

The gamma distribution has two parameters, t and λ. Recall that the mean and variance are given by t/λ and t/λ^2, respectively. Writing these in terms of the population moments, we have

$$E[X] = \frac{t}{\lambda} \qquad (3.29)$$

and

$$V(X) = E[X^2] - (E[X])^2 = \frac{t}{\lambda^2}. \qquad (3.30)$$

The next step is to solve Equations 3.29 and 3.30 for t and λ. From Equation 3.29, we have $t = \lambda E[X]$, and substituting this in the second equation yields

$$E[X^2] - (E[X])^2 = \frac{\lambda E[X]}{\lambda^2}. \tag{3.31}$$

Rearranging Equation 3.31 gives the following expression for λ:

$$\lambda = \frac{E[X]}{E[X^2] - (E[X])^2}. \tag{3.32}$$

We can now obtain the parameter t in terms of the population moments (substitute Equation 3.32 for λ in Equation 3.29) as

$$t = \frac{(E[X])^2}{E[X^2] - (E[X])^2}. \tag{3.33}$$

To get our estimates, we substitute the sample moments for $E[X]$ and $E[X^2]$ in Equations 3.32 and 3.33. This yields

$$\hat{t} = \frac{\overline{X}^2}{\frac{1}{n}\sum_{i=1}^{n} X_i^2 - \overline{X}^2}, \tag{3.34}$$

and

$$\hat{\lambda} = \frac{\overline{X}}{\frac{1}{n}\sum_{i=1}^{n} X_i^2 - \overline{X}^2}. \tag{3.35}$$

❏

   In Table 3.1, we provide some suggested point estimates for several of the distributions covered in Chapter 2. This table also contains the names of functions to calculate the estimators. In Section 3.6, we discuss the MATLAB code available in the Statistics Toolbox for calculating maximum likelihood estimates of distribution parameters. The reader is cautioned that the estimators discussed in this chapter are not necessarily the best in terms of bias, variance, etc.

TABLE 3.1

Suggested Point Estimators for Parameters

Distribution	Suggested Estimator	MATLAB Function
Binomial	$\hat{p} = \dfrac{X}{n}$	`csbinpar`
Exponential	$\hat{\lambda} = 1/\overline{X}$	`csexpar`
Gamma	$\hat{t} = \overline{X}^2 \Big/ \left(\dfrac{1}{n}\sum X_i^2 - \overline{X}^2\right)$ $\hat{\lambda} = \overline{X} \Big/ \left(\dfrac{1}{n}\sum X_i^2 - \overline{X}^2\right)$	`csgampar`
Normal	$\hat{\mu} = \overline{X}$ $\hat{\sigma}^2 = S^2$	`mean` `var`
Multivariate normal	$\hat{\mu}_j = \dfrac{1}{n}\sum_{i=1}^{n} X_{ij}$ $\hat{\Sigma}_{ij} = \dfrac{n\sum_{k=1}^{n} X_{ik}X_{jk} - \sum_{k=1}^{n} X_{ik}\sum_{k=1}^{n} X_{jk}}{n(n-1)}$	`mean` `cov`
Poisson	$\hat{\lambda} = \overline{X}$	`cspoipar`

3.5 Empirical Distribution Function

Recall from Chapter 2 that the cumulative distribution function is given by

$$F(x) = P(X \leq x) = \int_{-\infty}^{x} f(t)\,dt \tag{3.36}$$

for a continuous random variable and by

$$F(a) = \sum_{x_i \le a} f(x_i) \tag{3.37}$$

for a discrete random variable. In this section, we examine the sample analog of the cumulative distribution function called the *empirical distribution function*. When it is not suitable to assume a distribution for the random variable, then we can use the empirical distribution function as an estimate of the underlying distribution. One can call this a *nonparametric* estimate of the distribution function because we are not assuming a specific parametric form for the distribution that generates the random phenomena. In a *parametric* setting, we would assume a particular distribution generated the sample and estimate the cumulative distribution function by estimating the appropriate function parameters.

The empirical distribution function is based on the *order statistics*. The order statistics for a sample are obtained by putting the data in ascending order. Thus, for a random sample of size n, the order statistics are defined as

$$X_{(1)} \le X_{(2)} \le \ldots \le X_{(n)},$$

with $X_{(i)}$ denoting the i-th order statistic. The order statistics for a random sample can be calculated easily in MATLAB using the **sort** function.

The empirical distribution function $\hat{F}_n(x)$ is defined as the number of data points less than or equal to x divided by the sample size n. It can be expressed in terms of the order statistics as follows:

$$\hat{F}_n(x) = \begin{cases} 0; & x < X_{(1)} \\ j/n; & X_{(j)} \le x < X_{(j+1)} \\ 1; & x \ge X_{(n)}. \end{cases} \tag{3.38}$$

Figure 3.2 illustrates these concepts. We show the empirical cumulative distribution function for a standard normal and include the theoretical distribution function to verify the results. In the following section, we describe a descriptive measure for a population called a quantile, along with its corresponding estimate. Quantiles are introduced here because they are based on the cumulative distribution function.

Quantiles

Quantiles have a fundamental role in statistics. For example, they can be used as a measure of central tendency and dispersion, they provide the critical values in hypothesis testing (see Chapter 7), and they are used in exploratory data analysis for assessing distributions (see Chapters 5 and 6).

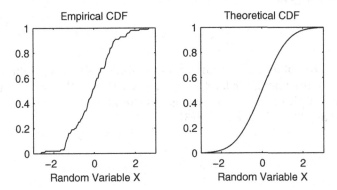

FIGURE 3.2
This shows the theoretical and empirical distribution functions for a standard normal distribution.

The **quantile** of a random variable (or equivalently of its distribution) is defined as a number q_p such that a proportion p of the distribution is less than or equal to q_p, where $0 < p < 1$. This can be calculated for a continuous random variable with density function $f(x)$ by solving

$$p = \int_{-\infty}^{q_p} f(x)dx \tag{3.39}$$

for q_p, or by using the inverse of the cumulative distribution function,

$$q_p = F^{-1}(p). \tag{3.40}$$

Stating this another way, the p-th quantile of a random variable X is the value q_p such that

$$F(q_p) = P(X \le q_p) = p \tag{3.41}$$

for $0 < p < 1$.

Some well-known examples of quantiles are the **quartiles**. These are denoted by $q_{0.25}$, $q_{0.5}$, and $q_{0.75}$. In essence, these divide the distribution into four equal (in terms of probability or area under the curve) segments. The second quartile is also called the **median** and satisfies

$$0.5 = \int_{-\infty}^{q_{0.5}} f(x)dx. \tag{3.42}$$

We can get a measure of the dispersion of the random variable by looking at the *interquartile range* (IQR) given by

$$IQR = q_{0.75} - q_{0.25}. \tag{3.43}$$

One way to obtain an estimate of the quantiles is based on the empirical distribution function. If we let $X_{(1)}, X_{(2)}, ..., X_{(n)}$ denote the order statistics for a random sample of size n, then $X_{(j)}$ is an estimate of the $(j - 0.5)/n$ quantile [Banks, et al., 2001; Cleveland, 1993]:

$$X_{(j)} \approx F^{-1}\left(\frac{j - 0.5}{n}\right). \tag{3.44}$$

We are not limited to a value of 0.5 in Equation 3.44. In general, we can estimate the p-th quantile using the following:

$$\hat{q}_p = X_{(j)}; \qquad \frac{j-1}{n} < p \le \frac{j}{n}; \qquad j = 1, ..., n. \tag{3.45}$$

As already stated, Equation 3.45 is not the only way to estimate quantiles. For more information on other methods, see Kotz and Johnson [Vol. 7, 1986]. The analyst should exercise caution when calculating quartiles (or other quantiles) using computer packages. Statistical software packages define them differently [Frigge, Hoaglin, and Iglewicz, 1989], so these statistics might vary depending on the formulas that are used.

Example 3.5

In this example, we will show one way to determine the sample quartiles. The second sample quartile $\hat{q}_{0.5}$ is the sample median of the data set. We can calculate this using the function **median**. We could calculate the first quartile $\hat{q}_{0.25}$ as the median of the ordered data that are at the median or below. The third quartile $\hat{q}_{0.75}$ would be calculated as the median of the data that are at $\hat{q}_{0.5}$ or above. The following MATLAB code illustrates these concepts.

```
% Generate the random sample and sort.
x = sort(rand(1,100));
% Find the median of the lower half - first quartile.
q1 = median(x(1:50));
% Find the median.
q2 = median(x);
% Find the median of the upper half - third quartile.
q3 = median(x(51:100));
```

The quartiles obtained from this random sample are

$$q1 = 0.29, \quad q2 = 0.53, \quad q3 = 0.79$$

The theoretical quartiles for the uniform distribution are $q_{0.25} = 0.25$, $q_{0.5} = 0.5$, and $q_{0.75} = 0.75$. So, we see that the estimates seem reasonable.
❑

Equation 3.44 provides one way to estimate the quantiles from a random sample. In some situations, we might need to determine an estimate of a quantile that does not correspond to $(j - 0.5)/n$. For instance, this is the case when we are constructing q-q plots (see Chapter 5), and the sample sizes differ. We can use interpolation to find estimates of quantiles that are not represented by Equation 3.44.

Example 3.6

The MATLAB function **interp1** (in the standard package) returns the interpolated value Y_I at a given X_I, based on some observed values X_{obs} and Y_{obs}. The general syntax is

```
yint = interp1(xobs, yobs, xint);
```

In our case, the argument of F^{-1} in Equation 3.44 represents the observed values X_{obs}, and the order statistics $X_{(j)}$ correspond to the Y_{obs}. The MATLAB code for this procedure is shown next.

```
% First generate some standard normal data.
x = randn(500,1);
% Now get the order statistics. These will serve
% as the observed values for the ordinate (Y_obs).
xs = sort(x);
% Now get the observed values for the abscissa (X_obs).
n=length(x);
phat = ((1:n)-0.5)/n;
% We want to get the quartiles.
p = [0.25, 0.5, 0.75];
% The following provides the estimates of the quartiles
% using linear interpolation.
qhat = interp1(phat,xs,p);
```

The resulting estimates are

$$\text{qhat} = -0.6928 \quad 0.0574 \quad 0.6453.$$

The reader is asked to explore this further in the exercises.
❑

3.6 MATLAB® Code

The MATLAB Statistics Toolbox has functions for calculating the maximum likelihood estimates for most of the common distributions, including the gamma and the Weibull distributions. It is important to remember that the parameters estimated for some of the distributions (e.g., exponential and gamma) are different from those defined in Chapters 2 and 3. Table 3.2 provides a partial list of MATLAB functions for calculating statistics, and Table 3.3 has relevant functions from the Computational Statistics Toolbox.

TABLE 3.2

List of MATLAB functions for calculating statistics

Purpose	MATLAB Function
These functions are available in the standard MATLAB package.	`mean` `var` `std` `cov` `median` `corrcoef` `max, min` `sort`
These functions for calculating descriptive statistics are available in the MATLAB Statistics Toolbox.	`harmmean` `iqr` `kurtosis` `mad` `moment` `prctile` `range` `skewness` `trimmean`
These MATLAB Statistics Toolbox functions provide the maximum likelihood estimates for distributions.	`betafit` `binofit` `expfit` `gamfit` `normfit` `poissfit` `weibfit` `unifit` `mle`

TABLE 3.3

List of Functions from Chapter 3 Included in the
Computational Statistics Toolbox

Purpose	MATLAB Function
These functions are used to obtain parameter estimates for a distribution.	`csbinpar`
	`csexpar`
	`csgampar`
	`cspoipar`
	`csunipar`
These functions return the quantiles.	`csbinoq`
	`csexpoq`
	`csunifq`
	`csweibq`
	`csnormq`
	`csquantiles`
Other descriptive statistics	`csmomentc`
	`cskewness`
	`cskurtosis`
	`csmoment`
	`csecdf`

3.7 Further Reading

Many books discuss sampling distributions and parameter estimation. These topics are covered at an undergraduate level in most introductory statistics books for engineers or nonstatisticians. For the advanced undergraduate and beginning graduate student, we recommend the text on mathematical statistics by Hogg and Craig [1978]. Another excellent introductory book on mathematical statistics that contains many applications and examples is written by Mood, Graybill, and Boes [1974]. Other texts at this same level include Bain and Engelhardt [1992], Bickel and Doksum [2001], and Lindgren [1993]. For the reader interested in the theory of point estimation on a more advanced graduate level, the books written by Lehmann and Casella [1998] and Lehmann [1994] are classics.

Most of the texts already mentioned include descriptions of other methods (Bayes methods, minimax methods, Pitman estimators, etc.) for estimating parameters. For an introduction to robust estimation methods, see the books by Wilcox [1997], Launer and Wilkinson [1979], Huber [1981], or Rousseeuw and Leroy [1987]. The survey paper by Hogg [1974] is another useful resource. Finally, the text by Keating, Mason, and Sen [1993] provides an introduction to Pitman's measure of closeness as a way to assess the performance of competing estimators.

Exercises

3.1.   Generate 500 random samples from the standard normal distribution for sample sizes of $n = 2, 15$, and 45. At each sample size, calculate the sample mean for all 500 samples. How are the means distributed as n gets large? Look at a histogram of the sample means to help answer this question. What is the mean and variance of the sample means for each n? Is this what you would expect from the Central Limit Theorem? Here is some MATLAB code to get you started.

For each n:

```
% Generate 500 random samples of size n:
x = randn(n, 500);
% Get the mean of each sample:
xbar = mean(x);
% Do a histogram with superimposed normal density.
% This function is in the MATLAB Statistics Toolbox.
% If you do not have this, then just use the
% function hist instead of histfit.
histfit(xbar);
```

3.2.   Repeat Problem 3.1 for random samples drawn from a uniform distribution. Use the MATLAB function **rand** to get the samples.

3.3.   We have two unbiased estimators T_1 and T_2 of the parameter θ. The variances of the estimators are given by $V(T_2) = 8$ and $V(T_1) = 4$. What is the MSE of the estimators? Which estimator is better and why? What is the relative efficiency of the two estimators?

3.4.   Repeat Example 3.1 using different sample sizes. What happens to the coefficient of skewness and kurtosis as the sample size gets large?

3.5.   Repeat Example 3.1 using samples generated from a standard normal distribution. You can use the MATLAB function **randn** to generate your samples. What happens to the coefficient of skewness and kurtosis as the sample size gets large?

3.6.   Generate a random sample that is uniformly distributed over the interval $(0, 1)$. Plot the empirical distribution function over the interval $(-0.5, 1.5)$. There is also a function in the Statistics Toolbox called **cdfplot** that will do this.

3.7. Generate a random sample of size 100 from a normal distribution with mean 10 and variance of 2. Use the following:

randn(1,100)*sqrt(2)+10)

Plot the empirical cumulative distribution function. What is the value of the empirical distribution function evaluated at a point less than the smallest observation in your random sample? What is the value of the empirical cumulative distribution function evaluated at a point that is greater than the largest observation in your random sample?

3.8. Generate a random sample of size 100 from a normal distribution. What are the estimated quartiles?

3.9. Generate a random sample of size 100 from a uniform distribution (use the MATLAB function **rand** to generate the samples). What are the sample quantiles for $p = 0.33, 0.40, 0.63, 0.90$? Is this what you would expect from theory?

3.10. Write a MATLAB function that will return the sample quartiles based on the general definition given for sample quantiles (Equation 3.44).

3.11. Repeat Examples 3.5 and 3.6 for larger sample sizes. Do your estimates for the quartiles get closer to the theoretical values?

3.12. Derive the median for an exponential random variable.

3.13. Calculate the quartiles for the exponential distribution.

3.14. Compare the values obtained for the estimated quartiles in Example 3.6 with the theoretical quantities. You can find the theoretical quantities using **norminv**. Increase the sample size to $n = 1000$. Does your estimate get better?

3.15. Another measure of skewness, called the *quartile coefficient of skewness*, for a sample is given by

$$\hat{\gamma}_{1_q} = \frac{\hat{q}_{0.75} - 2\hat{q}_{0.5} + \hat{q}_{0.25}}{\hat{q}_{0.75} - \hat{q}_{0.25}}.$$

Write a MATLAB function that returns this statistic.

3.16. Investigate the bias in the maximum likelihood estimate of the variance that is given in Equation 3.28. Generate a random sample from the standard normal distribution. You can use the **randn** function that is available in the standard MATLAB package. Calculate $\hat{\sigma}^2$ using Equation 3.28 and record the value in a vector. Repeat this process (generate a random sample from the standard normal distribution, estimate the variance, save the value) many times. Once you are done with this procedure, you should have many estimates for the variance. Take the mean of these estimates to get an estimate of the expected value of $\hat{\sigma}^2$. How does this compare with the known value of $\sigma^2 = 1$? Does this indicate that the maximum likelihood estimate for the variance is biased? What is the estimated bias from this procedure?

Chapter 4

Generating Random Variables

4.1 Introduction

Many of the methods in computational statistics require the ability to generate random variables from known probability distributions. This is at the heart of Monte Carlo simulation for statistical inference (Chapter 7), bootstrap and resampling methods (Chapters 7 and 8), and Markov chain Monte Carlo techniques (Chapter 14). In addition, we use simulated random variables to illustrate many other topics in this book.

There are many excellent books available that discuss techniques for generating random variables and the underlying theory; references will be provided in the last section. Our purpose in covering this topic is to give readers the tools they need to generate the types of random variables that often arise in practice and to provide examples illustrating the methods. We first discuss general techniques for generating random variables, such as the inverse transformation and acceptance-rejection methods. We then provide algorithms and MATLAB® code for generating random variables for some useful distributions.

4.2 General Techniques for Generating Random Variables

Uniform Random Numbers

Most methods for generating random variables start with random numbers that are uniformly distributed on the interval $(0, 1)$. We will denote these random variables by the letter U. With the advent of computers, we now have the ability to generate uniform random variables very easily. However, we have to caution the reader that the numbers generated by computers are really *pseudorandom* because they are generated using a deterministic algorithm. The techniques used to generate uniform random variables have been widely studied in the literature, and it has been shown that some generators have serious flaws [Gentle, 1998].

The basic MATLAB program has a three functions for generating random variables. The function **rand** generates uniform random numbers in the open interval $(0, 1)$. Use **randi(imax, n)** to get n random integers from the discrete uniform distribution ranging from 1 to **imax**. The **randn** function generates random variables from the standard normal distribution.

There are several optional arguments to **rand** and **randn**, and we take a moment to discuss them because they will be useful in simulations. The function **rand** with no arguments returns a single instance of the random variable U. To get an $m \times n$ array of uniform variates, you can use the syntax **rand(m, n)**. A note of caution: if you use **rand(n)**, then you get an $n \times n$ matrix. This can be very large and might cause memory errors in MATLAB.

The sequence of random numbers that is generated in MATLAB depends on the seed of the generator. The seed is reset to the default of zero when it starts up, so the same sequences of random variables are generated whenever you start MATLAB. This can sometimes be an advantage in situations where we would like to obtain a specific random sample, as we illustrate in the next example.

There is a function called **rng** that controls the seed of the generator and which random number generator is used by **rand**, **randn**, and **randi**. If you call the function with no input argument, then it returns information about the random number generator. For example, this is the information we get in the command window before we have generated any random numbers:

```
rng

ans =

    Type: 'twister'
    Seed: 0
    State: [625x1 uint32]
```

Note that it provides the type of generator, which is the Mersenne Twister and the seed. We can set the seed to a specified value using

```
rng(seed)
```

where **seed** is a number. We can reset the generator to the default values with

```
rng('default')
```

See the MATLAB **help** on **rng** for more information and options.

It should be noted that random numbers that are uniformly distributed over an interval a to b may be generated by a simple transformation, as follows:

$$X = (b - a) \cdot U + a. \tag{4.1}$$

Example 4.1

In this example, we illustrate the use of MATLAB's function **rand**.

```
% Obtain a vector of uniform random variables in (0,1).
x = rand(1,1000);
% Do a histogram to plot.
% First get the height of the bars.
[N,X] = hist(x,15);
% Use the bar function to see the distribution.
bar(X,N,1,'w')
title('Histogram of Uniform Random Variables')
xlabel('X')
ylabel('Frequency')
```

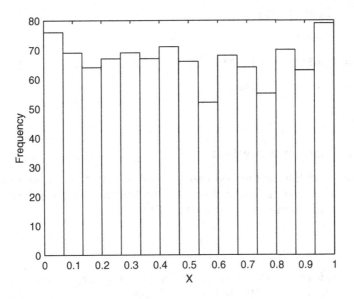

FIGURE 4.1
This figure shows a histogram of a random sample from the uniform distribution on the interval (0, 1).

The resulting histogram is shown in Figure 4.1. In some situations, the analyst might need to reproduce results from a simulation, say to verify a conclusion or to illustrate an interesting sample. To accomplish this, the state of the uniform random number generator should be specified at each iteration of the loop. This is can be done in MATLAB as shown next.

```
% Generate 3 random samples of size 5.
x = zeros(3,5);    % Allocate the memory.
for i = 1:3
    rng(i) % set the seed
    x(i,:) = rand(1,5);
end
```

The three sets of random variables are

```
x =
```

0.4170	0.7203	0.0001	0.3023	0.1468
0.4360	0.0259	0.5497	0.4353	0.4204
0.5508	0.7081	0.2909	0.5108	0.8929

We can easily recover the five random variables generated in the second sample by setting the state of the random number generator, as follows:

```
rng(2)
xt = rand(1,5);
```

From this, we get

```
xt = 0.4360    0.0259    0.5497    0.4353    0.4204
```

which is the same as before.

❏

Inverse Transform Method

The inverse transform method can be used to generate random variables from a continuous distribution. It uses the fact that the cumulative distribution function F is uniform $(0, 1)$ [Ross, 1997]:

$$U = F(X). \tag{4.2}$$

If U is a uniform $(0, 1)$ random variable, then we can obtain the desired random variable X from the following relationship:

$$X = F^{-1}(U). \tag{4.3}$$

We see an example of how to use the inverse transform method when we discuss generating random variables from the exponential distribution (see Example 4.6). The general procedure for the inverse transformation method is outlined here.

PROCEDURE – INVERSE TRANSFORM METHOD (CONTINUOUS)

1. Derive the expression for the inverse distribution function $F^{-1}(U)$.
2. Generate a uniform random number U.
3. Obtain the desired X from $X = F^{-1}(U)$.

This same technique can be adapted to the discrete case [Banks, et al., 2001]. Say we would like to generate a discrete random variable X that has a probability mass function given by

$$P(X = x_i) = p_i; \qquad x_0 < x_1 < x_2 < \dots; \qquad \sum_i p_i = 1. \qquad (4.4)$$

We get the random variables by generating a random number U and then deliver the random number X according to the following:

$$X = x_i, \qquad \text{if} \quad F(x_{i-1}) < U \le F(x_i). \qquad (4.5)$$

We illustrate this procedure using a simple example.

Example 4.2
We would like to simulate a discrete random variable X that has probability mass function given by

$$P(X = 0) = 0.3,$$
$$P(X = 1) = 0.2,$$
$$P(X = 2) = 0.5.$$

The cumulative distribution function is

$$F(x) = \begin{cases} 0; & x < 0 \\ 0.3; & 0 \le x < 1 \\ 0.5; & 1 \le x < 2 \\ 1.0; & 2 \le x. \end{cases}$$

We generate random variables for X according to the following scheme

$$X = \begin{cases} 0; & U \le 0.3 \\ 1; & 0.3 < U \le 0.5 \\ 2; & 0.5 < U \le 1. \end{cases}$$

This is easily implemented in MATLAB as we show in the next example. The procedure is illustrated in Figure 4.2, for the situation where a uniform random variable 0.73 was generated. Note that this would return the variate $x = 2$.
❑

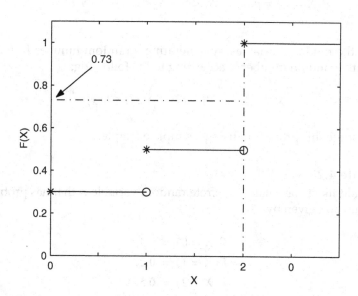

FIGURE 4.2

This figure illustrates the inverse transform procedure for generating discrete random variables. If we generate a uniform random number of u = 0.73, then this yields a random variable of x = 2.

We now outline the steps to implement this procedure. This will be useful when we describe a method for generating Poisson random variables.

PROCEDURE – INVERSE TRANSFORM (DISCRETE)

1. Define a probability mass function for x_i, $i = 1, ..., k$. Note that k could grow infinitely.
2. Generate a uniform random number U.
3. If $U \le p_0$, deliver $X = x_0$

4. Else if $U \leq p_0 + p_1$, deliver $X = x_1$
5. Else if $U \leq p_0 + p_1 + p_2$, deliver $X = x_2$
6. ... Else if $U \leq p_0 + ... + p_k$, deliver $X = x_k$.

Example 4.3

We repeat the previous example using this procedure and implement it in MATLAB. We generate 1000 variates from the desired probability mass function using the following commands.

```
% Set up storage space for the variables.
n = 1000;
X = zeros(1,n);
% These are the x's in the domain:
x = 0:2;
% These are the probability masses for each x:
pr = [0.3 0.2 0.5];
% Generate 1000 random variables from the desired
% distribution.
for i = 1:n
    u = rand;   % Generate the U.
    if u <= pr(1)
        X(i) = x(1);
    elseif u <= sum(pr(1:2))
        % It has to be between 0.3 and 0.5.
        X(i) = x(2);
    else
        X(i) = x(3); % It has to be between 0.5 and 1.
    end
end
```

One way to verify that our random variables are from the desired distribution is to look at the relative frequency of each x.

```
% Find the proportion of each number.
x0 = length(find(X==0))/n;
x1 = length(find(X==1))/n;
x2 = length(find(X==2))/n;
```

The resulting estimated probabilities are

$$\hat{P}(x = x_0) = 0.26$$

$$\hat{P}(x = x_1) = 0.21$$

$$\hat{P}(x = x_2) = 0.53.$$

These values are reasonable when compared with the desired probability mass values.
❏

Acceptance-Rejection Method

In some cases, we might have a simple method for generating a random variable from one density, say $g(y)$, instead of the density we are seeking. We can use this density to generate data from the desired continuous density $f(x)$. We first generate a random number Y from $g(y)$ and accept the value with a probability proportional to the ratio $(f(Y)/(g(Y)))$.

If we define c as a constant that satisfies

$$\frac{f(y)}{g(y)} \leq c; \qquad \text{for all } y, \tag{4.6}$$

then we can generate the desired variates using the procedure outlined next. The constant c is needed because we might have to adjust the height of $g(y)$ to ensure that it is above $f(y)$. We generate points from $cg(y)$, and those points that are inside the curve $f(y)$ are accepted as belonging to the desired density. Those that are outside are rejected. It is best to keep the number of rejected variates small for maximum efficiency.

PROCEDURE – ACCEPTANCE-REJECTION METHOD (CONTINUOUS)

1. Choose a density $g(y)$ that is easy to sample from.
2. Find a constant c such that Equation 4.6 is satisfied.
3. Generate a random number Y from $g(y)$.
4. Generate a uniform random number U.
5. If

$$U \leq \frac{f(Y)}{cg(Y)},$$

then accept $X = Y$, else go to step 3.

Example 4.4

We shall illustrate the acceptance-rejection method by generating random variables from the beta distribution with parameters $\alpha = 2$ and $\beta = 1$ [Ross, 1997]. This yields the following probability density function:

$$f(x) = 2x; \qquad 0 < x < 1. \tag{4.7}$$

Since the domain of this density is 0 to 1, we use the uniform distribution for our $g(y)$. We must find a constant that we can use to inflate the uniform so it is above the desired beta density. This constant is given by the maximum value of the density function, and from Equation 4.7, we see that $c = 2$. For more complicated functions, techniques from calculus or the MATLAB function **fminsearch** may be used. The following MATLAB code generates 100 random variates from the desired distribution. We save both the accepted and the rejected variates for display purposes only.

```
c = 2;    % constant
n = 100;   % Generate 100 random variables.
% Set up the arrays to store variates.
x = zeros(1,n);   % random variates
xy = zeros(1,n);% corresponding y values
rej = zeros(1,n);% rejected variates
rejy = zeros(1,n);  % corresponding y values
irv = 1;
irej = 1;
while irv <= n
    y = rand(1);   % random number from g(y)
    u = rand(1);   % random number for comparison
    if u <= 2*y/c;
        x(irv) = y;
        xy(irv) = u*c;
        irv = irv+1
    else
        rej(irej) = y;
        rejy(irej) = u*c; % really comparing u*c<=2*y
        irej = irej + 1
    end
end
```

In Figure 4.3, we show the accepted and rejected random variates that were generated in this process. Note that the accepted variates are those that are less than $f(x)$.
❑

We can easily adapt this method to generate random variables from a discrete distribution. Here we have a method for simulating a random variable with a probability mass function $q_i = P(Y = i)$, and we would like to obtain a random variable X having a probability mass function $p_i = P(X = i)$. As in the continuous case, we generate a random variable Y from q_i and accept this value with probability $(p_Y/(cq_Y))$.

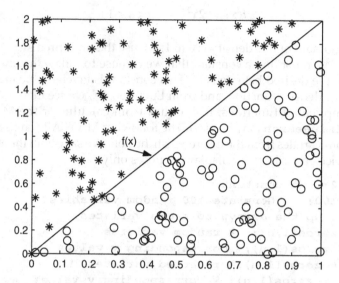

FIGURE 4.3
This shows the points that were accepted ("o") as being generated by $f(x) = 2x$ and those points that were rejected (""). The curve represents $f(x)$, so we see that the accepted variates are the ones below the curve.*

PROCEDURE – REJECTION METHOD (DISCRETE)

1. Choose a probability mass function q_i that is easy to sample from.
2. Find a constant c such that $p_Y < cq_Y$.
3. Generate a random number Y from the density q_i.
4. Generate a uniform random number U.

5. If

$$U \le \frac{p_Y}{cq_Y},$$

then deliver $X = Y$, else go to step 3.

Example 4.5
In this example, we use the discrete form of the acceptance-rejection method to generate random variables according to the probability mass function defined as follows:

$$P(X = 1) = 0.15,$$
$$P(X = 2) = 0.22,$$
$$P(X = 3) = 0.33,$$
$$P(X = 4) = 0.10,$$
$$P(X = 5) = 0.20.$$

We let q_Y be the discrete uniform distribution on $1, ..., 5,$ where the probability mass function is given by

$$q_y = \frac{1}{5}; \qquad y = 1, ..., 5.$$

We describe a method for generating random variables from the discrete uniform distribution in a later section. The value for c is obtained as the maximum value of p_y/q_y, which is 1.65. This quantity is obtained by taking the maximum p_y, which is $P(X = 3) = 0.33$, and dividing by $1/5$:

$$\frac{max(p_y)}{1/5} = 0.33 \times 5 = 1.65.$$

The steps for generating the variates are

1. Generate a variate Y from the discrete uniform density on $1, ..., 5$. One could use the base MATLAB function **randi**, the Statistics Toolbox function **unidrnd**, or **csdunrnd** from the Computational Statistics Toolbox.
2. Generate a uniform random number U.
3. If

$$U \le \frac{p_Y}{cq_Y} = \frac{p_Y}{1.65 \cdot 1/5} = \frac{p_Y}{0.33},$$

then deliver $X = Y$, else return to step 1.

The implementation of this example in MATLAB is left as an exercise.
❑

4.3 Generating Continuous Random Variables

Normal Distribution

The main MATLAB program has a function that will generate numbers from the standard normal distribution, so we do not discuss any techniques for generating random variables from the normal distribution. For the reader who is interested in how normal random variates can be generated, most of the references provided in Section 4.6 contain this information.

The MATLAB function for generating standard normal random variables is called **randn**, and its functionality is similar to the function **rand** that was discussed in the previous section. As with the uniform random variable U, we can obtain a normal random variable X with mean μ and variance σ^2 by means of a transformation. Letting Z represent a standard normal random variable (possibly generated from **randn**), we get the desired X from the relationship

$$X = Z \cdot \sigma + \mu. \tag{4.8}$$

Exponential Distribution

The inverse transform method can be used to generate random variables from the exponential distribution and serves as an example of this procedure. The distribution function for an exponential random variable with parameter λ is given by

$$F(x) = 1 - e^{-\lambda x}; \qquad 0 < x < \infty. \tag{4.9}$$

Letting

$$u = F(x) = 1 - e^{-\lambda x}, \tag{4.10}$$

we can solve for x, as follows:

$$u = 1 - e^{-\lambda x}$$
$$e^{-\lambda x} = 1 - u$$
$$-\lambda x = \log(1 - u)$$
$$x = -\frac{1}{\lambda}\log(1 - u).$$

By making note of the fact that $1 - u$ is also uniformly distributed over the interval (0,1), we can generate exponential random variables with parameter λ using the transformation

$$X = -\frac{1}{\lambda}\log(U).\tag{4.11}$$

Example 4.6

The following MATLAB code will generate exponential random variables for a given λ.

```
% Set up the parameters.
lam = 2;
n = 1000;
% Generate the random variables.
uni = rand(1,n);
X = -log(uni)/lam;
```

We can generate a set of random variables and plot them to verify that the function does yield exponentially distributed random variables. We plot a histogram of the results along with the theoretical probability density function in Figure 4.4. The MATLAB code given next shows how we did this.

```
% Get the values to draw the theoretical curve.
x = 0:.1:5;
% This is a function in the Statistics Toolbox.
y = exppdf(x,1/2);
% Get the information for the histogram.
[N,h] = hist(X,10);
% Change bar heights to make it correspond to
% the theoretical density - see Chapter 5.
N = N/(h(2)-h(1))/n;
% Do the plots.
bar(h,N,1,'w')
hold on
plot(x,y)
hold off
xlabel('X')
ylabel('f(x) - Exponential')
```

❑

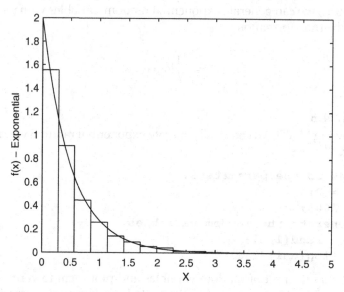

FIGURE 4.4
This shows a probability density histogram of the random variables generated in Example 4.6. We also superimpose the curve corresponding to the theoretical probability density function with $\lambda = 2$. The histogram and the curve match quite well.

Gamma

In this section, we present an algorithm for generating a gamma random variable with parameters (t, λ), where t is an integer. Recall that it has the following distribution function:

$$F(x) = \int_0^{\lambda x} \frac{e^{-y} y^{t-1}}{(t-1)!} dy. \tag{4.12}$$

The inverse transform method cannot be used in this case because a simple closed form solution for its inverse is not possible. It can be shown [Ross, 1997] that the sum of t independent exponentials with the same parameter λ is a gamma random variable with parameters t and λ. This leads to the following transformation based on t uniform random numbers:

$$X = -\frac{1}{\lambda} \log U_1 - \dots - \frac{1}{\lambda} \log U_t. \tag{4.13}$$

We can simplify this and compute only one logarithm by using a familiar relationship of logarithms. This yields the following:

$$X = -\frac{1}{\lambda}\log(U_1 \times \ldots \times U_t) = -\frac{1}{\lambda}\log\left(\prod_{i=1}^{t} U_i\right). \qquad (4.14)$$

Example 4.7

The MATLAB code given next implements the algorithm described above for generating gamma random variables, when the parameter t is an integer.

```
n = 1000;
t = 3;
lam = 2;
% Generate the uniforms needed. Each column
% contains the t uniforms for a realization of a
% gamma random variable.
U = rand(t,n);
% Transform according to Equation 4.13.
% See Example 4.8 for an illustration of Equation 4.14.
logU = -log(U)/lam;
X = sum(logU);
```

To see whether the implementation of the algorithm is correct, we plot them in a probability density histogram.

```
% Now do the histogram.
[N,h] = hist(X,10);
% Change bar heights.
N = N/(h(2)-h(1))/n;
% Now get the theoretical probability density.
% This is a function in the Statistics Toolbox.
x = 0:.1:6;
y = gampdf(x,t,1/lam);
bar(h,N,1,'w')
hold on
plot(x,y,'k')
hold off
```

The histogram and the corresponding theoretical probability density function are shown in Figure 4.5.
❏

Chi-Square

A chi-square random variable with v degrees of freedom is a special case of the gamma distribution, where $\lambda = 1/2$, $t = v/2$ and v is a positive integer. This can be generated using the gamma distribution method

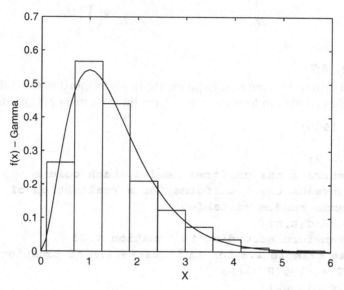

FIGURE 4.5
This shows the probability density histogram for a set of gamma random variables with
$t = 3$ *and* $\lambda = 2$.

described above with one change. We have to make this change because the method we presented for generating gamma random variables is for integer t, which works for even values of ν.

When ν is even, say $2k$, we can obtain a chi-square random variable from

$$X = -2\log\left(\prod_{i=1}^{k} U_i\right). \tag{4.15}$$

When ν is odd, say $2k + 1$, we can use the fact that the chi-square distribution with ν degrees of freedom is the sum of ν squared independent standard normals [Ross, 1997]. We obtain the required random variable by first simulating a chi-square with $2k$ degrees of freedom and adding a squared standard normal variate Z, as follows:

$$X = Z^2 - 2\log\left(\prod_{i=1}^{k} U_i\right). \tag{4.16}$$

Example 4.8

In this example, we provide a function that will generate chi-square random variables.

```
% function X = cschirnd(n,nu)
% This function will return n chi-square
% random variables with degrees of freedom nu.

function X = cschirnd(n,nu)
% Generate the uniforms needed.
rm = rem(nu,2);
k = floor(nu/2);
if rm == 0    % then even degrees of freedom
   U = rand(k,n);
   if k ~= 1
      X = -2*log(prod(U));
   else
      X = -2*log(U);
   end
else          % odd degrees of freedom
   U = rand(k,n);
   Z = randn(1,n);
   if k ~= 1
      X = Z.^2-2*log(prod(U));
   else
      X = Z.^2-2*log(U);
   end
end
```

The use of this function to generate random variables is left as an exercise.
❑

The chi-square distribution is useful in situations where we need to systematically investigate the behavior of a statistic by changing the skewness of the distribution. As the degrees of freedom for a chi-square increases, the distribution changes from being right skewed to one approaching normality and symmetry.

Beta

The beta distribution is useful in simulations because it covers a wide range of distribution shapes, depending on the values of the parameters α and β. These shapes include skewed, uniform, approximately normal, and a bimodal distribution with an interior dip.

First, we describe a simple approach for generating beta random variables with parameters α and β, when both are integers [Rubinstein, 1981; Gentle,

1998]. It is known [David, 1981] that the k-th order statistic of n uniform $(0,1)$ variates is distributed according to a beta distribution with parameters k and $n - k + 1$. This means that we can generate random variables from the beta distribution using the following procedure.

PROCEDURE – BETA RANDOM VARIABLES (INTEGER PARAMETERS)

   1. Generate $\alpha + \beta - 1$ uniform random numbers: $U_1, ..., U_{\alpha + \beta - 1}$
   2. Deliver $X = U_{(\alpha)}$ which is the α-th order statistic.

One simple way to generate random variates from the beta distribution is to use the following result from Rubinstein [1981]. If Y_1 and Y_2 are independent random variables, where Y_1 has a gamma distribution with parameters α and 1, and Y_2 follows a gamma distribution with parameters β and 1, then

$$X = \frac{Y_1}{Y_1 + Y_2} \tag{4.17}$$

is from a beta distribution with parameters α and β. This is the method that is used in the MATLAB Statistics Toolbox function **betarnd** that generates random variates from the beta distribution. We illustrate the use of **betarnd** in the following example.

Example 4.9
We use this example to illustrate the use of the MATLAB Statistics Toolbox function that generates beta random variables. In general, most of these toolbox functions for generating random variables use the following general syntax:

```
rvs = pdfrnd(par1,par2,nrow,ncol);
```

Here, *pdf* refers to the type of distribution (see Table 4.1). The first several arguments represent the appropriate parameters of the distribution, so the number of them might change. The last two arguments denote the number of rows and the number of columns in the array of random variables that are returned by the function. We use the function **betarnd** to generate random variables from two beta distributions with different parameters α and β. First we look at the case where $\alpha = 3$ and $\beta = 3$. To generate $n = 500$ beta random variables (that are returned in a row vector), we use the following commands:

```
% Let a = 3, b = 3
n = 500;
a = 3;
b = 3;
```

```
rvs = betarnd(a,b,1,n);
```

We can construct a histogram of the random variables and compare it to the corresponding beta probability density function. This is easily accomplished in MATLAB as shown next.

```
% Now do the histogram.
[N,h] = hist(rvs,10);
% Change bar heights.
N = N/(h(2)-h(1))/n;
% Now get the theoretical probability density.
x = 0:.05:1;
y = betapdf(x,a,b);
plot(x,y)
axis equal
bar(h,N,1,'w')
hold on
plot(x,y,'k')
hold off
```

The result is shown in the left plot of Figure 4.6. Notice that this density looks approximately bell-shaped. The beta density on the right has parameters $\alpha = 0.5$ and $\beta = 0.5$. We see that this curve has a dip in the middle with modes on either end. The reader is asked to construct this plot in the exercises.
❑

Multivariate Normal

In the following chapters, we will have many examples where we need to generate multivariate random variables in order to study the methods of computational statistics as they apply to multivariate distributions. Thus, we need some methods for generating multivariate random variables. The easiest distribution of this type to generate is the multivariate normal. We cover other methods for generating random variables from more general multivariate distributions in Chapter 14.

The method is similar to the one used to generate random variables from a univariate normal distribution. One starts with a d-dimensional vector of standard normal random numbers. These can be transformed to the desired distribution using

$$x = R^T z + \mu. \tag{4.18}$$

Here z is a $d \times 1$ vector of standard normal random numbers, μ is a $d \times 1$ vector representing the mean, and R is a $d \times d$ matrix such that $R^T R = \Sigma$. The matrix R can be obtained in several ways, one of which is the Cholesky

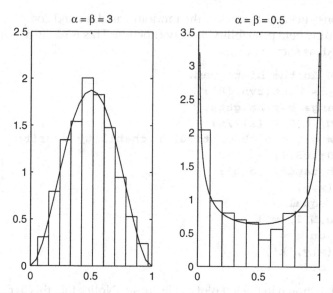

FIGURE 4.6
This figure shows two histograms created from random variables generated from the beta distribution. The beta distribution on the left has parameters α = 3 and β = 3, while the one on the right has parameters α = 0.5 and β = 0.5.

factorization of the covariance matrix Σ. This is the method we illustrate next. Another possibility is to factor the matrix using singular value decomposition, which will be shown in the examples provided in Chapter 5.

Example 4.10

The function **csmvrnd** generates multivariate normal random variables using the Cholesky factorization. Note that we are transposing the transformation given in Equation 4.18, yielding the following:

$$X = ZR + \mu^{T},$$

where X is an $n \times d$ matrix of d-dimensional random variables and Z is an $n \times d$ matrix of standard normal random variables.

```
% function X = csmvrnd(mu,covm,n);
% This function will return n multivariate random
% normal variables with d-dimensional mean mu and
% covariance matrix covm. Note that the covariance
% matrix must be positive definite (all eigenvalues
% are greater than zero), and the mean
% vector is a column
```

```
function X = csmvrnd(mu,covm,n)
d = length(mu);
% Get Cholesky factorization of covariance.
R = chol(covm);
% Generate the standard normal random variables.
Z = randn(n,d);
X = Z*R + ones(n,1)*mu';
```

We illustrate its use by generating some multivariate normal random variables with $\mu^T = (-2, 3)$ and covariance

$$\Sigma = \begin{bmatrix} 1 & 0.7 \\ 0.7 & 1 \end{bmatrix}.$$

```
% Generate the multivariate random normal variables.
mu = [-2;3];
covm = [1 0.7 ; 0.7 1];
X = csmvrnd(mu,covm,500);
```

To check the results, we plot the random variables in a scatterplot in Figure 4.7. We can also calculate the sample mean and sample covariance matrix to compare with what we used as input arguments to **csmvrnd**. By typing **mean(X)** at the command line, we get

$$-2.0629 \qquad 2.9394$$

Similarly, entering **corrcoef(X)** at the command line yields

```
1.0000    0.6957
0.6957    1.0000
```

We see that these values for the sample statistics correspond to the desired mean and covariance. We note that you could also use the **cov** function to compare the variances. The MATLAB Statistics Toolbox now has a function for generating multivariate normal random variables. It is called **mvnrnd**.
❑

Multivariate Student's *t* Distribution

As we discussed in Chapter 2, we might need to use a distribution that has fatter tails than the multivariate normal, in which case the multivariate *t* distribution is a viable alternative. We can generate multivariate *t* random variables using a generalization of the *t* as the ratio of a standard normal random variable and the square root of an independently distributed chi-square random variable divided by its degrees of freedom v. We illustrate the procedure in the next example.

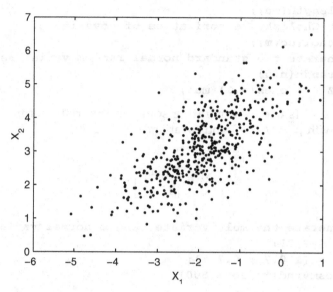

FIGURE 4.7
This shows the scatterplot of the random variables generated using the function **csmvrnd***.*

Example 4.11

The following steps show how to generate multivariate *t* random variables with the desired correlation.

```
% Generate multivariate normal random variables.
R = [1 -.8; -.8 1];  % Correlation matrix.
df = 5;
d = 2;
n = 500;
% Generate n 2-D multivariate normal random
% variables, centered at 0, with covariance C.
C = chol(R);
Xnorm = randn(n,d)*C;
% Generate chi-square random variables and
% divide by the degrees of freedom.
Xchi = sqrt(chi2rnd(df,n,1)./df);
% Divide to get the multivariate t random variables.
Xt = Xnorm./repmat(Xchi(:),1,d);
% Do a scatterplot.
plot(Xt(:,1),Xt(:,2),'.');
xlabel('X_1');ylabel('X_2');
title('Multivariate t Random Variables with \nu = 5')
```

The scatterplot of the random variables is given in Figure 4.8. We can check the correlation using the command:

```
corrcoef(Xt)
```

The result is

```
    1.0000    -0.7965
   -0.7965     1.0000
```

and we see that the correlation is close to what we specified. The MATLAB Statistics Toolbox includes a function **mvtrnd** that will generate random variables from a multivariate t distribution. The function takes either the correlation or the covariance matrix as an argument.
❑

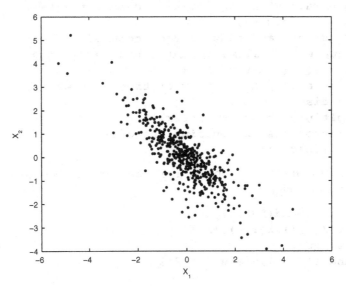

FIGURE 4.8
This is the scatterplot of multivariate t random variables with $v = 5$.

Generating Variates on a Sphere

In some applications, we would like to generate d-dimensional random variables that are distributed on the surface of the unit hypersphere S^d, $d = 2, \ldots$. Note that when $d = 2$ the surface is a circle, and for $d = 3$ the surface is a sphere. We will be using this technique in Chapter 6, where we present an algorithm for exploratory data analysis using projection pursuit. The easiest method is to generate d standard normal random variables and then to scale them such that the magnitude of the vector is one. This is illustrated in the following example.

Example 4.12

The following function **cssphrnd** generates random variables on a d-dimensional unit sphere. We illustrate its use by generating random variables that are on the unit circle S^2.

```
% function X = cssphrnd(n,d);
% This function will generate n d-dimensional
% random variates that are distributed on the
% unit d-dimensional sphere. d >= 2

function X = cssphrnd(n,d)
if d < 2
    error('ERROR - d must be greater than 1.')
    break
end
% Generate standard normal random variables.
tmp = randn(d,n);
% Find the magnitude of each column.
% Square each element, add, and take the square root.
mag = sqrt(sum(tmp.^2));
% Make a diagonal matrix of them - inverses.
dm = diag(1./mag);
% Multiply to scale properly.
% Transpose so X contains the observations.
X = (tmp*dm)';
```

We can use this function to generate a set of random variables for $d = 2$ and plot the result in Figure 4.9.

```
X = cssphrnd(500,2);
plot(X(:,1),X(:,2),'x')
axis equal
xlabel('X_1'),ylabel('X_2')
```

❑

4.4 Generating Discrete Random Variables

Binomial

A binomial random variable with parameters n and p represents the number of successes in n independent trials. We can obtain a binomial random variable by generating n uniform random numbers $U_1, U_2, ..., U_n$ and letting

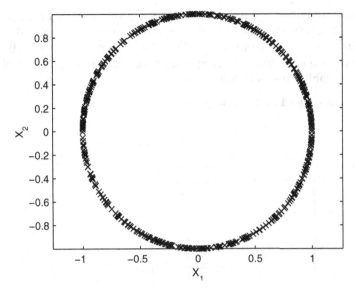

FIGURE 4.9
This is the scatterplot of the random variables generated in Example 4.12. These random variables are distributed on the surface of a 2D unit sphere (i.e., a unit circle).

X be the number of U_i that are less than or equal to p. This is easily implemented in MATLAB as illustrated in the following example.

Example 4.13

We implement this algorithm for generating binomial random variables in the function **csbinrnd**.

```
% function X = csbinrnd(n,p,N)
% This function will generate N binomial
% random variables with parameters n and p.

function X = csbinrnd(n,p,N)
X = zeros(1,N);
% Generate the uniform random numbers:
% N variates of n trials.
U = rand(N,n);
% Loop over the rows, finding the number
% less than p
for i = 1:N
    ind = find(U(i,:) <= p);
    X(i) = length(ind);
```

end

We use this function to generate a set of random variables that are distributed according to the binomial distribution with parameters $n = 6$ and $p = 0.5$. The histogram of the random variables is shown in Figure 4.10. Before moving on, we offer the following more efficient way to generate binomial random variables in MATLAB:

```
X = sum(rand(n,N) <= p);
```

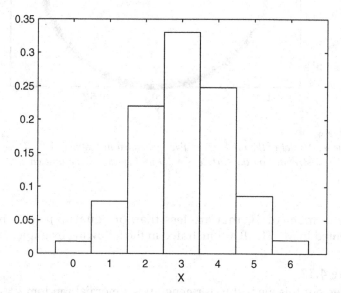

FIGURE 4.10
This is the histogram for the binomial random variables generated in Example 4.13. The parameters for the binomial are $n = 6$ and $p = 0.5$.

Poisson

We use the inverse transform method for discrete random variables as described in Ross [1997] to generate variates from the Poisson distribution. We need the following recursive relationship between successive Poisson probabilities

$$p_{i+1} = P(X = i) = \frac{\lambda}{i+1}p_i; \qquad i \geq 0.$$

This leads to the following method.

PROCEDURE – GENERATING POISSON RANDOM VARIABLES

1. Generate a uniform random number U.
2. Initialize the quantities: $i = 0$, $p_0 = e^{-\lambda}$, and $F_0 = p_0$.
3. If $U \leq F_i$, then deliver $X = i$. Return to step 1.
4. Else increment the values: $p_{i+1} = \lambda p_i/(i+1)$, $i = i+1$, and $F_{i+1} = F_i + p_{i+1}$.
5. Return to step 3.

This algorithm could be made more efficient when λ is large. The interested reader is referred to Ross [1997] for more details.

Example 4.14

The following shows how to implement the procedure for generating Poisson random variables in MATLAB.

```
% function X = cspoirnd(lam,n)
% This function will generate Poisson
% random variables with parameter lambda.
% The reference for this is Ross, 1997, page 50.

function x = cspoirnd(lam,n)
x = zeros(1,n);
j = 1;
while j <= n
    flag = 1;
    % initialize quantities
    u = rand(1);
    i = 0;
    p = exp(-lam);
    F = p;
    while flag % generate the variate needed
        if u <= F % then accept
            x(j) = i;
            flag = 0;
            j = j+1;
        else % move to next probability
            p = lam*p/(i+1);
            i = i+1;
            F = F + p;
        end
    end
end
```

We can use this to generate a set of Poisson random variables with $\lambda = 0.5$, and show a histogram of the data in Figure 4.11.

```
% Set the parameter for the Poisson.
lam = 0.5;
N = 5000; % Sample size
x = cspoirnd(lam,N);
edges = 0:max(x);
f = histc(x,edges);
bar(edges,f/N,1,'w')
```

As an additional check to ensure that our algorithm is working correctly, we can determine the observed relative frequency of each value of the random variable X and compare that to the corresponding theoretical values.

```
% Determine the observed relative frequencies.
% These are the estimated values.
relf = zeros(1,max(x)+1);
for i = 0:max(x)
    relf(i+1) = length(find(x==i))/N;
end
% Use the Statistics Toolbox function to get the
% theoretical values.
y = poisspdf(0:4,.5);
```

When we display these in the command window, we have the following:

```
% These are the estimated values.
relf = 0.5860    0.3080    0.0840    0.0200    0.0020
% These are the theoretical values.
y = 0.6065    0.3033    0.0758    0.0126    0.0016
```

❏

Discrete Uniform

When we implement some of the Monte Carlo methods in Chapter 7 (such as the bootstrap), we will need the ability to generate numbers that follow the discrete uniform distribution. This is a distribution where X takes on values in the set $\{1, 2, ..., N\}$, and the probability that X equals any of the numbers is $1/N$. This distribution can be used to randomly sample with replacement from a group of N objects.

We can generate from the discrete uniform distribution using the following transform:

$$X = \lceil NU \rceil,$$

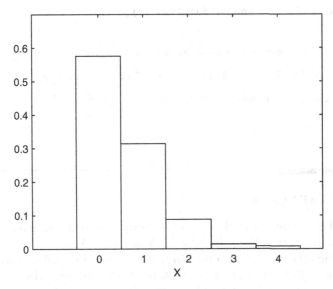

FIGURE 4.11
This is the histogram for random variables generated from the Poisson with $\lambda = 0.5$.

where the function $\lceil y \rceil$, $y \geq 0$ means to round up the argument y. The next example shows how to implement this in MATLAB.

Example 4.15
The method for generating discrete uniform is implemented in the function **csdunrnd**, given next.

```
% function X = csdunrnd(N,n)
% This function will generate random variables
% from the discrete uniform distribution. It picks
% numbers uniformly between 1 and N.

function X = csdunrnd(N,n)
X = ceil(N*rand(1,n));
```

To verify that we are generating the right random variables, we can look at the observed relative frequencies. Each should have relative frequency of $1/N$. This is shown next where $N = 5$ and the sample size is 500.

```
N = 5;
n = 500;
x = csdunrnd(N,n);
% Determine the estimated relative frequencies.
relf = zeros(1,N);
```

```
for i = 1:N
   relf(i) = length(find(x==i))/n;
end
```

Printing out the observed relative frequencies, we have

relf = 0.1820   0.2080   0.2040   0.1900   0.2160

which is close to the theoretical value of $1/N = 1/5 = 0.2$.
❑

4.5 MATLAB® Code

The MATLAB Statistics Toolbox has functions to generate random variables from all of the distributions discussed in Chapter 2. As we explained in that chapter, the analyst must keep in mind that probability distributions are often defined differently, so caution should be exercised when using any software package. Table 4.1 provides a partial list of the MATLAB functions that are available for random number generation. A complete list can be found in the Statistics Toolbox User's Guide. The reader should note that the **gamrnd, weibrnd,** and **exprnd** functions use the alternative definition for the given distribution (see Chapter 2).

Another function that might prove useful in implementing computational statistics methods is called **randperm**. This is provided with the standard MATLAB software package, and it generates random permutations of the integers 1 to n. The result can be used to permute the elements of a vector. For example, to permute the elements of a vector **x** of size **n**, use the following MATLAB statements:

```
% Get the permuted indices.
ind = randperm(n);
% Now re-order based on the permuted indices.
xperm = x(ind);
```

The observant reader might have noticed a (possibly) unfamiliar entry in Table 4.1 called *copulas*. Copulas are functions that link univariate marginals to their joint multivariate distribution. Copulas are useful in the area of computational statistics because they allow one to generate multivariate distributions with arbitrary marginal distributions. One does this by first specifying the marginals and then choosing a copula to impose the desired correlation structure. A more detailed description is beyond the scope of this text. However, the MATLAB Statistics Toolbox User's Guide has an excellent discussion of copulas and instructions on how to generate random variables using the **copularnd** function.

TABLE 4.1

Partial List of Functions in the MATLAB Statistics
Toolbox for Generating Random Variables

Distribution	MATLAB Function
Beta	`betarnd`
Binomial	`binornd`
Chi-square	`chi2rnd`
Discrete uniform	`unidrnd`
Exponential	`exprnd`
Gamma	`gamrnd`
Normal	`normrnd`
Poisson	`poissrnd`
Continuous uniform	`unifrnd`
Weibull	`weibrnd`
Multivariate normal	`mvnrnd`
Multivariate *t*	`mvtrnd`
Copulas	`copularnd`

We also provide some functions in the Computational Statistics Toolbox
for generating random variables. These are summarized in Table 4.2. Note
that these generate random variables using the distributions as defined in the
previous chapter.

TABLE 4.2

List of Functions from Chapter 4 Included in the
Computational Statistics Toolbox

Distribution	MATLAB Function
Beta	`csbetarnd`
Binomial	`csbinrnd`
Chi-square	`cschirnd`
Discrete uniform	`csdunrnd`
Exponential	`csexprnd`
Gamma	`csgamrnd`
Multivariate normal	`csmvrnd`
Poisson	`cspoirnd`
Points on a sphere	`cssphrnd`

4.6 Further Reading

In this text we do not attempt to assess the computational efficiency of the methods for generating random variables. If the statistician or engineer is performing extensive Monte Carlo simulations, then the time it takes to generate random samples becomes important. In these situations, the reader is encouraged to consult Gentle [1998] or Rubinstein [1981] for efficient algorithms. Our goal is to provide methods that are easily implemented using MATLAB or other software, in case the data analyst must write his or her own functions for generating random variables from nonstandard distributions.

There has been considerable research into methods for random number generation, and we refer the reader to the sources mentioned next for more information on the theoretical foundations. The book by Ross [1997] is an excellent resource and is suitable for advanced undergraduate students. He addresses simulation in general and includes a discussion of discrete event simulation and Markov chain Monte Carlo methods. Another text that covers the topic of random number generation and Monte Carlo simulation is Gentle [1998]. This book includes an extensive discussion of uniform random number generation and covers more advanced topics such as Gibbs sampling. Two other resources on random number generation are Rubinstein [1981] and Kalos and Whitlock [1986]. For a description of methods for generating random variables from more general multivariate distributions, see Johnson [1987]. The article by Deng and Lin [2000] offers improvements on some of the standard uniform random number generators.

There are several review articles on copulas. A very readable one is by Frees and Valdez [1998]. The authors discuss properties of copulas, their relationship to measures of dependence, and practical applications. The article also has an annotated bibliography. Another review article is one by Genest and Favre [2007]. They present an introduction to inference for copula models by working with a small numerical example making it easier to understand the concepts. Finally, Genest and Rivest [1993] provide a strategy for choosing the family of copulas that fit a given set of observations.

The MATLAB Statistics Toolbox User's Guide and the function reference documentation are also useful resources for information on ways to generate random variables, copulas, and more. Use the **Help** button on the desktop ribbon interface to access the documents. See Appendix A for information.

Exercises

4.1.   Repeat Example 4.3 using values of $n = 100, 300, 500, 700, 1000$. What happens to the estimated probability mass function (i.e., the relative frequencies from the random samples) as the sample size gets bigger?

4.2.   Write the MATLAB code to implement Example 4.5. Generate 500 random variables from this distribution and construct a histogram (**hist** function) to verify your code.

4.3.   Using the algorithm implemented in Example 4.3, write a MATLAB function that will take any probability mass function (i.e., a vector of probabilities) and return the desired number of random variables generated according to that probability function.

4.4.   Write a MATLAB function that will return random numbers that are uniformly distributed over the interval (a, b).

4.5.   Write a MATLAB function that will return random numbers from the normal distribution with mean μ and variance σ^2. The user should be able to set values for the mean and variance as input arguments.

4.6.   Write a function that will generate chi-square random variables with v degrees of freedom by generating v standard normals, squaring them and then adding them up. This uses the fact that

$$X = Z_1^2 + \dots + Z_v^2$$

is chi-square with v degrees of freedom. Generate some random variables and plot in a histogram. The degrees of freedom should be an input argument set by the user.

4.7.   An alternative method for generating beta random variables is described in Rubinstein [1981]. Generate two variates $Y_1 = U_1^{1/\alpha}$ and $Y_2 = U_2^{1/\beta}$, where the U_i are from the uniform distribution. If $Y_1 + Y_2 \leq 1$, then

$$X = \frac{Y_1}{Y_1 + Y_2},$$

is from a beta distribution with parameters α and β. Implement this algorithm.

4.8.   Run Example 4.4 and generate 1000 random variables. Determine the number of variates that were rejected and the total number generated to obtain the random sample. What percentage were rejected? How efficient was it?

4.9.   Implement Example 4.5 in MATLAB. Generate 100 random variables. What is the relative frequency of each value of the random variable 1, ..., 5? Does this match the probability mass function?

4.10.   Generate four sets of random variables with $v = 2, 5, 15, 20$, using the function **cschirnd**. Create histograms for each sample. How does the shape of the distribution depend on the degrees of freedom v?

4.11.   Repeat Example 4.14 for larger sample sizes. Is the agreement better between the observed relative frequencies and the theoretical values?

4.12.   Generate 1000 binomial random variables for $n = 5$ and $p = 0.3, 0.5, 0.8$. In each case, determine the observed relative frequencies and the corresponding theoretical probabilities. How is the agreement between them?

4.13.   The MATLAB Statistics Toolbox has a GUI called **randtool**. This is an interactive demo that generates random variables from distributions that are available in the toolbox. The user can change parameter values and see the results via a histogram. There are options to change the sample size and to output the results. To start the GUI, simply type **randtool** at the command line. Run the function and experiment with the distributions that are discussed in the text (normal, exponential, gamma, beta, etc.).

4.14.   The plot on the right in Figure 4.6 shows a histogram of beta random variables with parameters $\alpha = \beta = 0.5$. Construct a similar plot using the information in Example 4.9.

4.15.   Generate random variables from the multivariate t distribution for $v = 5, 15, 25, 35$. Use $d = 2$, the same correlation matrix from Example 4.11, and $n = 100$. Construct scatterplots and compare them to a scatterplot of random variables that are bivariate normal with the same correlation matrix and sample size.

Chapter 5

Exploratory Data Analysis

5.1 Introduction

Exploratory data analysis (EDA) is quantitative detective work according to John Tukey [1977]. EDA is the philosophy that data should first be explored without assumptions about probabilistic models, error distributions, number of groups, relationships between the variables, etc. for the purpose of discovering what they can tell us about the phenomena we are investigating. The goal of EDA is to explore the data to reveal patterns and features that will help the analyst better understand, analyze, and model the data. With the advent of powerful desktop computers and high resolution graphics capabilities, these methods and techniques are within the reach of every statistician, engineer, and data analyst.

EDA is a collection of techniques for revealing information about the data and methods for visualizing them to see what they can tell us about the underlying process that generated it. In most situations, exploratory data analysis should precede confirmatory analysis (e.g., hypothesis testing, ANOVA, etc.) to ensure that the analysis is appropriate for the data set. Some examples and goals of EDA are given next to help motivate the reader.

- If we have a time series, then we would plot the values over time to look for patterns such as trends, seasonal effects, or change points. In Chapter 14, we have an example of a time series that shows evidence of a change point in a Poisson process.

- We have observations that relate two characteristics or variables, and we are interested in how they are related. Is there a linear or a nonlinear relationship? Are there patterns that can provide insight into the process that relates the variables? We will see examples of this application in Chapters 8, 12, and 13.

- We need to provide some summary statistics that describe the data set. We should look for outliers or aberrant observations that might contaminate the results. If EDA indicates extreme observations are in the data set, then robust statistical methods might be more appropriate. In Chapter 13, we illustrate an example where a graphical look at the data indicates the presence of outliers, so we use a robust method of nonparametric regression.

- We have a random sample that will be used to develop a model. This model will be included in our simulation of a process (e.g., simulating a physical process such as a queue). We can use EDA techniques to help us determine how the data might be distributed and what model might be appropriate.

In this chapter, we will be discussing graphical EDA and how these techniques can be used to gain information and insights about the data. Some experts include techniques such as smoothing, probability density estimation, clustering, and principal component analysis in exploratory data analysis. We agree that these can be part of EDA, but we do not cover them in this chapter. Principal component analysis and other methods for finding structure are described in Chapter 6. Smoothing techniques are discussed in Chapter 13 where we present methods for nonparametric regression. Techniques for probability density estimation are presented in Chapter 9, but we do discuss simple histograms in this chapter. Methods for clustering are described in Chapter 11.

At this point, we need to inform the reader that there is a companion text to this one called *Exploratory Data Analysis with MATLAB®, Second Edition* [Martinez, Martinez, and Solka, 2010]. There is also an EDA GUI Toolbox that implements most of the functions described in the EDA text. We point this out now because several of these GUIs allow one to easily explore data sets using many of the techniques in this chapter. For example, there are GUIs for exploring univariate and bivariate distributions, conducting data tours, and visualizing high-dimensional data.

It is likely that some of the visualization methods in this chapter are familiar to statisticians, data analysts, and engineers. As we stated in Chapter 1, one of the goals of this book is to promote the use of MATLAB for statistical analysis. Some readers might not be familiar with the extensive graphics capabilities of MATLAB, so we endeavor to describe the most useful ones for data analysis. In Section 5.2, we consider techniques for visualizing univariate data. These include such methods as stem-and-leaf plots, box plots, histograms, and quantile plots. We turn our attention to techniques for visualizing bivariate data in Section 5.3 and include a description of surface plots, scatterplots, and bivariate histograms. Section 5.4 offers several methods for viewing multi-dimensional data, such as slices, isosurfaces, star plots, parallel coordinates, and Andrews curves.

5.2 Exploring Univariate Data

Two important goals of EDA are (1) to determine a reasonable model for the process that generated the data and (2) to locate possible outliers in the sample. For example, we might be interested in finding out whether the distribution that generated the data is symmetric or skewed. We might also like to know whether it has one mode or many modes. The univariate visualization techniques presented here will help us answer questions such as these.

Histograms

A *histogram* is a way to graphically represent the frequency distribution of a data set. Histograms are a good way to

- summarize a data set to understand general characteristics of the distribution such as shape, spread, or location,
- suggest possible probabilistic models, or
- determine unusual behavior.

In this chapter, we look only at the simple, basic histogram. Variants and extensions of the histogram are discussed in Chapter 9.

A *frequency histogram* is obtained by creating a set of bins or intervals that cover the range of the data set. It is important that these bins do not overlap and that they have equal width. We then count the number of observations that fall into each bin. To visualize this, we plot the frequency as the height of a bar, with the width of the bar representing the width of the bin. The histogram is determined by two parameters, the bin width and the starting point of the first bin. We discuss these issues in greater detail in Chapter 9. *Relative frequency histograms* are obtained by representing the height of the bin by the relative frequency of the observations that fall into the bin.

The basic MATLAB package has a function for calculating and plotting a univariate histogram. This function is illustrated in the example given next.

Example 5.1

In this example, we look at a histogram of the data in **forearm**. These data [Hand, et al., 1994; Pearson and Lee, 1903] consist of 140 measurements of the length in inches of the forearm of adult males. We can obtain a simple histogram in MATLAB using these commands:

```
load forearm
subplot(1,2,1)
```

```
% The hist function optionally returns the
% bin centers and frequencies.
[n,x] = hist(forearm);
% Plot and use the argument of width=1
% to produce bars that touch.
bar(x,n,1);
axis square
title('Frequency Histogram')
% Now create a relative frequency histogram.
% Divide each box by the total number of points.
subplot(1,2,2)
bar(x,n/140,1)
title('Relative Frequency Histogram')
axis square
```

These plots are shown in Figure 5.1. Notice that the shapes of the histograms are the same in both types of histograms, but the vertical axis is different. From the shape of the histograms, it seems reasonable to assume that the data are normally distributed.

❑

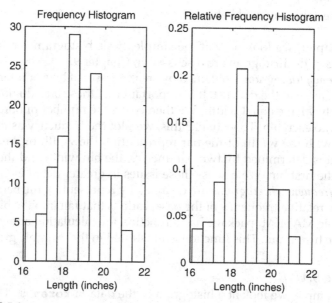

FIGURE 5.1
On the left is a frequency histogram of the **forearm** *data, and on the right is the relative frequency histogram. These indicate that the distribution is unimodal and that the normal distribution is a reasonable model.*

One problem with using a frequency or relative frequency histogram is that they do not represent meaningful probability densities. The heights of the bars represent a step function that does not integrate to one. This can be seen by superimposing a corresponding normal distribution over the relative frequency histogram as shown in Figure 5.2.

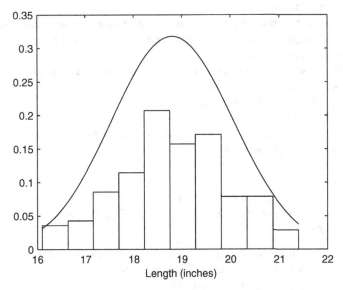

FIGURE 5.2
*This shows a relative frequency histogram of the **forearm** data. Superimposed on the histogram is the normal probability density function using parameters estimated from the data. Note that the curve is higher than the histogram, indicating that the heights of the histogram bars do not correspond to a valid probability density function.*

A *density histogram* is a histogram that has been normalized so it will integrate to one. This means that if we add up the *areas* represented by the bars, then they should add up to one. A density histogram is given by the following equation:

$$\hat{f}(x) = \frac{v_k}{nh} \qquad x \text{ in } B_k, \qquad (5.1)$$

where B_k denotes the k-th bin, v_k represents the number of data points that fall into the k-th bin, and h represents the width of the bins. In the following example, we reproduce the histogram of Figure 5.2 using the density histogram.

Example 5.2

Here we explore the **forearm** data using a density histogram. Assuming a normal distribution and estimating the parameters from the data, we can superimpose a smooth curve that represents an estimated density for the normal distribution.

```
% Get parameter estimates for the normal distribution.
mu = mean(forearm);
v = var(forearm);
% Obtain normal pdf based on parameter estimates.
xp = linspace(min(forearm),max(forearm));
yp = csnormp(xp,mu,v);
% Get the information needed for a histogram.
[nu,x] = hist(forearm);
% Get the widths of the bins.
h = x(2)-x(1);
% Plot as density histogram - Equation 5.1.
bar(x,nu/(140*h),1)
hold on
plot(xp,yp)
xlabel('Length (inches)')
title('Density Histogram and Density Estimate')
hold off
```

The results are shown in Figure 5.3. Note that the assumption of normality for the data is not unreasonable. The estimated density function and the density histogram match up quite well.

❑

Stem-and-Leaf

Stem-and-leaf plots were introduced by Tukey [1977] as a way of displaying data in a structured list. Presenting data in a table or an ordered list does not readily convey information about how the data are distributed, as is the case with histograms.

If we have data where each observation consists of at least two digits, then we can construct a stem-and-leaf diagram. To display these, we separate each measurement into two parts: the *stem* and the *leaf*. The stems are comprised of the leading digit or digits, and the remaining digit makes up the leaf. For example, if we had the number 75, then the stem is the 7, and the leaf is the 5. If the number is 203, then the stem is 20 and the leaf is 3.

The stems are listed to the left of a vertical line with all of the leaves corresponding to that stem listed to the right. If the data contain decimal places, then they can be rounded for easier display. An alternative is to move the decimal place to specify the appropriate leaf unit. We provide a function

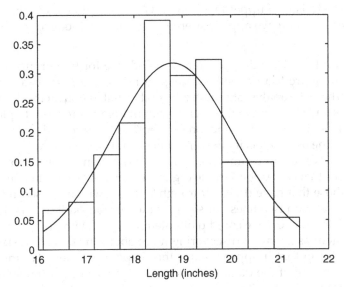

FIGURE 5.3
Density histogram for the **forearm** *data. The curve represents a normal probability density function with parameters given by the sample mean and sample variance of the data. From this we see that the normal distribution is a reasonable probabilistic model.*

with the text (**csstemleaf**) that will construct stem-and-leaf plots, and its use is illustrated in the next example.

Example 5.3

The heights of 32 Tibetan skulls [Hand, et al. 1994; Morant, 1923] measured in millimeters is given in the file **tibetan**. These data comprise two groups of skulls collected in Tibet. One group of 17 skulls comes from graves in Sikkim and nearby areas of Tibet and the other 15 skulls come from a battlefield in Lhasa. The original data contain five measurements, but for this example, we only use the fourth measurement. This is the upper face height, and we round to the nearest millimeter. We use the function **csstemleaf** that is provided with the text.

```
load tibetan
% This loads up all 5 measurements of the skulls.
% We use the fourth characteristic to illustrate
% the stem-and-leaf plot. We first round them.
x = round(tibetan(:,4));
csstemleaf(x)
title('Height (mm) of Tibetan Skulls')
```

The resulting stem-and-leaf is shown in Figure 5.4. From this plot, we see there is not much evidence that there are two groups of skulls, if we look only at the characteristic of upper face height. We will explore these data further in Chapter 10, where we apply pattern recognition methods to the problem. ❏

It is possible that we do not see much evidence for two groups of skulls because there are too few stems. Keep in mind that EDA is an iterative process, where the analyst should try several visualization methods in search of patterns and structure in the data. An alternative approach is to plot more than one line per stem. The function **csstemleaf** has an optional argument that allows the user to specify two lines per stem. The default value is one line per stem, as we saw in Example 5.3 (Figure 5.4). When we plot two lines per stem, leaves that correspond to the digits 0 through 4 are plotted on the first line and those that have digits 5 through 9 are shown on the second line. A stem-and-leaf with two lines per stem for the Tibetan skull data is shown in Figure 5.5. In practice, one could plot a stem-and-leaf with one and with two lines per stem as a way of discovering more about the data. The stem-and-leaf is useful in that it approximates the shape of the density, and it also provides a listing of the data. One can usually recover the original data set from the stem-and-leaf (if it has not been rounded), unlike the histogram. A disadvantage of the stem-and-leaf plot is that it is not useful for large data sets, while a histogram is very effective in reducing and displaying massive data sets.

Quantile-Based Plots — Continuous Distributions

If we need to compare two distributions, then we can use the quantile plot to visually compare them. This is also applicable when we want to compare a distribution and a sample or to compare two samples. In comparing the distributions or samples, we are interested in knowing how they are shifted relative to each other. In essence, we want to know if they are distributed in the same way. This is important when we are trying to determine the distribution that generated our data, possibly with the goal of using that information to generate data for Monte Carlo simulation. Another application where this is useful is in checking model assumptions, such as normality, before we conduct our analysis.

In this part, we discuss several versions of quantile-based plots. These include *quantile-quantile plots* (q-q plots) and *quantile plots* (sometimes called a *probability plot*). Quantile plots for discrete variables are discussed next. The quantile plot is used to compare a sample with a theoretical distribution. Typically, a q-q plot (sometimes called an *empirical quantile plot*) is used to determine whether two random samples are generated by the same distribution. It should be noted that the q-q plot can also be used to

```
6 │ 2 3 5 5 6 8 9

7 │ 0 0 1 1 1 2 2 3 4 4 4 4 5 6 6 7 7 7 8 9 9

8 │ 0 1 2 3
  │
```

FIGURE 5.4
This shows the stem-and-leaf plot for the upper face height of 32 Tibetan skulls. The data have been rounded to the nearest millimeter.

```
6 │ 2 3

6 │ 5 5 6 8 9

7 │ 0 0 1 1 1 2 2 3 4 4 4 4

7 │ 5 6 6 7 7 7 8 9 9

8 │ 0 1 2 3

8 │
  │
```

FIGURE 5.5
This shows a stem-and-leaf plot for the upper face height of 32 Tibetan skulls where we now have two lines per stem. Note that we see approximately the same information (a unimodal distribution) as in Figure 5.4.

compare a random sample with a theoretical distribution by generating a sample from the theoretical distribution as the second sample.

Q-Q Plot

The q-q plot was originally proposed by Wilk and Gnanadesikan [1968] to visually compare two distributions by graphing the quantiles of one versus the quantiles of the other. Say we have two data sets consisting of univariate measurements. We denote the order statistics for the first data set by

$$x_{(1)}, x_{(2)}, \ldots, x_{(n)}.$$

Let the order statistics for the second data set be

$$y_{(1)}, y_{(2)}, \ldots, y_{(m)},$$

with $m \leq n$.

We look first at the case where the sizes of the data sets are equal, so $m = n$. In this case, we plot as points the sample quantiles of one data set versus the other data set. This is illustrated in Example 5.4. If the data sets come from the same distribution, then we would expect the points to approximately follow a straight line.

A major strength of the quantile-based plots is that they do not require the two samples (or the sample and theoretical distribution) to have the same location and scale parameter. If the distributions are the same, but differ in location or scale, then we would still expect the quantile-based plot to produce a straight line.

Example 5.4

We will generate two sets of normal random variables and construct a q-q plot. The q-q plot (Figure 5.6) follows a straight line (approximately), indicating that the samples come from the same distribution.

```
% Generate the random variables.
x = randn(1,75);
y = randn(1,75);
% Find the order statistics.
xs = sort(x);
ys = sort(y);
% Now construct the q-q plot.
plot(xs,ys,'o')
xlabel('X - Standard Normal')
ylabel('Y - Standard Normal')
axis equal
```

If we repeat the above MATLAB commands using a data set generated from an exponential distribution and one that is generated from the standard normal, then we have the plot shown in Figure 5.7. Note that the points in this q-q plot do not follow a straight line, leading us to conclude that the data are not generated from the same distribution.

❏

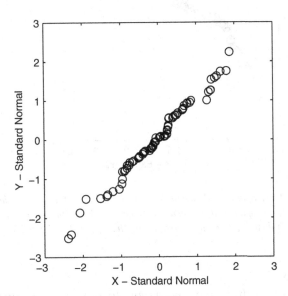

FIGURE 5.6
This is a q-q plot of X and Y where both data sets are generated from a standard normal distribution. Note that the points follow a line, as expected.

We now look at the case where the sample sizes are not equal. Without loss of generality, we assume that $m < n$. To obtain the q-q plot, we graph the $y_{(i)}$, $i = 1, ..., m$ against the $(i - 0.5)/m$ quantile of the other data set. Note that this definition is not unique, as values other than 0.5 can be used [Cleveland, 1993]. The $(i - 0.5)/m$ quantiles of the x data are usually obtained via interpolation, and we show in the next example how to use the function **csquantiles** to get the desired plot.

Users should be aware that q-q plots provide a rough idea of how similar the distribution is between two random samples. If the sample sizes are small, then a lot of variation is expected, so comparisons might be suspect. To help aid the visual comparison, some q-q plots include a reference line. These are lines that are estimated using the first and third quartiles $(q_{0.25}, q_{0.75})$ of each data set. The line is added to the graphic and plotted to cover the range of the data. The MATLAB Statistics Toolbox provides a function called **qqplot** that displays this type of plot. We show next how to add the line.

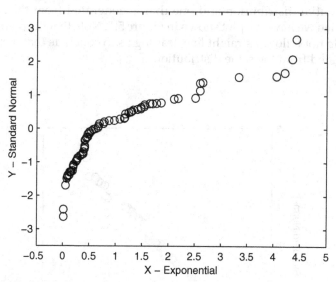

FIGURE 5.7
This is a q-q plot where one random sample is generated from the exponential distribution and one is generated by a standard normal distribution. Note that the points do not follow a straight line, indicating that the distributions that generated the random variables are not the same.

Example 5.5

This example shows how to do a q-q plot when the samples do not have the same number of points. We use the function **csquantiles** to get the required sample quantiles from the data set that has the larger sample size. We then plot these versus the order statistics of the other sample, as we did in the previous examples. Note that we add a reference line based on the first and third quartiles of each data set, using the function **polyfit** (see Chapter 8 for more information on this function).

```
% Generate the random variables.
m = 50;
n = 75;
x = randn(1,n);
y = randn(1,m);
% Find the order statistics for y.
ys = sort(y);
% Now find the associated quantiles using the x.
% Probabilities for quantiles:
p = ((1:m) - 0.5)/m;
xs = csquantiles(x,p);
% Construct the plot.
plot(xs,ys,'ko')
```

```
% Get the reference line.
% Use the 1st and 3rd quartiles of each set to
% get a line.
qy = csquantiles(y,[0.25,0.75]);
qx = csquantiles(x,[0.25,0.75]);
[pol, s] = polyfit(qx,qy,1);
% Add the line to the figure.
yhat = polyval(pol,xs);
hold on
plot(xs,yhat,'k')
xlabel('Sample Quantiles - X'),
ylabel('Sorted Y Values')
hold off
```

From Figure 5.8, the assumption that each data set is generated according to the same distribution seems reasonable.
❏

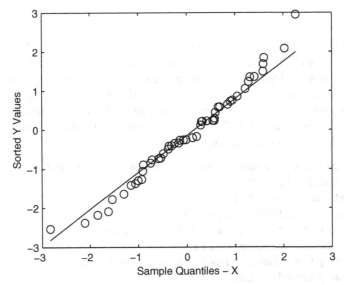

FIGURE 5.8
Here we show the q-q plot of Example 5.5. In this example, we also show the reference line estimated from the first and third quartiles of the data sets. The q-q plot shows that the data do seem to come from the same distribution.

Quantile Plot

A *quantile plot* or *probability plot* is one where the theoretical quantiles are plotted against the order statistics for the sample. Thus, on one axis we plot the $x_{(i)}$ and on the other axis we plot

$$F^{-1}\left(\frac{i-0.5}{n}\right),$$

where $F^{-1}(.)$ denotes the inverse of the cumulative distribution function for the hypothesized distribution. As before, the 0.5 in the above argument can be different [Cleveland, 1993]. A well-known example of a quantile plot is the *normal probability plot*, where the ordered sample versus the quantiles of the normal distribution are plotted.

The MATLAB Statistics Toolbox has several functions for obtaining quantile plots. One is called **normplot**, and it produces a normal probability plot. So, if one would like to assess the assumption that a data set comes from a normal distribution, then this is the one to use. Another function is called **probplot**. Using this function, one can construct probability plots for several distributions, such as the exponential, lognormal, normal, and others. There is also a function for constructing a quantile plot that compares a data set to the Weibull distribution. This is called **wblplot**. For quantile plots with other theoretical distributions, one can use the MATLAB code given next, substituting the appropriate function to get the theoretical quantiles.

Example 5.6

This example illustrates how you can display a quantile plot in MATLAB. We first generate a random sample from the standard normal distribution as our data set. The sorted sample is an estimate of the $(i-0.5)/n$ quantile, so we next calculate these probabilities and get the corresponding theoretical quantiles. Finally, we use the function **norminv** from the Statistics Toolbox to get the theoretical quantiles for the normal distribution. The resulting quantile plot is shown in Figure 5.9.

```
% Generate a random sample from a standard normal.
x = randn(1,100);
% Get the probabilities.
prob = ((1:100)-0.5)/100;
% Now get the theoretical quantiles.
qp = norminv(prob,0,1);
% Now plot theoretical quantiles versus
% the sorted data.
plot(sort(x),qp,'ko')
xlabel('Sorted Data')
ylabel('Standard Normal Quantiles')
```

To further illustrate these concepts, let's see what happens when we generate a random sample from a uniform (0, 1) distribution and check it against the normal distribution. The MATLAB code is given next, and the quantile plot is shown in Figure 5.10. As expected, the points do not lie on a line, and we see that the data are not from a normal distribution.

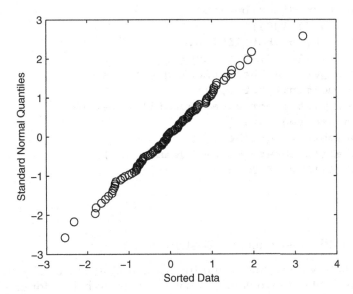

FIGURE 5.9
This is a quantile plot or normal probability plot of a random sample generated from a standard normal distribution. Note that the points approximately follow a straight line, indicating that the normal distribution is a reasonable model for the sample.

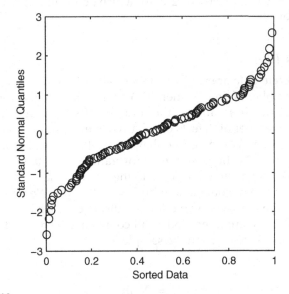

FIGURE 5.10
Here we have a quantile plot where the sample is generated from a uniform distribution, and the theoretical quantiles are from the normal distribution. The shape of the curve verifies that the sample is not from a normal distribution.

```
% Generate a random sample from a
% uniform distribution.
x = rand(1,100);
% Get the probabilities.
prob = ((1:100)-0.5)/100;
% Now get the theoretical quantiles.
qp = norminv(prob,0,1);
% Now plot theoretical quantiles versus
% the sorted data.
plot(sort(x),qp,'ko')
ylabel('Standard Normal Quantiles')
xlabel('Sorted Data')
```

❏

Quantile Plots — Discrete Distributions

Previously, we described quantile plots that are primarily used for continuous data. We would like to have a similar technique for graphically comparing the shapes of discrete distributions. Hoaglin and Tukey [1985] developed several plots to accomplish this. We present two of them here: the *Poissonness plot* and the *binomialness plot*. These will enable us to search for evidence that our discrete data follow a Poisson or a binomial distribution. They also serve to highlight which points might be incompatible with the model.

Poissonness Plot

Typically, discrete data are whole number values that are often obtained by counting the number of times something occurs. For example, these might be the number of traffic fatalities, the number of school-age children in a household, the number of defects on a hard drive, or the number of errors in a computer program. We sometimes have the data in the form of a frequency distribution that lists the possible count values (e.g., 0, 1, 2, ...) and the number of observations that are equal to the count values.

The counts will be denoted as k, with $k = 0, 1, ..., L$. We will assume that L is the maximum observed value for our discrete variable or counts in the data set and that we are interested in all counts between 0 and L. Thus, the total number of observations in the sample is

$$N = \sum_{k=0}^{L} n_k,$$

where n_k represents the number of observations that are equal to the count k.

A basic Poissonness plot is constructed by plotting the count values k on the horizontal axis and

$$\phi(n_k) = \ln(k!n_k/N) \tag{5.2}$$

on the vertical axis. These are plotted as symbols, similar to the quantile plot. If a Poisson distribution is a reasonable model for the data, then this should follow a straight line. Systematic curvature in the plot would indicate that these data are not consistent with a Poisson distribution. The values for $\phi(n_k)$ tend to have more variability when n_k is small, so Hoaglin and Tukey [1985] suggest plotting a special symbol or a "1" to highlight these points.

Example 5.7

This example is taken from Hoaglin and Tukey [1985]. In the late 1700s, Alexander Hamilton, John Jay, and James Madison wrote a series of 77 essays under the title of *The Federalist*. These appeared in the newspapers under a pseudonym. Most analysts accept that John Jay wrote 5 essays, Alexander Hamilton wrote 43, Madison wrote 14, and 3 were jointly written by Hamilton and Madison. Later, Hamilton and Madison claimed that they each solely wrote the remaining 12 papers. To verify this claim, Mosteller and Wallace [1964] used statistical methods, some of which were based on the frequency of words contained in blocks of text. Table 5.1 gives the frequency distribution for the word *may* in papers that were known to be written by Madison.

TABLE 5.1

Frequency distribution of the word *may* in essays known to be written by James Madison.[a]

Number of Occurrences of the Word *may* (k)	Number of Blocks (n_k)
0	156
1	63
2	29
3	8
4	4
5	1
6	1

[a] The n_k represent the number of blocks of text that contained k occurrences of the word *may* [Hoaglin and Tukey, 1985]

We are not going to repeat the analysis of Mosteller and Wallace, we are simply using the data to illustrate a Poissonness plot. The following MATLAB code produces the Poissonness plot shown in Figure 5.11.

```
k = 0:6;  % vector of counts
n_k = [156 63 29 8 4 1 1];
N = sum(n_k);
% Get vector of factorials.
fact = zeros(size(k));
for i = k
    fact(i+1) = factorial(i);
end
% Get phi(n_k) for plotting.
phik = log(fact.*n_k/N);
% Find the counts that are equal to 1.
% Plot these with the symbol 1.
% Plot rest with a symbol.
ind = find(n_k~=1);
plot(k(ind),phik(ind),'o')
ind = find(n_k==1);
if ~isempty(ind)
    text(k(ind),phik(ind),'1')
end
% Add some white space to see better.
axis([-0.5 max(k)+1 min(phik)-1 max(phik)+1])
xlabel('Number of Occurrences - k')
ylabel('\phi (n_k)')
```

The Poissonness plot has significant curvature indicating that the Poisson distribution is not a good model for these data. There are also a couple of points with a frequency of 1 that seem incompatible with the rest of the data. Thus, if a statistical analysis of these data relies on the Poisson model, then any results are suspect.
❏

Hoaglin and Tukey [1985] suggest a modified Poissonness plot that is obtained by changing the n_k, which helps account for the variability of the individual values. They propose the following change:

$$
n_k^* = \begin{cases} n_k - 0.67 - 0.8n_k/N; & n_k \geq 2 \\ 1/e; & n_k = 1 \\ \text{undefined}; & n_k = 0. \end{cases} \tag{5.3}
$$

As we will see in the following example where we apply the modified Poissonness plot to the word frequency data, the main effect of the modified plot is to highlight those data points with small counts that do not behave contrary to the other observations. Thus, if a point that is displayed as a 1 in a modified Poissonness plot seems different from the rest of the data, then it should be investigated.

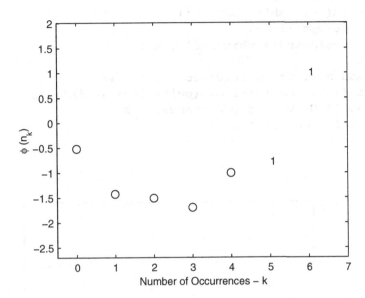

FIGURE 5.11
This is a basic Poissonness plot using the data in Table 5.1. The symbol 1 indicates that
$n_k = 1$.

Example 5.8

We return to the word frequency data in Table 5.1 and show how to get a
modified Poissonness plot. In this modified version shown in Figure 5.12, we
see that the points where $n_k = 1$ do not seem so different from the rest of the
data.

```
% Poissonness plot - modified
k = 0:6;   % vector of counts
% Find n*_k.
n_k = [156 63 29 8 4 1 1];
N = sum(n_k);
phat = n_k/N;
nkstar = n_k-0.67-0.8*phat;
% Get vector of factorials.
fact = zeros(size(k));
for i = k
    fact(i+1) = factorial(i);
end
% Find the frequencies that are 1; nkstar=1/e.
ind1 = find(n_k==1);
nkstar(ind1)= 1/2.718;
% Get phi(n_k) for plotting.
phik = log(fact.*nkstar/N);
```

```
ind = find(n_k~=1);
plot(k(ind),phik(ind),'o')
if ~isempty(ind1)
    text(k(ind1),phik(ind1),'1')
end
% Add some white space to see better.
axis([-0.5 max(k)+1 min(phik)-1 max(phik)+1])
xlabel('Number of Occurrences - k')
ylabel('\phi (n^*_k)')
```

❑

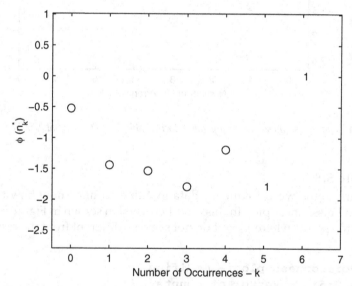

FIGURE 5.12
This is a modified Poissonness plot for the word frequency data in Table 5.1. Here the counts where $n_k = 1$ do not seem radically different from the rest of the observations.

Binomialness Plot

A binomialness plot is obtained by plotting k along the horizontal axis and plotting

$$\varphi(n_k^*) = \ln\left\{\frac{n_k^*}{N \times \binom{n}{k}}\right\}, \tag{5.4}$$

along the vertical axis. Recall that n represents the number of trials, and n_k^* is given by Equation 5.3. As with the Poissonness plot, we are looking for an approximate linear relationship between k and $\varphi(n_k^*)$. An example of the binomialness plot is given in Example 5.9.

TABLE 5.2

Frequency Distribution for the Number of Females in a Queue of Size 10 [Hoaglin and Tukey, 1985]

Number of Females (k)	Number of Blocks (n_k)
0	1
1	3
2	4
3	23
4	25
5	19
6	18
7	5
8	1
9	1
10	0

Example 5.9

Hoaglin and Tukey [1985] provide a frequency distribution representing the number of females in 100 queues of length 10. These data are given in Table 5.2. The MATLAB code to display a binomialness plot for $n = 10$ is given next. Note that we cannot display $\varphi(n_k^*)$ for $k = 10$ (in this example) because it is not defined for $n_k = 0$. The resulting binomialness plot is shown in Figure 5.13, and it indicates a linear relationship. Thus, the binomial model for these data seems adequate.

```
% Binomialness plot.
k = 0:9;
n = 10;
n_k = [1 3 4 23 25 19 18 5 1 1];
N = sum(n_k);
nCk = zeros(size(k));
for i = k
    nCk(i+1) = nchoosek(n,i);
% nchoosek is a function in basic MATLAB.
end
phat = n_k/N;
nkstar = n_k-0.67-0.8*phat;
% Find the frequencies that are 1; nkstar=1/e.
ind1 = find(n_k==1);
```

```
nkstar(ind1) = 1/2.718;
% Get phi(n_k) for plotting.
phik = log(nkstar./(N*nCk));
% Find the counts that are equal to 1.
ind = find(n_k~=1);
plot(k(ind),phik(ind),'o')
if ~isempty(ind1)
    text(k(ind1),phik(ind1),'1')
end
% Add some white space to see better.
axis([-0.5 max(k)+1 min(phik)-1 max(phik)+1])
xlabel('Number of Females - k')
ylabel('\phi (n^*_k)')
```

❑

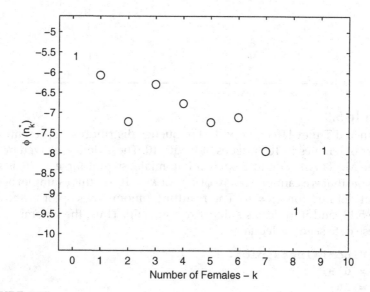

FIGURE 5.13
This shows the binomialness plot for the data in Table 5.2. From this it seems reasonable to use the binomial distribution to model the data.

Box Plots

Box plots (sometimes called box-and-whisker diagrams) have been in use for many years [Tukey, 1977]. As with other visualization techniques we have just discussed, they are used to display the distribution of a sample. Five values from a data set are used to construct the basic version of the box plot.

These are the three sample quartiles $(\hat{q}_{0.25}, \hat{q}_{0.5}, \hat{q}_{0.75})$, the minimum value in the sample, and the maximum value.

There are many variations of the box plot, and it is important to note that they are defined differently depending on the software package that is used. Frigge, Hoaglin, and Iglewicz [1989] describe a study on how box plots are implemented in some popular statistics programs such as Minitab, S, SAS, SPSS, and others. The main difference lies in how outliers and quartiles are defined. Therefore, depending on how the software calculates these, different plots might be obtained [Frigge, Hoaglin, and Iglewicz, 1989].

Before we describe the box plot, we need to define some terms. Recall from Chapter 3, that the *interquartile range* (IQR) is the difference between the first and the third sample quartiles. This gives the range of the middle 50% of the data. It is estimated from the following:

$$IQR = \hat{q}_{0.75} - \hat{q}_{0.25}.$$
(5.5)

Two limits are also defined: a lower limit (LL) and an upper limit (UL). These are calculated from the estimated IQR as follows:

$$LL = \hat{q}_{0.25} - 1.5 \cdot IQR$$
$$UL = \hat{q}_{0.75} + 1.5 \cdot IQR.$$
(5.6)

The idea is that observations that lie outside these limits are possible outliers. *Outliers* are data points that lie away from the rest of the data. This might mean that the data were incorrectly measured or recorded. On the other hand, it could mean that they represent extreme points that arise naturally according to the distribution. In any event, they are sample points that are suitable for further investigation.

Adjacent values are the most extreme observations in the data set that are within the lower and the upper limits. If there are no potential outliers, then the adjacent values are simply the maximum and the minimum data points.

To construct a vertical box plot, we place horizontal lines at each of the three quartiles and draw vertical lines to create a box. We then extend a line from the first quartile to the smallest adjacent value and do the same for the third quartile and largest adjacent value. These lines are sometimes called the whiskers. Finally, any possible outliers are shown as an asterisk or some other plotting symbol. An example of a box plot is shown in Figure 5.14.

Box plots for different samples can be plotted together for visually comparing the corresponding distributions. The MATLAB Statistics Toolbox contains a function called **boxplot** for creating this type of display. It displays one box plot for each column of data. When we want to compare data sets, it is better to display a box plot with notches. These notches represent the uncertainty in the locations of central tendency and provide a rough measure of the significance of the differences between the values. If the

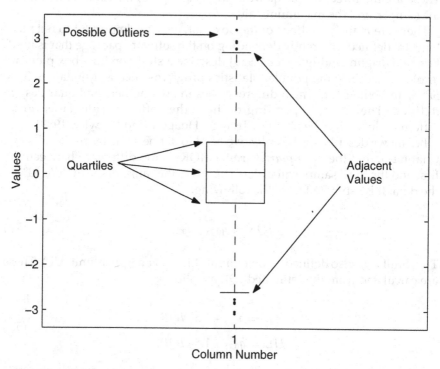

FIGURE 5.14
An example of a box plot with possible outliers shown as points.

notches do not overlap, then there is evidence that the medians are significantly different. The length of the whisker is easily adjusted using optional input arguments to **boxplot**. For more information on this function and to find out what other options are available, type **help boxplot** at the MATLAB command line.

Example 5.10

In this example, we first generate random variables from a uniform distribution on the interval (0, 1), a standard normal distribution, and an exponential distribution. We will then display the box plots corresponding to each sample using the MATLAB function **boxplot**.

```
% Generate a sample from the uniform distribution.
xunif = rand(100,1);
% Generate sample from the standard normal.
xnorm = randn(100,1);
% Generate a sample from the exponential distribution.
% NOTE: this function is from the Statistics Toolbox.
xexp = exprnd(1,100,1);
```

```
% Construct a boxplot with notches.
boxplot([xunif,xnorm,xexp],'notch','on')
```

It can be seen in Figure 5.15 that the box plot readily conveys the shape of the distribution. A symmetric distribution will have whiskers with approximately equal lengths, and the two sides of the box will also be approximately equal. This would be the case for the uniform or normal distribution. A skewed distribution will have one side of the box and whisker longer than the other. This can be seen in Figure 5.15 for the exponential distribution. If the interquartile range is small, then the data in the middle are packed around the median. Conversely, if it is large, then the middle 50% of the data are widely dispersed.
❑

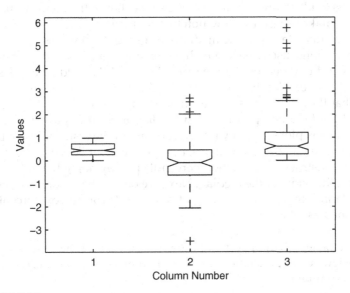

FIGURE 5.15
Here we have three box plots. The one on the left is for a sample from the uniform distribution. The data for the middle box plot came from a standard normal distribution, while the data for the box plot on the right came from an exponential. Notice that the shape of each distribution is apparent from the information contained in the box plots.

Several enhancements and variations to the basic box plot have been proposed. We already discussed one such variation (shown in Figure 5.15): the notched boxplot [McGill, Tukey, and Larsen, 1978]. The notches represent the uncertainty in the location of the median and provide a measure of the significance of the difference. If the intervals represented by the notches do not overlap, then there is evidence that the medians differ at the 5% significance level. See Chapter 7 for information on hypothesis testing.

One could also incorporate a sense of the sample size by making the width of the box proportional to some function of n. McGill, Tukey, and Larsen [1978] recommend using widths that are proportional to the square root of n. Other possibilities include making the widths directly proportional to n or using a logit scale.

Benjamini [1988] connects the width of the box plot to the density of the data, and he describes two types of box plots. One is called the *histplot*, where the width of the box at the three quartiles is proportional to an estimate of the density at those points. His implementation uses the density histogram to obtain these, but other estimation methods can be used instead. A *vaseplot* extends this idea to include all quantiles, so the width of the box at each point is proportional to the corresponding estimated density at the point. This tends to produce a vase-like shape, hence the name of the plot. It no longer makes sense to draw notches, so Benjamini uses shaded bars to indicate the uncertainty in the medians, as in regular box plots.

The extensions discussed so far change the width of the box plot only; the whiskers and outliers remain the same. Esty and Banfield [2003] describe a variant of the box plot that no longer has the whiskers or the outliers, which means that the ambiguities in their definition are no longer of concern. The Esty and Banfield box plot is called the *box-percentile plot*. This uses the sides of the box plot to convey information about the distribution of the data over the entire range of data values, not just the inner 50% of the data.

We can construct a vertical box-percentile plot by doing the following. Let w indicate the width at the median. Next, we obtain the order statistics of our observed random sample, $x_{(1)}, ..., x_{(n)}$. Then the sides of the box-percentile plot are obtained as follows:

1. For $x_{(k)}$ less than or equal to the median, we plot the observation at height $x_{(k)}$ at a distance $kw/(n + 1)$ on either side of a vertical axis of symmetry.

2. For $x_{(k)}$ greater than the median, we plot the point at height $x_{(k)}$ at a distance $(n + 1 - k)w/(n + 1)$ on either side of a vertical axis of symmetry.

The sides of the box-percentile plot provide information about the empirical cumulative distribution function. We provide two MATLAB functions that implement these box plots and show how to apply them to the data set in the previous example.

Example 5.11

We will only show how to use the functions (**boxp** and **boxprct**) for the box plot variants that are provided in the Computational Statistics Toolbox, rather than the actual code. The interested reader can consult the function files for details on how they are constructed. We note that the Statistics Toolbox function (**boxplot**) does not handle different sample sizes. The functions used in this example can deal with variable sample sizes, but we have to put the samples in a cell array rather than a matrix.

```
% Use data similar to that in Example 5.10, but
% but make it different sample sizes. We need to
% store in a cell array.
X{1} = rand(50,1);    % uniform
X{2} = randn(100,1);  % normal
X{3} = exprnd(1,500,1); % exponential
% Construct a histplot. This function is
% included in the Computational Statistics Toolbox.
boxp(X,'hp')
title('Histplot of Uniform, Normal, Exponential')
```

The side-by-side box plots are shown in Figure 5.16. Note that we still have whiskers and potential outliers. Also, we did not construct variable width box plots in this example. Variable width box plots can be obtained using the same **boxp** function, and the reader is asked to construct that plot in the exercises. The code given next will create a box-percentile plot.

```
% Construct a box-percentile plot. This function is
% included in the Computational Statistics Toolbox.
boxprct(X)
title('Box-Percentile Plot')
```

The **boxprct** function will also do a variable width version of the box-percentile plot, where the width at the median is a function of the sample size n. The box plot functions from this example are shown in Figure 5.16.
❑

5.3 Exploring Bivariate and Trivariate Data

Using Cartesian coordinates, we can view up to three dimensions. For example, we could view bivariate data as points or trivariate data as a point cloud. We could also view a bivariate function, $z = f(x, y)$ as a surface. Visualizing anything more than three dimensions is very difficult, but we do offer some techniques in the next section. In this section, we present several methods for visualizing 2D and 3D data, looking first at bivariate data. Most of the techniques that we discuss are readily available in the basic MATLAB program.

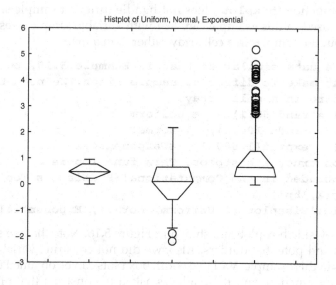

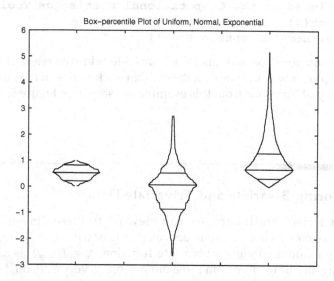

FIGURE 5.16

The top figure shows the histplot of Example 5.11. The sides of the box at the three quartiles are proportional to the density at that point. The bottom figure illustrates the box-percentile plot for the same data set. Here the sides of the box plots are similar to the empirical distribution function. Note that we no longer have whiskers and outliers with the box-percentile plot.

Scatterplots

Perhaps one of the easiest ways to visualize bivariate data is with the *scatterplot*. A scatterplot is obtained by displaying the ordered pairs as points using some plotting symbol. This type of plot conveys useful information such as how the data are distributed in the two dimensions and how the two variables are related (e.g., a linear or a nonlinear relationship). Before any modeling, such as regression, is done using bivariate data, the analyst should always look at a scatterplot to see what type of relationship is reasonable. We will explore this further in Chapters 8 and 12.

A scatterplot can be obtained easily in MATLAB using the **plot** command. One just enters the marker style or plotting symbol as one of the arguments. See the **help** on **plot** for more information on what characters are available. By entering a marker (or line) style, you tell MATLAB that you do *not* want to connect the points with a straight line, which is the default. We have already seen many examples of how to use the **plot** function in this way when we constructed the quantile and q-q plots.

An alternative function for scatterplots that is available with MATLAB is the function called **scatter**. This function takes the input vectors **x** and **y** and plots them as symbols. There are optional arguments that will plot the markers as different colors and sizes. These alternatives are explored in Example 5.12.

Example 5.12

We first generate a set of bivariate normal random variables using the technique described in Chapter 4. However, it should be noted that we find the matrix **R** in Equation 4.19 using singular value decomposition rather than Cholesky factorization. We then create a scatterplot using the **plot** function and the **scatter** function. The resulting plots are shown in Figure 5.17 and Figure 5.18.

```
% Create a positive definite covariance matrix.
vmat = [2, 1.5; 1.5, 9];
% Create mean at (2,3).
mu = [2 3];
[u,s,v] = svd(vmat);
vsqrt = ( v*(u'.*sqrt(s)))';
% Get standard normal random variables.
td = randn(250,2);
% Use x=z*sigma+mu to transform - see Chapter 4.
data = td*vsqrt+ones(250,1)*mu;
% Create a scatterplot using the plot function.
% Figure 5.16.
plot(data(:,1),data(:,2),'x')
axis equal
% Create a scatterplot using the scatter function.
```

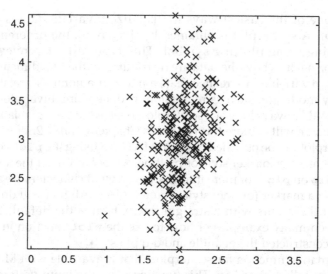

FIGURE 5.17
This is a scatterplot of the sample in Example 5.12 using the **plot** *function. We can see that the data seem to come from a bivariate normal distribution. Here we use* **'x'** *as an argument to the* **plot** *function to plot the symbols as x's.*

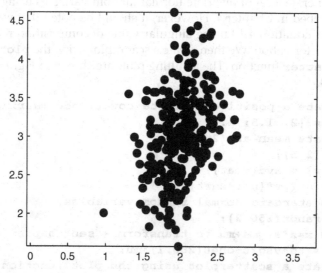

FIGURE 5.18
This is a scatterplot of the sample in Example 5.12 using the **scatter** *function with filled markers.*

```
% Figure 5.17.
% Use filled-in markers.
scatter(data(:,1),data(:,2),'filled')
axis equal
box on
```

❑

Surface Plots

If we have data that represent a function defined over a bivariate domain, such as $z = f(x, y)$, then we can view our values for z as a surface. MATLAB provides two functions that display a matrix of z values as a surface: **mesh** and **surf**.

The **mesh** function displays the values as points above a rectangular grid in the x-y plane and connects adjacent points with straight lines. The mesh lines can be colored using various options, but the default method maps the height of the surface to a color.

The **surf** function is similar to **mesh**, except that the open spaces between the lines are filled in with color, with lines shown in black. Other options available with the **shading** command remove the lines or interpolate the color across the patches. An example of where the ability to display a surface can be used is in visualizing a probability density function (see Chapter 9).

Example 5.13

In this example, we begin by generating a grid over which we evaluate a bivariate normal density function. We then calculate the z values that correspond to the function evaluated at each x and y. We can display this as a surface using **surf**, which is shown in Figure 5.19.

```
% Create a bivariate standard normal.
% First create a grid for the domain.
[x,y] = meshgrid(-3:.1:3,-3:.1:3);
% Evaluate using the bivariate standard normal.
z = (1/(2*pi))*exp(-0.5*(x.^2+y.^2));
% Do the plot as a surface.
surf(x,y,z)
```

❑

Special effects can be achieved by changing color maps and using lighting. For example, lighting and color can help highlight structure or features on functions that have many bumps or a jagged surface. We will see some examples of how to use these techniques in the next section and in the exercises at the end of the chapter.

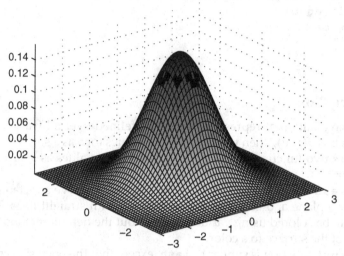

FIGURE 5.19
*This shows a **surf** plot of a bivariate normal probability density function.*

Contour Plots

We can also use contour plots to view our surface. Contour plots show lines of constant surface values, similar to topographical maps. Two functions are available in MATLAB for creating 2D and 3D contour plots. These are called **contour** and **contour3**.

The **pcolor** function shows the same information that is in a contour plot by mapping the surface height to a set of colors. It is sometimes useful to combine the two on the same plot. MATLAB also has the **contourf** function that will create a combination **pcolor** and **contour** plot. The various options that are available for creating contour plots are illustrated in Example 5.14.

Example 5.14

MATLAB has a function called **peaks** that returns a surface with peaks and depressions that can be used to illustrate contour plots. We show how to use the **peaks** function in this example. The following MATLAB code demonstrates how to create the 2D contour plot in Figure 5.20.

```
% Get the data for plotting.
[x,y,z] = peaks;
% Create a 2D contour plot with labels.
% This returns the information for the labels.
```

```
c = contour(x,y,z);
% Add the labels to the plot.
clabel(c)
```

A filled contour plot, which is a combination of **pcolor** and **contour**, is given in Figure 5.21. The MATLAB command needed to get this plot is given here.

```
% Create a 2D filled contour plot.
contourf(x,y,z,15)
```

Finally, a 3D contour plot is easily obtained using the **contour3** function as shown next. The resulting contour plot is shown in Figure 5.22.

```
contour3(x,y,z,15)
```
❑

Bivariate Histogram

In the previous section, we described the univariate density histogram as a way of viewing how our data are distributed over the range of the data. We can extend this to any number of dimensions over a partition of the space [Scott, 2015]. However, in this section, we restrict our attention to the bivariate histogram given by

$$\hat{f}(\mathbf{x}) = \frac{v_k}{nh_1h_2} \qquad \mathbf{x} \text{ in } B_k, \tag{5.7}$$

where v_k represents the number of observations falling into the bivariate bin B_k and h_i is the width of the bin for the x_i coordinate axis. Example 5.15 shows how to get the bivariate density histogram in MATLAB.

Example 5.15

We generate bivariate standard normal random variables and use them to illustrate how to get the bivariate density histogram. We use the optimal bin width for data generated from a standard bivariate normal given in Scott [2015]. We postpone discussion of the optimal bin width and how to obtain it until Chapter 9. A scatterplot of the data and the resulting histogram are shown in Figure 5.23.

```
% Generate sample that is
% standard normal in each dimension.
n = 1000;
d = 2;
x = randn(n,d);
% Need bin origins.
```

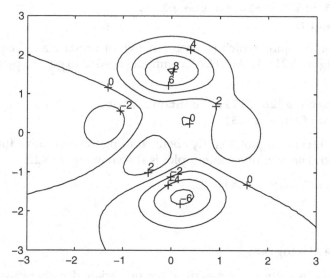

FIGURE 5.20
This is a labeled contour plot of the **peaks** *function. The labels make it easier to understand the hills and valleys in the surface.*

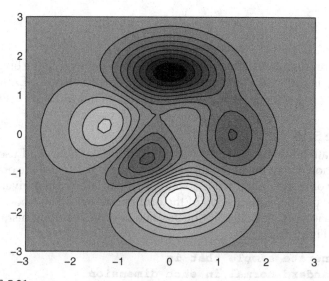

FIGURE 5.21
This is a filled contour plot of the **peaks** *surface. It is created using the* **contourf** *function.*

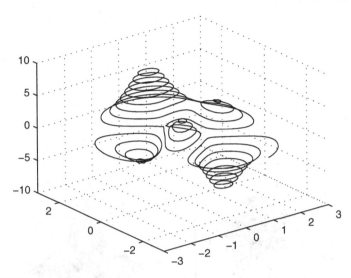

FIGURE 5.22
This is a 3D contour plot of the **peaks** *function.*

```
bin0 = [floor(min(x(:,1))) floor(min(x(:,2)))];
% The bin widths - h - are covered later.
h = 3.504*n^(-0.25)*ones(1,2);
% find the number of bins
nb1 = ceil((max(x(:,1))-bin0(1))/h(1));
nb2 = ceil((max(x(:,2))-bin0(2))/h(2));
% find the mesh
t1 = bin0(1):h(1):(nb1*h(1)+bin0(1));
t2 = bin0(2):h(2):(nb2*h(2)+bin0(2));
[X,Y] = meshgrid(t1,t2);
% Find bin frequencies.
[nr,nc] = size(X);
vu = zeros(nr-1,nc-1);
for i = 1:(nr-1)
    for j = 1:(nc-1)
        xv = [X(i,j) X(i,j+1) X(i+1,j+1) X(i+1,j)];
        yv = [Y(i,j) Y(i,j+1) Y(i+1,j+1) Y(i+1,j)];
        in = inpolygon(x(:,1),x(:,2),xv,yv);
        vu(i,j) = sum(in(:));
    end
end
Z = vu/(n*h(1)*h(2));
% Get some axes that make sense.
[XX,YY] = meshgrid(linspace(-3,3,nb1),...
    linspace(-3,3,nb2));
```

```
surf(XX,YY,Z)
```

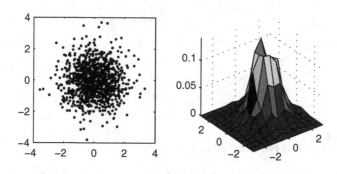

FIGURE 5.23
On the left is a scatterplot of the data. A surface plot of the bivariate density histogram is on the right. Compare the estimated density given by the surface with the one shown in Figure 5.19.

We displayed the resulting bivariate histogram using the **surf** plot in MATLAB. The matrix **Z** in Example 5.14 contains the bin heights. When MATLAB constructs a **mesh** or **surf** plot, the elements of the **Z** matrix represent heights above the x-y plane. The surface is obtained by plotting the points and joining adjacent points with straight lines. Therefore, a **surf** or **mesh** plot of the bivariate histogram bin heights is a linear interpolation between adjacent bins. In essence, it provides a *smooth* version of a histogram. In the next example, we offer another method for viewing the bivariate histogram.

Example 5.16

In this example, we first show the bin heights of the bivariate histogram as bars using the MATLAB function **bar3**. The colors are mapped to the column number of the **Z** matrix, not to the heights of the bins. The resulting histogram is shown in Figure 5.24.

```
% The Z matrix was obtained in Example 5.15.
bar3(Z,1)
% Use some Handle Graphics.
set(gca,'YTickLabel',' ','XTickLabel',' ')
set(gca,'YTick',0,'XTick',0)
grid off
```

MATLAB now has a function in the Statistics Toolbox that will construct bivariate histograms. This function is called **hist3**. This function works similarly to the univariate function, **hist**. The following fragment of code will construct a bivariate histogram using the data generated in Example 5.15.

```
% Make a histogram with see-through bars. The
% hist3 function is in the Statistics Toolbox.
% We use the data (x) generated in Example 5.15.
hist3(x,[10 10],'FaceAlpha',0.35);
set(gcf,'renderer','opengl');
axis tight
```

This plot is shown in the bottom of Figure 5.24. The following MATLAB code constructs a plot that displays the distribution in a different way. We can use the **scatter** plotting function with arguments that relate the marker size and color to the height of the bins. We add the **colorbar** to map the heights of the bins to the color.

```
% Plot the 2D histogram as a scatterplot with
% heights proportional to marker size.
% Find the bin centers to use in the scatterplot.
n1 = length(t1);
n2 = length(t2);
tt1 = linspace((t1(1)+t1(2))/2,...
    (t1(n1-1)+t1(n1))/2,nb1);
tt2 = linspace((t2(1)+t2(2))/2,...
    (t2(n2-1)+t2(n2))/2,nb2);
[xxs,yys] = meshgrid(tt1,tt2);
scatter(xxs(:),yys(:),(Z(:)+eps)*1000,...
    (Z(:)+eps)*1000,'filled')
% Create a colorbar and set the axis
% to the correct scale
h_ax = colorbar;
% Get the current labels.
temp = get(h_ax,'Yticklabel');
[nr,nc] = size(temp);
% Convert from strings to numbers.
newlab = cell(nr,1);
tempcell = cellstr(temp);
% Re-scale and convert back to numbers.
```

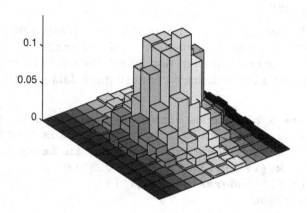

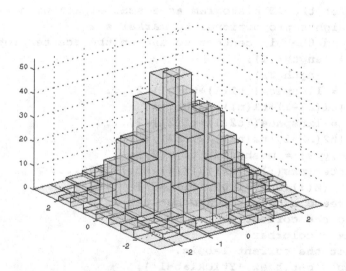

FIGURE 5.24
This top plot shows the same bivariate histogram of Figure 5.23, where the heights of the bars are plotted using the MATLAB function **bar3**. *The bottom histogram was created using the Statistics Toolbox function* **hist3**.

```
for i=1:nr
    newlab{i}=num2str((str2num(tempcell{i})/1000));
end
set(h_ax,'Yticklabel',newlab)
```

This graphic is given in Figure 5.25. Note that we still see the same bivariate normal distribution. The reader might want to compare this plot with the scatterplot of the sample shown in Figure 5.23.
❑

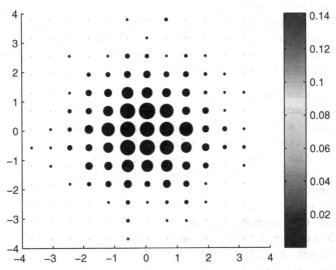

FIGURE 5.25
Here is a different display of the bivariate histogram of Example 5.16. The size and color of the markers indicate the heights of the bins.

3D Scatterplot

As with 2D data, one way we can view trivariate data is with a 3D scatterplot. This is the 3D analog of the bivariate scatterplot. In this case, the ordered triples (x, y, z) are plotted as points. MATLAB provides a function called **scatter3** that will create a 3D scatterplot. As in the bivariate case, you can also use the **plot3** function using a symbol for the marker style to obtain a 3D scatterplot.

A helpful MATLAB command when visualizing anything in 3D is **rotate3d**. Simply type this in at the command line, and you will be able to rotate your graphic using the mouse. There is also a toolbar button that activates the same capability. One reason for looking at scatterplots of the data is to look for interesting structures. The ability to view these structures for 3D data is dependent on the viewpoint or projection to the screen. When looking at 3D scatterplots, the analyst should rotate them to search for patterns or structure that might be hidden by the viewpoint.

Example 5.17
Three variables were measured on 10 insects from each of three species [Hand, et al.,1994]. The variables correspond to the width of the first joint of the first tarsus, the width of the first joint of the second tarsus, and the maximal width of the aedeagus. All widths are measured in microns. These data were originally used in cluster analysis [Lindsey, Herzberg, and Watts, 1987]. What we would like to see from the scatterplot is whether the data for each species can be separated from the others. In other words, is there clear separation or clustering between the species using these variables? The 3D scatterplot for these data is shown in Figure 5.26. This view of the scatterplot indicates that using these variables for pattern recognition or clustering (see Chapters 10 and 11) is reasonable.

```
% Load the insect data
load insect
% Create a 3D scatter plot using a different color
% and marker for each class of insect.
% Plot the first class and hold the plot.
plot3(insect(1:10,1),insect(1:10,2),...
    insect(1:10,3),'ro')
hold on
% Plot the second class.
plot3(insect(11:20,1),insect(11:20,2),...
    insect(11:20,3),'kx')
% Plot the third class.
plot3(insect(21:30,1),insect(21:30,2),...
    insect(21:30,3),'b*')
grid on, hold off
```
□

5.4 Exploring Multi-Dimensional Data

Several methods have been developed to address the problem of visualizing multi-dimensional data. Here we consider applications where we are trying to explore data that has more than three dimensions ($d > 3$). We discuss several ways of visualizing multi-dimensional data. These include the scatterplot matrix, slices, 3D contours, star plots, Andrews curves, and parallel coordinates.

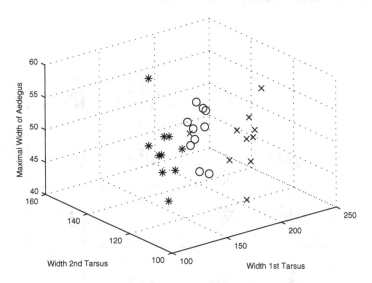

FIGURE 5.26
This is a 3D scatterplot of the **insect** *data. Each species is plotted using a different symbol. This plot indicates that we should be able to identify (with reasonable success) the species based on these three variables.*

Scatterplot Matrix

In the previous sections, we presented the scatterplot as a way of looking at 2D and 3D data. We can extend this to multi-dimensional data by looking at 2D scatterplots of all possible pairs of variables. This allows one to view pairwise relationships and to look for interesting structures in two dimensions. MATLAB provides a function called **plotmatrix** that will create a scatterplot matrix. Its use is illustrated next.

Example 5.18

The **iris** data are well known to statisticians and are often used to illustrate classification, clustering, or visualization techniques. The data were collected by Anderson [1935] and were analyzed by Fisher [1936], so the data are often called *Fisher's iris data* by statisticians. The data consist of 150 observations containing four measurements based on the petals and sepals of three species of iris. These three species are *Iris setosa, Iris virginica,* and *Iris versicolor.* We apply the **plotmatrix** function to the iris data set.

```
load iris
% This loads up three matrices, one for each species.
% Get the plotmatrix display of the Iris setosa data.
```

```
[H,ax,bigax,P] = plotmatrix(setosa);
axes(bigax),title('Iris Setosa')
```

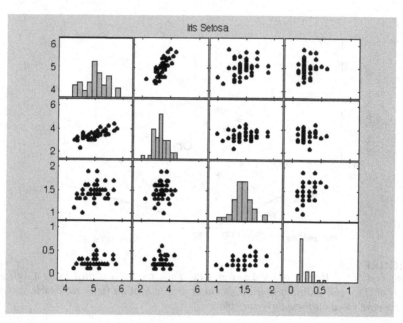

FIGURE 5.27
This is the scatterplot matrix for the Iris setosa data using the **plotmatrix** *function.*

The results are shown in Figure 5.27. Several argument options are available for the **plotmatrix** function. If the first two arguments are matrices, then MATLAB plots a column of the first matrix versus the corresponding column of the other second matrix. In our example, we use a single matrix argument, and MATLAB creates scatterplots of all possible pairs of variables (or columns of **X**). Histograms of each variable or column are shown along the diagonal of the scatterplot matrix. Optional output arguments allow one to add a title or change the plot as shown in the following MATLAB commands. Here we replace the histograms with text that identifies the variable names and display the result in Figure 5.28.

```
% Create the labels as a cell array of strings.
labs = {'Sepal Length','Sepal Width',...
    'Petal Length', 'Petal Width'};
[H,ax,bigax,P] = plotmatrix(virginica);
axes(bigax)
title('Virginica')
% Delete the histograms.
delete(P)
%Put the labels in - the positions might have
```

```
% to be adjusted depending on the text.
for i = 1:4
   txtax = axes('Position',get(ax(i,i),'Position'),...
        'units','normalized');
   text(.1, .5,labs{i})
   set(txtax,'xtick',[],'ytick',[],...
        'xgrid','off','ygrid','off','box','on')
end
```
❑

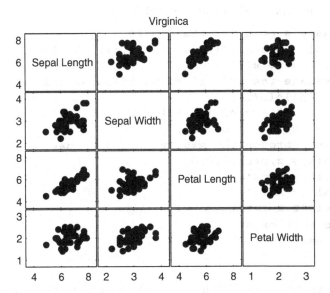

FIGURE 5.28
By using MATLAB's Handle Graphics, we can add text for the variable name to the diagonal boxes.

Slices and Isosurfaces

If we have a function defined over a volume, $f(x, y, z)$, then we can view it using the MATLAB **slice** function or the **isosurface** function. This situation could arise in cases where we have a probability density function defined over a volume. The **slice** capability allows us to view the distribution of our data on slices through a volume. The **isosurface** function allows us to view 3D contours through our volume. These are illustrated in the following examples.

Example 5.19

To illustrate the **slice** function, we need $f(x, y, z)$ values that are defined over a 3D grid or volume. We will use a trivariate normal distribution centered at the origin with covariance equal to the identity matrix. The following MATLAB code displays slices through the $x = 0$, $y = 0$, and $z = 0$ planes, and the resulting display is shown in Figure 5.29. A standard normal bivariate density is given in Figure 5.30 to help the reader understand what the **slice** function is showing. The density or height of the surface defined over the volume is mapped to a color. Therefore, in the **slice** plot, you can see that the maximum density or surface height is at the origin with the height decreasing at the edges of the slices. The color at each point is obtained by interpolation into the volume $f(x, y, z)$.

```
% Create a grid for the domain.
[x,y,z] = meshgrid(-3:.5:3,-3:.5:3,-3:.5:3);
[n,d] = size(x(:));
% Evaluate the trivariate standard normal.
a = (2*pi)^(3/2);
arg = (x.^2 + y.^2 + z.^2);
prob = exp((-.5)*arg)/a;
% Slice through the x=0, y=0, z=0 planes.
slice(x,y,z,prob,0,0,0)
xlabel('X Axis')
ylabel('Y Axis')
zlabel('Z Axis')
% Add a colorbar.
colorbar
```

❑

Isosurfaces are a way of viewing contours through a volume. An isosurface is a surface where the function values $f(x, y, z)$ are constant. These are similar to α-level contours [Scott, 2015], which are defined by

$$S_\alpha = \{\mathbf{x}: f(\mathbf{x}) = \alpha f_{max}\}; \qquad 0 \le \alpha \le 1, \tag{5.8}$$

where $\mathbf{x}$ is a d-dimensional vector. Generally, the α-level contours are nested surfaces.

The function **isosurface(X,Y,Z,V,isosvalue)** determines the contour from the volume data **V** at the value given by **isovalue**. The arrays in **X**, **Y**, and **Z** define the coordinates for the volume. The outputs from this function are the faces and vertices corresponding to the isosurface and can be passed directly into the **patch** function for displaying.

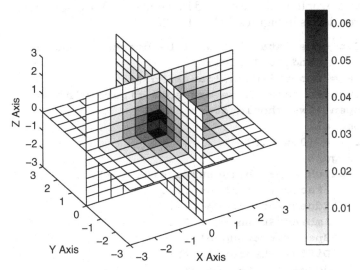

FIGURE 5.29
These are slices through the $x = 0, y = 0, z = 0$ planes for a standard trivariate normal distribution. Each of these planes slice through the volume, and the value of the volume (in this case, the height of the trivariate normal density) is represented by the color. The mode at the origin is clearly seen.

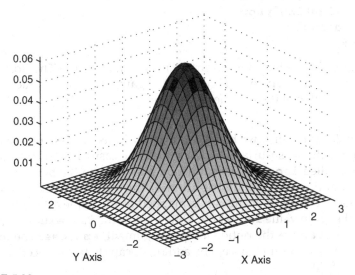

FIGURE 5.30
This is the surface plot for a standard normal bivariate distribution to help the reader understand what is shown in Figure 5.29.

Example 5.20

We illustrate several isosurfaces of 3D contours for data that is uniformly distributed over the volume defined by a unit cube. We display two contours of different levels in Figures 5.31 and 5.32.

```
% Get some data that will be between 0 and 1.
data = rand(10,10,10);
data = smooth3(data,'gaussian');
% Just in case there are some figure windows
% open - we should start anew.
close all
for i = [0.4 0.6]
    figure
    hpatch = patch(isosurface(data,i),...
        'Facecolor','blue',...
        'Edgecolor','none',...
        'AmbientStrength',.2,...
        'SpecularStrength',.7,...
        'DiffuseStrength',.4);
    isonormals(data,hpatch)
    title(['f(x,y,z) = ' num2str(i)])
    daspect([1,1,1])
    axis tight
    axis off
    view(3)
    camlight right
    camlight left
    lighting phong
    drawnow
end
```

In Figure 5.31, we have the isosurface for $f(x, y, z) = 0.4$. The isosurface for $f(x, y, z) = 0.6$ is given in Figure 5.32. Again, these are surface contours where the value of the volume is the same.
❑

It would be better if we had a context to help us understand what we are viewing with the isosurfaces. This can be done easily in MATLAB using the function called **isocaps**. This function puts caps on the boundaries of the domain and shows the distribution of the volume $f(x, y, z)$ above the isosurface. The color of the cap is mapped to the values $f(x, y, z)$ that are above the given value **isovalue**. Values below the **isovalue** can be shown on the **isocap** via the optional input argument, **enclose**. The following example illustrates this concept by adding isocaps to the surfaces obtained in Example 5.20.

f(x,y,z) = 0.4

FIGURE 5.31
This is the isosurface of Example 5.20 for $f(x, y, z) = 0.4$.

f(x,y,z) = 0.6

FIGURE 5.32
This is the isosurface of Example 5.20 for $f(x, y, z) = 0.6$.

Example 5.21

These MATLAB commands show how to add **isocaps** to the isosurfaces in the previous example.

```
for i=[0.4 0.6]
```

```
figure
hpatch = patch(isosurface(data,i),...
   'Facecolor','blue',...
   'Edgecolor','none',...
   'AmbientStrength',.2,...
   'SpecularStrength',.7,...
   'DiffuseStrength',.4);
isonormals(data,hpatch)
patch(isocaps(data,i),...
   'Facecolor','interp',...
   'EdgeColor','none')
colormap hsv
title(['f(x,y,z) = ' num2str(i)])
daspect([1,1,1])
axis tight, axis off
view(3), camlight right, camlight left
lighting phong
drawnow
end
```

Figure 5.33 shows the **isosurface** of Figure 5.31 with the **isocaps**. It is easier now to see what values are *"inside"* the isosurface or contour. Figure 5.34 shows the **isocaps** added to the **isosurface** corresponding to Figure 5.32.
❑

Glyphs

Several methods for visualizing high-dimensional data using a glyph were developed in the 1970s. These were proposed before computers became widely available, so they are suitable for drawing by hand. Thus, they tend to be more appropriate for very small data sets. Some examples of glyphs that can be used are stars, faces, and profiles [du Toit, Steyn, and Stumpf, 1986]. We discuss two of these methods: stars and faces.

Chernoff [1973] developed a method where d-dimensional observations are represented by a cartoon face, with features of the face reflecting the values of the measurements. The size and shape of the nose, eyes, mouth, outline of the face and eyebrows, etc. would be determined by the value of the measurements. Chernoff faces can be used to determine simple trends in the data, but they can be hard to interpret in many cases. The MATLAB Statistics Toolbox includes a function called **glyphplot** that will create Chernoff faces. We illustrate its use in the next example.

Star diagrams were developed by Fienberg [1979] as a way of viewing multi-dimensional observations as a glyph or star. Each observed data point in the sample is plotted as a star, with the value of each measurement shown

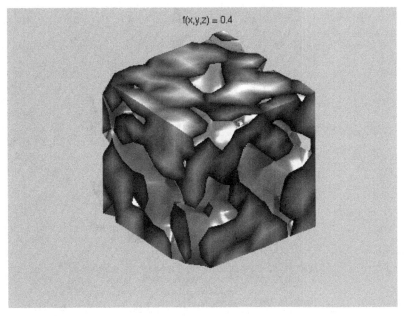

FIGURE 5.33
This is the **isosurface** *of Figure 5.31 with* **isocaps** *added. Note that the color of the edges is mapped to the volume. The default is to map all values above $f(x, y, z) = 0.4$ to the color on the* **isocaps**. *This can be changed by an input argument to* **isocaps**.

as a radial line from a common center point. Thus, each measured value for an observation is plotted as a spoke that is proportional to the size of the measured variable with the ends of the spokes connected with line segments to form a star. Star plots are a nice way to view the entire data set over all dimensions, but they are not suitable when there is a large number of observations ($n > 10$) or many dimensions (e.g., $d > 15$).

The next example applies this technique to data obtained from ratings of eight brands of cereal [Chakrapani and Ehrenberg, 1981; Venables and Ripley, 1994].

Example 5.22

The later versions of the MATLAB Statistics Toolbox have a function called **glyphplot** for constructing star plots and Chernoff faces. We will use that function in this example. To illustrate the concept, we use the **cereal** data. This **.mat** file contains a matrix where each row corresponds to an observation and each column represents one of the variables or the percent agreement with the following statements about the cereal:

- Come back to
- Tastes nice

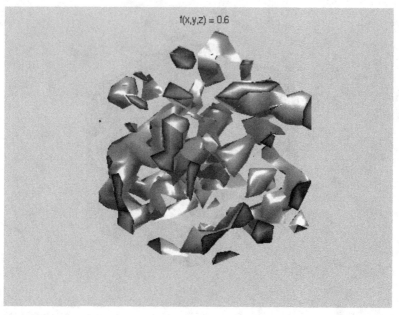

FIGURE 5.34
This is the **isosurface** *of Figure 5.32 with* **isocaps** *added. Note that the color of the edges is mapped to the volume.*

- Popular with all the family
- Very easy to digest
- Nourishing
- Natural flavor
- Reasonably priced
- A lot of food value
- Stays crispy in milk
- Helps to keep you fit
- Fun for children to eat

First, we load the **cereal** data.

```
load cereal
% This file contains the labels and
% the matrix of 8 observations.
```

The default syntax of **glyphplot(X)** will construct star plots. The syntax for a plot using faces is

```
glyphplot(X, 'Glyph','face')
```

Type **help glyphplot** on the command line for information on how the variables are mapped to 17 features of the face. The following code constructs the Chernoff face plot for the **cereal** data.

```
% Construct the stars and face plot using the
% Statistics Toolbox function.
% The following produces a star plot.
glyphplot(cereal)
glyphplot(cereal,'glyph','face','Obslabels',labs)
box on
title('Chernoff Face Plot of Cereal Data')
```

The plots are shown in Figure 5.35. It is hard to get a sense of the variables and their values from looking at these plots. Additionally, it is not clear what values are mapped to the various features of the faces.
❑

Andrews Curves

Andrews curves [Andrews, 1972] were developed as a method for visualizing multi-dimensional data by mapping each observation onto a function. This is similar to star plots or Chernoff faces in that each observation or sample point is represented by a glyph, except that in this case the glyph is a curve. The function for Andrews curves is defined as

$$f_x(t) = x_1/\sqrt{2} + x_2\sin t + x_3\cos t + x_4\sin 2t + x_5\cos 2t + \dots, \qquad (5.9)$$

where the range of t is given by $-\pi \le t \le \pi$. Each observation is projected onto a set of orthogonal basis functions represented by sines and cosines and then plotted. Thus, each sample point is now represented by a curve given by Equation 5.9. We illustrate how to get the Andrews curves in the following example.

Example 5.23
We use a simple example to show how to get Andrews curves. The data we have are the following 3D observations:

$$\mathbf{x}_1 = (2, 6, 4)$$
$$\mathbf{x}_2 = (5, 7, 3)$$
$$\mathbf{x}_3 = (1, 8, 9).$$

Using Equation 5.9, we construct three curves, one corresponding to each data point. The Andrews curves for the data are

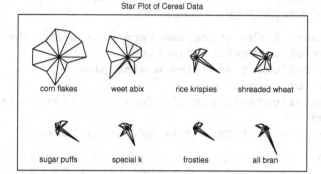

FIGURE 5.35

The top graphic shows a star plot for the **cereal** *data. The Chernoff face plot for the same data is shown in the bottom picture. We used the* **glyphplot** *function in the Statistics Toolbox to create both of these pictures. Note that plots are not suitable for large data sets.*

$$f_{x_1}(t) = 2/\sqrt{2} + 6\sin t + 4\cos t$$

$$f_{x_2}(t) = 5/\sqrt{2} + 7\sin t + 3\cos t$$

$$f_{x_3}(t) = 1/\sqrt{2} + 8\sin t + 9\cos t.$$

We plot these three functions in MATLAB using the following commands. The Andrews curves for these data are shown in Figure 5.36.

```
% Get the domain.
t = linspace(-pi,pi);
% Evaluate function values for each observation.
f1 = 2/sqrt(2)+6*sin(t)+4*cos(t);
```

```
f2 = 5/sqrt(2)+7*sin(t)+3*cos(t);
f3 = 1/sqrt(2)+8*sin(t)+9*cos(t);
plot(t,f1,'.',t,f2,'*',t,f3,'o')
legend('F1','F2','F3')
xlabel('t')
```

❑

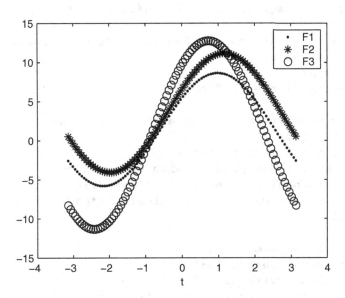

FIGURE 5.36
Andrews curves for the three data points in Example 5.23.

It has been shown [Andrews, 1972; Embrechts and Herzberg, 1991] that because of the mathematical properties of the trigonometric functions, the Andrews curves preserve means, distance (up to a constant), and variances. One consequence of this is that Andrews curves showing functions close together suggest that the corresponding data points will also be close together. Thus, one use of Andrews curves is to look for clustering of the data points.

Example 5.24

We show how to construct Andrews curves for the **iris** data, using only the observations for *Iris setosa* and *Iris virginica* observations. We plot the curves for each species in a different line style to see if there is evidence that we can distinguish between the species using these variables.

```
load iris
% This defines the domain that will be plotted.
theta = (-pi+eps):0.1:(pi-eps);
```

```
n = 50;
p = 4;
ysetosa = zeros(n,p);
% There will n curves plotted,
% one for each data point.
yvirginica = zeros(n,p);
% Take dot product of each row with observation.
ang = zeros(length(theta),p);
fstr = '[1/sqrt(2) sin(i) cos(i) sin(2*i)]';
k = 0;
% Evaluate sin and cos functions at each angle theta.
for i = theta
   k = k+1;
   ang(k,:) = eval(fstr);
end
% Now generate a 'y' for each observation.
for i = 1:n
  for j = 1:length(theta)
     % Find dot product with observation.
     ysetosa(i,j)=setosa(i,:)*ang(j,:)';
     yvirginica(i,j)=virginica(i,:)*ang(j,:)';
  end
end
% Do all of the plots.
plot(theta,ysetosa(1,:),'r',...
     theta,yvirginica(1,:),'b-.')
legend('Iris Setosa','Iris Virginica')
hold
for i = 2:n
  plot(theta,ysetosa(i,:),'r',...
     theta,yvirginica(i,:),'b-.')
end
hold off
title('Andrews Plot')
xlabel('t')
ylabel('Andrews Curve')
```

The curves are shown in Figure 5.37. By plotting the two groups with different line styles, we can gain some insights about whether these two species of iris can be distinguished based on these features. From the Andrews curves, we see that the observations exhibit similarity within each class and that they show differences between the classes. Thus, we might get reasonable discrimination using these features.
❑

Andrews curves are dependent on the order of the variables. Lower frequency terms exert more influence on the shape of the curves, so

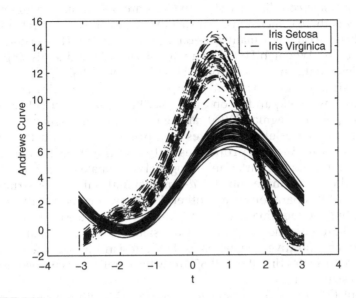

FIGURE 5.37
These are the Andrews curves for the Iris setosa and Iris virginica data. The curves corresponding to each species are plotted with different line styles. Note that the observations within each group show similar curves, and that we seem to be able to separate these two species.

reordering the variables and viewing the resulting plot might provide insights about the data. By lower frequency terms, we mean those that are first in the sum given in Equation 5.9. Embrechts and Herzberg [1991] also suggest that the data be rescaled so they are centered at the origin and have covariance equal to the identity matrix. Andrews curves can be extended by using orthogonal bases other than sines and cosines. For example, Embrechts and Herzberg [1991] illustrate Andrews curves using Legendre polynomials and Chebychev polynomials.

The MATLAB Statistics Toolbox has a function called **andrewsplot** that will construct an Andrews curves plot. The basic syntax is

andrewsplot(X)

It has several useful features. For example, it will plot the curves using different colors for given groups; it will plot the curves for given quantiles (useful for large samples); and it will plot curves for various standardizations.

Parallel Coordinates

In the Cartesian coordinate system the axes are orthogonal, so the most we can view is three dimensions. If instead we draw the axes parallel to each other, then we can view many axes on the same display. This technique was developed by Wegman [1986] as a way of viewing and analyzing multi-dimensional data and was introduced by Inselberg [1985] in the context of computational geometry and computer vision. Parallel coordinate techniques were expanded on and described in a statistical setting by Wegman [1990]. Wegman [1990] also gave a rigorous explanation of the properties of parallel coordinates as a projective transformation and illustrated the duality properties between the parallel coordinate representation and the Cartesian orthogonal coordinate representation.

A parallel coordinate plot for d-dimensional data is constructed by drawing d lines parallel to each other. We draw d copies of the real line representing the coordinates for $x_1, x_2, ..., x_d$. The lines are the same distance apart and are perpendicular to the Cartesian y axis. Additionally, they all have the same positive orientation as the Cartesian x axis. Some versions of parallel coordinates [Inselberg, 1985] draw the parallel axes perpendicular to the Cartesian x axis.

A point $C = (c_1, ..., c_4)$ is shown in Figure 5.38 with the MATLAB code that generates it given in Example 5.25. We see that the point is a polygonal line with vertices at $(c_i, i - 1), i = 1, ..., d$ in Cartesian coordinates on the x_i parallel axis. Thus, a point in Cartesian coordinates is represented in parallel coordinates as a series of connected line segments.

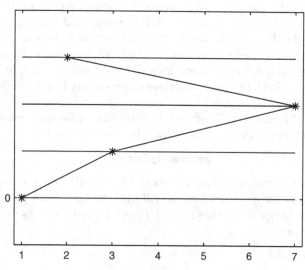

FIGURE 5.38
This shows the parallel coordinate representation for the 4D point (1,3,7,2).

Example 5.25

We now plot the point $C = (1, 3, 7, 2)$ in parallel coordinates using these MATLAB commands.

```
c = [1 3 7 2];
% Get range of parallel axes.
x = [1 7];
% Plot the 4 parallel axes.
plot(x,zeros(1,2),x,ones(1,2),x, ...
    2*ones(1,2),x,3*ones(1,2))
hold on
% Now plot point c as a polygonal line.
plot(c,0:3,c,0:3,'*')
ax = axis;
axis([ax(1) ax(2) -1 4 ])
set(gca,'ytick',0)
hold off
```

❑

If we plot observations in parallel coordinates with colors designating what class they belong to, then the parallel coordinate display can be used to determine whether the variables will enable us to separate the classes. This is similar to the Andrews curves in Example 5.24, where we used the Andrews curves to view the separation between two species of iris. The next example shows how parallel coordinates can display the correlation between two variables.

Example 5.26

We first generate a set of 20 bivariate normal random variables with correlation given by 1. We plot the data using the function called **csparallel** to show how to recognize various types of correlation in parallel coordinate plots.

```
% Get a covariance matrix with correlation 1.
covmat = [1 1; 1 1];
% Generate the bivariate normal random variables.
% Note: you could use csmvrnd to get these.
[u,s,v] = svd(covmat);
vsqrt = (v*(u'.*sqrt(s)))';
subdata = randn(20,2);
data = subdata*vsqrt;
% Create parallel plot using CS Toolbox function.
csparallel(data)
title('Correlation of 1')
```

This is shown in Figure 5.39.The direct linear relationship between the first variable and the second variable is readily apparent. We can generate data

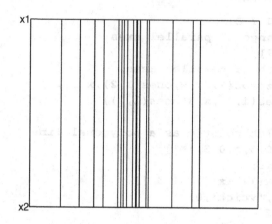

FIGURE 5.39
This is a parallel coordinate plot for bivariate data that have a correlation coefficient of 1.

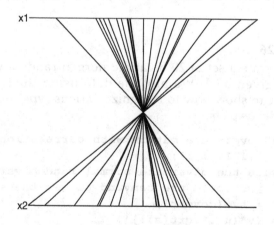

FIGURE 5.40
The data shown in this parallel coordinate plot are negatively correlated.

that are correlated differently by changing the covariance matrix. For example, to obtain a random sample for data with a correlation of 0.2, we can use

```
covmat = [4 1.2; 1.2, 9];
```

In Figure 5.40, we show the parallel coordinates plot for data that have a correlation coefficient of -1. Note the different structure that is visible in the parallel coordinates plot.
❑

In the previous example, we showed how parallel coordinates can indicate the relationship between variables. To provide further insight, we illustrate how parallel coordinates can indicate clustering of variables in a dimension. Figure 5.41 shows data that can be separated into clusters in both of the dimensions. This is indicated on the parallel coordinate representation by separation or groups of lines along the x_1 and x_2 parallel axes. In Figure 5.42, we have data that are separated into clusters in only one dimension, x_1, but not in the x_2 dimension. This appears in the parallel coordinates plot as a gap in the x_1 parallel axis.

As with Andrews curves, the order of the variables makes a difference. Adjacent parallel axes provide some insights about the relationship between consecutive variables. To see other pairwise relationships, we must permute the order of the parallel axes. Wegman [1990] provides a systematic way of finding all permutations such that all adjacencies in the parallel coordinate display will be visited. This idea is also discussed in more detail in Martinez et al., 2010].

Before we proceed to other topics, we provide an example applying parallel coordinates to the **iris** data. In Example 5.27, we illustrate a parallel coordinates plot of the two classes: *Iris setosa* and *Iris virginica*.

Example 5.27

First we load the **iris** data. An optional input argument of the **csparallel** function is the line style for the lines. This usage is shown next, where we plot the *Iris setosa* observations as dot-dash lines and the *Iris virginica* as solid lines. The parallel coordinate plot is given in Figure 5.43.

```
load iris
figure
csparallel(setosa,'-.')
hold on
csparallel(virginica,'-')
hold off
```

From this plot, we see evidence of groups or separation in coordinates x_2 and x_3.
❑

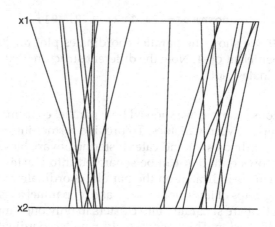

FIGURE 5.41
Clustering in two dimensions produces gaps in both parallel axes.

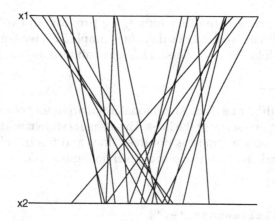

FIGURE 5.42
Clustering in only one dimension produces a gap in the corresponding parallel axis, x_1.

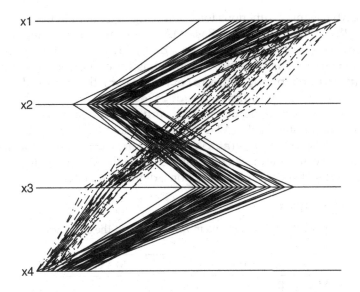

FIGURE 5.43
*Here we see an example of a parallel coordinate plot for the **iris** data. The Iris setosa is shown as dot-dash lines and the Iris virginica as solid lines. There is evidence of groups in two of the coordinate axes, indicating that reasonable separation between these species could be made based on these features.*

The Statistics Toolbox has a function that will construct parallel coordinate plots. This function is called **parallelcoords**. The basic syntax is

parallelcoords(X)

As with Andrews plots, one can have parallel coordinate plots for grouped data, where each group is shown with a different color. One can also create a parallel coordinates plot using various types of standardization or plot just the median and given quantiles.

5.5 MATLAB® Code

MATLAB has many functions for visualizing data, both in the main package and in the Statistics Toolbox. Base MATLAB has functions for scatterplots (**scatter**), histograms (**hist, bar**), and scatterplot matrices (**plotmatrix**). The Statistics Toolbox has functions for constructing q-q plots (**normplot, probplot, qqplot, wblplot**), the empirical cumulative distribution function (**cdfplot, ecdf, ecdfhist**), grouped versions of plots (**gscatter, gplotmatrix**), and others. Some other graphing functions in the standard MATLAB package that might be of interest include pie charts (**pie**), stair plots (**stairs**), error bars (**errorbar**), and stem plots (**stem**). Some are summarized in Table 5.3. See the MATLAB documentation for a complete list of graphics functions, demos, and GUIs.

TABLE 5.3

Visualization Functions in MATLAB and the Statistics Toolbox

Purpose	MATLAB Function
MATLAB	
Discrete data plots	**stem, stairs**
Contour plots	**contour**
Histogram	**hist**
Plots	**plot, plot3**
Scatterplot	**scatter, scatter3**
Scatterplot matrix	**plotmatrix**
MATLAB Statistics Toolbox	
Andrews curves	**andrewsplot**
Boxplot	**boxplot**
Empirical CDF	**cdfplot**
Glyph plot (faces and stars)	**glyphplot**
Grouped scatterplot	**gscatter**
Grouped scatterplot matrix	**gplotmatrix**
Histogram of empirical CDF	**ecdfhist**
Histogram — bivariate	**hist3**
Normal probability plot	**normplot**
Parallel coordinate plot	**parallelcoords**
Probability plots	**probplot**
Quantile–quantile plot	**qqplot**
Scatterplot with histograms	**scatterhist**
Weibull probability plot	**wblplot**

There are two EDA Toolboxes that have more functionality for exploratory data analysis as described in Martinez et al. [2010]. Both of these toolboxes are available for download from the CRC Press website for the referenced text or from

http://www.pi-sigma.info/

The methods for statistical graphics described in Cleveland's *Visualizing Data* [1993] have been implemented in MATLAB. They are available for download at

http://www.datatool.com/resources.html

This book contains many useful techniques for visualizing data. Since MATLAB code is available for these methods, we urge the reader to refer to this highly readable text for more information on statistical visualization.

Rousseeuw, Ruts, and Tukey [1999] describe a bivariate generalization of the univariate boxplot called a *bagplot*. This type of plot displays the location, spread, correlation, skewness, and tails of the data set. Rousseeuw, Ruts, and Tukey provide MATLAB code for creating a bagplot. We include this code in the EDA Toolbox.

Finally, we include several functions in the Computational Statistics Toolbox that implement some of the algorithms and graphics covered in Chapter 5. These are summarized in Table 5.4.

TABLE 5.4

List of Functions from Chapter 5 Included in the Computational Statistics Toolbox

Purpose	MATLAB Function
Star plot	csstars
Stem-and-leaf plot	csstemleaf
Box Plot — variable width and histplot	boxp
Box-percentile plot	boxprct
Parallel coordinates plot	csparallel
Q-Q plot	csqqplot
Poissonness plot	cspoissplot
Andrews curves	csandrews
Exponential probability plot	csexpoplot
Binomial plot	csbinoplot

5.6 Further Reading

One of the first treatises on graphical exploratory data analysis is John Tukey's *Exploratory Data Analysis* [1977]. In this book, he explains many aspects of EDA, including smoothing techniques, graphical techniques, and others. The material in this book is practical and is readily accessible to readers with rudimentary knowledge of data analysis. Another excellent book on this subject is *Graphical Exploratory Data Analysis* [du Toit, Steyn, and Stumpf, 1986], which includes several techniques (e.g., profiles) that we do not cover. For texts that emphasize the visualization of technical data, see Fortner and Meyer [1997] and Fortner [1995]. The paper by Wegman, Carr, and Luo [1993] discusses many of the methods we present, along with others such as stereoscopic displays, generalized nonlinear regression using skeletons, and a description of d-dimensional grand tour. This paper and Wegman [1990] provide an excellent theoretical treatment of parallel coordinates.

The Grammar of Graphics by Wilkinson [1999] describes a foundation for producing graphics for scientific journals, the Internet, statistical packages, or any visualization system. It looks at the rules for producing pie charts, bar charts, scatterplots, maps, function plots, and many others.

For the reader who is interested in visualization and information design, the four books by Edward Tufte are recommended. His first book (now in its second edition), *The Visual Display of Quantitative Information* [Tufte, 2001], shows how to depict numbers. The second in the series is called *Envisioning Information* [Tufte, 1990], and illustrates how to deal with pictures of nouns (e.g., maps, aerial photographs, weather data). The third book is entitled *Visual Explanations* [Tufte, 1997], and it discusses how to illustrate pictures of verbs. The fourth book examines how visualizing data can convey complex information and knowledge. These books also provide many examples of good graphics and bad graphics.

We highly recommend the book by Wainer [1997] for any statistician, engineer, or data analyst. Wainer also discusses the subject of good and bad graphics in a way that is accessible to the general reader. Wainer has a wonderful text called *Graphic Discovery: A Trout in the Milk and Other Visual Adventures* [2004] that takes the reader on a tour of data displays through history.

Another text that discusses ways to create better graphics for a wide range of applications is *Creating More Effective Graphs* by Robbins [2013]. This text is easy to read, so it is accessible to anyone who needs to understand how to construct effective graphs for communicating information about data. It also includes many types of plots that we do not discuss, such as mosaic plots, trellis displays, month plots, and more.

We did not discuss methods for visualizing categorical data. Some methods for doing this can be found in Robbins [2005]. Another source of information on this topic is *Visualizing Categorical Data* by Friendly [2000]. It uses the SAS software to present the ideas, but it has some good examples of mosaic plots and other methods.

High-dimensional data can also be viewed using color histograms or data images. Color histograms are described in Wegman [1990]. Data images are discussed in Minotte and West [1998] and are a special case of color histograms.

For more information on the graphical capabilities of MATLAB, we refer the reader to the MATLAB documentation. Another excellent resource is the book called *Graphics and GUI's with MATLAB, Third Edition* by Marchand and Holland [2003]. This text goes into more detail on the graphics capabilities in MATLAB for data analysis such as lighting, use of the camera, animation, etc.

Various versions and extensions of the stem-and-leaf plot are available. We show an ordered stem-and-leaf plot in this book, but ordering is not required. Another version shades the leaves. Most introductory applied statistics books have information on stem-and-leaf plots (e.g., Montgomery, et al. [1998]). Hunter [1988] proposes an enhanced stem-and-leaf called the *digidot plot*. This combines a stem-and-leaf with a time sequence plot. As data are collected they are plotted as a sequence of connected dots and a stem-and-leaf is created at the same time.

Hoaglin and Tukey [1985] provide similar plots for other discrete distributions. These include the negative binomial, the geometric, and the logarithmic series. They also discuss graphical techniques for plotting confidence intervals instead of points.

Scatterplot techniques are discussed in Carr, et al. [1987]. The methods presented in this paper are especially pertinent to the situation facing analysts today, where the typical data set that must be analyzed is often very large ($n = 10^3, 10^6, ...$). They recommend various forms of binning (including hexagonal binning) and representing the value by gray scale or symbol area.

Other techniques for visualizing multi-dimensional data have been proposed in the literature and are also implemented in base MATLAB. One example is called *brushing*. Brushing [Venables and Ripley, 1994; Cleveland, 1993] is an interactive technique where the user can highlight data points on a scatterplot, and the same points are highlighted on all other plots. For example, in a scatterplot matrix, highlighting a point in one plot shows up as highlighted in all of the others. This helps illustrate interesting structure across plots. Martinez et al., [2010] discuss brushing and linking in more detail.

Exercises

5.1.   Generate a sample of 1000 univariate standard normal random variables using **randn**. Construct a frequency histogram, relative frequency histogram, and density histogram. For the density histogram, superimpose the corresponding theoretical probability density function. How well do they match?

5.2.   Repeat Problem 5.1 for random samples generated from the exponential, gamma, and beta distributions.

5.3.   Do a quantile plot of the Tibetan skull data (**tibetan**) of Example 5.3 using the standard normal quantiles. Is it reasonable to assume the data follow a normal distribution?

5.4.   Try the following MATLAB code using the 3D multivariate normal as defined in Example 5.19. This will create a slice through the volume at an arbitrary angle. Notice that the colors indicate a normal distribution centered at the origin with the covariance matrix equal to the identity matrix.

```
% Draw a slice at an arbitrary angle
hs = surf(linspace(-3,3,20),...
   linspace(-3,3,20),zeros(20));
% Rotate the surface :
rotate(hs,[1,-1,1],30)
% Get the data that will define the
% surface at an arbitrary angle.
xd = get(hs,'XData');
yd = get(hs,'YData');
zd = get(hs,'ZData');
delete(hs)
% Draw slice:
slice(x,y,z,prob,xd,yd,zd)
axis tight
% Now plot this using the peaks surface as the slice.
% Try plotting against the peaks surface
[xd,yd,zd] = peaks;
slice(x,y,z,prob,xd,yd,zd)
axis tight
```

5.5.   Repeat Example 5.24 using the data for *Iris virginica* and *Iris versicolor*. Do the Andrews curves indicate separation between the classes? Do you think it will be difficult to separate these classes based on these features?

5.6.   Repeat Example 5.4, where you generate random variables such that

a. $X \sim N(0, 2)$ and $Y \sim N(0, 1)$
b. $X \sim N(5, 1)$ and $Y \sim N(0, 1)$

How can you tell from the q-q plot that the scale and the location parameters are different?

5.7.   Write a MATLAB program that permutes the axes in a parallel coordinates plot. Apply it to the **iris** data and comment on the results.

5.8.   Write a MATLAB program that permutes the order of the variables and plots the resulting Andrews curves. Apply it to the **iris** data and comment on the results.

5.9.   Implement Andrews curves using a different set of basis functions as suggested in the text.

5.10. Repeat Example 5.17 and use **rotate3d** (or the rotate toolbar button) to rotate about the axes. Do you see any separation of the different types of insects?

5.11. Do a scatterplot matrix of the *Iris versicolor* data (see Example 5.18).

5.12. Define a trivariate normal as your volume, $f(x, y, z)$. Use the MATLAB functions **isosurface** and **isocaps** to obtain contours of constant volume or probability (in this case).

5.13. Construct a quantile plot using the **forearm** data, comparing the sample to the quantiles of a normal distribution. Is it reasonable to model the data using the normal distribution?

5.14. The **moths** data represent the number of moths caught in a trap over 24 consecutive nights [Hand, et al., 1994]. Use the stem-and-leaf to explore the shape of the distribution.

5.15. The **biology** data set contains the number of research papers for 1534 biologists [Tripathi and Gupta, 1988; Hand, et al., 1994]. Construct a binomial plot of these data. Analyze your results.

5.16. In the **counting** data set, we have the number of scintillations in 72-second intervals arising from the radioactive decay of polonium [Rutherford and Geiger, 1910; Hand, et al., 1994]. Construct a Poissonness plot. Does this indicate agreement with the Poisson distribution?

5.17. Use the MATLAB Statistics Toolbox function **boxplot** to compare box plots of the features for each species of **iris** data.

5.18.   The **thrombos** data set contains measurements of urinary-throm-boglobulin excretion in 12 normal and 12 diabetic patients [van Oost, et al.; 1983; Hand, et al., 1994]. Put each of these into a column of a matrix and use the **boxplot** function to compare normal versus diabetic patients.

5.19.   To explore the **shading** options in MATLAB, try the following code from the documentation:

```
% The ezsurf function is in base MATLAB.
% First get a surface.
ezsurf('sin(sqrt(x^2+y^2))/sqrt(x^2+y^2)',...
    [-6*pi,6*pi])
% Now add some lighting effects:
view(0,75)
shading interp
lightangle(-45,30)
set(findobj('type','surface'),...
    'FaceLighting','phong',...
    'AmbientStrength',0.3,'DiffuseStrength',0.8,...
    'SpecularStrength',0.9,'SpecularExponent',25,...
    'BackFaceLighting','unlit')
axis off
```

5.20.   The **bank** data contains two matrices comprised of measurements made on genuine money and forged money. Combine these two matrices into one and use some of the visualization techniques mentioned in this chapter to explore the data.

5.21.   Repeat Example 5.11 and construct a variable width box plot. You can use the MATLAB command: **boxp(X,'vw')**. Obtain a variable width box-percentile plot using: **boxp(X,'vw')**. Compare these with the other box plots from Examples 5.10 and 5.11.

Chapter 6

Finding Structure

6.1 Introduction

In the last chapter, we presented several graphical methods for exploring data sets, but all of them involved static displays. In this chapter, we will be discussing graphical methods for EDA that are dynamic and sometimes allow the user to interact with the display to uncover structure. The power and usefulness of these methods become apparent when used to explore high-dimensional data sets.

Structure can mean many things and is often not defined in advance. In other words, we will know structure when we see it. For example, we might be interested in visually locating groups or possibly finding anomalies (e.g., holes or nonlinear structures or patterns). However, in some of the techniques we will discuss in this chapter, structure means non-Gaussian or some sort of departure from normality.

We will also see that finding structure sometimes involves projecting our data to a lower-dimensional space. Therefore, we start off with a brief introduction to the mechanics of projecting data, along with a discussion of a well-known method for transforming data called principal component analysis. This is followed by a methodology called projection pursuit EDA, where one can find interesting projections of the data, with *interesting* being defined in several ways. The next section has a discussion of independent component analysis and its connection with principal component analysis and projection pursuit. This is followed by dynamic or animated methods that are known as grand tours. We conclude with two techniques for reducing the dimensionality belonging to the family of nonlinear methods: multidimensional scaling and isometric feature mapping.

6.2 Projecting Data

The Andrews curves and parallel coordinate plots we presented in the previous chapter allow us to visualize all of the data points and all of the dimensions at once. An Andrews curve accomplishes this by mapping a data point to a curve. Parallel coordinate displays accomplish this by mapping each observation to a polygonal line with vertices on parallel axes. Another option is to tackle the problem of visualizing multi-dimensional data by reducing the dimension via a suitable projection. These methods reduce the data to 1D or 2D by projecting onto a line or a plane and then displaying each point in some suitable graphic, such as a scatterplot. Once the data are reduced to something that can be easily viewed, then exploring the data for patterns or interesting structure is possible. This process is called *dimensionality reduction.*

The concept of projecting data is important, and several of the methods described in this chapter project the data to some other space to search for structure or to reduce the dimensionality of the data. For example, principal component analysis, independent component analysis, projection pursuit, and the grand tour all involve projections. So, we will first briefly describe this concept before moving on to specific techniques for finding structure.

Example 6.1
We illustrate how projection works for a small 2D data set, where we will reduce the dimensionality by projecting to a line. In this case, the projection matrix is given by

$$\mathbf{P} = \begin{bmatrix} (\cos\theta)^2 & \cos\theta\sin\theta \\ \cos\theta\sin\theta & (\sin\theta)^2 \end{bmatrix}.$$

The coordinates of the projected observations are a weighted sum of the original variables, where $\mathbf{P}$ contains the weights. We project the observations using the matrix notation

$$\mathbf{y}_i = \mathbf{P}^T\mathbf{x}_i; \qquad i = 1, \ldots, n. \tag{6.1}$$

The following code shows how to project all of the data at once using the projection matrix.

```
% Specify some data.
X = [-2 4; 2 4;6 1;8 10;7 5;11 8];
% Create the projection matrix.
```

```
theta = pi/4;
c2 = cos(theta)^2;
cs = cos(theta)*sin(theta);
s2 = sin(theta)^2;
P = [c2 cs; cs s2];
% Now project the data onto the line.
Xp = X*P;
% Create the plot shown in Figure 6.1.
plot(Xp(:,1),Xp(:,2),'d',X(:,1),X(:,2),'o')
hold on
plot([0  12 ], [0 12 ], ':')
hold off
axis([-4 12 0 12])
```

The results are shown in Figure 6.1, where one can see that the projected observations now fall on the line. The original data points are shown as circles, and the transformed observations are shown as diamonds.
❑

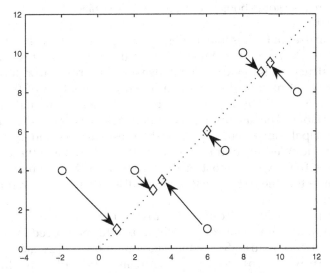

FIGURE 6.1
This illustrates the projection of 2D data (circles) onto a line (diamonds). See Example 6.1 for details.

The example above did not reduce the dimensionality of the data, since the projected points in the matrix **Xp** are still 2D. We show in the next section how to use a particular projection to obtain data with fewer dimensions. For more information on projections and other matrix transformations, please see Strang [2005] or any linear algebra text.

6.3 Principal Component Analysis

One well-known method for reducing dimensionality is *principal component analysis* (PCA) [Jackson, 1991]. This method uses the eigenvector decomposition of the covariance (or the correlation) matrix. The data are then projected onto the eigenvector corresponding to the maximum eigenvalue (sometimes known as the first principal component) to reduce the data to one dimension. In this case, the eigenvector corresponds to one that follows the direction of the maximum variation in the data. Therefore, if we project onto the first principal component, then we will be using the direction that accounts for the maximum amount of variation using only one dimension. Alternatively, we could project onto two dimensions using the eigenvectors corresponding to the largest and second largest eigenvalues. This would project onto the plane spanned by these eigenvectors.

As we see shortly, PCA can be thought of in terms of projection pursuit, where the interesting structure is the variance of the projected data. Principal component analysis is discussed in many linear algebra texts [Strang, 2005], and there are also some books dedicated to this subject alone [Jackson, 1991; Joliffe, 2002].

One of the uses of PCA is to map observations to a space such that they are uncorrelated. However, our main interest in the method, as it is used in the search for structure, is to reduce the dimensionality from d to k, where $k < d$, and the new set of coordinates are a linear combination of the original variables. We do this by finding k orthogonal linear combinations of the original variables. The first principal component has the largest variance. The second principal component has the next largest variance and is orthogonal to the first one. We can continue in this way until we have the k principal components that explain most of the variance in the original data. The assumption is that the rest of the PCs can be discarded without losing a lot of information.

As noted previously, PCA can start with either the covariance matrix or the correlation matrix. In this text, we will illustrate the procedure using the sample covariance matrix, **S**. For information on other approaches, such as the decomposition of the correlation matrix and singular value decomposition, see Joliffe [2002] or Jackson [1991].

We showed how to calculate the sample covariance matrix **S** in Chapter 3, but we repeat the definition for ease of reading. We first subtract the d-dimensional sample mean from each observation to center the data at the sample mean. We denote this centered data matrix as $\mathbf{X}_c$. Note that one should get a vector of zeros (approximately) when using the command **mean(Xc)**. The sample covariance matrix is found as

$$\mathbf{S} = \frac{1}{n-1}\mathbf{X}_c^T\mathbf{X}_c,$$

with the jk-th element of $\mathbf{S}$ given by

$$s_{jk} = \frac{1}{n-1}\sum_{i=1}^{n}(x_{ij} - \bar{x}_j)(x_{ik} - \bar{x}_k); \qquad j, k = 1, ..., d.$$

The next step is to calculate the eigenvectors and the eigenvalues of the sample covariance matrix. The eigenvalues are found by solving the following equation for each λ_j, $j = 1, ..., d$:

$$|\mathbf{S} - \lambda\mathbf{I}| = 0, \tag{6.2}$$

where $\mathbf{I}$ is a $d \times d$ identity matrix and $|*|$ denotes the matrix determinant. If one uses the definition of the determinant to expand Equation 6.2, then the result is a polynomial of degree d.

Once we have the eigenvalues, we can then find the eigenvectors by solving the following set of equations for vectors $\mathbf{a}_j$:

$$(\mathbf{S} - \lambda_j\mathbf{I})\mathbf{a}_j = 0, \qquad j = 1, ..., d,$$

subject to the condition that the set of eigenvectors are orthonormal. This means that the magnitude of each eigenvector is one, and they are orthogonal to each other. Mathematically, this can be expressed as

$$\mathbf{a}_i^T\mathbf{a}_i = 1$$
$$\mathbf{a}_i^T\mathbf{a}_j = 0.$$

We know from matrix algebra that *any* square, symmetric, nonsingular matrix can be transformed to a diagonal matrix using the eigenvectors. In our description here, we use the sample covariance matrix $\mathbf{S}$, so the transformation is given by

$$\mathbf{L} = \mathbf{A}^T\mathbf{S}\mathbf{A},$$

where each eigenvector ($\mathbf{a}_j$) of $\mathbf{S}$ is a column in the matrix $\mathbf{A}$, and $\mathbf{L}$ is a diagonal matrix with the eigenvalues along the diagonal. The eigenvalues are typically ordered as $\lambda_1 \geq \lambda_2 \geq ... \geq \lambda_d$, with the eigenvectors following the same order. This means that the eigenvector $\mathbf{a}_1$ that goes with λ_1 is the first column of $\mathbf{A}$.

The eigenvectors of **S** can be used to obtain new transformed variables called *principal components* (PCs). The j-th PC is given by

$$z_j = \mathbf{a}_j^T(\mathbf{x} - \bar{\mathbf{x}}), \qquad (6.3)$$

for $j = 1, ..., d$. We see from Equation 6.3 that the PCs are a linear combination of the original variables where the elements of $\mathbf{a}_j$ provide the coefficients. We transform the data to the principal component coordinate system using the following matrix multiplication:

$$\mathbf{Y} = \mathbf{X}_c\mathbf{A}. \qquad (6.4)$$

The matrix **Y** in Equation 6.4 contains the *principal component scores*. The PC scores have zero mean because we first centered the data about the mean, and they are uncorrelated. We could also transform the original observations, but the resulting PC scores would have a different mean. To summarize, the transformed *variables* (Equation 6.3) are the *principal components* and the transformed *observations* (Equation 6.4) are the *PC scores*.

So far, we have not addressed the issue of reducing the dimensionality using PCA, since the PC scores in Equation 6.4 are still d-dimensional. We now show how to accomplish this using the eigenvectors that correspond to the k ($k < d$) largest eigenvalues. It can be shown [Joliffe, 2002; Jackson, 1991] that the sum of the eigenvalues is equal to the total variance of the original variables. Therefore, one could include in the projection only those eigenvectors that correspond to the highest eigenvalues. In this way, we are accounting for the largest amount of variation with smaller dimensionality.

We can reduce the dimensionality using the following

$$\mathbf{Y}_k = \mathbf{X}_c\mathbf{A}_k, \qquad (6.5)$$

where $\mathbf{A}_k$ contains the first k eigenvectors or columns of **A**. Note that the matrix $\mathbf{Y}_k$ has n rows and k columns, so each data point is now reduced to k variables.

The issue of what value to use for k is an important one, and several methods have been proposed over the years. We briefly mention just a few of them here and refer the reader to Joliffe [2002], Jackson [1991], and Martinez et al. [2010] for more details. The methods presented here are the scree plot and the cumulative percentage of variance explained.

The *scree plot* [Cattell, 1966] is perhaps the most popular method, and as the title suggests, it is graphical rather than quantitative. A scree plot is a plot of the eigenvalues versus the index of the eigenvalue. Depending on the size of the eigenvalues, it might be more informative to plot the log of the eigenvalues. This type of plot is then called a *log-eigenvalue plot*. To find a value for k using the scree plot, one looks for the elbow in the curve. This is the part of the curve where it levels off and becomes almost flat. One could also calculate the change in the slope of the lines connecting the points. When the slopes become less steep, then this is a reasonable value for k. We will illustrate a scree plot in Example 6.2.

A popular method for determining the number of dimensions k is the *cumulative percentage of variance explained*. In this method, one selects the k PCs that contribute a cumulative percentage of total variation in the data. This is calculated using

$$t_k = 100 \times \left(\sum_{j=1}^{k} \lambda_j \right) \div \left(\sum_{j=1}^{d} \lambda_j \right).$$

The user must choose a value for t_d, but typical values range between 70% and 95%.

Example 6.2

We illustrate the procedure for PCA using the glass identification data set, downloaded from the UCI Machine Learning Repository (Newman, et al., 1998]. These data were originally used in criminal investigations, where the goal was to classify glass for use as evidence. The data set consists of 214 observations and 9 attributes. We will use the scree plot and the cumulative percentage of variance explained to choose the number of dimensions to keep. First, we load the data, center it, and find the covariance matrix.

```
load glassdata
[n,d] = size(glassdata);
% Center the data.
Xc = glassdata - repmat(sum(glassdata)/n,n,1);
% Find the covariance matrix.
covm = cov(Xc);
```

Next, we use the **eig** function that is part of the main MATLAB® package. The Statistics Toolbox has more extensive functionality for PCA via the **princomp** function.

```
[v,d]= eig(covm);
% Get the eigenvalues and make sure they
% are ordered.
% Also get the indexes of the sort,
% to re-order the columns of the eigenvector matrix.
[eigvals,inds] = sort(diag(d),'descend');
```

We construct the scree plot using the following code, and show the results in Figure 6.2.

```
% Create the scree plot.
plot(1:9, eigvals ,'ko-')
title('Scree Plot - Glass Data')
xlabel('Index')
ylabel('Eigenvalue')
```

We see an elbow at around $k = 3$. The following MATLAB commands will calculate the cumulative percentage of variance.

```
% Now calculate the cumulative percentage of variance.
pervar = 100*cumsum(eigvals)/sum(eigvals);
```

The results are shown here:

   47.6205   73.9398   84.7198   94.9223   98.2290

We see that approximately 95% of the variance is explained by the first four eigenvalues, indicating that $k = 4$ is a reasonable estimate using this method. Using the estimated number of dimensions of $k = 3$ from the scree plot method, we can project the data to the first three PCs as follows.

```
eigvecs = v(:,inds);
P3 = eigvecs(:,1:3);
Xp3 = Xc*P3;
```

The reader will be asked to explore the principal component scores in the exercises.
❑

6.4 Projection Pursuit EDA

There are an infinite number of planes we can use to reduce the dimensionality of our data. As we just mentioned, the first two principal components in PCA span one such plane, providing a projection such that the variation in the projected data is maximized over all possible 2D projections. However, this might not be the best plane for highlighting interesting and informative structure in the data. Here, *structure* is defined to be departure from normality and includes such things as clusters, linear structures, holes, outliers, etc. Thus, the objective is to find a projection plane that provides a 2D view of our data such that the structure (or departure from normality) is maximized over all possible 2D projections.

We can use the Central Limit Theorem to explain why we are interested in departures from normality. Linear combinations of data, including non-normal data, look normal. So, in most of the low-dimensional projections, one observes a Gaussian. If there is something interesting (e.g., clusters, etc.), then it has to be in the nonnormal projections.

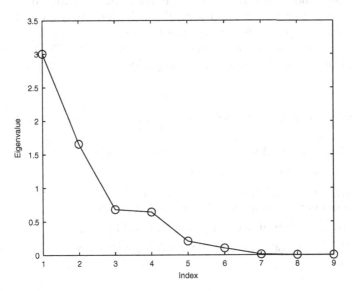

FIGURE 6.2
Here is the scree plot for the glass data. There is an obvious elbow in the curve at three, indicating we should keep three principal components.

Friedman and Tukey [1974] describe projection pursuit as a way of searching for and exploring nonlinear structure in multi-dimensional data by examining many 2D projections of the data. The idea is that 2D orthogonal projections of the data should reveal structure that is in the original data. The projection pursuit technique can also be used to obtain 1D projections; here we look only at the 2D case. Extensions to this method are also described in the literature by Friedman [1987], Posse [1995a, 1995b], Huber [1985], and Jones and Sibson [1987]. In our presentation of projection pursuit exploratory data analysis, we follow the method of Posse [1995a, 1995b].

Projection pursuit exploratory data analysis (PPEDA) is accomplished by visiting many projections to find an interesting one, where *interesting* is measured by an index. In most cases, our interest is in nonnormality, so the projection pursuit index usually measures the departure from normality. The index we use is known as the ***chi-square index*** and is developed in Posse [1995a, 1995b]. For completeness, other projection indexes are given in Appendix B, and the interested reader is referred to Posse [1995b] for a simulation analysis of the performance of these indexes.

PPEDA consists of two parts:

1. A projection pursuit index that measures the degree of the structure or departure from normality, and

2. A method for finding the projection that yields the highest value for the index.

Posse [1995a, 1995b] uses a random search to locate the global optimum of the projection index and combines it with the structure removal of Friedman [1987] to get a sequence of interesting 2D projections. Each projection found shows a structure that is less important (in terms of the projection index) than the previous one. Before we describe this method for PPEDA, we give a summary of the notation that we use in projection pursuit exploratory data analysis.

NOTATION – PROJECTION PURSUIT EXPLORATORY DATA ANALYSIS

- **Z** is the sphered version of the data matrix **X**.
- **S** is the sample covariance matrix.
- α, β are orthonormal ($\alpha^T\alpha = 1 = \beta^T\beta$ and $\alpha^T\beta = 0$) d-dimensional vectors that span the projection plane.
- $P(\alpha, \beta)$ is the projection plane spanned by α and β.
- (z_i^α, z_i^β) are the sphered observations projected onto the vectors α and β:

$$z_i^\alpha = z_i^T\alpha$$
$$z_i^\beta = z_i^T\beta$$

- (α^*, β^*) denotes the plane where the index is maximum.
- $PI_{\chi^2}(\alpha, \beta)$ denotes the chi-square projection index evaluated using the data projected onto the plane spanned by α and β.
- ϕ_2 is the standard bivariate normal density.
- B_k is a box in the projection plane.
- c_k is the probability evaluated over the k-th region B_k using the standard bivariate normal,

$$c_k = \iint\limits_{B_k} \phi_2 dz_1 dz_2 .$$

- I_{B_k} is the indicator function for region B_k.
- $\eta_j = \pi j/36, j = 0, ..., 8$ is the angle by which the data are rotated in the plane before being assigned to regions B_k.
- $\alpha(\eta_j)$ and $\beta(\eta_j)$ are given by

$$\alpha(\eta_j) = \alpha\cos\eta_j - \beta\sin\eta_j$$
$$\beta(\eta_j) = \alpha\sin\eta_j + \beta\cos\eta_j$$

- *c* is a scalar that determines the size of the neighborhood around (α^*, β^*) that is visited in the search for planes that provide better values for the projection pursuit index.
- **v** is a vector uniformly distributed on the unit *d*-dimensional sphere.
- *half* specifies the number of steps without an increase in the projection index, at which time the value of the neighborhood is halved.
- *m* represents the number of searches or random starts to find the best plane.

Projection Pursuit Index

The ***Posse chi-square index*** is developed by first dividing the plane into 48 regions or boxes B_k distributed in rings [1995a, 1995b]. See Figure 6.3 for an illustration of how the plane is partitioned. All regions have the same angular width of 45 degrees and the inner regions have the same radial width of $(2\log 6)^{1/2}/5$. This choice for the radial width provides regions with approximately the same probability for the standard bivariate normal distribution. The regions in the outer ring have probability 1/48. The regions are constructed in this way to account for the radial symmetry of the bivariate normal distribution.

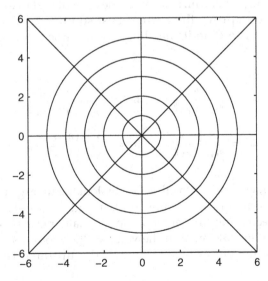

FIGURE 6.3
This shows the layout of the regions B_k for the chi-square projection index [Posse, 1995a].

Posse [1995a, 1995b] provides the population version of the projection index. We present only the empirical version here because that is the one that must be implemented on the computer. The projection index is given by

$$PI_{\chi^2}(\alpha, \beta) = \frac{1}{9} \sum_{j=1}^{8} \sum_{k=1}^{48} \frac{1}{c_k} \left[\frac{1}{n} \sum_{i=1}^{n} I_{B_k}(z_i^{\alpha(\eta_j)}, z_i^{\beta(\eta_j)}) - c_k \right]^2.$$

The chi-square projection index is not affected by the presence of outliers. This means that an interesting projection obtained using this index will not be one that is interesting solely because of outliers, unlike some of the other indexes (see Appendix B). It is sensitive to distributions that have a hole in the core, and it will also yield projections that contain clusters. The chi-square projection pursuit index is fast and easy to compute, making it appropriate for large sample sizes. Posse [1995a] provides a formula to approximate the percentiles of the chi-square index so the analyst can assess the significance of the observed value of the projection index.

Finding the Structure

The second part of PPEDA requires a method for optimizing the projection index over all possible projections onto 2D planes. Posse [1995a] shows that his optimization method outperforms the steepest-ascent techniques [Friedman and Tukey, 1974]. The Posse algorithm starts by randomly selecting a starting plane, which becomes the current best plane (α^*, β^*). The method seeks to improve the current best solution by considering two candidate solutions within its neighborhood. These candidate planes are given by

$$\mathbf{a}_1 = \frac{\alpha^* + c\mathbf{v}}{\|\alpha^* + c\mathbf{v}\|} \qquad \mathbf{b}_1 = \frac{\beta^* - (\mathbf{a}_1^T \beta^*)\mathbf{a}_1}{\|\beta^* - (\mathbf{a}_1^T \beta^*)\mathbf{a}_1\|}$$

$$\mathbf{a}_2 = \frac{\alpha^* - c\mathbf{v}}{\|\alpha^* - c\mathbf{v}\|} \qquad \mathbf{b}_2 = \frac{\beta^* - (\mathbf{a}_2^T \beta^*)\mathbf{a}_2}{\|\beta^* - (\mathbf{a}_2^T \beta^*)\mathbf{a}_2\|}. \tag{6.6}$$

In this approach, we start a global search by looking in large neighborhoods of the current best solution plane (α^*, β^*) and gradually focus in on a maximum by decreasing the neighborhood by half after a specified number of steps with no improvement in the value of the projection pursuit index. When the neighborhood is small, then the optimization process is terminated.

A summary of the steps for the exploratory projection pursuit algorithm is given here. Details on how to implement these steps are provided in Example 6.3. The complete search for the best plane involves repeating steps

2 through 9 of the procedure m times, using m random starting planes. Keep in mind that the best plane (α^*, β^*) is the plane where the projected data exhibit the greatest departure from normality.

PROCEDURE – PROJECTION PURSUIT EXPLORATORY DATA ANALYSIS

1. Sphere the data using the following transformation:

$$\mathbf{Z}_i = \Lambda^{-1/2} \mathbf{Q}^T (\mathbf{X}_i - \bar{\mathbf{x}}) \qquad i = 1, \ldots, n,$$

   where the columns of $\mathbf{Q}$ are the eigenvectors obtained from $\mathbf{S}$, Λ is a diagonal matrix of corresponding eigenvalues, and $\mathbf{X}_i$ is the i-th observation.

2. Generate a random starting plane, (α_0, β_0). This is the current best plane, (α^*, β^*).

3. Evaluate the projection index $PI_{\chi^2}(\alpha_0, \beta_0)$ for the starting plane.

4. Generate two candidate planes $(\mathbf{a}_1, \mathbf{b}_1)$ and $(\mathbf{a}_2, \mathbf{b}_2)$ according to Equation 6.6.

5. Find the value of the projection index for these planes, $PI_{\chi^2}(\mathbf{a}_1, \mathbf{b}_1)$ and $PI_{\chi^2}(\mathbf{a}_2, \mathbf{b}_2)$.

6. If one of the candidate planes yields a higher value of the projection pursuit index, then that one becomes the current best plane (α^*, β^*).

7. Repeat steps 4 through 6 while there are improvements in the projection pursuit index.

8. If the index does not improve for *half* times, then decrease the value of c by half.

9. Repeat steps 4 through 8 until c is some small number set by the analyst.

It is important to note that in PPEDA we are working with sphered or standardized versions of the original data. Some researchers in this area [Huber, 1985] discuss the benefits and the disadvantages of this approach.

Structure Removal

In PPEDA, we locate a projection that provides a maximum of the projection index. We have no reason to assume that there is only one interesting projection, and there might be other views that reveal insights about our data. To locate other views, Friedman [1987] devised a method called *structure removal*. The overall procedure is to perform projection pursuit as outlined above, remove the structure found at that projection, and repeat the projection pursuit process to find a projection that yields another maximum value of the projection pursuit index. Proceeding in this manner will provide a sequence of projections providing informative views of the data.

Structure removal in two dimensions is an iterative process. The procedure repeatedly transforms data that are projected to the current solution plane (the one that maximized the projection pursuit index) to standard normal until they stop becoming more normal. We can measure *more normal* using the projection pursuit index.

We start with a $d \times d$ matrix $\mathbf{U}^*$, where the first two rows of the matrix are the vectors of the projection obtained from PPEDA. The rest of the rows of $\mathbf{U}^*$ have ones on the diagonal and zero elsewhere. For example, if $d = 4$, then

$$\mathbf{U}^* = \begin{bmatrix} \alpha_1^* & \alpha_2^* & \alpha_3^* & \alpha_4^* \\ \beta_1^* & \beta_2^* & \beta_3^* & \beta_4^* \\ 0 & 0 & 1 & 0 \\ 0 & 0 & 0 & 1 \end{bmatrix}.$$

We use the Gram-Schmidt process [Strang, 2005] to make $\mathbf{U}^*$ orthonormal. We denote the orthonormal version as $\mathbf{U}$.

The next step in the structure removal process is to transform the $\mathbf{Z}$ matrix using the following

$$\mathbf{T} = \mathbf{U}\mathbf{Z}^T. \tag{6.7}$$

In Equation 6.7, $\mathbf{T}$ is $d \times n$, so each column of the matrix corresponds to a d-dimensional observation. With this transformation, the first two dimensions (the first two rows of $\mathbf{T}$) of every transformed observation are the projection onto the plane given by (α^*, β^*).

We now remove the structure that is represented by the first two dimensions. We let Θ be a transformation that transforms the first two rows of $\mathbf{T}$ to a standard normal and the rest remain unchanged. This is where we actually remove the structure, making the data normal in that projection (the first two rows). Letting $\mathbf{T}_1$ and $\mathbf{T}_2$ represent the first two rows of $\mathbf{T}$, we define the transformation as follows:

$$\Theta(\mathbf{T}_1) = \Phi^{-1}[F(\mathbf{T}_1)]$$

$$\Theta(\mathbf{T}_2) = \Phi^{-1}[F(\mathbf{T}_2)] \tag{6.8}$$

$$\Theta(\mathbf{T}_i) = \mathbf{T}_i; \qquad i = 3, ..., d \ ,$$

where Φ^{-1} is the inverse of the standard normal cumulative distribution function, and F is a function defined next (see Equations 6.9 and 6.10). We see from Equation 6.8 that we will be changing only the first two rows of $\mathbf{T}$.

We now describe the transformation of Equation 6.8 in more detail, working only with $\mathbf{T}_1$ and $\mathbf{T}_2$. First, we note that $\mathbf{T}_1$ can be written as

$$\mathbf{T}_1 = (z_1^{\alpha^*}, ..., z_j^{\alpha^*}, ..., z_n^{\alpha^*}),$$

and $\mathbf{T}_2$ as

$$\mathbf{T}_2 = (z_1^{\beta^*}, ..., z_j^{\beta^*}, ..., z_n^{\beta^*}).$$

Recall that $z_j^{\alpha^*}$ and $z_j^{\beta^*}$ would be coordinates of the j-th observation projected onto the plane spanned by (α^*, β^*).

Next, we define a rotation about the origin through the angle γ as follows

$$
\begin{aligned}
\tilde{z}_j^{1(t)} &= z_j^{1(t)} \cos\gamma + z_j^{2(t)} \sin\gamma \\
\tilde{z}_j^{2(t)} &= z_j^{2(t)} \cos\gamma - z_j^{1(t)} \sin\gamma,
\end{aligned}
\tag{6.9}
$$

where $\gamma = 0, \pi/4, \pi/8, 3\pi/8$ and $z_j^{1(t)}$ represents the j-th element of $\mathbf{T}_1$ at the t-th iteration of the process. We now apply the following transformation to the rotated points:

$$z_j^{1(t+1)} = \Phi^{-1} \left\{ \frac{r(\tilde{z}_j^{1(t)}) - 0.5}{n} \right\} \qquad z_j^{2(t+1)} = \Phi^{-1} \left\{ \frac{r(\tilde{z}_j^{2(t)}) - 0.5}{n} \right\}, \tag{6.10}$$

where $r(\tilde{z}_j^{1(t)})$ represents the rank (position in the ordered list) of $\tilde{z}_j^{1(t)}$.

This transformation *replaces* each rotated observation by its normal score in the projection. With this procedure, we are deflating the projection index by making the data more normal. It is evident in the procedure given next that this is an iterative process. Friedman [1987] states that during the first few iterations, the projection index should decrease rapidly. After approximate normality is obtained, the index might oscillate with small changes. Usually, the process takes between 5 to 15 complete iterations to remove the structure.

Once the structure is removed using this process, we must transform the data back using

$$\mathbf{Z}' = \mathbf{U}^T \Theta (\mathbf{U} \mathbf{Z}^T). \tag{6.11}$$

In other words, we transform back using the transpose of the orthonormal matrix $\mathbf{U}$. From matrix theory [Strang, 2005], we see that all directions orthogonal to the structure (i.e., all rows of **T** other than the first two) have not been changed. Whereas, the structure has been spherized and then transformed back.

PROCEDURE – STRUCTURE REMOVAL

1. Create the orthonormal matrix **U**, where the first two rows of **U** contain the vectors α^*, β^*.
2. Transform the data **Z** using Equation 6.7 to get **T**.
3. Using only the first two rows of **T**, rotate the observations using Equation 6.9.
4. Normalize each rotated point according to Equation 6.10.
5. For angles of rotation $\gamma = 0, \pi/4, \pi/8, 3\pi/8$, repeat steps 3 through 4.
6. Evaluate the projection index using $z_j^{1(t+1)}$ and $z_j^{2(t+1)}$, after going through an entire cycle of rotation (Equation 6.9) and normalization (Equation 6.10).
7. Repeat steps 3 through 6 until the projection pursuit index stops changing.
8. Transform the data back using Equation 6.11.

Example 6.3

We use a synthetic data set to illustrate the MATLAB functions used for PPEDA. These data contain two structures, both of which are clusters. So we will search for two planes that maximize the projection pursuit index. First we load the data set that is contained in the file called **ppdata**. This loads a matrix **X** containing 400 six-dimensional observations. We also set up the constants we need for the algorithm.

```
% First load up a synthetic data set.
% This has structure - clusters - in two planes.
% Note that the data is in
% ppdata.mat
load ppdata
% For m random starts, find the best projection plane
% using N structure removal procedures.
% Find two structures:
N = 2;
% Four random starts:
m = 4;
c = tan(80*pi/180);
% Number of steps with no increase.
half = 30;
```

We now set up some arrays to store the results of projection pursuit.

```
% To store the N structures:
[n,d] = size(X);
astar = zeros(d,N);
bstar = zeros(d,N);
ppmax = zeros(1,N);
```

Next we have to sphere the data.

```
% Sphere the data.
muhat = mean(X);
[V,D] = eig(cov(X));
Xc = X-ones(n,1)*muhat;
Z = ((D)^(-1/2)*V'*Xc')';
```

We use the sphered data as input to the function **csppeda**. The outputs from this function are the vectors that span the plane containing the structure and the corresponding value of the projection pursuit index. We can find multiple interesting projections by removing the structure at each iteration of the loop using the function **csppstrtrem**.

```
% Now do the PPEDA.
% Find a structure, remove it,
% and look for another one.
Zt = Z;
for i = 1:N
    [astar(:,i),bstar(:,i),ppmax(i)] =...
            csppeda(Zt,c,half,m);
    % Now remove the structure.
    Zt = csppstrtrem(Zt,astar(:,i),bstar(:,i));
end
```

Note that each column of **astar** and **bstar** contains the projections for a structure, each one found using m random starts of the Posse algorithm. To see the first structure and second structures, we project onto the best planes as follows:

```
% Now project and see the structure.
proj1 = [astar(:,1), bstar(:,1)];
proj2 = [astar(:,2), bstar(:,2)];
Zp1 = Z*proj1;
Zp2 = Z*proj2;
figure
plot(Zp1(:,1),Zp1(:,2),'k.'),title('Structure 1')
xlabel('\alpha^*'),ylabel('\beta^*')
figure
plot(Zp2(:,1),Zp2(:,2),'k.'),title('Structure 2')
xlabel('\alpha^*'),ylabel('\beta^*')
```

The results are shown in Figure 6.4 and Figure 6.5, where we see that projection pursuit did find two structures. The first structure has a projection pursuit index of 2.67, and the second structure has an index equal to 0.572.
❑

6.5 Independent Component Analysis

Independent component analysis (ICA) is another tool that can be used to help extract information from our data sets. In particular, ICA belongs to a class of techniques called *blind source separation methods*, where one would like to separate an observed mixture of signals into source signals. For example, one might have two radio stations at different locations, with both of them broadcasting on the same frequency. The mixed signals are picked up by two radio receivers at different geographical locations, and ICA can be used to recover the original source signals.

ICA can be applied to different data such as images, communications channels, stock prices, and others. The method is appropriate when the underlying phenomena or source signals are contained in an observed mixture of noisy signals. As we will see, we do not have to make many assumptions about the source signals to perform ICA.

We will use the example given above of two radio signals from different physical sources to explain ICA. If the two signals are broadcast by two different stations and using a fine time scale, then we can assume that the amplitude of one signal at a particular point in time is unrelated to the amplitude of the other signal at the same point. So, to separate the signals from the mixture, we might look for time-varying signals that are unrelated. The final assumption being that if we find such signals, then they should arise from different physical processes (or broadcasts in our case). Stone [2004] points out that this last assumption going in the reverse direction is not necessarily correct, but it does work in most applications.

Two signals being unrelated can be expressed in terms of statistical independence. As we know from the concepts in Chapter 2, if signals are statistically independent, then the value of one of the signals would not give us any information about the value of other source signals in the mixture. So, ICA seeks to separate the data into a set of statistically independent (or unrelated) component signals, which are then assumed to be from some meaningful source.

We illustrate this idea in Figure 6.6 [Stone, 2004], where we show two signals that are included in the main MATLAB package: a gong and a chirp. The top plot in Figure 6.6 shows the signals over time. If we plot the amplitude of the gong against the amplitude of the chirp (the bottom graphic in Figure 6.6), then we see that there is no obvious pattern or dependency between the two signals.

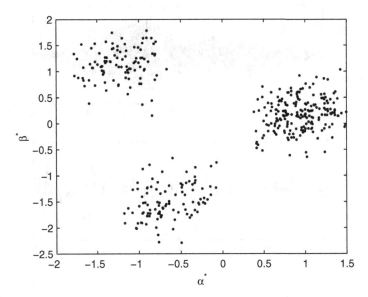

FIGURE 6.4
Here we see the first structure that was found using PPEDA. This structure yields a value of 2.67 for the chi-square projection pursuit index.

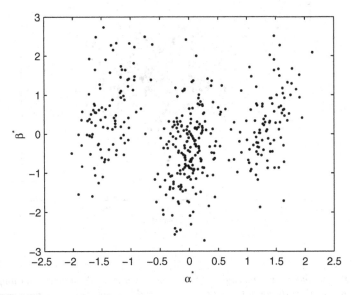

FIGURE 6.5
Here is the second structure we found using PPEDA. This structure has a value of 0.572 for the chi-square projection pursuit index.

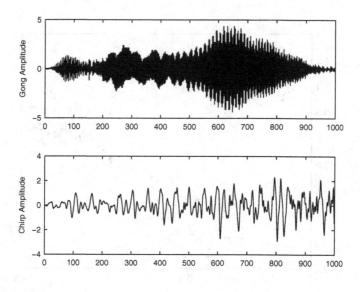

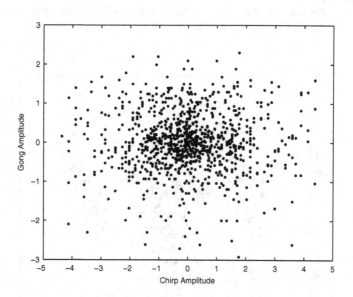

FIGURE 6.6

The top graphic shows the amplitudes of two signals over time. We then plot the amplitude of one signal against the amplitude of the other signal to graphically illustrate the concept of independence between the signals. The result is a cloud of points with no obvious pattern or dependency between the signals [Stone, 2004].

Note that PCA and ICA are somewhat related in that they are methods that can be used to transform the data. However, a PCA mapping produces *uncorrelated* data (or signals), while ICA finds *nonnormal* and *independent* source signals. Because this is an important issue, we take a moment to discuss the difference between statistical independence and the correlation of random variables [Hyvärinen, 1999].

Without loss of generality, we will illustrate these concepts using only two random variables X and Y, both with a mean of zero. We know from Chapter 2 that random variables are independent if the joint probability density function can be written as a product of the marginal probability functions:

$$f(X, Y) = f_X(X) \times f_Y(Y). \tag{6.12}$$

Equation 6.12 implies that

$$E[X^p, Y^q] = E[X^p] \times E[Y^q], \tag{6.13}$$

for positive integers, p and q. If we take $p = q = 1$, then we have the covariance between X and Y. Therefore, Equation 6.13 gives

$$E[X, Y] = E[X] \times E[Y]. \tag{6.14}$$

We know that correlation is a normalized version of the covariance, so if X and Y are uncorrelated, then Equation 6.14 is applicable [Stone, 2004].

To summarize, uncorrelated random variables X and Y imply that the first moment of the joint probability density function is equal to the product of the first moments of the marginals. Independence is a much stricter condition, and it depends on *all* moments of the joint probability density function (Equation 6.13). In general, independence always implies uncorrelatedness, but it does not always go the other way. That is, uncorrelatedness does not imply independence. There is one special case, though, where this is true. If the random variables have a joint multivariate normal distribution, then independence and uncorrelatedness are equivalent.

Formal definitions of ICA can be found in the literature, and Hyvärinen [1999] notes that there are several definitions of ICA. We are interested in the general definition because it has been shown that ICA is a special case of projection pursuit when this definition is used. The general definition simply states that ICA consists of finding a linear transformation so that the resulting components are as independent as possible, and we do this by optimizing a function that measures independence.

We already learned that projection pursuit looks for transformations where the projected data are nonnormal (or interesting), according to some index measuring the degree of the departure from normality. In the case of ICA, the interesting structure we are looking for is independent components. We will use the kurtosis of the probability density function of a signal as the way to measure the degree of independence or nonnormality in projection pursuit.

We now provide a more mathematical representation of the problem. For simplicity, we will focus on the case of two sound signals and two mixtures (i.e., two receivers). A set of two sound signals can be represented in matrix form as

$$\mathbf{G} = \begin{bmatrix} \mathbf{g}_1 & \mathbf{g}_2 \end{bmatrix},$$

where $\mathbf{g}_i$ is an $n \times 1$ signal, and n represents the number of observations in each signal.

If we have two sound signals that are received by one sensor, then the output is a *signal mixture*. This mixture is a weighted sum of the source signals, where the relative proportion of each signal in the mixture depends on several factors. So, for receivers $\mathbf{x}_1$ and $\mathbf{x}_2$, we have signal mixtures given by

$$\mathbf{x}_1 = a\,\mathbf{g}_1 + b\,\mathbf{g}_2$$
$$\mathbf{x}_2 = c\,\mathbf{g}_1 + d\,\mathbf{g}_2$$

The goal of ICA is to unmix these signals and recover the individual source signals:

$$\mathbf{g}_1 = \alpha\,\mathbf{x}_1 + \beta\,\mathbf{x}_2$$
$$\mathbf{g}_2 = \gamma\,\mathbf{x}_1 + \delta\,\mathbf{x}_2$$

by estimating the unmixing coefficients. That is, we wish to find a vector $\mathbf{w}$ that maximizes nonnormality (i.e., kurtosis) to extract *estimated signals* $\mathbf{y}$ from a mixture $\mathbf{x}$. We illustrate these concepts in the next example.

Example 6.4

The following example was adapted from the projection pursuit MATLAB code found in Stone [2004], and it uses a gradient ascent method to find the weight vector that maximizes the kurtosis. The idea is that a projection with maximum kurtosis would show nonnormal structure. It should be noted that we are looking for a 1D projection, and only one signal will be found at each application of the procedure. To find more than one source signal, we would need to apply something similar to the structure removal process, as we explain shortly. We already created a mixture of signals based on the **chirp** and **gong** signals that are included in the basic MATLAB software. The data are saved in the file called **icaexample.mat**.

```
% Load the data.
load icaexample
```

This file has several signal arrays, along with the frequency **N**. One is the matrix of signal mixtures **x**, where each column corresponds to amplitudes recorded at a receiver. We show one of the signal mixtures at the top of Figure 6.7. The file also contains the original unmixed signal matrix **G**, where the first column is the **chirp** signal and the second corresponds to the **gong**. Finally, as in projection pursuit, we spherized the data, producing the matrix **z**. See the online version of this example for the MATLAB code that was used to create the mixture of signals and the matrix **z**. The steps to extract a source signal from **z** are shown here:

```
% Initialize weight vector and normalize.
w = randn(1,2);
w = w/norm(w);
% Specify the step size.
step = 0.02;
% Do projection pursuit using gradient ascent.
for iter=1:100
      % Get estimated source signal, y. The matrix z is
% the spherized signal (see the file for details).
      y = z*w';
      % Get estimated kurtosis. Note that
      % we can use a simplified form because
      % the weight vector has unit length.
      K = mean(y.^4)-3;
      % Find gradient of kurtosis.
      Ycube = repmat(y.^3,1,2);
      grad = mean(Ycube.*z);
      % Update w and normalize.
      w = w + step*grad;
      w = w/norm(w);
end
```

The extracted signal is shown at the bottom of Figure 6.7. We can verify that the extracted signal matches the **chirp** by listening to it. This is done using the following commands.

```
% Play back both signals to compare them.
soundsc(G(:,1),N); % This is the original signal.
soundsc(y,N); % This is the extracted signal.
```

❏

We extracted just one signal in the previous example, but we probably want to find more. After all, this is the idea behind blind source separation. We can extract multiple source signals by first removing the current extracted signal from each of the remaining mixtures using Gram-Schmidt

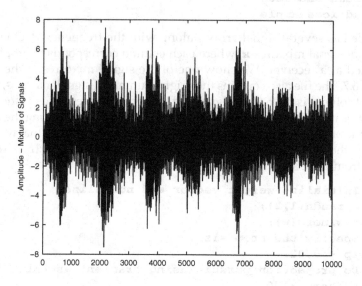

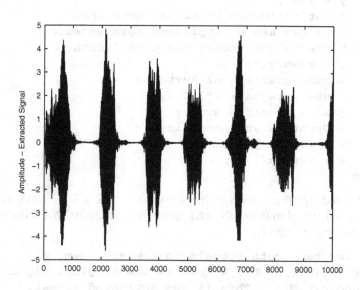

FIGURE 6.7
Th upper plot shows one of the signal mixtures. This is a weighted sum of the **gong** *and* **chirp** *signals. The lower plot is a picture of the first signal we extracted using projection pursuit with kurtosis measuring the interesting structure.*

orthogonalization [Strang, 2005]. This process is sometimes known as *deflation*. We then apply the procedure outlined in the previous example to get the next source signal.

We now describe Gram-Schmidt orthogonalization for our two signal, two mixture situation. We denote the original mixtures as $\mathbf{x}_1(0)$ and $\mathbf{x}_2(0)$, and we represent the first extracted signal as $\mathbf{y}_1$. The signal $\mathbf{y}_1$ is removed from the mixtures as follows

$$\mathbf{x}_1(1) = \mathbf{x}_1(0) - \frac{\mathbf{y}_1^T \mathbf{x}_1(0)\mathbf{y}_1}{\mathbf{y}_1^T \mathbf{y}_1}$$

$$\mathbf{x}_2(1) = \mathbf{x}_2(0) - \frac{\mathbf{y}_1^T \mathbf{x}_2(0)\mathbf{y}_1}{\mathbf{y}_1^T \mathbf{y}_1},$$

where $\mathbf{x}_i(1)$ means that one signal has been extracted from the original mixture. This is easily extended to M extracted signals from M signal mixtures. The reader will be asked to implement this in the exercises and to extract the second signal of our mixture.

The independent component analysis methodology has gained widespread use and has become an important tool for analyzing large data sets. Also, there are many versions of ICA, along with various algorithms to implement them. We provide some references for the interested reader at the end of the chapter.

6.6 Grand Tour

The grand tour of Asimov [1985] is an interactive visualization technique that enables the analyst to look for interesting structure embedded in multi-dimensional data. The idea is to project the d-dimensional data onto a plane and then rotate the plane through all possible angles, searching for structure in the data. As with projection pursuit, structure is defined as departure from normality, such as clusters, spirals, linear relationships, etc.

In this procedure, we first determine a plane, project the data onto it, and then view it as a 2D scatterplot. This process is repeated for a sequence of planes. If the sequence of planes is smooth (in the sense that the orientation of the plane changes slowly), then the result is a movie that shows the data points moving in a continuous manner. Asimov [1985] describes two methods for conducting a grand tour, called the *torus algorithm* and the *random interpolation algorithm*. Neither of these methods is ideal. With the torus method we may end up spending too much time in certain regions, and it is computationally intensive. The random interpolation method is better computationally, but cannot be reversed easily (to recover the projection) unless the set of random numbers used to generate the tour is retained. Thus, this method requires a lot of computer storage. Because of these limitations, we describe the *pseudo grand tour* described in Wegman and Shen [1993].

One of the important aspects of the torus grand tour is the need for a continuous space-filling path through the manifold of planes. This requirement satisfies the condition that the tour will visit *all* possible orientations of the projection plane. Here, we do not follow a space-filling curve, which is why we call it a pseudo grand tour. In spite of this, the pseudo grand tour has many benefits:

- It can be calculated easily;
- It does not spend a lot of time in any one region;
- It still visits an ample set of orientations; and
- It is easily reversible.

The fact that the pseudo grand tour is easily reversible enables the analyst to recover the projection for further analysis. Two versions of the pseudo grand tour are available: one that projects onto a line and one that projects onto a plane.

As with projection pursuit, we need unit vectors that comprise the desired projection. In the 1D case, we require a unit vector $\alpha(t)$ such that

$$\|\alpha(t)\|^2 = \sum_{i=1}^{d} \alpha_i^2(t) = 1$$

for every t, where t represents a point in the sequence of projections. For the pseudo grand tour, $\alpha(t)$ must be a continuous function of t and should produce *all* possible orientations of a unit vector.

We obtain the projection of the data using

$$z_i^{\alpha(t)} = \alpha^T(t)x_i, \tag{6.15}$$

where x_i is the i-th d-dimensional data point. To get the movie view of the pseudo grand tour, we plot the values $z_i^{\alpha(t)}$ on a fixed 1D coordinate system, redisplaying the projected points as t increases.

The pseudo grand tour in two dimensions is similar. We need a second unit vector $\beta(t)$ that is orthogonal to $\alpha(t)$,

$$\|\beta(t)\|^2 = \sum_{i=1}^{d} \beta_i^2(t) = 1 \qquad \alpha^T(t)\beta(t) = 0.$$

We project the data onto the second vector using

$$z_i^{\beta(t)} = \beta^T(t)\mathbf{x}_i. \tag{6.16}$$

To obtain the movie view of the 2D pseudo grand tour, we display $z_i^{\alpha(t)}$ and $z_i^{\beta(t)}$ in a 2D scatterplot, replotting the points as t increases.

In general, the basic idea of grand tours is to project the data onto a 1D or 2D space and plot the projected data, repeating this process many times to provide many views of the data. It is important for viewing purposes to make the time steps small to provide a nearly continuous path and to provide smooth motion of the points. The reader should note that grand tours are an interactive approach to EDA. The analyst must stop the tour when an interesting projection is found.

Asimov [1985] contends that we are viewing more than one or two dimensions in these tours because the speed vectors provide further information. For example, the further away a point is from the computer screen, the faster the point rotates. We believe that the extra dimension conveyed by the speed is difficult to understand unless the analyst has experience looking at grand tour movies.

In order to implement the pseudo grand tour, we need a way of obtaining the projection vectors $\alpha(t)$ and $\beta(t)$. First we consider the data vector $\mathbf{x}$. If d is odd, then we augment each data point with a zero, to get an even number of elements. In this case,

$$\mathbf{x} = (x_1, ..., x_d, 0); \qquad \text{for } d \text{ odd.}$$

This will not affect the projection. So, without loss of generality, we present the method with the understanding that d is even. We take the vector $\alpha(t)$ to be

$$\alpha(t) = \sqrt{2/d}(\sin\omega_1 t, \cos\omega_1 t, ..., \sin\omega_{d/2} t, \cos\omega_{d/2} t), \tag{6.17}$$

and the vector $\beta(t)$ as

$$\beta(t) = \sqrt{2/d}(\cos\omega_1 t, -\sin\omega_1 t, ..., \cos\omega_{d/2} t, -\sin\omega_{d/2} t). \tag{6.18}$$

We choose ω_i and ω_j such that the ratio ω_i/ω_j is irrational for every i and j. Additionally, we must choose these such that no ω_i/ω_j is a rational multiple of any other ratio. It is also recommended that the time step Δt be a small positive irrational number. One way to obtain irrational values for ω_i is to let $\omega_i = \sqrt{P_i}$, where P_i is the i-th prime number.

The steps for implementing the 2D pseudo grand tour are given here. The details on how to implement this in MATLAB are given in Example 6.5.

PROCEDURE – PSEUDO GRAND TOUR

1. Set each ω_i to an irrational number.
2. Find vectors $\alpha(t)$ and $\beta(t)$ using Equations 6.17 and 6.18.
3. Project the data onto the plane spanned by these vectors using Equations 6.15 and 6.16.
4. Display the projected points, $z_i^{\alpha(t)}$ and $z_i^{\beta(t)}$, in a 2D scatterplot.
5. Using Δt irrational, increment the time, and repeat steps 2 through 4.

Before we illustrate this in an example, we note that once we stop the tour at an interesting projection, we can easily recover the projection by knowing the time step.

Example 6.5

In this example, we use the **iris** data to illustrate the grand tour. First we load up the data and set up some preliminaries.

```
% This is for the iris data.
load iris
% Put data into one matrix.
x = [setosa;virginica;versicolor];
% Set up vector of frequencies.
th = sqrt([2 3]);
% Set up other constants.
[n,d] = size(x);
% This is a small irrational number:
delt = eps*10^14;
% Do the tour for some specified time steps.
maxit = 1000;
cof = sqrt(2/d);
% Set up storage space for projection vectors.
a = zeros(d,1);
b = zeros(d,1);
z = zeros(n,2);
```

We now do some preliminary plotting, just to get the handles we need to use MATLAB's Handle Graphics for plotting. This enables us to update the points that are plotted rather than replotting the entire figure.

```
% Get an initial plot, so the tour can be implemented
% using Handle Graphics.
Hlin1 = plot(z(1:50,1),z(1:50,2),'ro');
set(gcf,'backingstore','off')
set(gca,'Drawmode','fast')
hold on
```

```
Hlin2 = plot(z(51:100,1),z(51:100,2),'go');
Hlin3 = plot(z(101:150,1),z(101:150,2),'bo');
hold off
axis equal
axis vis3d
axis off
```

Now we do the actual pseudo grand tour, where we use a maximum number of iterations given by **maxit**.

```
for t = 0:delt:(delt*maxit)
  % Find the transformation vectors.
  for j = 1:d/2
    a(2*(j-1)+1) = cof*sin(th(j)*t);
    a(2*j) = cof*cos(th(j)*t);
    b(2*(j-1)+1) = cof*cos(th(j)*t);
    b(2*j) = cof*(-sin(th(j)*t));
  end
  % Project onto the vectors.
  z(:,1) = x*a;
  z(:,2) = x*b;
  set(Hlin1,'xdata',z(1:50,1),'ydata',z(1:50,2))
  set(Hlin2,'xdata',z(51:100,1),'ydata',z(51:100,2))
  set(Hlin3,'xdata',z(101:150,1),'ydata',z(101:150,2))
  drawnow
end
```

❑

6.7 Nonlinear Dimensionality Reduction

We have already seen several methods for reducing the dimensionality of our data as a way of searching for structure in a data set. These include projection pursuit, the grand tour, principal component analysis, and independent component analysis. While there are many methods for nonlinear dimensionality reduction, we present only two methods. These are multidimensional scaling and isometric feature mapping. Other types of nonlinear dimensionality reduction are locally linear embedding, Hessian eigenmaps, and nonlinear forms of PCA and ICA. We will provide references for these methods at the end of the chapter.

Multidimensional Scaling

Multidimensional scaling (MDS) encompasses techniques for dimensionality reduction based on proximities or similarities measured on a set of objects. The purpose of MDS is to find a configuration of the data in a low-dimensional space, such that objects that are close together in the full-dimensional space are close together after we reduce the dimensionality. MDS is now widely available in most statistical packages, including the MATLAB Statistics Toolbox.

We will first discuss some notation and definitions before we go on to describe the various categories of MDS [Cox and Cox, 2001]. In general, MDS starts with measures of proximity that quantify how similar or dissimilar observations are to one another. The dissimilarity between observations x_r and x_s is denoted by δ_{rs} and the similarity is denoted by s_{rs}. The following is true for *most* definitions of these measures

$$\delta_{rs} \geq 0 \qquad \delta_{rr} = 0$$

$$0 \leq s_{rs} \leq 1 \qquad s_{rr} = 1 .$$

This implies that for a dissimilarity measure, small values indicate observations that are close together. The opposite is true for similarity measures; large values indicate observations are similar. These two types of proximity measures can be converted to the other when necessary. Therefore, we can assume the proximities are dissimilarities without loss of generality. We also assume that they are arranged in matrix form, denoted by $\mathbf{D}$. The size of this matrix is $n \times n$, and it is a symmetric matrix. Because of its symmetry, it is sometimes given in upper (or lower) triangular form, where only the diagonal and upper (or lower) elements of the matrix are given.

We refer to the distance between observation r and s in the lower-dimensional space by d_{rs}. In the MDS literature, the configuration of points in the lower-dimensional space is often denoted by $\mathbf{X}$. We follow that convention here, but the reader should be careful not to confuse this with the data matrix. To summarize, with MDS, we start with a dissimilarity matrix $\mathbf{D}$ and finish with k-dimensional transformed observations entered as rows in a matrix $\mathbf{X}$.

As with PCA, we must also choose a value for k. In MDS, the value for k often depends on the application. If the transformed observations will be used in further analysis, such as classification and clustering, then we might want to use something similar to the scree plot to help us choose k. However, when they are used in graphical exploratory data analysis, we might choose $k = 2$ or $k = 3$ for visualizing in scatterplots.

There are many different techniques and algorithms for MDS, most of which fall into two main categories known as metric MDS and nonmetric MDS. The main difference between them comes from the assumptions of how the dissimilarities δ_{rs} are mapped to the configuration of distances in the lower dimensional space.

Metric MDS assumes that the dissimilarities δ_{rs} and distances d_{rs} are related as follows:

$$d_{rs} \approx f(\delta_{rs}), \qquad (6.19)$$

where f is a continuous monotonic function. We can see from Equation 6.19, that metric MDS seeks a configuration of points in a lower-dimensional space where distances in that space are approximately the same as some function of the dissimilarities in the full space. Metric MDS is often used in the literature to denote *classical MDS*, which will be described next. However, metric MDS includes more than this one technique, such as least squares scaling and others [Cox and Cox, 2001].

Nonmetric MDS relaxes the metric properties of Equation 6.19 and requires only that the rank order of the dissimilarities be preserved. Thus, the scaling function must obey the monotonicity constraint:

$$\delta_{rs} < \delta_{ab} \Rightarrow f(\delta_{rs}) \le f(\delta_{ab}),$$

for all observations. Nonmetric MDS is also known as *ordinal MDS*.

Most metric and nonmetric MDS solutions require the iterative optimization of an objective function based on the squared discrepancies between d_{rs} and δ_{rs}. However, if the proximity measure in the original d-dimensional space and the distance d_{rs} are taken to be Euclidean, then a closed form solution is available to find the configuration of points in a k-dimensional space. This is the *classical MDS* approach, where the scaling function relating the dissimilarities and distances is the identity function. Thus, we look for a mapping such that

$$d_{rs} = \delta_{rs}.$$

This technique originated with Young and Householder [1938], Torgerson [1952], and Gower [1966]. Gower showed the importance of classical scaling, and he gave it the name *principal coordinate analysis* because it has similarities to PCA. We describe the steps of the method next. The reader interested in the derivation can consult Cox and Cox [2001], Borg and Groenen [1997], or Seber [1984].

PROCEDURE – CLASSICAL MDS

1. Using the matrix of dissimilarities **D**, find **Q**, where each element of **Q** is given by

$$q_{rs} = -\frac{1}{2}\delta_{rs}^2.$$

2. Find the centered matrix **H** using

$$\mathbf{H} = \mathbf{I} - n^{-1}\mathbf{1}\mathbf{1}^T,$$

where **I** is the $n \times n$ identity matrix, and **1** is a vector of n ones.

3. Find the matrix B, as follows:

$$\mathbf{B} = \mathbf{HQH}.$$

4. Determine the eigenvectors and eigenvalues of **B**.

5. The coordinates in the lower-dimensional space are given by

$$\mathbf{X} = \mathbf{A}_k\mathbf{L}_k^{1/2},$$

where $\mathbf{A}_k$ holds the eigenvectors corresponding to the k largest eigenvalues, and $\mathbf{L}_k^{1/2}$ contains the square root of the k largest eigenvalues along the diagonal.

In some cases, the matrix **B** might have negative eigenvalues. One could ignore the negative eigenvalues and proceed to step 5; other options are given in Cox and Cox [2001]. If the dissimilarities are actually Euclidean distances, then this problem should not come up. It can also be shown that PCA and classical MDS provide equivalent results when the dissimilarities are Euclidean distances.

Example 6.6

We use the **cereal** data set in this example. Recall that this data set has 8 observations and 11 variables. First we have to load the data and find the interpoint distance (dissimilarity) matrix, **D**. We will use the **pdist** function that is part of the Statistics Toolbox.

```
load cereal
[n,d] = size(cereal);
% Get the matrix of dissimilarities.
D = squareform(pdist(cereal,'cityblock'));
```

The city block metric is given by

$$\delta_{rs} = \sum_{j=1}^{d} |x_{rj} - x_{sj}|.$$

We are now ready to implement the steps for classical MDS, as follows:

```
% Now implement the steps for classical MDS.
Q = -0.5*D.^2;
H = eye(n) - n^(-1)*ones(n,1)*ones(1,n);
B = H*Q*H;
[A,L] = eig(B);
[vals, inds] = sort(diag(L),'descend');
A = A(:,inds);
% Reduce the dimensionality to 2D.
X = A(:,1:2)*diag(sqrt(vals(1:2)));
```

We reduced the data to 2D to make it easier to visualize. We can construct a scatterplot using these steps:

```
% Plot the points and label them.
plot(X(:,1),X(:,2),'o')
text(X(:,1),X(:,2),labs)
title('Classical MDS - City Block Metric')
```

The plot is shown in Figure 6.8. We see from the labels that the sweet cereals are close together, as are some of the non-sweet cereals in the upper right corner. It is interesting to note that corn flakes is off by itself and is not close to any cereal. These results are consistent with the glyph plots we showed in Figure 5.35.
❑

An advantage of using multi-dimensional scaling methods is that one uses the interpoint distance matrix as input to the process rather than the entire data set. If the data set has large d (many variables) and small n (few observations), with $d \gg n$, then the interpoint distance matrix is a smaller matrix than the covariance or correlation matrix that would be used in PCA. So, in these cases, MDS is a viable option for dimensionality reduction, when computer memory constraints prohibit the use of something like PCA.

Isometric Feature Mapping (ISOMAP)

Isometric feature mapping (ISOMAP) belongs to a general class of methods that have come to be known as *manifold learning*. As we will see shortly, ISOMAP is essentially an application of classical MDS, where estimates of the geodesic distance between points is used to measure the dissimilarity between observations. These manifold learning approaches assume the data lie on a submanifold M of smaller dimension than the full space (see Figure 6.9). The techniques we briefly mention here produce coordinates in a lower dimensional space, such that the neighborhood structure of the submanifold is preserved.

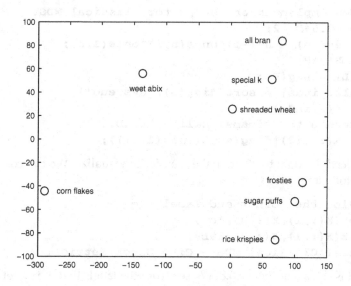

FIGURE 6.8

Here is a scatterplot of the **cereal** *data where the observations have been reduced to 2D using classical MDS with the city block distance used for the input. Note that similar cereals (e.g., sweet cereals) are close together in this space, indicating that the observations are similar. The reader should compare this with the star and Chernoff face plots (see Chapter 5) of the same data set.*

Several methods for manifold learning have been published in the literature. One of these methods is called *locally linear embedding* (LLE) that was developed by Roweis and Saul [2000]. LLE is an eigenvector-based method, and the results do not depend on iteratively solving an optimization problem, as is the case in most versions of MDS and projection pursuit. Another related approach is called *Hessian eigenmaps* (HLLE) [Donoho and Grimes, 2003]. An assumption of ISOMAP is that the manifold *M* is globally isometric to a convex subset of a low-dimensional Euclidean space. Hessian eigenmaps will recover a low-dimensional parametrization for data lying on a manifold that is locally isometric to an open, connected subset of Euclidean space. This expands the class of data sets where isometry principles can be applied to manifold learning.

In our experience, ISOMAP is more stable than these other methods, so we will explain this technique in more detail. ISOMAP was developed by Tenenbaum, de Silva, and Langford [2000] as an application of, or enhancement to, classical MDS. Before explaining ISOMAP in more detail, we first want to illustrate the idea of a submanifold. We use an example found in Tenenbaum, et al. called the Swiss roll. This is illustrated in Figure 6.9. Looking at this picture, we can think of the Swiss roll as a loosely rolled up (infinitely thin) sheet. Thus, it is really a 2D manifold that is embedded in a 3D space, and the goal of the manifold learning methods is to recover this structure.

The innovation in ISOMAP has to do with the idea that one should use something other than a straight-line Euclidean distance to measure the dissimilarity between points in this type of problem. This concept is illustrated in Figure 6.10, where this time we show a data set that lies along the Swiss roll manifold. We see that points might be close together using Euclidean distance (the two points connected by the line), but they are really far apart when looking at their distance when traveling on the surface of the submanifold. So, Euclidean distance is not a good indication of closeness in terms of the neighborhood structure given by the manifold.

The basic idea, then, of ISOMAP is to use distances along a geodesic path, hopefully measured along the submanifold M, as measures of dissimilarity. The ISOMAP method assumes the data lie on an unknown manifold M that is embedded in a high dimensional space. It seeks a mapping that preserves the distances between observations (as in classical MDS), where the distance is given by the geodesic path between points.

The input to the ISOMAP procedure is the interpoint dissimilarity matrix **D**. First one finds the neighboring points, where the neighborhood is specified by the number of nearest neighbors or a radius ε. The neighborhood relations are then represented as a weighted graph, where the edges of the graph have weights equal to the input distance. The second step of ISOMAP provides estimates of the geodesic distances between all pairs of points i and j by computing their shortest path distance using the neighborhood graph. In the final step, classical MDS is applied to the geodesic distances and an embedding is found in a lower dimensional space. The steps for ISOMAP are outlined next.

PROCEDURE – ISOMAP

1. Using the matrix of dissimilarities **D**, construct the neighborhood graph over all observations by connecting the ij-th point if point i and j are in the same neighborhood. Set the lengths of the edges equal to the distance between i and j.

2. Calculate the shortest paths between points as given by the graph.

3. Obtain the lower dimensional embedding by applying classical MDS to the geodesic paths found in step 2.

Different embeddings from ISOMAP can be obtained when other neighborhoods or interpoint distances are used as inputs. By changing these inputs, users can explore their data to find different structures. We also note that, depending on the neighborhood and interpoint distance, there might be a zero path length between points. In this case, there will be separate unconnected subsets of points in the lower dimensional embeddings

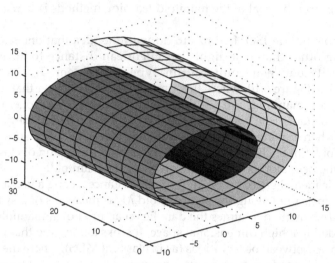

FIGURE 6.9
This shows the submanifold for the Swiss roll. This is a 2D submanifold embedded in a 3D space [Martinez et al., 2010].

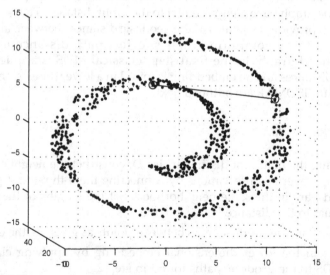

FIGURE 6.10
Here is a data set that was randomly generated according to the Swiss roll parametrization [Tenenbaum, de Silva, and Langford, 2000]. The Euclidean distance between two points is given by the straight line shown here. If we are seeking the neighborhood structure as given by the submanifold M, then it would be better to use the distance between the points if one travels along the manifold [Martinez et al., 2010].

returned by ISOMAP. If the MATLAB function demonstrated next returns fewer than n points, then one should try increasing the neighborhood size or change the measure of dissimilarity used.

Example 6.7

We return to the glass data of Example 6.2 to illustrate the use of the ISOMAP MATLAB code provided by Tenenbaum, et al. [2000], which is also included with the Computational Statistics Toolbox. In this example, we show only how to use the ISOMAP function and how to extract the lower dimensional data sets. The input to **isomap** is an $n \times n$ interpoint distance (or dissimilarity) matrix. The MATLAB Statistics Toolbox has two useful functions for calculating dissimilarities: **pdist** and **squareform**. Several dissimilarity metrics are available through **pdist** function, such as Euclidean, standardized Euclidean, Mahalanobis, Minkowski, and others. Some of the metrics are similarity measures that have been converted to dissimilarity (e.g., cosine of the angle between vectors, correlation, Spearman's rank correlation, and Jaccard). The output from **pdist** function is a $1 \times (n \cdot (n-1)/2)$ vector representing the $(n \cdot (n-1)/2)$ pairs of observations. The **isomap** function requires the square interpoint dissimilarity matrix, so we use the **squareform** function to convert the output of **pdist**.

```
% First load the data.
load glassdata
% Then get the interpoint dissimilarity matrix.
% We will use standardized Euclidean distance.
tmp = pdist(glassdata,'seuclidean');
% Now put it into a square matrix.
D = squareform(tmp);
```

The basic syntax for **isomap** is

```
[Y, R, E] = isomap(D, n_fcn, n_size, options);
```

where **D** is the dissimilarity matrix, **n_fcn** specifies the type of neighborhood (**'epsilon'** or **'k'**), and **n_size** is the neighborhood size. See the command line **help** on **isomap** for more information on the **options** argument. The output **Y** is a structure, where the cell array **Y.coords** contains the coordinates for the lower-dimensional embeddings. The default settings have **isomap** returning embedding dimensions of one to ten, so one can get the observations in 1D to 10D spaces. These values are easily changed via the **options** argument.

```
% Now we do ISOMAP. Code is available from
% Tenenbaum, de Silva, and Langford (2000)
% and is included in the Computational Statistics
% Toolbox.We will define the neighborhood using the
% number of nearest neighbors, k = 5.
```

```
[Y,R,E] = isomap(D,'k',5);
```

If default settings are used, then the **isomap** function will produce two plots: a scree plot where the horizontal axis represents the ISOMAP dimensionality (see Figure 6.11) and a scatterplot of the observations in 2D with the neighborhood graph (Figure 6.12). There is a nice elbow in the scree plot for ISOMAP dimensionality of three. The following commands show how to extract the data for a given lower dimensional embedding. It is also important to note that the data sets returned in the cell array **Y.coords** are given with the n columns representing the observations, so we have to take the transpose.

```
% We need to extract the data from structure Y.
% Scree plot shows that a value of 3 is a good one.
X3 = Y.coords{3}';
% Plot the data in a scatterplot matrix.
plotmatrix(X3);
```

We display the 3D ISOMAP embedding in Figure 6.13, where we can see some interesting non-Gaussian structures.
❑

6.8 MATLAB® Code

There is an open-source EDA Toolbox that has MATLAB functions and GUIs that implement many of the procedures described in this chapter. The EDA Toolbox contains more functions for dimensionality reduction and data tours than what we described here. These include a permutation tour for Andrews' curves and parallel coordinate plots, the pseudo grand tour, the k-dimensional grand tour, LLE, HLLE, and others. The toolbox is a companion to the text *Exploratory Data Analysis with MATLAB®, Second Edition* [Martinez, et al., 2010] and can be downloaded from CRC Press.

We include several functions in the Computational Statistics Toolbox that implement some of the algorithms and graphics covered in this chapter. These are summarized in Table 6.1.

Several of the nonlinear dimensionality reduction methods mentioned in this chapter were implemented in MATLAB by the original authors. These include LLE, HLLE, and ISOMAP. Websites for ISOMAP and LLE are

```
http://isomap.stanford.edu/
http://www.cs.toronto.edu/~roweis/lle/
```

There are some useful user-contributed toolboxes for manifold learning and dimensionality reduction available on The MathWorks, Inc. website. Go to the MATLAB Central file exchange at

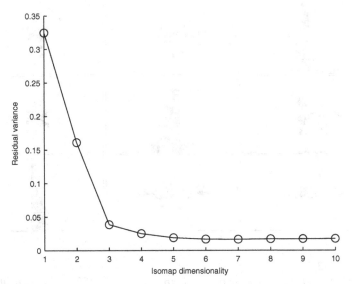

FIGURE 6.11

This shows the scree plot output of **isomap** *of the* **glassdata**. *The horizontal axis represents the lower-dimensional embedding size. As with the PCA scree plot, we look for the elbow in the curve. We see from the plot above that 3D seems a reasonable choice for the lower-dimensional embedding.*

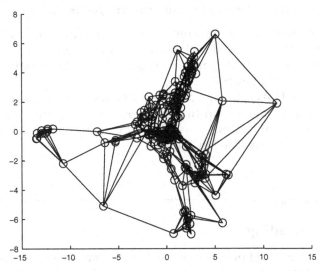

FIGURE 6.12

This is a scatterplot of the 2D ISOMAP coordinates for the **glassdata**. *The neighborhood map is also shown as edges.*

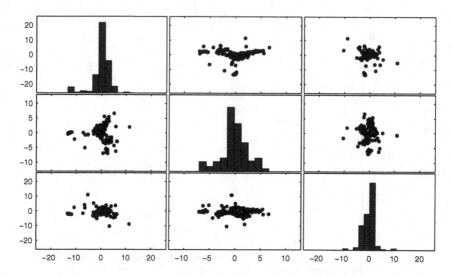

FIGURE 6.13
This scatterplot matrix shows the **glassdata** *using the 3D ISOMAP coordinates. We see some non-Gaussian structure.*

http://www.mathworks.com/matlabcentral/

and search using appropriate keywords.

For more information on independent component analysis, see the following website:

http://www.cis.hut.fi/projects/ica/

This website has papers, references, other useful links to ICA, and downloadable MATLAB code. In particular, a wonderful (free) MATLAB toolbox for independent component analysis called FastICA can be downloaded from that website.

The main MATLAB package and the Statistics Toolbox have functions relevant to topics covered in this chapter. These include functions for principal component analysis, classical MDS, metric MDS, and nonmetric MDS. The functions are listed in Table 6.2.

6.9 Further Reading

For a general treatment of linear and matrix algebra, we recommend any of the texts by Strang. Some examples are *Linear Algebra and its Applications* [2005] and *Introduction to Linear Algebra* [2003]. These texts also include a discussion of the theory behind principal component analysis, as do most

TABLE 6.1

List of Functions from Chapter 6 Included in the
Computational Statistics Toolbox

Purpose	MATLAB Function
ISOMAP	`isomap`
PPEDA	`csppeda`
	`csppstrtrem`
	`csppind`
Grand Tour	`torustour`
Independent Component Analysis	`csica`

TABLE 6.2

Functions from Base MATLAB and the Statistics Toolbox

Purpose	MATLAB Function
MATLAB	
Find eigenvectors and eigenvalues	`eig`
Statistics Toolbox	
Principal Component Analysis	`princomp, pcacov`
Classical MDS	`cmdscale`
Nonmetric and metric MDS	`mdscale`

texts on multivariate data analysis. Some examples of books on multivariate analysis are Manly [2004] and Seber [1984]. As mentioned previously in the chapter, there are also some texts devoted to principal component analysis by Joliffe [2002] and Jackson [1991]. Both of these texts are easy to read and offer many useful examples, in addition to the underlying theory. In our opinion, the Joliffe text tends to be more for the statistician or mathematician, while the Jackson book is more appropriate for engineers.

Multi-dimensional scaling methods are described in most multivariate data analysis texts [Manly 2004]. There are also some books dedicated to this subject. One of these is by Cox and Cox [2001], which is a highly readable text. The other reference is by Borg and Groenen [1997]. Both books describe many of the algorithms used for MDS as a way to aid understanding of the concepts.

We can recommend two books on independent component analysis. First is a book by Stone [2004]. This text is a tutorial introduction to the topic, and it includes many examples and code in MATLAB. Another excellent resource is a book by Hyvärinen, Karhunen, and Oja [2001]. This text is very readable and serves as a good source on theory and applications of ICA. Several review and tutorial articles have been published on independent component analysis. These include Hyvärinen [1999], Hyvärinen and Oja [2000], and Fodor [2002].

The book by Cook and Swayne [2007] discusses several of the visualization methods discussed in this chapter, such as data tours and multi-dimensional scaling, where they use R and GGobi as their implementation software. For information on R and GGobi, see

http://www.r-project.org/

http://www.ggobi.org/

For another approach to interactive graphics, see *Visual Statistics* by Young, Valero-Mora, and Friendly [2006]. This book is accessible to those with little statistical training, and it uses a free visual statistics system called ViSta. For information on downloading ViSta, see

http://www.visualstats.org/

The paper by Wegman, Carr, and Luo [1993] discusses many of the data tour methods we present in this chapter. This paper and Wegman [1990] also discuss the permutation tour for parallel coordinate plots.

There are several interesting papers on projection pursuit for EDA. Jones and Sibson [1987] describe a steepest-ascent algorithm that starts from either principal components or random starts. Friedman [1987] combines steepest-ascent with a stepping search to look for a region of interest. Crawford [1991] uses genetic algorithms to optimize the projection index. Perisic and Posse [2005] propose other indices for projection pursuit EDA that are based on the empirical distribution function and do not have any extra parameters that need to be tuned.

Other uses for projection pursuit have been proposed. These include projection pursuit probability density estimation [Friedman, Stuetzle, and Schroeder, 1984], projection pursuit regression [Friedman and Stuetzle, 1981], robust estimation [Li and Chen, 1985], and projection pursuit for pattern recognition [Flick, et al., 1990]. A 3D projection pursuit algorithm is given in Nason [1995].

For a theoretical and comprehensive description of projection pursuit, the reader is directed to Huber [1985]. This invited paper with discussion also presents applications of projection pursuit to computer tomography and to the deconvolution of time series. Another paper that provides applications of projection pursuit is Jones and Sibson [1987]. Not surprisingly, projection pursuit has been combined with the grand tour by Cook, et al. [1995]. Montanari and Lizzani [2001] apply projection pursuit to the variable selection problem. Bolton and Krzanowski [1999] describe the connection between projection pursuit and principal component analysis.

Exercises

6.1.   Generate some multivariate normal random variables using the following code (or use the corresponding function included in the Computational Statistics Toolbox):

```
mu = zeros(1,2);
sigma = [1 .8; .8 1];
X = mvnrnd(mu, sigma, 100);
```

Verify the covariance by using **corrcoef(X)**. We see that the data points are correlated, as expected. Now, find the eigenvectors of the covariance matrix using

```
eig(cov(X))
```

Use the eigenvectors to project the data as shown in Example 6.1. Verify that the projected observations are uncorrelated.

6.2.   Generate 1000, 10D random variables using **randn**. Construct a scree plot and find the cumulative percentage of variance as shown in Example 6.2. Is there any evidence that one could reduce the dimensionality of these data? If so, what would be a value for k?

6.3.   Looking at Figure 6.2, there appears to be another elbow in the curve at $k = 5$. Reduce the dimensionality of the data to 5D using

```
P5 = eigvecs(:,1,5);
Xp5 = Xc*P5;
```

Do a scatterplot matrix of the 3D and 5D data and compare them. Look at the first observation in 3D and 5D by using **Xp3(1,:)** and **Xp5(1,:)** and compare them.

6.4.   The **bank** data contains two matrices comprised of measurements made on genuine money and forged money. Combine these two matrices into one and use PPEDA to discover any clusters or groups in the data. Compare your results with the known groups in the data.

6.5.   Using the data in Example 6.3, do a scatterplot matrix of the original sphered data set. Note the structures in the first four dimensions. Get the first structure and construct another scatterplot matrix of the sphered data after the first structure has been removed. Repeat this process after both structures are removed.

6.6.   Apply PPEDA to the **glassdata** (Example 6.2). Describe any interesting structure that you find.

6.7.   Use Gram-Schmidt orthogonalization to remove the extracted **chirp** signal from the signal mixtures (i.e., the columns of **z**) in Example 6.4. First, listen to the remaining signal mixture. Does it sound like the **gong**? Apply the ICA steps of Example 6.4 to extract the second signal and listen to it using **soundsc**.

6.8.   Compare the estimated mixing coefficients of Example 6.4 with the true values in **mixmat**. Do the same for the estimate of the second signal found in the previous problem.

6.9.   Apply classical MDS to the **cereal** data using Euclidean distance. Reduce the data to 2D and construct a scatterplot. Compare your results with Figure 6.8.

6.10.   Use **gplotmatrix** to do a grouped scatterplot of the ISOMAP results for the **glassdata** (Example 6.7). Does this mapping provide good separation of the groups? Use various graphical techniques (e.g., Andrews curves, parallel coordinate plots, scatterplot matrix, etc.) to explore the other configurations from ISOMAP (i.e., in **Y.coords**).

6.11.   Verify that the two vectors used in Equations 6.17 and 6.18 are orthonormal.

6.12.   Load the data sets in **posse**. These contain several data sets from Posse [1995b]. Apply the PPEDA method to these data.

Chapter 7

Monte Carlo Methods for Inferential Statistics

7.1 Introduction

Methods in inferential statistics are used to draw conclusions about a population and to measure the reliability of these conclusions using information obtained from a random sample. Inferential statistics involves techniques such as estimating population parameters using point estimates, calculating confidence interval estimates for parameters, hypothesis testing, and modeling (e.g., regression and density estimation). To measure the reliability of the inferences that are made, the statistician must understand the distribution of any statistics that are used in the analysis. In situations where we use a well-understood statistic, such as the sample mean, this is easily done analytically. However, in many applications, we do not want to be limited to using such simple statistics or to making simplifying assumptions. The goal of this chapter is to explain how simulation or Monte Carlo methods can be used to make inferences when the traditional or analytical statistical methods fail.

According to Murdoch [2000], the term *Monte Carlo* originally referred to simulations that involved random walks and was first used by Jon von Neumann and S. M. Ulam in the 1940s. Today, the *Monte Carlo method* refers to any simulation that involves the use of random numbers. In the following sections, we show that Monte Carlo simulations (or experiments) are an easy and inexpensive way to understand the phenomena of interest [Gentle, 1998]. To conduct a simulation experiment, you need a model that represents your population or phenomena of interest and a way to generate random numbers (according to your model) using a computer. The data that are generated from your model can then be studied as if they were observations. As we will see, one can use statistics based on the simulated data (means, medians, modes, variance, skewness, etc.) to gain understanding about the population.

In Section 7.2, we give a short overview of methods used in classical inferential statistics, covering such topics as hypothesis testing, power, and confidence intervals. The reader who is familiar with these may skip this section. In Section 7.3, we discuss Monte Carlo simulation methods for hypothesis testing and for evaluating the performance of the tests. The bootstrap method for estimating the bias and variance of estimates is presented in Section 7.4. Finally, Sections 7.5 and 7.6 conclude the chapter with information about available MATLAB® code and references on Monte Carlo simulation and the bootstrap.

7.2 Classical Inferential Statistics

In this section, we will cover two of the main methods in inferential statistics: hypothesis testing and calculating confidence intervals. With confidence intervals, we are interested in obtaining an interval of real numbers that we expect (with specified confidence) contains the true value of a population parameter. In hypothesis testing, our goal is to make a decision about not rejecting or rejecting some statement about the population based on data from a random sample. We give a brief summary of the concepts in classical inferential statistics, endeavoring to keep the theory to a minimum. There are many books available that contain more information on these topics. We recommend Casella and Berger [1990], Walpole and Myers [1985], Bickel and Doksum [2001], Lindgren [1993], Montgomery, Runger, and Hubele [1998], and Mood, Graybill, and Boes [1974].

Hypothesis Testing

In hypothesis testing, we start with a *statistical hypothesis*, which is a conjecture about one or more populations. Some examples of these are

- A transportation official in the Washington, D.C. area thinks that the mean travel time to work for northern Virginia residents has increased from the average time it took in 1995.

- A medical researcher would like to determine whether aspirin decreases the risk of heart attacks.

- A pharmaceutical company needs to decide whether a new vaccine is superior to the one currently in use.

- An engineer has to determine whether there is a difference in accuracy between two types of instruments.

We generally formulate our statistical hypotheses in two parts. The first is the *null hypothesis* represented by H_0, which denotes the hypothesis we would like to test. Usually, we are searching for departures from this statement. Using one of the examples given above, the engineer would have the null hypothesis that there is no difference in the accuracy between the two instruments.

There must be an *alternative hypothesis* such that we would decide in favor of one or the other, and this is denoted by H_1. If we reject H_0, then this leads to the acceptance of H_1. Returning to the engineering example, the alternative hypothesis might be that there is a difference in the instruments or that one is more accurate than the other. When we perform a statistical hypothesis test, we can never know with certainty which hypothesis is true. For ease of exposition, we will use the terms *accept the null hypothesis* and *reject the null hypothesis* for our decisions resulting from statistical hypothesis testing.

To clarify these ideas, let's look at the example of the transportation official who wants to determine whether the average travel time to work has increased from the time it took in 1995. The mean travel time to work for northern Virginia residents in 1995 was 45 minutes. Since he wants to determine whether the mean travel time has increased, the statistical hypotheses are given by:

$$H_0: \qquad \mu = 45 \text{ minutes}$$
$$H_1: \qquad \mu > 45 \text{ minutes.}$$

The logic behind statistical hypothesis testing is summarized next, with details and definitions given after.

STEPS OF HYPOTHESIS TESTING

1. Determine the null and alternative hypotheses, using mathematical expressions if applicable. Usually, this is an expression that involves a characteristic or descriptive measure of a population.

2. Take a random sample from the population of interest.

3. Calculate a statistic from the sample that provides information about the null hypothesis. We use this to make our decision.

4. If the value of the statistic is consistent with the null hypothesis, then do not reject H_0.

5. If the value of the statistic is not consistent with the null hypothesis, then reject H_0 and accept the alternative hypothesis.

The problem then becomes one of determining when a statistic is consistent with the null hypothesis. Recall from Chapter 3 that a statistic is itself a random variable and has a probability distribution associated with it. So, in order to decide whether an observed value of the statistic is consistent with the null hypothesis, we must know the distribution of the statistic when the null hypothesis is true. The statistic used in step 3 is called a *test statistic*.

Let's return to the example of the travel time to work for northern Virginia residents. To perform the analysis, the transportation official takes a random sample of 100 residents in northern Virginia and measures the time it takes them to travel to work. He uses the sample mean to help determine whether there is sufficient evidence to reject the null hypothesis and conclude that the mean travel time has increased. The sample mean that he calculates is 47.2 minutes. This is slightly higher than the mean of 45 minutes for the null hypothesis. However, the sample mean is a random variable and has some variation associated with it. If the variance of the sample mean under the null hypothesis is large, then the observed value of $\bar{x} = 47.2$ minutes might not be inconsistent with H_0. This is explained further in Example 7.1.

Example 7.1

We continue with the transportation example. We need to determine whether the value of the statistic obtained from a random sample drawn from the population is consistent with the null hypothesis. Here we have a random sample comprised of $n = 100$ commute times. The sample mean of these observations is $\bar{x} = 47.2$ minutes. If the transportation official assumes that the travel times to work are normally distributed with $\sigma = 15$ minutes (one might know a reasonable value for σ based on previous experience with the population), then we know from Chapter 3 that $\bar{x}$ is approximately normally distributed with mean μ_X and standard deviation $\sigma_{\bar{x}} = \sigma_X/\sqrt{n}$. Standardizing the observed value of the sample mean, we have

$$z_0 = \frac{\bar{x} - \mu_0}{\sigma_X/\sqrt{n}} = \frac{\bar{x} - \mu_0}{\sigma_{\bar{x}}} = \frac{47.2 - 45}{15/\sqrt{100}} = \frac{2.2}{1.5} = 1.47, \qquad (7.1)$$

where z_0 is the observed value of the test statistic, and μ_0 is the mean under the null hypothesis. Thus, we have that the value of $\bar{x} = 47.2$ minutes is 1.47 standard deviations away from the mean, if the null hypothesis is really true. (This is why we use μ_0 in Equation 7.1.) We know that approximately 95% of normally distributed random variables fall within two standard deviations either side of the mean. Thus, $\bar{x} = 47.2$ minutes is not inconsistent with the null hypothesis.
❏

In hypothesis testing, the rule that governs our decision might be of the form: *if the observed statistic is within some region, then we reject the null hypothesis*. The **critical region** is an interval for the test statistic over which we would reject H_0. This is sometimes called the **rejection region**. The **critical value** is that value of the test statistic that divides the domain of the test statistic into a region where H_0 will be rejected and one where H_0 will be accepted. We need to know the distribution of the test statistic under the null hypothesis to find the critical value(s).

The critical region depends on the distribution of the statistic under the null hypothesis, the alternative hypothesis, and the amount of error we are willing to tolerate. Typically, the critical regions are areas in the tails of the distribution of the test statistic when H_0 is true. It could be in the lower tail, the upper tail, or both tails, and which one is appropriate depends on the alternative hypothesis. For example:

- If a large value of the test statistic would provide evidence for the alternative hypothesis, then the critical region is in the upper tail of the distribution of the test statistic. This is sometimes referred to as an ***upper tail test***.

- If a small value of the test statistic provides evidence for the alternative hypothesis, then the critical region is in the lower tail of the distribution of the test statistic. This is sometimes referred to as a ***lower tail test***.

- If small or large values of the test statistic indicate evidence for the alternative hypothesis, then the critical region is in the lower and upper tails. This is sometimes referred to as a ***two-tail test***.

There are two types of errors that can occur when we make a decision in statistical hypothesis testing. The first is a ***Type I error***, which arises when we reject H_0 when it is really true. The other error is called ***Type II error***, and this happens when we fail to detect that H_0 is actually false. These errors are summarized in Table 7.1.

TABLE 7.1

Types of Error in Statistical Hypothesis Testing

Type of Error	Description	Probability of Error
Type I Error	Rejecting H_0 when it is true	α
Type II Error	Not rejecting H_0 when it is false	β

Recall that we are usually searching for significant evidence that the alternative hypothesis is valid, and we do not want to change from the status quo (i.e., reject H_0) unless there is sufficient evidence in the data to lead us in that direction. So, when setting up a hypothesis test we ensure that the probability of wrongly rejecting H_0 is controlled. The probability of making a Type I error is denoted by α and is sometimes called the ***significance level*** of the test. The α is set by the analyst, and it represents the maximum probability of Type I error that will be tolerated. Typical values of α are $\alpha = 0.01, 0.05, 0.10$. The critical value is found as the quantile (under the null hypothesis) that gives a significance level of α.

The specific procedure for conducting a hypothesis test using these ideas is given next. This is called the **critical value approach** because the decision is based on whether the value of the test statistic falls in the rejection region. We will discuss an alternative method later in this section. The concepts of hypothesis testing using the critical value approach are illustrated in Example 7.2.

PROCEDURE – HYPOTHESIS TESTING (CRITICAL VALUE APPROACH)

1. Determine the null and alternative hypotheses.

2. Find a test statistic T that will provide evidence that H_0 should be accepted or rejected (e.g, a large value of the test statistic indicates H_0 should be rejected).

3. Obtain a random sample from the population of interest and compute the observed value of the test statistic t_o using the sample.

4. Using the sampling distribution of the test statistic under the null hypothesis and the significance level, find the critical value(s). That is, find the t such that

    <u>Upper Tail Test</u>: $P_{H_0}(T \leq t) = 1 - \alpha$

    <u>Lower Tail Test</u>: $P_{H_0}(T \leq t) = \alpha$

    <u>Two-Tail Test</u>: $P_{H_0}(T \leq t_1) = \alpha/2$ and $P_{H_0}(T \leq t_2) = 1 - \alpha/2$,

    where $P_{H_0}(.)$ denotes the probability under the null hypothesis.

5. If the value of the test statistic t_o falls in the critical region, then reject the null hypothesis.

Example 7.2

Here, we illustrate the critical value approach to hypothesis testing using the transportation example. Our test statistic is given by

$$z = \frac{\bar{x} - \mu_0}{\sigma_{\bar{x}}},$$

and we observed a value of $z_o = 1.47$ based on the random sample of $n = 100$ commute times. We want to conduct the hypothesis test at a significance level given by $\alpha = 0.05$. Since our alternative hypothesis is that the commute times have increased, a large value of the test statistic provides evidence for H_1. We can find the critical value using the MATLAB Statistics Toolbox as follows:

$$\text{cv = norminv(0.95,0,1);}$$

This yields a critical value of 1.645. Thus, if $z_o \geq 1.645$, then we reject H_0. Since the observed value of the test statistic is less than the critical value, we do not reject H_0. The regions corresponding to this hypothesis test are illustrated in Figure 7.1.
❑

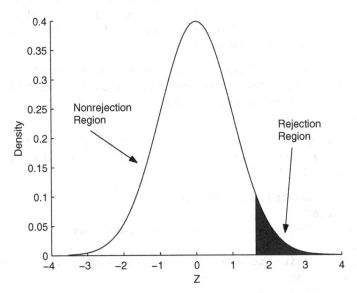

FIGURE 7.1
This shows the critical region (shaded region) for the hypothesis test of Examples 7.1 and 7.2. If the observed value of the test statistic falls in the shaded region, then we reject the null hypothesis. Note that this curve reflects the distribution for the test statistic under the null hypothesis.

The probability of making a Type II error is represented by β, and it depends on the sample size, the significance level of the test, and the alternative hypothesis. The last part is important to remember: *the probability that we will not detect a departure from the null hypothesis depends on the distribution of the test statistic under the alternative hypothesis.* Recall that the alternative hypothesis allows for many different possibilities, yielding many distributions under H_1. So, we must determine the Type II error for every alternative hypothesis of interest.

A more convenient measure of the performance of a hypothesis test is to determine the probability of not making a Type II error. This is called the *power* of a test. We can consider this to be the probability of rejecting H_0

when it is really false. Roughly speaking, one can think of the power as the ability of the hypothesis test to detect a false null hypothesis. The power is given by

$$\text{Power} = 1 - \beta. \tag{7.2}$$

As we see in Example 7.3, the power of the test to detect departures from the null hypothesis depends on the true value of μ.

Example 7.3

Returning to the transportation example, we illustrate the concepts of Type II error and power. It is important to keep in mind that these values depend on the true mean μ, so we have to calculate the Type II error for different values of μ. First we get a vector of values for μ:

```
% Get several values for the mean under the alternative
% hypothesis. Note that we are getting some values
% below the null hypothesis.
mualt = 40:60;
```

It is actually easier to understand the power when we look at a test statistic based on $\bar{x}$ rather than z_o. So, we convert the critical value to its corresponding $\bar{x}$ value:

```
% Note the critical value:
cv = 1.645;
% Note the standard deviation for x-bar:
sig = 1.5;
% It's easier to use the non-standardized version,
% so convert:
ct = cv*1.5 + 45;
```

We find the area under the curve to the left of the critical value (the nonrejection region) for each of these values of the true mean. That would be the probability of not rejecting the null hypothesis.

```
% Get a vector of critical values that is
% the same size as mualt.
ctv = ct*ones(size(mualt));
% Now get the probabilities to the left of this value.
% These are the probabilities of the Type II error.
beta = normcdf(ctv,mualt,sig);
```

Note that the variable **beta** contains the probability of Type II error (the area to the left of the critical value **ctv** under a normal curve with mean **mualt** and standard deviation **sig**) for every μ. To get the power, simply subtract all of the values for **beta** from one.

```
% To get the power: 1-beta
pow = 1 - beta;
```

We plot the power against the true value of the population mean in Figure 7.2. Note that as $\mu > \mu_0$, the power (or the likelihood that we can detect the alternative hypothesis) increases.

```
plot(mualt,pow);
xlabel('True Mean \mu')
ylabel('Power')
axis([40 60 0 1.1])
```

We leave it as an exercise for the reader to plot the probability of making a Type II error.
❑

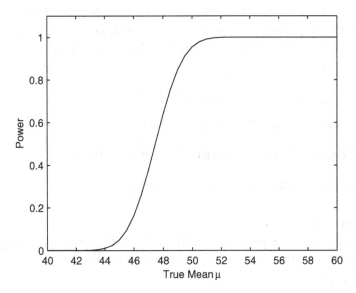

FIGURE 7.2
This shows the power (or probability of not making a Type II error) as a function of the true value of the population mean μ. Note that as the true mean gets larger, then the likelihood of not making a Type II error increases.

There is an alternative approach to hypothesis testing, which uses a quantity called a *p*-value. A ***p-value*** is defined as the probability of observing a value of the test statistic as extreme as or more extreme than the one that is observed, when the null hypothesis H_0 is true. The word *extreme* refers to the direction of the alternative hypothesis. For example, if a small value of the test statistic (a lower tail test) indicates evidence for the alternative hypothesis, then the *p*-value is calculated as

$$p\text{-value} = P_{H_0}(T \leq t_o),$$

where t_o is the observed value of the test statistic T, and $P_{H_0}(.)$ denotes the probability under the null hypothesis. The p-value is sometimes referred to as the ***observed significance level***.

In the p-value approach, a small value indicates evidence for the alternative hypothesis and would lead to rejection of H_0. Here small refers to a p-value that is less than or equal to α. The steps for performing hypothesis testing using the p-value approach are given next and are illustrated in the next example.

PROCEDURE – HYPOTHESIS TESTING (P-VALUE APPROACH)

1. Determine the null and alternative hypotheses.
2. Find a test statistic T that will provide evidence about H_0.
3. Obtain a random sample from the population of interest and compute the value of the test statistic t_o from the sample.
4. Calculate the p-value:

<u>Lower Tail Test</u>: $p - \text{value} = P_{H_0}(T \leq t_o)$

<u>Upper Tail Test</u>: $p - \text{value} = P_{H_0}(T \geq t_o)$

5. If the p-value $\leq \alpha$, then reject the null hypothesis.

For a two-tail test, the p-value is determined similarly.

Example 7.4

In this example, we repeat the hypothesis test of Example 7.2 using the p-value approach. First we set some of the values we need:

```
mu = 45;
sig = 1.5;
xbar = 47.2;
% Get the observed value of test statistic.
zobs = (xbar - mu)/sig;
```

The p-value is the area under the curve greater than the value for **zobs**. We can find it using the following command:

```
pval = 1-normcdf(zobs,0,1);
```

We get a p-value of 0.071. If we are doing the hypothesis test at the 0.05 significance level, then we would not reject the null hypothesis. This is consistent with the results we had previously.

❑

Note that in each approach, knowledge of the distribution of T under the null hypothesis H_0 is needed. How to tackle situations where we do not know the distribution of our statistic is the focus of the rest of the chapter.

Confidence Intervals

In Chapter 3, we discussed several examples of estimators for population parameters such as the mean, the variance, moments, and others. We call these *point estimates*. It is unlikely that a point estimate obtained from a random sample will exactly equal the true value of the population parameter. Thus, it might be more useful to have an interval of numbers that we expect will contain the value of the parameter. This type of estimate is called an *interval estimate*. An understanding of confidence intervals is needed for the bootstrap methods covered in Section 7.4.

Let θ represent a population parameter that we wish to estimate, and let T denote a statistic that we will use as a point estimate for θ. The observed value of the statistic is denoted as $\hat{\theta}$. An interval estimate for θ will be of the form

$$\hat{\theta}_{Lo} < \theta < \hat{\theta}_{Up}, \tag{7.3}$$

where $\hat{\theta}_{Lo}$ and $\hat{\theta}_{Up}$ depend on the observed value $\hat{\theta}$ and the distribution of the statistic T.

If we know the sampling distribution of T, then we are able to determine values for $\hat{\theta}_{Lo}$ and $\hat{\theta}_{Up}$ such that

$$P(\hat{\theta}_{Lo} < \theta < \hat{\theta}_{Up}) = 1 - \alpha, \tag{7.4}$$

where $0 < \alpha < 1$. Equation 7.4 indicates that we have a probability of $1 - \alpha$ that we will select a random sample that produces an interval that contains θ. This interval (Equation 7.3) is called a $(1 - \alpha) \cdot 100\%$ confidence interval. The philosophy underlying confidence intervals is the following. Suppose we repeatedly take samples of size n from the population and compute the random interval given by Equation 7.3. Then the relative frequency of the intervals that contain the parameter θ would approach $1 - \alpha$. It should be noted that one-sided confidence intervals can be defined similarly [Mood, Graybill, and Boes, 1974].

To illustrate these concepts, we use Equation 7.4 to get a confidence interval for the population mean μ. Recall from Chapter 3 that we know the distribution for $\overline{X}$. We define $z^{(\alpha/2)}$ as the z value that has an area under the standard normal curve of size $\alpha/2$ to the left of it. In other words, we use $z^{(\alpha/2)}$ to denote the value of z such that

$$P(Z < z^{(\alpha/2)}) = \alpha/2.$$

Therefore, the area between $z^{(\alpha/2)}$ and $z^{(1-\alpha/2)}$ is $1-\alpha$. This is shown in Figure 7.3.

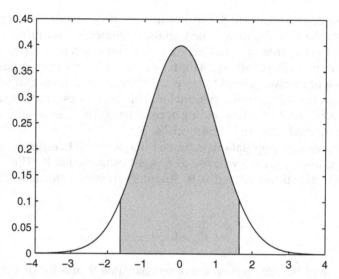

FIGURE 7.3
The left vertical line corresponds to $z^{(\alpha/2)}$, and the right vertical line is at $z^{(1-\alpha/2)}$. So, the non-shaded areas in the tails each have an area of $(\alpha/2)$, and the shaded area in the middle is $1-\alpha$.

We can see from this that the shaded area has probability $1-\alpha$, and

$$P(z^{(\alpha/2)} < Z < z^{(1-\alpha/2)}) = 1-\alpha, \tag{7.5}$$

where

$$Z = \frac{\overline{X} - \mu}{\sigma/\sqrt{n}}. \tag{7.6}$$

If we substitute this into Equation 7.5, then we have

$$P\left(z^{(\alpha/2)} < \frac{\overline{X} - \mu}{\sigma/\sqrt{n}} < z^{(1-\alpha/2)}\right) = 1-\alpha. \tag{7.7}$$

Rearranging the inequalities in Equation 7.7, we obtain

$$P\left(\overline{X} - z^{(1-\alpha/2)}\frac{\sigma}{\sqrt{n}} < \mu < \overline{X} - z^{(\alpha/2)}\frac{\sigma}{\sqrt{n}}\right) = 1 - \alpha. \qquad (7.8)$$

Comparing Equations 7.8 and 7.4, we see that

$$\hat{\theta}_{Lo} = \overline{X} - z^{(1-\alpha/2)}\frac{\sigma}{\sqrt{n}} \qquad \hat{\theta}_{Up} = \overline{X} - z^{(\alpha/2)}\frac{\sigma}{\sqrt{n}}.$$

Example 7.5

We provide an example of finding a 95% confidence interval, using the transportation application of before. Recall that $n = 100$, $\bar{x} = 47.2$ minutes, and the standard deviation of the travel time to work is $\sigma = 15$ minutes. Since we want a 95% confidence interval, $\alpha = 0.05$.

```
mu = 45;
sig = 15;
n = 100;
alpha = 0.05;
xbar = 47.2;
```

We can get the endpoints for a 95% confidence interval as follows:

```
% Get the 95% confidence interval.
% Get the value for z_alpha/2.
zlo = norminv(1-alpha/2,0,1);
zhi = norminv(alpha/2,0,1);
thetalo = xbar - zlo*sig/sqrt(n);
thetaup = xbar - zhi*sig/sqrt(n);
```

We get a value of $\hat{\theta}_{Lo} = 44.26$ and $\hat{\theta}_{Up} = 50.14$.
□

We return to confidence intervals later, where we discuss bootstrap methods for obtaining them. First, however, we look at Monte Carlo methods for hypothesis testing.

7.3 Monte Carlo Methods for Inferential Statistics

The sampling distribution is known for many statistics. However, these are typically derived using assumptions about the underlying population under study or for large sample sizes. In many cases, we do not know the sampling distribution for the statistic, or we cannot be sure that the assumptions are satisfied. We can address these cases using Monte Carlo simulation methods, which is the topic of this section. Some of the uses of Monte Carlo simulation for inferential statistics are the following:

- Performing inference when the distribution of the test statistic is not known analytically,
- Assessing the performance of inferential methods when parametric assumptions are violated,
- Testing the null and alternative hypotheses under various conditions,
- Evaluating the performance (e.g., power) of inferential methods, and
- Comparing the quality of estimators.

In this section, we cover situations in inferential statistics where we do know something about the distribution of the population or we are willing to make assumptions about the distribution. In the next section, we discuss bootstrap methods that can be used when no assumptions are made about the underlying distribution of the population.

Basic Monte Carlo Procedure

The fundamental idea behind *Monte Carlo simulation* for inferential statistics is that insights regarding the characteristics of a statistic can be gained by repeatedly drawing random samples from the same population or distribution of interest and observing the behavior of the statistic over the samples. In other words, we estimate the distribution of the *statistic* by randomly sampling from the population or distribution and recording the value of the statistic for each sample. The observed values of the statistic for these samples are used to estimate the distribution.

The first step is to decide on a pseudo-population or distribution that the analyst assumes represents the real population in all relevant aspects. We use the word *pseudo* here to emphasize the fact that we obtain our samples using a computer and pseudo random numbers. For example, we might assume that the underlying population is exponentially distributed if the random variable represents the time before a part fails, or we could assume the random variable comes from a normal distribution if we are measuring IQ scores. In this text, we consider this type of Monte Carlo simulation to be a parametric technique because we sample from a known (or assumed) distribution.

The basic Monte Carlo procedure is outlined here. Later, we provide procedures illustrating some specific uses of Monte Carlo simulation as applied to statistical hypothesis testing.

PROCEDURE – BASIC MONTE CARLO SIMULATION

1. Determine the pseudo-population or model that represents the true population of interest.
2. Use a sampling procedure to sample from the pseudo-population or distribution.
3. Calculate a value for the statistic of interest and store it.
4. Repeat steps 2 and 3 for M trials.
5. Use the M values found in step 4 to study the distribution of the statistic.

It is important to keep in mind that when sampling from the pseudo-population, the analyst should ensure that all relevant characteristics reflect the statistical situation. For example, the same sample size and sampling strategy should be used when trying to understand the performance of a statistic using simulation. This means that the distribution for the statistic obtained via Monte Carlo simulation is relevant only for the conditions of the sampling procedure and the assumptions about the pseudo-population.

Note that in the last step of the Monte Carlo simulation procedure, the analyst can use the estimated distribution of the statistic to study characteristics of interest. For example, one could use this information to estimate the skewness, bias, standard deviation, kurtosis, and many other characteristics.

Monte Carlo Hypothesis Testing

Recall that in statistical hypothesis testing, we have a test statistic that provides evidence that the null hypothesis should be rejected or not. Once we observe the value of the test statistic, we decide whether that particular value is consistent with the null hypothesis. To make that decision, we must know the distribution of the statistic when the null hypothesis is true. Estimating the distribution of the test statistic under the null hypothesis is one of the goals of Monte Carlo hypothesis testing. We discuss and illustrate the Monte Carlo method as applied to the critical value and p-value approaches to hypothesis testing.

Recall that in the critical value approach to hypothesis testing, we are given a significance level α. We then use this significance level to find the appropriate critical region in the distribution of the test statistic when the null hypothesis is true. Using the Monte Carlo method, we determine the critical value using the estimated distribution of the test statistic. The basic procedure is to randomly sample many times from the pseudo-population representing the null hypothesis, calculate the value of the test statistic at each trial, and use these values to estimate the distribution of the test statistic.

PROCEDURE – MONTE CARLO HYPOTHESIS TESTING (CRITICAL VALUE)

1. Using an available random sample of size n from the population of interest, calculate the observed value of the test statistic, t_o.

2. Decide on a pseudo-population that reflects the characteristics of the true population under the null hypothesis.

3. Obtain a random sample of size n from the pseudo-population.

4. Calculate the value of the test statistic using the random sample in step 3 and record it.

5. Repeat steps 3 and 4 for M trials. We now have values $t_1, ..., t_M$, that serve as an estimate of the distribution of the test statistic, T, when the null hypothesis is true.

6. Obtain the critical value for the given significance level α:

   <u>Lower Tail Test</u>: get the α-th sample quantile, $\hat{q}_\alpha$, from the $t_1, ..., t_M$.

   <u>Upper Tail Test</u>: get the $(1 - \alpha)$-th sample quantile, $\hat{q}_{1-\alpha}$, from the $t_1, ..., t_M$.

   <u>Two-Tail Test</u>: get the sample quantiles $\hat{q}_{\alpha/2}$ and $\hat{q}_{1-\alpha/2}$ from the $t_1, ..., t_M$.

7. If t_o falls in the critical region, then reject the null hypothesis.

The critical values in step 6 can be obtained using the estimate of a sample quantile that we discussed in Chapter 3. The function **csquantiles** from the Computational Statistics Toolbox is also available to find these values.

In the examples given next, we apply the Monte Carlo method to a familiar hypothesis testing situation where we are testing a hypothesis about the population mean. As we saw earlier, we can use analytical approaches for this type of test. We use this simple application in the hope that the reader will better understand the ideas of Monte Carlo hypothesis testing and then easily apply them to more complicated problems.

Example 7.6

This toy example illustrates the concepts of Monte Carlo hypothesis testing. The **mcdata** data set contains 25 observations. We are interested in using these data to test the following null and alternative hypotheses:

$$H_0: \quad \mu = 454$$
$$H_1: \quad \mu < 454.$$

We will perform our hypothesis test using simulation to get the critical values. We decide to use the following as our test statistic

$$z = \frac{\bar{x} - 454}{\sigma / \sqrt{n}}.$$

First, we take care of some preliminaries.

```
% Load up the data.
load mcdata
n = length(mcdata);
% Population sigma is known.
sigma = 7.8;
sigxbar = sigma/sqrt(n);
% Get the observed value of the test statistic.
Tobs = (mean(mcdata)-454)/sigxbar;
```

The observed value of the test statistic is $t_o = -2.56$. The next step is to decide on a model for the population that generated our data. We suspect that the normal distribution with $\sigma = 7.8$ is a good model, and we check this assumption using a normal probability plot. The resulting plot in Figure 7.4 shows that we can use the normal distribution as the pseudo-population.

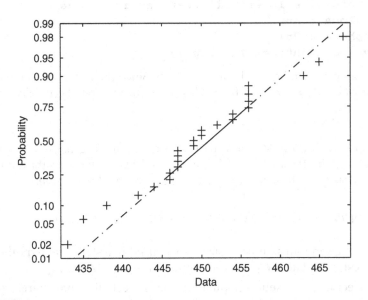

FIGURE 7.4
This normal probability plot for the **mcdata** *data shows that assuming a normal distribution for the data is reasonable.*

```
% This command generates the normal probability plot.
% It is a function in the MATLAB Statistics Toolbox.
normplot(mcdata)
```

We are now ready to implement the Monte Carlo simulation. We use 1000 trials in this example. At each trial, we randomly sample from the distribution of the test statistic under the null hypothesis (the normal distribution with $\mu = 454$ and $\sigma = 7.8$) and record the value of the test statistic.

```
M = 1000;% Number of Monte Carlo trials
% Storage for test statistics from the MC trials.
Tm = zeros(1,M);
% Start the simulation.
for i = 1:M
    % Generate a random sample under H_0
    % where n is the sample size.
    xs = sigma*randn(1,n) + 454;
    Tm(i) = (mean(xs) - 454)/sigxbar;
end
```

Now that we have the estimated distribution of the test statistic contained in the variable **Tm**, we can use that to estimate the critical value for a lower tail test.

```
% Get the critical value for alpha.
% This is a lower-tail test, so it is the
% alpha quantile.
alpha = 0.05;
cv = csquantiles(Tm,alpha);
```

We get an estimated critical value of -1.75. Since the observed value of our test statistic is $t_o = -2.56$, which is less than the estimated critical value, we reject H_0.
❑

The procedure for Monte Carlo hypothesis testing using the p-value approach is similar. Instead of finding the critical value from the simulated distribution of the test statistic, we use it to estimate the p-value.

PROCEDURE – MONTE CARLO HYPOTHESIS TESTING (P-VALUE)

1. For a random sample of size n to be used in a statistical hypothesis test, calculate the observed value of the test statistic, t_o.
2. Decide on a pseudo-population that reflects the characteristics of the population under the null hypothesis.
3. Obtain a random sample of size n from the pseudo-population.
4. Calculate the value of the test statistic using the random sample in step 3 and record it as t_i.

5. Repeat steps 3 and 4 for M trials. We now have values $t_1, ..., t_M$, that serve as an estimate of the distribution of the test statistic, T, when the null hypothesis is true.

6. Estimate the p-value using the distribution found in step 5, using the following.

Lower Tail Test:

$$\hat{p}\text{-value} = \frac{\#(t_i \leq t_o)}{M}; \qquad i = 1, ..., M$$

Upper Tail Test:

$$\hat{p}\text{-value} = \frac{\#(t_i \geq t_o)}{M}; \qquad i = 1, ..., M$$

7. If $\hat{p}$-value $\leq \alpha$, then reject the null hypothesis.

Example 7.7

We return to the situation in Example 7.6 and apply Monte Carlo simulation to the p-value approach to hypothesis testing. Just to change things a bit, we use the sample mean as our test statistic.

```
% Let's change the test statistic to xbar.
Tobs = mean(mcdata);
% Number of Monte Carlo trials.
M = 1000;
% Start the simulation.
Tm = zeros(1,M);
for i = 1:M
    % Generate a random sample under H_0.
    xs = sigma*randn(1,n) + 454;
    Tm(i) = mean(xs);
end
```

We find the estimated p-value by counting the number of observations in **Tm** that are below the value of the observed value of the test statistic and dividing by M.

```
% Get the p-value. This is a lower tail test.
% Find all of the values from the simulation that are
% below the observed value of the test statistic.
ind = find(Tm <= Tobs);
pvalhat = length(ind)/M;
```

We have an estimated *p*-value given by 0.007. If the significance level of our test is $\alpha = 0.05$, then we would reject the null hypothesis.
❑

Monte Carlo Assessment of Hypothesis Testing

Monte Carlo simulation can be used to evaluate the performance of an inference model or hypothesis test in terms of the Type I error and the Type II error. For some statistics, such as the sample mean, these errors can be determined analytically. However, what if we have an inference test where the assumptions of the standard methods might be violated or the analytical methods cannot be applied? For instance, suppose we choose the critical value by using a normal approximation (when our test statistic is *not* normally distributed), and we need to assess the results of doing that? In these situations, we can use Monte Carlo simulation to estimate the Type I and the Type II error.

We first outline the procedure for estimating the Type I error. Because the Type I error occurs when we reject the null hypothesis test when it is true, we must sample from the pseudo-population that represents H_0.

PROCEDURE – MONTE CARLO ASSESSMENT OF TYPE I ERROR

1. Determine the pseudo-population or distribution when the null hypothesis is *true*.
2. Generate a random sample of size *n* from this pseudo-population.
3. Perform the hypothesis test using the critical value.
4. Determine whether a Type I error has been committed. In other words, was the null hypothesis rejected? We know that it should
   · not be rejected because we are sampling from the distribution according to the null hypothesis. Record the result for this trial as

$$I_i = \begin{cases} 1; & \text{Type I error is made} \\ 0; & \text{Type I error is not made.} \end{cases}$$

5. Repeat steps 2 through 4 for *M* trials.
6. The probability of making a Type I error is

$$\hat{\alpha} = \frac{1}{M}\sum_{i=1}^{M} I_i. \tag{7.9}$$

Note that in step 6, this is the same as calculating the proportion of times the null hypothesis is falsely rejected out of M trials. This provides an estimate of the significance level of the test for a given critical value.

The procedure is similar for estimating the Type II error of a hypothesis test. However, this error is determined by sampling from the distribution when the null hypothesis is false. There are many possibilities for the Type II error, and the analyst should investigate the Type II error for those alternative hypotheses that are of interest.

PROCEDURE – MONTE CARLO ASSESSMENT OF TYPE II ERROR

1. Determine a pseudo-population or distribution of interest where the null hypothesis is *false*.
2. Generate a random sample of size n from this pseudo-population.
3. Perform the hypothesis test using the significance level α and corresponding critical value.
4. Note whether a Type II error has been committed; i.e., was the null hypothesis *not* rejected? Record the result for this trial as

$$I_i = \begin{cases} 1; & \text{Type II error is made} \\ 0; & \text{Type II error is not made.} \end{cases}$$

5. Repeat steps 2 through 4 for M trials.
6. The probability of making a Type II error is

$$\hat{\beta} = \frac{1}{M} \sum_{i=1}^{M} I_i. \tag{7.10}$$

The Type II error rate is estimated using the proportion of times the null hypothesis is not rejected (when it should be) out of M trials.

Example 7.8

For the hypothesis test in Example 7.6, we had a critical value (from theory) of -1.645. We can estimate the significance level of the test using the following steps:

```
M = 1000;
alpha = 0.05;
% Get the critical value, using z as test statistic.
cv = norminv(alpha,0,1);
% Start the simulation.
Im = 0;
```

```
for i = 1:M
  % Generate a random sample under H_0.
  xs = sigma*randn(1,n) + 454;
  Tm = (mean(xs)-454)/sigxbar;
  if Tm <= cv    % then reject H_0
    Im = Im +1;
  end
end
alphahat = Im/M;
```

A critical value of -1.645 in this situation corresponds to a desired probability of Type I error of 0.05. From this simulation, we get an estimated value of 0.045, which is very close to the theoretical value. We now check the Type II error in this test. Note that we now have to sample from the alternative hypotheses of interest.

```
% Now check the probability of Type II error.
% Get some alternative hypotheses:
mualt = 445:458;
betahat = zeros(size(mualt));
for j = 1:length(mualt)
    Im = 0;
    % Get the true mean.
    mu = mualt(j);
    for i = 1:M
        % Generate a sample from H_1.
        xs = sigma*randn(1,n) + mu;
        Tm = (mean(xs)-454)/sigxbar;
        if Tm > cv    % Then did not reject H_0.
            Im = Im +1;
        end
    end
    betahat(j) = Im/M;
end
% Get the estimated power.
powhat = 1-betahat;
```

We plot the estimated power as a function of μ in Figure 7.5. As expected, as the true value for μ gets closer to 454 (the mean under the null hypothesis), the power of the test decreases.
❑

An important point to keep in mind about the Monte Carlo simulations discussed in this section is that the experiment is applicable only for the situation that has been simulated. For example, when we assess the Type II error in Example 7.8, it is appropriate only for those alternative hypotheses, sample size, and critical value. What would be the probability of Type II error if some other departure from the null hypothesis is used in the simulation?

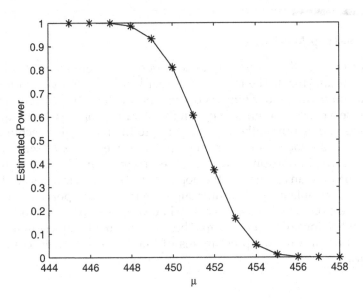

FIGURE 7.5
Here is the curve for the estimated power corresponding to the hypothesis test of Example 7.8.

In other cases, we might need to know whether the distribution of the statistic changes with sample size or skewness in the population or some other characteristic of interest. These variations are easily investigated using multiple Monte Carlo experiments.

One quantity that the researcher must determine is the number of trials that are needed in Monte Carlo simulations. This often depends on the computing assets that are available. If time and computer resources are not an issue, then M should be made as large as possible. Hope [1968] showed that results from a Monte Carlo simulation are unbiased for any M, under the assumption that the programming is correct.

Mooney [1997] states that there is no general theory that governs the number of trials in Monte Carlo simulation. However, he recommends the following general guidelines. The researcher should first use a small number of trials and ensure that the program is working properly. Once the code has been checked, the simulation or experiments can be run for very large M. Most simulations would have $M > 1000$, but M between 10,000 and 25,000 is not uncommon. One important guideline for determining the number of trials is the purpose of the simulation. If the tail of the distribution is of interest (e.g., estimating Type I error, getting p-values, etc.), then more trials are needed to ensure that there will be a good estimate of that area.

7.4 Bootstrap Methods

The treatment of the bootstrap methods described here comes from Efron and Tibshirani [1993]. The interested reader is referred to that text for more information on the underlying theory behind the bootstrap. There does not seem to be a consistent terminology in the literature for what techniques are considered bootstrap methods. Some refer to the resampling techniques of the previous section as bootstrap methods. Here, we use *bootstrap* to refer to Monte Carlo simulations that treat the original sample as the pseudo-population or as an estimate of the population. Thus, we now resample from the original sample instead of sampling from the pseudo-population.

In this section, we discuss the general bootstrap methodology, followed by some applications of the bootstrap. These include bootstrap estimates of the standard error, bootstrap estimates of bias, and bootstrap confidence intervals.

General Bootstrap Methodology

The bootstrap is a method of Monte Carlo simulation where no parametric assumptions are made about the underlying population that generated the random sample. Instead, we use the sample as an estimate of the population. This estimate is called the empirical distribution $\hat{F}$ where each x_i has probability mass $1/n$. Thus, each x_i has the same likelihood of being selected in a new sample taken from $\hat{F}$.

When we use $\hat{F}$ as our pseudo-population, then we resample *with replacement* from the original sample $\mathbf{x} = (x_1, ..., x_n)$. We denote the new sample obtained in this manner by $\mathbf{x}^* = (x_1^*, ..., x_n^*)$. Since we are sampling with replacement from the original sample, there is a possibility that some points x_i will appear more than once in $\mathbf{x}^*$ or maybe not at all. We are looking at the univariate situation, but the bootstrap concepts can also be applied in the d-dimensional case.

A small example serves to illustrate these ideas. Let's say that our random sample consists of the four numbers $\mathbf{x} = (5, 8, 3, 2)$. The following are possible samples $\mathbf{x}^*$ when we sample with replacement from $\mathbf{x}$:

$$\mathbf{x}^{*1} = (x_4, x_4, x_2, x_1) = (2, 2, 8, 5)$$

$$\mathbf{x}^{*2} = (x_4, x_2, x_3, x_4) = (2, 8, 3, 2).$$

We use the notation $\mathbf{x}^{*b}$, $b = 1, ..., B$ for the b-th bootstrap data set.

In many situations, the analyst is interested in estimating some parameter θ by calculating a statistic from the random sample. We denote this estimate by

$$\hat{\theta} = T = t(x_1, ..., x_n). \qquad (7.11)$$

We might also like to determine the standard error in the estimate $\hat{\theta}$ and the bias. The bootstrap method can provide an estimate of this when analytical methods fail. The method is also suitable for situations when the estimator $\hat{\theta} = t(x)$ is complicated.

To get estimates of bias or standard error of a statistic, we obtain B bootstrap samples by sampling with replacement from the original sample. We calculate the same statistic for every bootstrap sample to obtain the **bootstrap replications** of $\hat{\theta}$, as follows

$$\hat{\theta}^{*b} = t(x^{*b}); \qquad b = 1, ..., B. \qquad (7.12)$$

These B bootstrap replicates provide us with an estimate of the distribution of $\hat{\theta}$. This is similar to what we did in the previous section, except that we are not making any assumptions about the distribution for the original sample. Once we have the bootstrap replicates in Equation 7.12, we can use them to understand the distribution of the estimate.

The steps for the basic bootstrap methodology are given here, with detailed procedures for finding specific characteristics of $\hat{\theta}$ provided later. The issue of how large to make B is addressed later with each application of the bootstrap that we present.

PROCEDURE – BASIC BOOTSTRAP

1. Given a random sample, $x = (x_1, ..., x_n)$, calculate $\hat{\theta}$.
2. Sample with replacement from the original sample to get $x^{*b} = (x_1^{*b}, ..., x_n^{*b})$.
3. Calculate the same statistic using the bootstrap sample in step 2 to get $\hat{\theta}^{*b}$.
4. Repeat steps 2 through 3, B times.
5. Use this estimate of the distribution of $\hat{\theta}$ (i.e., the bootstrap replicates) to obtain the desired characteristic (e.g., standard error, bias or confidence interval).

Efron and Tibshirani [1993] discuss a method called the **parametric bootstrap**. In this case, the data analyst makes an assumption about the distribution that generated the original sample. Parameters for that distribution are estimated from the sample, and resampling (in step 2) is done using the assumed distribution and the estimated parameters. The parametric bootstrap is closer to the Monte Carlo methods described in the previous section.

For instance, say we have reason to believe that the data come from an exponential distribution with parameter λ. We need to estimate the variance and use

$$\hat{\theta} = \frac{1}{n}\sum_{i=1}^{n}(x_i - \bar{x})^2 \tag{7.13}$$

as the estimator. We can use the parametric bootstrap as outlined above to understand the behavior of $\hat{\theta}$. Since we assume an exponential distribution for the data, we estimate the parameter λ from the sample to get $\hat{\lambda}$. We then resample from an exponential distribution with parameter $\hat{\lambda}$ to get the bootstrap samples. The reader is asked to implement the parametric bootstrap in the exercises.

Bootstrap Estimate of Standard Error

When our goal is to estimate the standard error of $\hat{\theta}$ using the bootstrap method, we proceed as outlined in the previous procedure. Once we have the estimated distribution for $\hat{\theta}$, we use it to estimate the standard error for $\hat{\theta}$. This estimate is given by

$$\hat{SE}_B(\hat{\theta}) = \left\{\frac{1}{B-1}\sum_{b=1}^{B}\left(\hat{\theta}^{*b} - \overline{\hat{\theta}^{*}}\right)^2\right\}^{\frac{1}{2}}, \tag{7.14}$$

where

$$\overline{\hat{\theta}^{*}} = \frac{1}{B}\sum_{b=1}^{B}\hat{\theta}^{*b}. \tag{7.15}$$

Note that Equation 7.14 is just the sample standard deviation of the bootstrap replicates, and Equation 7.15 is the sample mean of the bootstrap replicates.

Efron and Tibshirani [1993] show that the number of bootstrap replicates B should be between 50 and 200 when estimating the standard error of a statistic. Often the choice of B is dictated by the computational complexity of $\hat{\theta}$, the sample size n, and the computer resources that are available. Even using a small value of B, say $B = 25$, the analyst will gain information about the variability of $\hat{\theta}$. In most cases, taking more than 200 bootstrap replicates to estimate the standard error is unnecessary.

The procedure for finding the bootstrap estimate of the standard error is given here and is illustrated in Example 7.9.

PROCEDURE – BOOTSTRAP ESTIMATE OF THE STANDARD ERROR

1. Given a random sample, $\mathbf{x} = (x_1, ..., x_n)$, calculate the statistic $\hat{\theta}$.
2. Sample with replacement from the original sample to get $\mathbf{x}^{*b} = (x_1^{*b}, ..., x_n^{*b})$.
3. Calculate the same statistic using the sample in step 2 to get the bootstrap replicates, $\hat{\theta}^{*b}$.
4. Repeat steps 2 through 3, B times.
5. Estimate the standard error of $\hat{\theta}$ using Equations 7.14 and 7.15.

Example 7.9

The lengths of the forearm (in inches) of 140 adult males are contained in the file **forearm** [Hand, et al., 1994]. We use these data to estimate the skewness of the population. We then estimate the standard error of this statistic using the bootstrap method. First we load the data and calculate the skewness.

```
load forearm
% Sample with replacement from this.
% First get the sample size.
n = length(forearm);
B = 100;% number of bootstrap replicates
% Get the value of the statistic of interest.
theta = skewness(forearm);
```

The estimated skewness in the **forearm** data is –0.11. To implement the bootstrap, we use the MATLAB Statistics Toolbox function **unidrnd** to sample with replacement from the original sample. The corresponding function from the Computational Statistics Toolbox can also be used. The output from this function will be indices from 1 to n that point to what observations have been selected for the bootstrap sample.

```
% Use unidrnd to get the indices to the resamples.
% Note that each column corresponds to indices
% for a bootstrap resample.
inds = unidrnd(n,n,B);
% Extract these from the data.
xboot = forearm(inds);
% We can get the skewness for each column using the
% MATLAB Statistics Toolbox function skewness.
thetab = skewness(xboot);
seb = std(thetab);
```

From this we get an estimated standard error in the skewness of 0.14. Efron and Tibshirani [1993] recommend that one look at histograms of the bootstrap replicates as a useful tool for understanding the distribution of $\hat{\theta}$. We show the histogram in Figure 7.6.

The MATLAB Statistics Toolbox has a function called **bootstrp** that returns the bootstrap replicates. We now show how to get the bootstrap estimate of standard error using this function.

```
% Now show how to do it with MATLAB Statistics Toolbox
% function: bootstrp.
Bmat = bootstrp(B,'skewness',forearm);
% What we get back are the bootstrap replicates.
% Get an estimate of the standard error.
sebmat = std(Bmat);
```

Note that one of the arguments to **bootstrp** is a string representing the function that calculates the statistic of interest. From this, we get an estimated standard error of 0.12.
❑

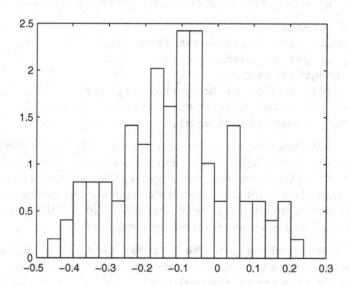

FIGURE 7.6
*This is a histogram for the bootstrap replicates in Example 7.9. This shows the estimated distribution of the sample skewness of the **forearm** data.*

Bootstrap Estimate of Bias

The standard error of an estimate is one measure of its performance. Bias is another quantity that measures the statistical accuracy of an estimate. From Chapter 3, the bias is defined as the difference between the expected value of the statistic and the parameter

$$\text{bias}(T) = E[T] - \theta. \tag{7.16}$$

The expectation in Equation 7.16 is taken with respect to the true distribution F. To get the bootstrap estimate of bias, we use the empirical distribution $\hat{F}$ as before. We resample from the empirical distribution and calculate the statistic using each bootstrap resample, yielding the bootstrap replicates $\hat{\theta}^{*b}$. We use these to estimate the bias from the following:

$$\widehat{\text{bias}}_B = \overline{\hat{\theta}^*} - \hat{\theta}, \tag{7.17}$$

where $\overline{\hat{\theta}^*}$ is given by the mean of the bootstrap replicates (Equation 7.15).

Presumably, one is interested in the bias in order to correct for it. The bias-corrected estimator is given by

$$\widehat{\theta} = \hat{\theta} - \widehat{\text{bias}}_B. \tag{7.18}$$

Using Equation 7.17 in Equation 7.18, we have

$$\widehat{\theta} = 2\hat{\theta} - \overline{\hat{\theta}^*}. \tag{7.19}$$

More bootstrap samples are needed to estimate the bias than are required to estimate the standard error. Efron and Tibshirani [1993] recommend that $B \geq 400$.

It is useful to have an estimate of the bias for $\hat{\theta}$, but caution should be used when correcting for the bias. Equation 7.19 will hopefully yield a less biased estimate, but it could turn out that $\widehat{\theta}$ will have a larger variation or standard error. It is recommended that if the estimated bias is small relative to the estimate of standard error (both of which can be estimated using the bootstrap method), then the analyst should not correct for the bias [Efron and Tibshirani, 1993]. However, if this is not the case, then perhaps some other, less biased, estimator should be used to estimate the parameter θ.

PROCEDURE – BOOTSTRAP ESTIMATE OF THE BIAS

1. Given a random sample, $\mathbf{x} = (x_1, ..., x_n)$, calculate the statistic $\hat{\theta}$.

2. Sample with replacement from the original sample to get
   $\mathbf{x}^{*b} = (x_1^{*b}, \ldots, x_n^{*b})$.

3. Calculate the same statistic using the sample in step 2 to get the
   bootstrap replicates, $\hat{\theta}^{*b}$.

4. Repeat steps 2 through 3, B times.

5. Using the bootstrap replicates, calculate $\overline{\hat{\theta}^*}$.

6. Estimate the bias of $\hat{\theta}$ using Equation 7.17.

Example 7.10

We return to the **forearm** data of Example 7.9, where now we want to
estimate the bias in the sample skewness. We use the same bootstrap
replicates as before, so all we have to do is to calculate the bias using Equation
7.17.

```
% Use the same replicates from before.
% Evaluate the mean using Equation 7.15.
meanb = mean(thetab);
% Now estimate the bias using Equation 7.17.
biasb = meanb - theta;
```

We have an estimated bias of −0.011. Note that this is small relative to the
standard error.
❑

In the next chapter, we discuss another method for estimating the bias and
the standard error of a statistic called the jackknife. The jackknife method is
related to the bootstrap. However, since it is based on the reuse or
partitioning of the original sample rather than resampling, we do not include
it here.

Bootstrap Confidence Intervals

There are several ways of constructing confidence intervals using the
bootstrap. We discuss three of them here: the standard interval, the
bootstrap-*t* interval, and the percentile method. Because it uses the jackknife
procedure, an improved bootstrap confidence interval called the BC_a will be
presented in the next chapter.

Bootstrap Standard Confidence Interval

The *bootstrap standard confidence interval* is based on the parametric form
of the confidence interval that was discussed in Section 7.2. We showed that
the $(1 - \alpha) \cdot 100\%$ confidence interval for the mean can be found using

$$P\left(\overline{X} - z^{(1-\alpha/2)}\frac{\sigma}{\sqrt{n}} < \mu < \overline{X} - z^{(\alpha/2)}\frac{\sigma}{\sqrt{n}}\right) = 1 - \alpha. \qquad (7.20)$$

Similar to this, the bootstrap standard confidence interval is given by

$$\left(\hat{\theta} - z^{(1-\alpha/2)}SE_{\hat{\theta}}, \hat{\theta} - z^{(\alpha/2)}SE_{\hat{\theta}}\right), \qquad (7.21)$$

where $SE_{\hat{\theta}}$ is the standard error for the statistic $\hat{\theta}$ obtained using the bootstrap [Mooney and Duval, 1993]. The confidence interval in Equation 7.21 can be used when the distribution for $\hat{\theta}$ is normally distributed or the normality assumption is plausible. This is easily coded in MATLAB using previous results and is left as an exercise for the reader.

Bootstrap-t Confidence Interval

The second type of confidence interval using the bootstrap is called the **bootstrap-t**. We first generate B bootstrap samples, and for each bootstrap sample the following quantity is computed:

$$z^{*b} = \frac{\hat{\theta}^{*b} - \hat{\theta}}{\hat{SE}^{*b}}. \qquad (7.22)$$

As before, $\hat{\theta}^{*b}$ is the bootstrap replicate of $\hat{\theta}$, but $\hat{SE}^{*b}$ is the estimated standard error of $\hat{\theta}^{*b}$ for that bootstrap sample. If a formula exists for the standard error of $\hat{\theta}^{*b}$, then we can use that to determine the denominator of Equation 7.22. For instance, if $\hat{\theta}$ is the sample mean, then we can calculate the standard error as explained in Chapter 3. However, in most situations where we have to resort to using the bootstrap, these formulas are not available. One option is to use the bootstrap method of finding the standard error, keeping in mind that you are estimating the standard error of $\hat{\theta}^{*b}$ using the bootstrap sample $\mathbf{x}^{*b}$. In other words, one resamples with replacement from the bootstrap sample $\mathbf{x}^{*b}$ to get an estimate of $\hat{SE}^{*b}$.

Once we have the B bootstrapped z^{*b} values from Equation 7.22, the next step is to estimate the quantiles needed for the endpoints of the interval. The $\alpha/2$-th quantile, denoted by $\hat{t}^{(\alpha/2)}$ of the z^{*b}, is estimated by

$$\alpha/2 = \frac{\#(z^{*b} \le \hat{t}^{(\alpha/2)})}{B}. \qquad (7.23)$$

This says that the estimated quantile is the $\hat{t}^{(\alpha/2)}$ such that $100 \cdot \alpha/2\%$ of the points z^{*b} are less than this number. For example, if $B = 100$ and

$\alpha/2 = 0.05$, then $\hat{t}^{(0.05)}$ could be estimated as the fifth largest value of the z^{*b} ($B \cdot \alpha/2 = 100 \cdot 0.05 = 5$). Alternatively, one could estimate the quantiles as discussed in Chapter 3.

We are now ready to calculate the bootstrap-t confidence interval. This is given by

$$(\hat{\theta} - \hat{t}^{(1-\alpha/2)} \times \hat{SE}_{\hat{\theta}}, \hat{\theta} - \hat{t}^{(\alpha/2)} \times \hat{SE}_{\hat{\theta}}), \tag{7.24}$$

where $\hat{SE}_{\hat{\theta}}$ is an estimate of the standard error of $\hat{\theta}$. The bootstrap-t interval is suitable for location statistics such as the mean or quantiles. However, its accuracy for more general situations is questionable [Efron and Tibshirani, 1993]. The next method based on the bootstrap percentiles is more reliable.

PROCEDURE – BOOTSTRAP-T CONFIDENCE INTERVAL

1. Given a random sample, $\mathbf{x} = (x_1, ..., x_n)$, calculate $\hat{\theta}$.
2. Sample with replacement from the original sample to get $\mathbf{x}^{*b} = (x_1^{*b}, ..., x_n^{*b})$.
3. Calculate the same statistic using the sample in step 2 to get $\hat{\theta}^{*b}$.
4. Use the bootstrap sample $\mathbf{x}^{*b}$ to get the standard error of $\hat{\theta}^{*b}$. This can be calculated using a formula or estimated by the bootstrap.
5. Calculate z^{*b} using the information found in steps 3 and 4.
6. Repeat steps 2 through 5, B times, where $B \geq 1000$.
7. Order the z^{*b} from smallest to largest. Find the quantiles $\hat{t}^{(1-\alpha/2)}$ and $\hat{t}^{(\alpha/2)}$.
8. Estimate the standard error $\hat{SE}_{\hat{\theta}}$ of $\hat{\theta}$ using the B bootstrap replicates of $\hat{\theta}^{*b}$ (from step 3).
9. Use Equation 7.24 to get the confidence interval.

The number of bootstrap replicates that are needed is quite large for confidence intervals. It is recommended that B should be 1000 or more. If no formula exists for calculating the standard error of $\hat{\theta}^{*b}$, then the bootstrap method can be used. This means that there are two levels of bootstrapping: one for finding the $\hat{SE}^{*b}$ and one for finding the z^{*b}, which can greatly increase the computational burden. For example, say that $B = 1000$, and we use 50 bootstrap replicates to find $\hat{SE}^{*b}$, then this results in a total of 50,000 resamples.

Example 7.11

Say we are interested in estimating the variance of the **forearm** data, and we decide to use the following statistic:

$$\hat{\sigma}^2 = \frac{1}{n}\sum_{i=1}^{n}(X_i - \bar{X})^2,$$

which is the sample second central moment. We write our own simple function called **mom** (included in the Computational Statistics Toolbox) to estimate this.

```
% This function will calculate the sample second
% central moment for a given sample vector x.
function mr = mom(x)
n = length(x);
mu = mean(x);
mr = (1/n)*sum((x-mu).^2);
```

We use this function as an input argument to **bootstrp** to get the bootstrap-*t* confidence interval. The MATLAB code given next also shows how to get the bootstrap estimate of standard error for each bootstrap sample. First we load the data and get the observed value of the statistic.

```
load forearm
n = length(forearm);
alpha = 0.1;
B = 1000;
thetahat = mom(forearm);
```

Now we get the bootstrap replicates using the function **bootstrp**. One of the optional output arguments from this function is a matrix of indices for the resamples. As shown next, each column of the output **bootsam** contains the indices to a bootstrap sample. We loop through all of the bootstrap samples to estimate the standard error of the bootstrap replicate using that resample.

```
% Get the bootstrap replicates and samples.
[bootreps, bootsam] = bootstrp(B,'mom',forearm);
% Set up some storage space for the SE's.
sehats = zeros(size(bootreps));
% Each column of bootsam contains indices
% to a bootstrap sample.
for i = 1:B
   % Extract the sample from the data.
     xstar = forearm(bootsam(:,i));
   bvals(i) = mom(xstar);
   % Do bootstrap using that sample to estimate SE.
   sehats(i) = std(bootstrp(25,'mom',xstar));
end
zvals = (bootreps - thetahat)./sehats;
```

Then we get the estimate of the standard error that we need for the endpoints of the interval.

```
% Estimate the SE using the bootstrap.
SE = std(bootreps);
```

Now we get the quantiles that we need for the interval given in Equation 7.24 and calculate the interval.

```
% Get the quantiles.
k = B*alpha/2;
szval = sort(zvals);
tlo = szval(k);
thi = szval(B-k);
% Get the endpoints of the interval.
blo = thetahat - thi*SE;
bhi = thetahat - tlo*SE;
```

The bootstrap-t interval for the variance of the **forearm** data is $(1.00, 1.57)$.
❏

Bootstrap Percentile Interval

An improved bootstrap confidence interval is based on the quantiles of the distribution of the bootstrap replicates. This technique has the benefit of being more stable than the bootstrap-t, and it also enjoys better theoretical coverage properties [Efron and Tibshirani, 1993]. The *bootstrap percentile confidence interval* is

$$(\hat{\theta}_B^{*(\alpha/2)}, \hat{\theta}_B^{*(1-\alpha/2)}), \qquad (7.25)$$

where $\hat{\theta}_B^{*(\alpha/2)}$ is the $\alpha/2$ quantile in the bootstrap distribution of $\hat{\theta}^*$. For example, if $\alpha/2 = 0.025$ and $B = 1000$, then $\hat{\theta}_B^{*(0.025)}$ is the $\hat{\theta}^{*b}$ in the 25th position of the ordered bootstrap replicates. Similarly, $\hat{\theta}_B^{*(0.975)}$ is the replicate in position 975. As discussed previously, some other suitable estimate for the quantile can be used.

The procedure is the same as the general bootstrap method, making it easy to understand and to implement. We outline the steps next.

PROCEDURE – BOOTSTRAP PERCENTILE INTERVAL

1. Given a random sample, $\mathbf{x} = (x_1, ..., x_n)$, calculate $\hat{\theta}$.
2. Sample with replacement from the original sample to get $\mathbf{x}^{*b} = (x_1^{*b}, ..., x_n^{*b})$.
3. Calculate the same statistic using the sample in step 2 to get the bootstrap replicates, $\hat{\theta}^{*b}$.

4. Repeat steps 2 through 3, B times, where $B \geq 1000$.

5. Order the $\hat{\theta}^{*b}$ from smallest to largest.

6. Calculate $B \cdot \alpha/2$ and $B \cdot (1 - \alpha/2)$.

7. The lower endpoint of the interval is given by the bootstrap replicate that is in the $B \cdot \alpha/2$-th position of the ordered $\hat{\theta}^{*b}$, and the upper endpoint is given by the bootstrap replicate that is in the $B \cdot (1 - \alpha/2)$-th position of the same ordered list. Alternatively, using quantile notation, the lower endpoint is the estimated quantile $\hat{q}_{\alpha/2}$ and the upper endpoint is the estimated quantile $\hat{q}_{1-\alpha/2}$, where the estimates are taken from the bootstrap replicates.

Example 7.12

Let's find the bootstrap percentile interval for the same **forearm** data. The confidence interval is easily found from the bootstrap replicates, as shown next.

```
% Use Statistics Toolbox function
% to get the bootstrap replicates.
bvals = bootstrp(B,'mom',forearm);
% Find the upper and lower endpoints
k = B*alpha/2;
sbval = sort(bvals);
blo = sbval(k);
bhi = sbval(B-k);
```

This interval is given by $(1.03, 1.45)$, which is slightly narrower than the bootstrap-*t* interval from Example 7.11.
□

So far, we discussed three types of bootstrap confidence intervals. The standard interval is the easiest and assumes that $\hat{\theta}$ is normally distributed. The bootstrap-*t* interval estimates the standardized version of $\hat{\theta}$ from the data, avoiding the normality assumptions used in the standard interval. The percentile interval is simple to calculate and obtains the endpoints directly from the bootstrap estimate of the distribution for $\hat{\theta}$ It has another advantage in that it is range-preserving. This means that if the parameter θ can take on values in a certain range, then the confidence interval will reflect that. This is not always the case with the other intervals.

According to Efron and Tibshirani [1993], the bootstrap-*t* interval has good coverage probabilities, but does not perform well in practice. The bootstrap percentile interval is more dependable in most situations, but does not enjoy the good coverage property of the bootstrap-*t* interval. There is another bootstrap confidence interval, called the BC_a interval, that has both good coverage and is dependable. This interval is described in the next chapter.

The bootstrap estimates of bias and standard error are also random variables, and they have their own error associated with them. So, how accurate are they? In the next chapter, we discuss how one can use the jackknife method to evaluate the error in the bootstrap estimates.

As with any method, the bootstrap is not appropriate in every situation. When analytical methods are available to understand the uncertainty associated with an estimate, then those are more efficient than the bootstrap. In what situations should the analyst use caution in applying the bootstrap? One important assumption that underlies the theory of the bootstrap is the notion that the empirical distribution function is representative of the true population distribution. If this is not the case, then the bootstrap will not yield reliable results. For example, this can happen when the sample size is small or the sample was not gathered using appropriate random sampling techniques. Chernick [1999] describes other examples from the literature where the bootstrap should not be used.

7.5 MATLAB® Code

We include several functions with the Computational Statistics Toolbox that implement some of the bootstrap techniques discussed in this chapter. These are listed in Table 7.2. Like **bootstrp**, these functions have an input argument that specifies a MATLAB function to calculate the statistic.

TABLE 7.2

List of MATLAB Functions for Chapter 7

Purpose	MATLAB Function
General bootstrap: resampling, estimates of standard error, and bias	**csboot** **bootstrp**
Constructing bootstrap confidence intervals	**csbootint** **csbooperint** **csbootbca** **bootci**

As we saw in the examples, the MATLAB Statistics Toolbox has a function called **bootstrp** that will return the bootstrap replicates from the input argument **bootfun** (e.g., **mean**, **std**, **var**, etc.). It takes an input data set, finds the bootstrap resamples, applies the **bootfun** to the resamples, and stores the replicate in the first row of the output argument. The user can get two outputs from the function: the bootstrap replicates and the indices that correspond to the points selected in the resample.

The Statistics Toolbox now has a function called **bootci** that will return the bootstrap confidence interval. Several types of intervals can be obtained using this function, including the basic percentile method and the normal approximation method. See the **help** documentation on **bootci** for more options.

Other software exists for Monte Carlo simulation as applied to statistics. The Efron and Tibshirani [1993] book describes S code for implementing the bootstrap. This code, written by the authors, can be downloaded from the statistics archive at Carnegie-Mellon University that was mentioned in Chapter 1. Another software package that has some of these capabilities is called Resampling Stats® [Simon, 1999], and information on this can be found at **www.resample.com**. Routines are available from Resampling Stats for [Kaplan, 1999] Excel.

7.6 Further Reading

Mooney [1997] describes Monte Carlo simulation for inferential statistics that is written in a way that is accessible to most data analysts. It has some excellent examples of using Monte Carlo simulation for hypothesis testing using multiple experiments, assessing the behavior of an estimator, and exploring the distribution of a statistic using graphical techniques. The text by Gentle [1998] has a chapter on performing Monte Carlo studies in statistics. He discusses how simulation can be considered as a scientific experiment and should be held to the same high standards. Hoaglin and Andrews [1975] provide guidelines and standards for reporting the results from computations. Efron and Tibshirani [1991] explain several computational techniques, written at a level accessible to most readers. Other articles describing Monte Carlo inferential methods can be found in Joeckel [1991], Hope [1968], Besag and Diggle [1977], Diggle and Gratton [1984], Efron [1979], Efron and Gong [1983], and Teichroew [1965].

There has been a lot of work in the literature on bootstrap methods. Perhaps the most comprehensive and easy to understand treatment of the topic can be found in Efron and Tibshirani [1993]. Efron's [1982] earlier monogram on resampling techniques describes the jackknife, the bootstrap, and cross-validation. A later book by Chernick [1999] gives an updated description of results in this area, and it also has an extensive bibliography on the bootstrap. Hall [1992] describes the connection between Edgeworth expansions and the bootstrap. A volume of papers on the bootstrap was edited by LePage and Billard [1992], where many applications of the bootstrap are explored. Politis, Romano, and Wolf [1999] present subsampling as an alternative to the bootstrap. A subset of articles that present the theoretical justification for the bootstrap are Efron [1981, 1985, 1987]. The paper by Boos and Zhang [2000] looks at a way to ease the computational burden of Monte Carlo estimation of the power of tests that uses resampling methods. For a nice discussion on the coverage of the bootstrap percentile confidence interval, see Polansky [1999]. A special issue of *Statistical Science* (Volume 18, Issue 2, 2003) was published to celebrate the silver anniversary of the bootstrap. The articles can be accessed at

http://projecteuclid.org/euclid.ss/1063994964

Exercises

7.1.   Repeat Example 7.1 where the population standard deviation for the travel times to work is $\sigma_X = 5$ minutes. Is $\bar{x} = 47.2$ minutes still consistent with the null hypothesis?

7.2.   Using the information in Example 7.3, plot the probability of Type II error as a function of μ. How does this compare with Figure 7.2?

7.3.   Would you reject the null hypothesis in Example 7.4 if $\alpha = 0.10$?

7.4.   Using the same value for the sample mean, repeat Example 7.3 for different sample sizes of $n = 50, 100, 200$. What happens to the curve showing the power as a function of the true mean as the sample size changes?

7.5.   Repeat Example 7.6 using a two-tail test. In other words, test for the alternative hypothesis that the mean is not equal to 454.

7.6.   Repeat Example 7.8 for larger M. Does the estimated Type I error get closer to the true value?

7.7.   Write MATLAB code that implements the parametric bootstrap. Test it using the **forearm** data. Assume that the normal distribution is a reasonable model for the data. Use your code to get a bootstrap estimate of the standard error and the bias of the coefficient of skewness and the coefficient of kurtosis. Get a bootstrap percentile interval for the sample central second moment using your parametric bootstrap approach.

7.8.   Write MATLAB code that will get the bootstrap standard confidence interval. Use it with the **forearm** data to get a confidence interval for the sample central second moment. Compare this interval with the ones obtained in the examples and in the previous problem.

7.9.   Use your program from Problem 7.8 and the **forearm** data to get a bootstrap confidence interval for the mean. Compare this to the theoretical one.

7.10. The **remiss** data set contains the remission times for 42 leukemia patients. Some of the patients were treated with the drug called 6-mercaptopurine (**mp**), and the rest were part of the control group (**control**). Use the techniques from Chapter 5 to help determine a suitable model (e.g., Weibull, exponential, etc.) for each group. Devise a Monte Carlo hypothesis test to test for the equality of means between the two groups [Hand, et al., 1994; Gehan, 1965]. Use the *p*-value approach.

7.11. Load the **lawpop** data set [Efron and Tibshirani, 1993]. These data contain the average scores on the LSAT (**lsat**) and the corresponding average undergraduate grade point average (**gpa**) for the 1973 freshman class at 82 law schools. Note that these data constitute the entire population. The data contained in **law** comprise a random sample of 15 of these classes. Obtain the true population variances for the **lsat** and the **gpa**. Use the sample in **law** to estimate the population variance using the sample central second moment. Get bootstrap estimates of the standard error and the bias in your estimate of the variance. Make some comparisons between the known population variance and the estimated variance.

7.12. Using the **lawpop** data, devise a test statistic to test for the significance of the correlation between the LSAT scores and the corresponding grade point averages. Get a random sample from the population, and use that sample to test your hypothesis. Do a Monte Carlo simulation of the Type I and Type II error of the test you devise.

7.13. In 1961, 16 states owned the retail liquor stores. In 26 others, the stores were owned by private citizens. The data contained in **whisky** reflect the price (in dollars) of a fifth of whisky from these 42 states. Note that this represents the population, not a sample. Use the **whisky** data to get an appropriate bootstrap confidence interval for the median price of whisky at the state-owned stores and the median price of whisky at the privately owned stores. First get the random sample from each of the populations, and then use the bootstrap with that sample to get the confidence intervals. Do a Monte Carlo study where you compare the confidence intervals for different sample sizes. Compare the intervals with the known population medians [Hand, et al., 1994].

7.14. The **quakes** data [Hand, et al., 1994] give the time in days between successive earthquakes. Use the bootstrap to get an appropriate confidence interval for the average time between earthquakes.

Chapter 8

Data Partitioning

8.1 Introduction

In this book, data partitioning refers to procedures where some observations from the sample are removed as part of the analysis. These techniques are used for the following purposes:

- To evaluate the accuracy of the model or classification scheme;
- To decide what is a reasonable model for the data;
- To find a smoothing parameter in density estimation;
- To estimate the bias and error in parameter estimation;
- And many others.

We start off with an example to motivate the reader. We have a sample where we measured the average atmospheric temperature and the corresponding amount of steam used per month [Draper and Smith, 1981]. Our goal in the analysis is to model the relationship between these variables. Once we have a model, we can use it to predict how much steam is needed for a given average monthly temperature. The model can also be used to gain understanding about the structure of the relationship between the two variables.

The problem then is deciding what model to use. To start off, one should always look at a scatterplot (or scatterplot matrix) of the data as discussed in Chapter 5. The scatterplot for these data is shown in Figure 8.1 and is examined in Example 8.3. We see from the plot that as the temperature increases, the amount of steam used per month decreases. It appears that using a line (i.e., a first degree polynomial) to model the relationship between the variables is not unreasonable. However, other models might provide a better fit. For example, a cubic or some higher degree polynomial might be a better model for the relationship between average temperature and steam usage.

So, how can we decide which model is better? To make that decision, we need to assess the accuracy of the various models. We could then choose the model that has the best accuracy or lowest error. In this chapter, we use the prediction error (see Equation 8.5) to measure the accuracy. One way to assess the error would be to observe new data (average temperature and corresponding monthly steam usage) and then determine what is the predicted monthly steam usage for the new observed average temperatures. We can compare this prediction with the true steam used and calculate the error. We do this for all of the proposed models and pick the model with the smallest error. The problem with this approach is that it is sometimes impossible to obtain new data, so all we have available to evaluate our models (or our statistics) is the original data set. In this chapter, we consider two methods that allow us to use the data already in hand for the evaluation of the models. These are ***cross-validation*** and the ***jackknife***.

Cross-validation is typically used to determine the classification error rate for pattern recognition applications or the prediction error when building models. In Chapter 10, we will see two applications of cross-validation where it is used to select the best classification tree and to estimate the misclassification rate. In this chapter, we show how cross-validation can be used to assess the prediction accuracy in a regression problem.

In the previous chapter, we covered the bootstrap method for estimating the bias and standard error of statistics. The jackknife procedure has a similar purpose and was developed prior to the bootstrap [Quenouille,1949]. The connection between the methods is well known and is discussed in the literature [Efron and Tibshirani, 1993; Efron, 1982; Hall, 1992]. We include the jackknife procedure here because it is more a data partitioning method than a simulation method such as the bootstrap. We return to the bootstrap at the end of this chapter, where we present another method of constructing bootstrap confidence intervals using the jackknife. In the last section, we show how the jackknife method can be used to assess the error in our bootstrap estimates.

8.2 Cross-Validation

Often, one of the jobs of a statistician or engineer is to create models using sample data, usually for the purpose of making predictions. For example, given a data set that contains the drying time and the tensile strength of batches of cement, can we model the relationship between these two variables? We would like to be able to predict the tensile strength of the cement for a given drying time that we will observe in the future. We must then decide what model best describes the relationship between the variables and estimate its accuracy.

Unfortunately, in many cases the naive researcher will build a model based on the data set and then use that same data to assess the performance of the model. The problem with this is that the model is being evaluated or tested with data it has already seen. Therefore, that procedure will yield an overly optimistic (i.e., low) prediction error (see Equation 8.5). Cross-validation is a technique that can be used to address this problem by iteratively partitioning the sample into two sets of data. One is used for building the model, and the other is used to test it.

We introduce cross-validation in a linear regression application, where we are interested in estimating the expected prediction error. We use linear regression to illustrate the cross-validation concept because it is a topic that most engineers and data analysts should be familiar with. However, before we describe the details of cross-validation, we briefly review the concepts in linear regression. We will return to this topic in Chapter 12.

Say we have a set of data, (X_i, Y_i), where X_i denotes a ***predictor variable*** and Y_i represents the corresponding ***response variable***. We are interested in modeling the dependency of Y on X. The easiest example of linear regression is in situations where we can fit a straight line between X and Y. In Figure 8.1, we show a scatterplot of 25 observed (X_i, Y_i) pairs [Draper and Smith, 1981]. The X variable represents the average atmospheric temperature measured in degrees Fahrenheit, and the Y variable corresponds to the pounds of steam used per month. From the scatterplot, we see that a straight line is a reasonable model for the relationship between these variables. We will use these data to illustrate linear regression.

The linear, first-order model is given by

$$Y = \beta_0 + \beta_1 X + \varepsilon, \tag{8.1}$$

where β_0 and β_1 are parameters that must be estimated from the data, and ε represents the error in the measurements. It should be noted that the word *linear* refers to the linearity of the parameters β_i. The *order* (or *degree*) of the model refers to the highest power of the predictor variable X. We know from elementary algebra that β_1 is the slope and β_0 is the *y*-intercept. As another example, we represent the linear, second-order model by

$$Y = \beta_0 + \beta_1 X + \beta_2 X^2 + \varepsilon. \tag{8.2}$$

To get the model in Equation 8.1, we need to estimate the parameters β_0 and β_1. Thus, the estimate of our model is

$$\hat{Y} = \hat{\beta}_0 + \hat{\beta}_1 X, \tag{8.3}$$

where $\hat{Y}$ denotes the predicted value of Y for some value of X, and $\hat{\beta}_0, \hat{\beta}_1$ are the estimated parameters. We do not go into the derivation of the estimators,

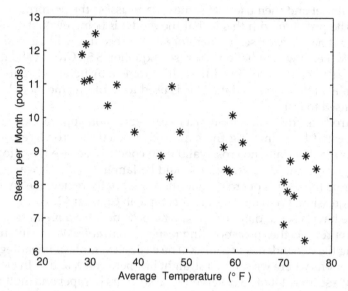

FIGURE 8.1
Scatterplot of a data set where we are interested in modeling the relationship between average temperature (the predictor variable) and the amount of steam used per month (the response variable). The scatterplot indicates that modeling the relationship with a straight line is reasonable.

since it can be found in most introductory statistics textbooks. Additionally, we are using the concept of model estimation in this chapter to illustrate data partitioning, so details are left for a later chapter.

Assume that we have a sample of observed predictor variables with corresponding responses. We denote these by (X_i, Y_i), $i = 1, ..., n$. The *least squares* fit is obtained by finding the values of the parameters that minimize the sum of the squared errors

$$RSE = \sum_{i=1}^{n} \varepsilon^2 = \sum_{i=1}^{n} (Y_i - (\beta_0 + \beta_1 X_i))^2, \tag{8.4}$$

where *RSE* denotes the *residual squared error*.

Estimates of the parameters $\hat{\beta}_0$ and $\hat{\beta}_1$ are easily obtained in MATLAB® using the function **polyfit**, and other methods available in MATLAB will be explored in Chapters 12 and 13. We use the function **polyfit** in Example 8.1 to model the linear relationship between the atmospheric temperature and the amount of steam used per month (see Figure 8.1).

Example 8.1

In this example, we show how to use the MATLAB function **polyfit** to fit a line to the **steam** data. The **polyfit** function takes three arguments: the observed **x** values, the observed **y** values, and the degree of the polynomial that we want to fit to the data. The following commands fit a polynomial of degree one to the steam data.

```
% Loads the vectors x and y.
load steam
% Fit a first degree polynomial to the data.
[p,s] = polyfit(x,y,1);
```

The output argument **p** is a vector of coefficients of the polynomial in decreasing order. So, in this case, the first element of **p** is the estimated slope $\hat{\beta}_1$, and the second element is the estimated y-intercept $\hat{\beta}_0$. The resulting model is

$$\hat{\beta}_0 = 13.62 \qquad \hat{\beta}_1 = -0.08.$$

We can use the **polyval** function to get predictions using this model. First, we have to get a domain of interest **xf**, and then we evaluate the estimated function over that domain.

```
% We can evaluate the polynomial over a
% domain of interest.
xf = linspace(min(x),max(x));
yf = polyval(p,xf);
```

Next, we produce a scatterplot of the original data with the fitted line superimposed on it.

```
% Now produce the plot for Figure 8.2
plot(x,y,'o',xf,yf,'-')
axis equal
xlabel('Average Temperature (\circ F)')
ylabel('Steam per Month (pounds)')
```

The predictions that would be obtained from the model (i.e., points on the line given by the estimated parameters) are shown in Figure 8.2, and we see that it seems to be a reasonable fit.
❑

The *prediction error* is defined as

$$PE = E[(Y - \hat{Y})^2], \tag{8.5}$$

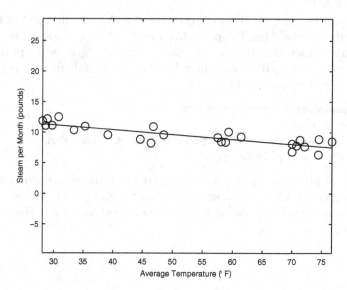

FIGURE 8.2
This figure shows a scatterplot of the **steam** *data along with the line obtained using* **polyfit**. *The estimate of the slope is* $\hat{\beta}_1 = -0.08$, *and the estimate of the y-intercept is* $\hat{\beta}_0 = 13.62$.

where the expectation is with respect to the true population. To estimate the error given by Equation 8.5, we need to test our model (obtained from **polyfit**) using an independent set of data that we denote by (x_i', y_i'). This means that we would take an observed (x_i', y_i') and obtain the estimate of $\hat{y}_i'$ using our model:

$$\hat{y}_i' = \hat{\beta}_0 + \hat{\beta}_1 x_i'. \tag{8.6}$$

We then compare $\hat{y}_i'$ with the true value of y_i'. Obtaining the outputs or $\hat{y}_i'$ from the model is easily done in MATLAB using the **polyval** function as shown in Example 8.2.

Say we have m independent observations (x_i', y_i') that we can use to test the model. We estimate the prediction error (Equation 8.5) using

$$\hat{PE} = \frac{1}{m}\sum_{i=1}^{m}(y_i' - \hat{y}_i')^2. \tag{8.7}$$

Equation 8.7 measures the average squared error between the predicted response obtained from the model and the true measured response. Other measures of error can be used, such as the absolute difference between the observed and predicted responses.

Example 8.2

We now show how to estimate the prediction error using Equation 8.7. We first choose some points from the **steam** data set and put them aside to use as an independent test sample. The rest of the observations are then used to obtain the model.

```
load steam
% Get the set that will be used to
% estimate the line.
indtest = 2:2:20; % Just pick some points.
xtest = x(indtest);
ytest = y(indtest);
% Now get the observations that will be
% used to fit the model.
xtrain = x;
ytrain = y;
% Remove the test observations.
xtrain(indtest) = [];
ytrain(indtest) = [];
```

The next step is to fit a first degree polynomial:

```
% Fit a first degree polynomial (the model)
% to the data.
[p,s] = polyfit(xtrain,ytrain,1);
```

We can use the MATLAB function **polyval** to get the predictions at the x values in the testing set and compare these to the observed y values in the testing set.

```
% Now get the predictions using the model and the
% testing data that was set aside.
yhat = polyval(p,xtest);
% The residuals are the difference between the true
% and the predicted values.
r = (ytest - yhat);
```

Finally, the estimate of the prediction error (Equation 8.7) is obtained as follows:

```
pe = mean(r.^2);
```

The estimated prediction error is $\hat{PE} = 0.91$. The reader is asked to explore this further in the exercises.
❏

What we just illustrated in Example 8.2 was a situation where we partitioned the data into one set for building the model and another for estimating the prediction error. This is perhaps not the best use of the data because we have all of the data available for evaluating the error in the model. We could repeat the above procedure, repeatedly partitioning the data into *many* training and testing sets. This is the fundamental idea underlying **cross-validation**.

The most general form of this procedure is called K-fold cross-validation. The basic concept is to split the data into K partitions of approximately equal size. One partition is reserved for testing, and the rest of the data are used for fitting the model. The test set is used to calculate the squared error $(y_i - \hat{y}_i)^2$. Note that the prediction $\hat{y}_i$ is from the model obtained using the current training set (one *without* the i-th observation in it). This procedure is repeated until all K partitions have been used as a test set. Note that we have n squared errors because each observation will be a member of one testing set. The average of these errors is the estimated expected prediction error.

In most situations, where the size of the data set is relatively small, the analyst can set $K = n$, so the size of the testing set is one. Since this requires fitting the model n times, this can be computationally expensive if n is large. We note, however, that there are efficient ways of doing this [Gentle 1998; Hjorth, 1994]. We outline the steps for cross-validation next and demonstrate this approach in Example 8.3.

PROCEDURE – CROSS-VALIDATION

1. Partition the data set into K partitions. For simplicity, we assume that $n = r \cdot K$, so there are r observations in each set.
2. Leave out one of the partitions for testing purposes.
3. Use the remaining $n - r$ data points for training (e.g., fit the model, build the classifier, estimate the probability density function, etc.).
4. Use the test set with the model and determine the error between the observed and predicted response.
5. Repeat steps 2 through 4 until all K partitions have been used as a test set.
6. Determine the average of the n errors.

Note that the error mentioned in step 4 depends on the application and the goal of the analysis [Hjorth, 1994]. For instance, in pattern recognition applications, this might be the cost of misclassifying a case. In the following example, we apply the cross-validation technique to help decide what type of model should be used for the **steam** data.

Example 8.3
In this example, we apply cross-validation to the modeling problem of Example 8.1. We fit linear, quadratic (degree 2) and cubic (degree 3) models to the data and compare their accuracy using the estimates of prediction error obtained from cross-validation.

```
% Set up the array to store the prediction errors.
n = length(x);
r1 = zeros(1,n);% store error - linear fit
r2 = zeros(1,n);% store error - quadratic fit
r3 = zeros(1,n);% store error - cubic fit
% Loop through all of the data. Remove one point at a
% time as the test point.
for i = 1:n
    xtest = x(i);% Get the test point.
    ytest = y(i);
    xtrain = x;% Get the points to build model.
    ytrain = y;
    xtrain(i) = [];% Remove test point.
    ytrain(i) = [];
    % Fit a first degree polynomial to the data.
    [p1,s] = polyfit(xtrain,ytrain,1);
    % Fit a quadratic to the data.
    [p2,s] = polyfit(xtrain,ytrain,2);
    % Fit a cubic to the data
    [p3,s] = polyfit(xtrain,ytrain,3);
    % Get the errors
    r1(i) = (ytest - polyval(p1,xtest)).^2;
    r2(i) = (ytest - polyval(p2,xtest)).^2;
    r3(i) = (ytest - polyval(p3,xtest)).^2;
end
```

We obtain the estimated prediction error of both models as follows:

```
% Get the prediction error for each one.
pe1 = mean(r1);
pe2 = mean(r2);
pe3 = mean(r3);
```

From this, we see that the estimated prediction error for the linear model is 0.86; the corresponding error for the quadratic model is 0.88; and the error for the cubic model is 0.95. Thus, between these three models, the first-degree polynomial is the best in terms of minimum expected prediction error.
❑

8.3 Jackknife

The *jackknife* is a data partitioning method like cross-validation, but the goal of the jackknife is more in keeping with that of the bootstrap. The jackknife method is used to estimate the bias and the standard error of statistics.

Let's say that we have a random sample of size n, and we denote our estimator of a parameter θ as

$$\hat{\theta} = T = t(x_1, x_2, ..., x_n). \tag{8.8}$$

So, $\hat{\theta}$ might be the mean, the variance, the correlation coefficient, or some other statistic of interest. Recall from Chapters 3 and 6 that T is also a random variable, and it has some error associated with it. We would like to get an estimate of the bias and the standard error of the estimate T, so we can assess the accuracy of the results.

When we cannot determine the bias and the standard error using analytical techniques, then methods such as the bootstrap or the jackknife may be used. The jackknife is similar to the bootstrap in that no parametric assumptions are made about the underlying population that generated the data, and the variation in the estimate is investigated by looking at the sample data.

The jackknife method is similar to cross-validation in that we leave out one observation x_i from our sample to form a *jackknife sample* as follows

$$x_1, ..., x_{i-1}, x_{i+1}, ..., x_n.$$

This says that the i-th jackknife sample is the original sample with the i-th data point removed. We calculate the value of the estimate using this reduced jackknife sample to obtain the i-th *jackknife replicate*. This is given by

$$T^{(-i)} = t(x_1, ..., x_{i-1}, x_{i+1}, ..., x_n).$$

This means that we leave out one point at a time and use the rest of the sample to calculate our statistic. We continue to do this for the entire sample, leaving out one observation at a time, and the end result is a sequence of n jackknife replications of the statistic.

The estimate of the bias of T obtained from the jackknife technique is given by [Efron and Tibshirani, 1993]

$$\widehat{\text{Bias}}_{Jack}(T) = (n-1)(\overline{T^{(J)}} - T), \tag{8.9}$$

where

$$\overline{T^{(J)}} = \frac{\sum_{i=1}^{n} T^{(-i)}}{n}. \tag{8.10}$$

We see from Equation 8.10 that $\overline{T^{(J)}}$ is simply the average of the jackknife replications of T.

The estimated standard error using the jackknife is defined as follows:

$$\hat{SE}_{Jack}(T) = \left[\frac{n-1}{n} \sum_{i=1}^{n} (T^{(-i)} - \overline{T^{(J)}})^2 \right]^{1/2}. \tag{8.11}$$

Equation 8.11 is essentially the sample standard deviation of the jackknife replications with a factor $(n-1)/n$ in front of the summation instead of $1/(n-1)$. Efron and Tibshirani [1993] show that this factor ensures that the jackknife estimate of the standard error of the sample mean, $\hat{SE}_{Jack}(\bar{x})$, is an unbiased estimate.

PROCEDURE – JACKKNIFE

1. Leave out an observation.
2. Calculate the value of the statistic using the remaining sample points to obtain $T^{(-i)}$.
3. Repeat steps 1 and 2, leaving out one point at a time, until all n $T^{(-i)}$ are recorded.
4. Calculate the jackknife estimate of the bias of T using Equation 8.9.
5. Calculate the jackknife estimate of the standard error of T using Equation 8.11.

The following two examples show how this is used to obtain jackknife estimates of the bias and standard error for an estimate of the correlation coefficient.

Example 8.4

In this example, we use a data set that has been examined in Efron and Tibshirani [1993]. These data consist of measurements collected on the freshman class of 82 law schools in 1973. The average score for the entering class on a national law test (**lsat**) and the average undergraduate grade point average (**gpa**) were recorded. A random sample of size $n = 15$ was taken from the population. We would like to use these sample data to estimate the correlation coefficient ρ between the test scores (**lsat**) and the grade point average (**gpa**). We start off by finding the statistic of interest.

```
% Loads up a matrix - law.
load law
% Estimate the desired statistic from the sample.
lsat = law(:,1);
gpa = law(:,2);
tmp = corrcoef(gpa,lsat);
% Recall from Chapter 3 that the corrcoef function
```

```
% returns a matrix of correlation coefficients. We
% want the one in the off-diagonal position.
T = tmp(1,2);
```

We get an estimated correlation coefficient of $\hat{\rho} = 0.78,$ and we would like to get an estimate of the bias and the standard error of this statistic. The following MATLAB code implements the jackknife procedure for estimating these quantities.

```
% Set up memory for jackknife replicates.
n = length(gpa);
reps = zeros(1,n);
for i = 1:n
    % Store as temporary vector:
    gpat = gpa;
    lsatt = lsat;
    % Leave i-th point out:
    gpat(i) = [];
    lsatt(i) = [];
    % Get correlation coefficient:
    % In this example, we want off-diagonal element.
    tmp = corrcoef(gpat,lsatt);
    reps(i) = tmp(1,2);
end
mureps = mean(reps);
sehat = sqrt((n-1)/n*sum((reps-mureps).^2));
% Get the estimate of the bias:
biashat = (n-1)*(mureps-T);
```

Our estimate of the standard error of the sample correlation coefficient is

$$\hat{SE}_{Jack}(\hat{\rho}) = 0.14,,$$

and our estimate of the bias is

$$\hat{Bias}_{Jack}(\hat{\rho}) = -0.0065.$$

This data set will be explored further in the exercises.
❑

Example 8.5

We provide a MATLAB function called **csjack** that implements the jackknife procedure. This will work with any MATLAB function that takes the random sample as the argument and returns a statistic. This function can be one that comes with MATLAB, such as **mean** or **var**, or it can be one written by the user. We illustrate its use with a user-written function called **corr** that returns the single correlation coefficient between two univariate random variables.

```
function r = cscorr(data)
% This function returns the single correlation
% coefficient between two variables.
tmp = corrcoef(data);
r = tmp(1,2);
```

The data used in this example are taken from Hand, et al. [1994]. They were originally from Anscombe [1973], where they were created to illustrate the point that even though an observed value of a statistic is the same for data sets ($\hat{\rho} = 0.82$),, that does not tell the entire story. He also used them to show the importance of looking at scatterplots because it is obvious from the plots that the relationships between the variables are not similar. The scatterplots are shown in Figure 8.3.

```
% Here is another example.
% We have 4 data sets with essentially the same
% correlation coefficient.
% The scatterplots look very different.
% When this file is loaded, you get four sets
% of x and y variables.
load anscombe
% Do the scatterplots.
subplot(2,2,1),plot(x1,y1,'k*');
subplot(2,2,2),plot(x2,y2,'k*');
subplot(2,2,3),plot(x3,y3,'k*');
subplot(2,2,4),plot(x4,y4,'k*');
```

We now determine the jackknife estimate of bias and standard error for $\hat{\rho}$ using **csjack**.

```
% Note that 'corr' is something we wrote.
[b1,se1,jv1] = csjack([x1,y1],'cscorr');
[b2,se2,jv2] = csjack([x2,y2],'cscorr');
[b3,se3,jv3] = csjack([x3,y3],'cscorr');
[b4,se4,jv4] = csjack([x4,y4],'cscorr');
```

The jackknife estimates of bias are

```
b1 = -0.0052
b2 =  0.0008
b3 =  0.1514
b4 =  NaN
```

The jackknife estimates of the standard error are

```
se1 = 0.1054
se2 = 0.1026
se3 = 0.1730
se4 = NaN
```

The jackknife procedure does not work for the fourth data set because when we leave out the last data point, the correlation coefficient is undefined for the remaining points.

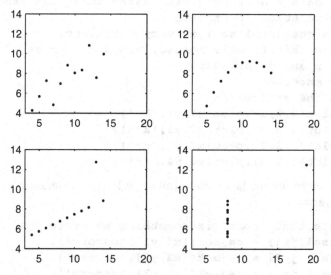

FIGURE 8.3
This shows the scatterplots of the four data sets discussed in Example 8.5. These data were created to show the importance of looking at scatterplots [Anscombe, 1973]. All data sets have the same estimated correlation coefficient of $\hat{\rho} = 0.82$, but it is obvious that the relationship between the variables is very different.

The jackknife method is also described in the literature using pseudo-values. The **jackknife pseudo-values** are given by

$$\widehat{T}_i = nT - (n-1)T^{(-i)} \qquad i = 1, \ldots, n, \qquad (8.12)$$

where $T^{(-i)}$ is the value of the statistic computed on the sample with the *i*-th data point removed.

We take the average of the pseudo-values given by

$$J(T) = \frac{\sum_{i=1}^{n} \widehat{T}_i}{n}, \qquad (8.13)$$

and use this to get the jackknife estimate of the standard error, as follows:

$$\hat{SE}_{JackP}(T) = \left[\frac{1}{n(n-1)} \sum_{i=1}^{n} (\widehat{T}_i - J(T))^2 \right]^{1/2}. \qquad (8.14)$$

PROCEDURE – PSEUDO-VALUE JACKKNIFE

1. Leave out an observation.
2. Calculate the value of the statistic using the remaining sample points to obtain $T^{(-i)}$.
3. Calculate the pseudo-value $\widehat{T}_i$ using Equation 8.12.
4. Repeat steps 2 and 3 for the remaining data points, yielding n values of $\widehat{T}_i$.
5. Determine the jackknife estimate of the standard error of T using Equation 8.14.

Example 8.6

We now repeat Example 8.4 using the jackknife pseudo-value approach and compare estimates of the standard error of the correlation coefficient for these data. The following MATLAB code implements the pseudo-value procedure.

```
% Loads up a matrix.
load law
lsat = law(:,1);
gpa = law(:,2);
% Get the statistic from the original sample
tmp = corrcoef(gpa,lsat);
T = tmp(1,2);
% Set up memory for jackknife replicates
n = length(gpa);
reps = zeros(1,n);
for i = 1:n
    % store as temporary vector
    gpat = gpa;
```

```
        lsatt = lsat;
        % leave i-th point out
        gpat(i) = [];
        lsatt(i) = [];
        % get correlation coefficient
        tmp = corrcoef(gpat,lsatt);
        % In this example, is off-diagonal element.
        % Get the jackknife pseudo-value for the i-th point.
        reps(i) = n*T-(n-1)*tmp(1,2);
    end
    JT = mean(reps);
    sehatpv = sqrt(1/(n*(n-1))*sum((reps - JT).^2));
```

We obtain an estimated standard error of $\hat{SE}_{JackP}(\hat{\rho}) = 0.14$, which is the same result we had before.
❑

Efron and Tibshirani [1993] describe a situation where the jackknife procedure does not work and suggest that the bootstrap be used instead. These are applications where the statistic is not smooth. An example of this type of statistic is the median. Here *smoothness* refers to statistics where small changes in the data set produce small changes in the value of the statistic. We illustrate this situation in the next example.

Example 8.7

Researchers collected data on the weight gain of rats that were fed four different diets based on the amount of protein (high and low) and the source of the protein (beef and cereal) [Snedecor and Cochran, 1967; Hand, et al., 1994]. We will use the data collected on the rats who were fed a low protein diet of cereal. The sorted data are

```
    x = [58, 67, 74, 74, 80, 89, 95, 97, 98, 107];
```

The median of this data set is $\hat{q}_{0.5} = 84.5$. To see how the median changes with small changes of x, we increment the fourth observation $x = 74$ by one. The change in the median is zero because it is still at $\hat{q}_{0.5} = 84.5$. In fact, the median does not change until we increment the fourth observation by 7, at which time the median becomes $\hat{q}_{0.5} = 85$. Let's see what happens when we use the jackknife approach to get an estimate of the standard error in the median.

```
    % Set up memory for jackknife replicates.
    n = length(x);
    reps = zeros(1,n);
    for i = 1:n
        % Store as temporary vector.
        xt = x;
        % Leave i-th point out.
```

```
    xt(i) = [];
    % Get the median.
    reps(i) = median(xt);
end
mureps = mean(reps);
sehat = sqrt((n-1)/n*sum((reps-mureps).^2));
```

The jackknife replicates are

$$89 \quad 89 \quad 89 \quad 89 \quad 89 \quad 80 \quad 80 \quad 80 \quad 80 \quad 80.$$

These give an estimated standard error of the median of $\hat{SE}_{Jack}(\hat{q}_{0.5}) = 13.5$. Because the median is not a smooth statistic, we have only a few distinct values of the statistic in the jackknife replicates. To understand this further, we now estimate the standard error using the bootstrap.

```
% Now get the estimate of standard error using
% the bootstrap.
[bhat,seboot,bvals]=csboot(x','median',500);
```

From this, we get an estimated standard error of the median of 8.05. The reader is asked in the exercises to see what happens when the statistic is the mean and should find that the jackknife and bootstrap estimates of the standard error of the mean are similar.
□

It can be shown [Efron and Tibshirani, 1993] that the jackknife estimate of the standard error of the median does not converge to the true standard error as $n \to \infty$. For the data set of Example 8.7, we had only two distinct values of the median in the jackknife replicates. This gives a poor estimate of the standard error of the median. On the other hand, the bootstrap produces data sets that are not as similar to the original data, so it yields reasonable results. The delete-d jackknife [Efron and Tibshirani, 1993; Shao and Tu, 1995] deletes d observations at a time instead of only one. This method addresses the problem of inconsistency with nonsmooth statistics.

8.4 Better Bootstrap Confidence Intervals

In Chapter 7, we discussed three types of confidence intervals based on the bootstrap: the bootstrap standard interval, the bootstrap-t interval and the bootstrap percentile interval. Each of them is applicable under more general assumptions and is superior in some sense (e.g., coverage performance, range-preserving, etc.) to the previous one. The bootstrap confidence interval that we present in this section is an improvement on the bootstrap percentile interval. This is called the BC_a **interval,** which stands for bias-corrected and accelerated.

Recall that the upper and lower endpoints of the $(1 - \alpha) \cdot 100\%$ bootstrap percentile confidence interval are given by

$$\text{Percentile Interval: } (\hat{\theta}_{Lo}, \hat{\theta}_{Hi}) = (\hat{\theta}_B^{*(\alpha/2)}, \hat{\theta}_B^{*(1-\alpha/2)}). \tag{8.15}$$

Say we have $B = 100$ bootstrap replications of our statistic, which we denote as $\hat{\theta}^{*b}$, $b = 1, ..., 100$. To find the percentile interval, we sort the bootstrap replicates in ascending order. If we want a 90% confidence interval, then one way to obtain $\hat{\theta}_{Lo}$ is to use the bootstrap replicate in the 5th position of the ordered list. Similarly, $\hat{\theta}_{Hi}$ is the bootstrap replicate in the 95th position. As discussed in the previous chapter, the endpoints could also be obtained using other quantile estimates.

The BC_a interval adjusts the endpoints of the interval based on two parameters, $\hat{a}$ and $\hat{z}_0$. The $(1 - \alpha) \cdot 100\%$ confidence interval using the BC_a method is

$$BC_a \text{ Interval: } (\hat{\theta}_{Lo}, \hat{\theta}_{Hi}) = (\hat{\theta}_B^{*(\alpha_1)}, \hat{\theta}_B^{*(\alpha_2)}), \tag{8.16}$$

where

$$\alpha_1 = \Phi\left(\hat{z}_0 + \frac{\hat{z}_0 + z^{(\alpha/2)}}{1 - \hat{a}(\hat{z}_0 + z^{(\alpha/2)})}\right) \tag{8.17}$$

$$\alpha_2 = \Phi\left(\hat{z}_0 + \frac{\hat{z}_0 + z^{(1-\alpha/2)}}{1 - \hat{a}(\hat{z}_0 + z^{(1-\alpha/2)})}\right).$$

Let's first look a little closer at α_1 and α_2 given in Equation 8.17. Since Φ denotes the standard normal cumulative distribution function, we know that $0 \le \alpha_1 \le 1$ and $0 \le \alpha_2 \le 1$. So we see from Equation 8.16 and 8.17 that instead of basing the endpoints of the interval on the confidence level of $1 - \alpha$, they are adjusted using information from the distribution of bootstrap replicates.

We now turn our attention to how we determine the parameters $\hat{a}$ and $\hat{z}_0$. The **bias-correction** is given by $\hat{z}_0$, and it is based on the proportion of bootstrap replicates $\hat{\theta}^{*b}$ that are less than the statistic $\hat{\theta}$ calculated from the original sample. It is given by

$$\hat{z}_0 = \Phi^{-1}\left(\frac{\#(\hat{\theta}^{*b} < \hat{\theta})}{B}\right), \tag{8.18}$$

where Φ^{-1} denotes the inverse of the standard normal cumulative distribution function.

The *acceleration parameter* $\hat{a}$ is obtained using the jackknife procedure as follows:

$$\hat{a} = \frac{\sum_{i=1}^{n} \left\{ \overline{\hat{\theta}^{(J)}} - \hat{\theta}^{(-i)} \right\}^3}{6 \left\{ \sum_{i=1}^{n} \left(\overline{\hat{\theta}^{(J)}} - \hat{\theta}^{(-i)} \right)^2 \right\}^{3/2}}, \tag{8.19}$$

where $\hat{\theta}^{(-i)}$ is the value of the statistic using the sample with the i-th data point removed (the i-th jackknife sample) and

$$\overline{\hat{\theta}^{(J)}} = \frac{1}{n} \sum_{i=1}^{n} \hat{\theta}^{(-i)}. \tag{8.20}$$

We can see from Equation 8.17 that if $\hat{a}$ and $\hat{z}_0$ are both equal to zero, then the BC_a is the same as the bootstrap percentile interval. For example,

$$\alpha_1 = \Phi\left\{ 0 + \frac{0 + z^{(\alpha/2)}}{1 - 0(0 + z^{(\alpha/2)})} \right\} = \Phi(z^{(\alpha/2)}) = \alpha/2,$$

with a similar result for α_2. Thus, when we do not account for the bias $\hat{z}_0$ and the acceleration $\hat{a}$, then Equation 8.16 reduces to the bootstrap percentile interval.

According to Efron and Tibshirani [1993], $\hat{z}_0$ is a measure of the difference between the median of the bootstrap replicates and $\hat{\theta}$ in normal units. If half of the bootstrap replicates are less than or equal to $\hat{\theta}$, then there is no median bias and $\hat{z}_0$ is zero. The parameter $\hat{a}$ measures the rate acceleration of the standard error of $\hat{\theta}$. For more information on the theoretical justification for these corrections, see Efron and Tibshirani [1993] and Efron [1987].

PROCEDURE – BC_a INTERVAL

1. Given a random sample, $\mathbf{x} = (x_1, \ldots, x_n)$, calculate the statistic of interest $\hat{\theta}$.
2. Sample with replacement from the original sample to get the bootstrap sample

$$\mathbf{x}^{*b} = (x_1^{*b}, \ldots, x_n^{*b}).$$

3. Calculate the same statistic as in step 1 using the sample found in step 2. This yields a bootstrap replicate $\hat{\theta}^{*b}$.

4. Repeat steps 2 through 3, B times, where $B \geq 1000$.

5. Calculate the bias correction (Equation 8.18) and the acceleration factor (Equation 8.19).

6. Determine the adjustments for the interval endpoints using Equation 8.17.

7. The lower endpoint of the confidence interval is the α_1 quantile $\hat{q}_{\alpha_1}$ of the bootstrap replicates, and the upper endpoint of the confidence interval is the α_2 quantile $\hat{q}_{\alpha_2}$ of the bootstrap replicates.

Example 8.8

We use an example from Efron and Tibshirani [1993] to illustrate the BC_a interval. Here we have a set of measurements of 26 neurologically impaired children who took a test of spatial perception called test A. We are interested in finding a 90% confidence interval for the variance of a random score on test A. We use the following estimate for the variance:

$$\hat{\theta} = \frac{1}{n}\sum_{i=1}^{n}(x_i - \bar{x})^2,$$

where x_i represents one of the test scores. This is a biased estimator of the variance, and when we calculate this statistic from the sample we get a value of $\hat{\theta} = 171.5$. We provide a function called **csbootbca** that will determine the BC_a interval. Because it is somewhat lengthy, we do not include the MATLAB code here, but the reader can view it in the M-file. However, before we can use the function **csbootbca**, we have to write an M-file function that will return the estimate of the second sample central moment using only the sample as an input. It should be noted that MATLAB Statistics Toolbox has a function (**moment**) that will return the sample central moments of any order. We do not use this with the **csbootbca** function because the function specified as an input argument to **csbootbca** can only use the sample as an input. Note that the function **mom** is the same function used in Chapter 7. We can get the bootstrap BC_a interval with the following command.

```
% First load the data.
load spatial
% Now find the BC-a bootstrap interval.
alpha = 0.10;
B = 2000;
```

```
% Use the function we wrote to get the
% 2nd sample central moment - 'mom'.
[blo,bhi,bvals,z0,ahat] = ...
csbootbca(spatial','mom',B,alpha);
```

From this function, we get a bias correction of $\hat{z}_0 = 0.16$ and an acceleration factor of $\hat{a} = 0.061$. The endpoints of the interval from **csbootbca** are $(115.97, 258.54)$. In the exercises, the reader is asked to compare this to the bootstrap-*t* interval and the bootstrap percentile interval.
❑

8.5 Jackknife-After-Bootstrap

In the previous chapter, we presented the bootstrap method for estimating the statistical accuracy of estimates. However, the bootstrap estimates of standard error and bias are also estimates, so they too have error associated with them. This error arises from two sources, one of which is the usual sampling variability because we are working with the sample instead of the population. The other variability comes from the fact that we are working with a finite number B of bootstrap samples.

We now turn our attention to estimating this variability using the jackknife-after-bootstrap technique. The characteristics of the problem are the same as in Chapter 7. We have a random sample $\mathbf{x} = (x_1, ..., x_n)$, from which we calculate our statistic $\hat{\theta}$. We estimate the distribution of $\hat{\theta}$ by creating B bootstrap replicates $\hat{\theta}^{*b}$. Once we have the bootstrap replicates, we estimate some feature of the distribution of $\hat{\theta}$ by calculating the corresponding feature of the distribution of bootstrap replicates. We will denote this feature or bootstrap estimate as $\hat{\gamma}_B$. As we saw before, $\hat{\gamma}_B$ could be the bootstrap estimate of the standard error, the bootstrap estimate of a quantile, the bootstrap estimate of bias, or some other quantity.

To obtain the jackknife-after-bootstrap estimate of the variability of $\hat{\gamma}_B$, we leave out one data point x_i at a time and calculate $\hat{\gamma}_B^{(-i)}$ using the bootstrap method on the remaining $n - 1$ data points. We continue in this way until we have the n values of $\hat{\gamma}_B^{(-i)}$. We estimate the variance of $\hat{\gamma}_B$ using the $\hat{\gamma}_B^{(-i)}$ values, as follows:

$$\hat{\text{var}}_{Jack}(\hat{\gamma}_B) = \frac{n-1}{n} \sum_{i=1}^{n} \left(\hat{\gamma}_B^{(-i)} - \overline{\hat{\gamma}}_B\right)^2, \tag{8.21}$$

where

$$\overline{\hat{\gamma}_B} = \frac{1}{n}\sum_{i=1}^{n}\hat{\gamma}_B^{(-i)}.$$

Note that this is just the jackknife estimate for the variance of a statistic, where the statistic that we have to calculate for each jackknife replicate is a bootstrap estimate.

This can be computationally intensive because we would need a new set of bootstrap samples when we leave out each data point x_i. There is a shortcut method for obtaining $\hat{\text{var}}_{Jack}(\hat{\gamma}_B)$ where we use the original B bootstrap samples. There will be some bootstrap samples where the i-th data point does not appear. Efron and Tibshirani [1993] show that if $n \geq 10$ and $B \geq 20$, then the probability is low that every bootstrap sample contains a given point x_i. We estimate the value of $\hat{\gamma}_B^{(-i)}$ by taking the bootstrap replicates for samples that do not contain the data point x_i. These steps are outlined next.

PROCEDURE – JACKKNIFE-AFTER-BOOTSTRAP

1. Given a random sample $x = (x_1, ..., x_n)$, calculate a statistic of interest $\hat{\theta}$.
2. Sample with replacement from the original sample to get a bootstrap sample $x^{*b} = (x_1^*, ..., x_n^*)$.
3. Using the sample obtained in step 2, calculate the same statistic that was determined in step one and denote by $\hat{\theta}^{*b}$.
4. Repeat steps 2 through 3, B times to estimate the distribution of $\hat{\theta}$.
5. Estimate the desired feature of the distribution of $\hat{\theta}$ (e.g., standard error, bias, etc.) by calculating the corresponding feature of the distribution of $\hat{\theta}^{*b}$. Denote this bootstrap estimated feature as $\hat{\gamma}_B$.
6. Now get the error in $\hat{\gamma}_B$. For $i = 1, ..., n$, find all samples $x^{*b} = (x_1^*, ..., x_n^*)$ that do not contain the point x_i. These are the bootstrap samples that can be used to calculate $\hat{\gamma}_B^{(-i)}$.
7. Calculate the estimate of the variance of $\hat{\gamma}_B$ using Equation 8.21.

Example 8.9

In this example, we show how to implement the jackknife-after-bootstrap procedure. For simplicity, we will use the MATLAB Statistics Toolbox function called **bootstrp** because it returns the indices for each bootstrap sample and the corresponding bootstrap replicate $\hat{\theta}^{*b}$. We return now to the **law** data where our statistic is the sample correlation coefficient. Recall that we wanted to estimate the standard error of the correlation coefficient, so $\hat{\gamma}_B$ will be the bootstrap estimate of the standard error.

```
% Use the law data.
load law
lsat = law(:,1);
gpa = law(:,2);

% Use the example in MATLAB documentation.
B = 1000;
[bootstat,bootsam] = bootstrp(B,'corrcoef',lsat,gpa);
```

The output argument **bootstat** contains the *B* bootstrap replicates of the statistic we are interested in, and the columns of **bootsam** contain the indices to the data points that were in each bootstrap sample. We can loop through all of the data points and find the columns of **bootsam** that do not contain that point. We then find the corresponding bootstrap replicates.

```
% Find the jackknife-after-bootstrap.
n = length(gpa);
% Set up storage space.
jreps = zeros(1,n);
% Loop through all points,
% Find the columns in bootsam that
% do not have that point in it.
for i = 1:n
    % Note that the columns of bootsam are
    % the indices to the samples.
    % Find all columns with the point.
    [I,J] = find(bootsam==i);
    % Find all columns without the point.
    jacksam = setxor(J,1:B);
    % Find the correlation coefficient for
    % each of the bootstrap samples that
    % do not have the point in them.
    bootrep = bootstat(jacksam,2);
    % In this case it is col 2 that we need.
    % Calculate the feature (gamma_b) we want.
    jreps(i) = std(bootrep);
end
% Estimate the error in gamma_b.
varjack = (n-1)/n*sum((jreps-mean(jreps)).^2);
% The original bootstrap estimate of error is:
gamma = std(bootstat(:,2));
```

We see that the estimate of the standard error of the correlation coefficient for this simulation is $\hat{\gamma}_B = \hat{SE}_{Boot}(\hat{\rho}) = 0.14$, and our estimated standard error in this bootstrap estimate is $\hat{SE}_{Jack}(\hat{\gamma}_B) = 0.088$.

❑

Efron and Tibshirani [1993] point out that the jackknife-after-bootstrap works well when the number of bootstrap replicates B is large. Otherwise, it overestimates the variance of $\hat{\gamma}_B$.

8.6 MATLAB® Code

The MATLAB Statistics Toolbox has a function for the jackknife, which is called **jackknife**. This function draws jackknife samples, calculates statistics on each sample using an input argument **jackfun**, and returns the results. The **bootci** function mentioned in the previous chapter for finding bootstrap confidence intervals will return the bias-corrected and accelerated interval.

As described earlier, we provide a function (**csjack**) that will implement the jackknife procedure for estimating the bias and standard error in an estimate. We also provide a function called **csjackboot** that will implement the jackknife-after-bootstrap. These functions are summarized in Table 8.1.

The cross-validation method is application specific, so users must write their own code for each situation. For example, we showed in this chapter how to use cross-validation to choose a model in regression by estimating the prediction error. We will revisit cross-validation in Chapters 10 and 13, where we use it to estimate the classification error and to choose the best tree in classification and regression trees.

TABLE 8.1

List of Functions from Chapter 8 Included in the Computational Statistics Toolbox.

Purpose	MATLAB Function
Implements the jackknife and returns the jackknife estimate of standard error and bias.	**csjack**
Returns the bootstrap BC_a confidence interval.	**csbootbca**
Implements the jackknife-after-bootstrap and returns the jackknife estimate of the error in the bootstrap.	**csjackboot**

8.7 Further Reading

There are very few books available where the cross-validation technique is the main topic, although Hjorth [1994] comes the closest. In that book, he discusses the cross-validation technique and the bootstrap and describes their use in model selection. Other sources on the theory and use of cross-validation are Efron [1982, 1983, 1986] and Efron and Tibshirani [1991, 1993]. Cross-validation is usually presented along with the corresponding applications. For example, to see how cross-validation can be used to select the smoothing parameter in probability density estimation, see Scott [2015]. Breiman, et al. [1984] and Webb [2011] describe how cross-validation is used to choose the right size classification tree.

The initial jackknife method was proposed by Quenouille [1949, 1956] to estimate the bias of an estimate. This was later extended by Tukey [1958] to estimate the variance using the pseudo-value approach. Efron [1982] is an excellent resource that discusses the underlying theory and the connection between the jackknife, the bootstrap, and cross-validation. A more recent text by Shao and Tu [1995] provides a guide to using the jackknife and other resampling plans. Many practical examples are included. They also present the theoretical properties of the jackknife and the bootstrap, examining them in an asymptotic framework. Efron and Tibshirani [1993] show the connection between the bootstrap and the jackknife through a geometrical representation. For a reference on the jackknife that is accessible to readers at the undergraduate level, we recommend Mooney and Duval [1993]. This text also gives a description of the delete-*d* jackknife procedure.

The use of jackknife-after-bootstrap to evaluate the error in the bootstrap is discussed in Efron and Tibshirani [1993] and Efron [1992]. Applying another level of bootstrapping to estimate this error is given in Loh [1987], Tibshirani [1988], and Hall and Martin [1988]. For other references on this topic, see Chernick [1999].

Exercises

8.1.   The **insulate** data set [Hand, et al., 1994] contains observations corresponding to the average outside temperature in degrees Celsius and the amount of weekly gas consumption measured in 1000 cubic feet. Do a scatterplot of the data corresponding to the measurements taken before insulation was installed. What is a good model for this? Use cross-validation with $K = 1$ to estimate the prediction error for your model. Use cross-validation with $K = 4$.

Does your error change significantly? Repeat the process for the data taken after insulation was installed.

8.2. Using the same procedure as in Example 8.2, use a quadratic (degree is 2) and a cubic (degree is 3) polynomial to build the model. What is the estimated prediction error from these models? Which one seems best: linear, quadratic, or cubic?

8.3. The **peanuts** data set [Hand, et al., 1994; Draper and Smith, 1981] contains measurements of the alfatoxin (X) and the corresponding percentage of noncontaminated peanuts in the batch (Y). Do a scatterplot of these data. What is a good model for these data? Use cross-validation to choose the best model.

8.4. Generate $n = 25$ random variables from a standard normal distribution that will serve as the random sample. Determine the jackknife estimate of the standard error for $\bar{x}$, and calculate the bootstrap estimate of the standard error. Compare these to the theoretical value of the standard error (see Chapter 3).

8.5. Using a sample size of $n = 15$, generate random variables from a uniform $(0,1)$ distribution. Determine the jackknife estimate of the standard error for $\bar{x}$, and calculate the bootstrap estimate of the standard error for the same statistic. Let's say we decide to use $s/\sqrt{n}$ as an estimate of the standard error for $\bar{x}$. How does this compare to the other estimates?

8.6. Use Monte Carlo simulation to compare the performance of the bootstrap and the jackknife methods for estimating the standard error and bias of the sample second central moment. For every Monte Carlo trial, generate 100 standard normal random variables and calculate the bootstrap and jackknife estimates of the standard error and bias. Show the distribution of the bootstrap estimates (of bias and standard error) and the jackknife estimates (of bias and standard error) in a histogram or a box plot. Make some comparisons of the two methods.

8.7. Repeat Problem 8.4 and use Monte Carlo simulation to compare the bootstrap and jackknife estimates of bias for the sample coefficient of skewness statistic and the sample coefficient of kurtosis (see Chapter 3).

8.8. Using the **law** data set in Example 8.4, find the jackknife replicates of the median. How many different values are there? What is the jackknife estimate of the standard error of the median? Use the bootstrap method to get an estimate of the standard error of the median. Compare the two estimates of the standard error of the median.

8.9. For the data in Example 8.7, use the bootstrap and the jackknife to estimate the standard error of the mean. Compare the two estimates.

8.10. Using the data in Example 8.8, find the bootstrap-*t* interval and the bootstrap percentile interval. Compare these to the BC_a interval found in Example 8.8.

Chapter 9

Probability Density Estimation

9.1 Introduction

We discussed several techniques for graphical exploratory data analysis in Chapters 5 and 6. One purpose of these exploratory techniques is to obtain information and insights about the distribution of the underlying population. For instance, we would like to know if the distribution is multimodal, skewed, symmetric, etc. Another way to gain understanding about the distribution of the data is to estimate the probability density function from the random sample, possibly using a nonparametric probability density estimation technique.

Estimating probability density functions is required in many areas of computational statistics. One of these is in the modeling and simulation of physical phenomena. We often have measurements from our process, and we would like to use those measurements to determine the probability distribution so we can generate random variables for a Monte Carlo simulation (Chapter 7). Another application where probability density estimation is used is in statistical pattern recognition (Chapter 10). In supervised learning, which is one approach to pattern recognition, we have measurements where each one is labeled with a class membership tag. We could use the measurements for each class to estimate the class-conditional probability density functions, which are then used in a Bayesian classifier. In other applications, we might need to determine the probability that a random variable will fall within some interval, so we would need to evaluate the cumulative distribution function. If we have an estimate of the probability density function, then we can easily estimate the required probability by integrating under the estimated curve. Finally, in Chapter 13, we show how to use density estimation techniques for nonparametric regression.

In this chapter, we cover *semi-parametric* and *nonparametric* techniques for probability density estimation. By these, we mean techniques where we make few or no assumptions about what functional form the probability density takes. This is in contrast to a *parametric* method, where the density is estimated by assuming a distribution and then estimating the parameters (see Chapter 3). We present three main methods of semi-parametric and nonparametric density estimation and their variants: histograms, kernel density estimates, and finite mixtures.

In the remainder of this section, we cover some ways to measure the estimation error in functions as background to what follows. Then, in Section 9.2, we present various histogram-based methods for probability density estimation. There we cover optimal bin widths for univariate and multivariate histograms, the frequency polygons, and averaged shifted histograms. Section 9.3 contains a discussion of kernel density estimation, both univariate and multivariate. In Section 9.4, we describe methods that model the probability density as a finite (less than n) weighted sum of component densities. Section 9.5 includes a discussion of how one can generate random variables from a finite mixture probability density function. As usual, we conclude with descriptions of available MATLAB® code and references to the topics covered in the chapter.

Before we can describe the various density estimation methods, we need to provide a little background on measuring the error in our estimated functions. We briefly present two ways to measure the error between the true function and the estimate of the function. These are called the mean integrated squared error (MISE) and the mean integrated absolute error (MIAE). Much of the underlying theory for choosing optimal parameters for probability density estimation is based on these concepts.

We start off by describing the mean squared error at a given point in the domain of the function. We can find the *mean squared error* (MSE) of the estimate $\hat{f}(x)$ at a point x from the following:

$$\text{MSE}[\hat{f}(x)] = E[(\hat{f}(x) - f(x))^2]. \tag{9.1}$$

Alternatively, we can determine the error over the domain for x by integrating. This gives us the *integrated squared error* (ISE):

$$\text{ISE} = \int (\hat{f}(x) - f(x))^2 dx. \tag{9.2}$$

The ISE is a random variable that depends on the true function $f(x)$, the estimator $\hat{f}(x)$, and the particular random sample that was used to obtain the estimate. Therefore, it makes sense to look at the expected value of the ISE or *mean integrated squared error*, which is given by

$$\text{MISE} = E\left[\int (\hat{f}(x) - f(x))^2 dx\right]. \tag{9.3}$$

To get the *mean integrated absolute error*, we simply replace the integrand with the absolute difference between the estimate and the true function. Thus, we have

$$\text{MIAE} = E\left[\int \left| \hat{f}(x) - f(x) \right| dx \right]. \tag{9.4}$$

These concepts are easily extended to the multivariate case.

9.2 Histograms

Histograms were introduced in Chapter 5 as a graphical way of summarizing or describing a data set. A histogram visually conveys how a data set is distributed, reveals modes and bumps, and provides information about relative frequencies of observations. Histograms are easy to create and are computationally feasible. Thus, they are well suited for summarizing large data sets. We revisit histograms here and examine optimal bin widths and where to start the bins. We also offer several extensions of the histogram, such as the frequency polygon and the averaged shifted histogram.

1D Histograms

Most introductory statistics textbooks expose students to the frequency histogram and the relative frequency histogram. The problem with these is that the total area represented by the bins does not sum to 1. Thus, these are not valid probability density estimates. The reader is referred to Chapter 5 for more information on this and an example illustrating the difference between a frequency histogram and a density histogram. Since our goal is to estimate a *bona fide* probability density, we want to have a function $\hat{f}(x)$ that is nonnegative and satisfies the constraint that

$$\int \hat{f}(x)dx = 1. \tag{9.5}$$

The histogram is calculated using a random sample $X_1, X_2, ..., X_n$. The analyst must choose an origin t_0 for the bins and a bin width h. These two parameters define the mesh over which the histogram is constructed. In what follows, we will see that it is the bin width that determines the smoothness of the histogram. Small values of h produce histograms with a lot of variation, while larger bin widths yield smoother histograms. This phenomenon is illustrated in Figure 9.1, where we show histograms with different bin

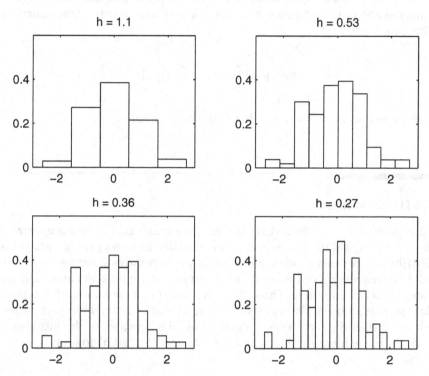

FIGURE 9.1
These are histograms for normally distributed random variables. Notice that for the larger bin widths, we have only one bump as expected. As the smoothing parameter gets smaller, the histogram displays more variation and spurious bumps appear in the histogram estimate.

widths. Because of this relationship, the bin width h is sometimes referred to as the ***smoothing parameter***.

Let $B_k = [t_k, t_{k+1})$ denote the k-th bin, where $t_{k+1} - t_k = h$, for all k. We represent the number of observations that fall into the k-th bin by v_k. The 1D histogram at a point x is defined as

$$\hat{f}_{Hist}(x) = \frac{v_k}{nh} = \frac{1}{nh}\sum_{i=1}^{n} I_{B_k}(X_i); \qquad x \text{ in } B_k, \qquad (9.6)$$

where $I_{B_k}(X_i)$ is the indicator function

$$I_{B_k}(X_i) = \begin{cases} 1, & X_i \text{ in } B_k \\ 0, & X_i \text{ not in } B_k. \end{cases}$$

This means that if we need to estimate the value of the probability density for a given x, then we obtain the value $\hat{f}_{Hist}(x)$ by taking the number of observations in the data set that fall into the same bin as x and multiplying by $(1/(nh))$.

Example 9.1

In this example, we illustrate MATLAB code that calculates the estimated value $\hat{f}_{Hist}(x)$ for a given x. We first generate random variables from a standard normal distribution.

```
n = 1000;
x = randn(n,1);
```

We then compute the histogram using MATLAB's **hist** function, using the default value of 10 bins. The issue of the bin width (or alternatively the number of bins) will be addressed shortly.

```
% Get the histogram-default is 10 bins.
[vk,bc] = hist(x);
% Get the bin width.
h = bc(2)- bc(1);
```

We can now obtain our histogram estimate at a point using the following code. Note that we have to adjust the output from **hist** to ensure that our estimate is a *bona fide* density. Let's get the estimate of our function at a point $x_0 = 0$.

```
% Now return an estimate at a point xo.
xo = 0;
% Find all of the bin centers less than xo.
ind = find(bc < xo);
% xo should be between these two bin centers.
b1 = bc(ind(end));
b2 = bc(ind(end)+1);
% Put it in the closer bin.
if (xo-b1) < (b2-xo)    % then put it in the 1st bin
    fhat = vk(ind(end))/(n*h);
else
    fhat = vk(ind(end)+1)/(n*h);
end
```

Our result is $\hat{f}(0) = 0.3851$. The true value for the standard normal evaluated at 0 is $1/\sqrt{2\pi} \approx 0.3989$, so we see that our estimate is close, but not equal to the true value.
❏

We now look at how we can choose the bin width h. Using some assumptions, Scott [2015] provides the following upper bound for the MSE (Equation 9.1) of $\hat{f}_{Hist}(x)$:

$$\mathrm{MSE}(\hat{f}_{Hist}(x)) \leq \frac{f(\xi_k)}{nh} + \gamma_k^2 h^2; \qquad x \text{ in } B_k, \tag{9.7}$$

where

$$hf(\xi_k) = \int_{B_k} f(t)dt; \qquad \text{for some } \xi_k \text{ in } B_k. \tag{9.8}$$

This is based on the assumption that the probability density function $f(x)$ is Lipschitz continuous over the bin interval B_k. A function is **Lipschitz continuous** if there is a positive constant γ_k such that

$$|f(x) - f(y)| < \gamma_k |x - y|; \qquad \text{for all } x, y \text{ in } B_k. \tag{9.9}$$

The first term on the right-hand side of Equation 9.7 is an upper bound for the variance of the density estimate, and the second term is an upper bound for the squared bias of the density estimate. These bounds show what happens to the density estimate when the bin width h is varied. We can try to minimize the MSE by varying the bin width h. We could set h very small to reduce the bias, but this also increases the variance. The increased variance in our density estimate is evident in Figure 9.1, where we see more spikes as the bin width gets smaller. Equation 9.7 illustrates a common problem in some density estimation methods: the trade-off between variance and bias as h is changed. Most of the optimal bin widths presented here are obtained by trying to minimize the squared error.

A rule for bin width selection that is often presented in introductory statistics texts is called **Sturges' Rule**. In reality, it is a rule that provides the number of bins in the histogram and is given by the following formula.

STURGES' RULE (HISTOGRAM)

$$k = 1 + \log_2 n.$$

Here k is the number of bins. The bin width h is obtained by taking the range of the sample data and dividing it into the requisite number of bins, k.

Some improved values for the bin width h can be obtained by assuming the existence of derivatives of the probability density function $f(x)$. We include the following results (without proof) because they are the basis for many of the univariate bin width rules presented in this chapter. The interested reader is referred to Scott [2015] for more details. Most of what we present here follows his excellent treatment of the subject.

Equation 9.7 provides a measure of the squared error at a point x. If we want to measure the error in our estimate for the entire function, then we can integrate over all values of x. Let's assume $f(x)$ has an absolutely continuous and a square-integrable first derivative. If we let n get very large $(n \to \infty)$, then the asymptotic MISE is

$$\text{AMISE}_{Hist}(h) = \frac{1}{nh} + \frac{1}{12}h^2 R(f'), \tag{9.10}$$

where $R(g) \equiv \int g^2(x)dx$ is used as a measure of the roughness of the function, and f' is the first derivative of $f(x)$. The first term of Equation 9.10 indicates the asymptotic integrated variance, and the second term refers to the asymptotic integrated squared bias. These are obtained as approximations to the integrated squared bias and integrated variance [Scott, 2015]. Note, however, that the form of Equation 9.10 is similar to the upper bound for the MSE in Equation 9.7 and indicates the same trade-off between bias and variance, as the smoothing parameter h changes.

The optimal bin width h_{Hist}^* for the histogram is obtained by minimizing the AMISE (Equation 9.10), so it is the h that yields the smallest MISE as n gets large. This is given by

$$h_{Hist}^* = \left(\frac{6}{nR(f')} \right)^{1/3}. \tag{9.11}$$

For the case of data that is normally distributed, we have a roughness of

$$R(f') = \frac{1}{4\sigma^3 \sqrt{\pi}}.$$

Using this in Equation 9.11, we obtain the following expression for the optimal bin width for normal data.

NORMAL REFERENCE RULE – 1D HISTOGRAM

$$h_{Hist}^* = \left(\frac{4\sigma^3 \sqrt{\pi}}{n} \right)^{1/3} \approx 3.5\sigma n^{-1/3}. \tag{9.12}$$

Scott [1979, 2015] proposed the sample standard deviation as an estimate of σ in Equation 9.12 to get the following bin width rule.

SCOTT'S RULE

$$\hat{h}^*_{Hist} = 3.5 \times s \times n^{-1/3}.$$

A robust rule was developed by Freedman and Diaconis [1981]. This uses the interquartile range (IQR) instead of the sample standard deviation.

FREEDMAN-DIACONIS RULE

$$\hat{h}^*_{Hist} = 2 \times IQR \times n^{-1/3}.$$

It turns out that when the data are skewed or heavy-tailed, the bin widths are too large using the normal reference rule. Scott [1979, 2015] derived the following correction factor for skewed data:

$$\text{skewness factor }_{Hist} = \frac{2^{1/3}\sigma}{e^{5\sigma^2/4}(\sigma^2+2)^{1/3}(e^{\sigma^2}-1)^{1/2}}. \tag{9.13}$$

The bin width obtained from Equation 9.12 should be multiplied by this factor when there is evidence that the data come from a skewed distribution. A factor for heavy-tailed distributions can be found in Scott [2015]. If one suspects the data come from a skewed or heavy-tailed distribution, as indicated by calculating the corresponding sample statistics (Chapter 3) or by graphical exploratory data analysis (Chapters 5 and 6), then the normal reference rule bin widths should be multiplied by these factors. Scott [2015] shows that the modification to the bin widths is greater for skewness and is not so critical for kurtosis.

Example 9.2

Data representing the waiting times (in minutes) between eruptions of the Old Faithful geyser at Yellowstone National Park were collected [Hand, et al., 1994]. These data are contained in the file **geyser**. In this example, we use an alternative MATLAB function (available in the standard MATLAB package) for finding a histogram, called **histc**. This takes the *bin edges* as one of the arguments. This is in contrast to the **hist** function that takes the *bin centers* as an optional argument. The following MATLAB code will construct a histogram density estimate for the Old Faithful geyser data.

```
load geyser
n = length(geyser);
% Use normal reference rule for bin width.
h = 3.5*std(geyser)*n^(-1/3);
% Get the bin mesh.
```

```
t0 = min(geyser)-1;
tm = max(geyser)+1;
rng = tm - t0;
nbin = ceil(rng/h);
bins = t0:h:(nbin*h + t0);
% Get the bin counts vk.
vk = histc(geyser,bins);
% Normalize to make it a bona fide density.
% We do not need the last count in fhat.
fhat(end) = [];
fhat = vk/(n*h);
```

We have to use the following to create a plot of our histogram density. The MATLAB **bar** function takes the bin centers as the argument, so we convert our mesh to bin centers before plotting. The plot is shown in Figure 9.2, and the existence of two modes is apparent.

```
% To plot this, use bar with the bin centers.
tm = max(bins);
bc = (t0+h/2):h:(tm-h/2);
bar(bc,fhat,1,'w')
```

❑

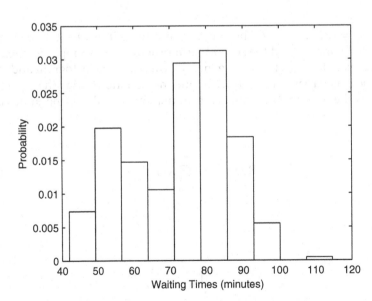

FIGURE 9.2
Histogram of Old Faithful **geyser** *data (waiting time between eruptions). Here we are using Scott's Rule for the bin widths.*

Multivariate Histograms

Given a data set that contains d-dimensional observations $\mathbf{X}_i$, we would like to obtain the probability density estimate $\hat{f}(\mathbf{x})$. We can extend the univariate histogram to d dimensions in a straightforward way. We first partition the d-dimensional space into hyper-rectangles of size $h_1 \times h_2 \times \ldots \times h_d$. We denote the k-th bin by B_k and the number of observations falling into that bin by v_k, with $\sum v_k = n$. The multivariate histogram is then defined as

$$\hat{f}_{Hist}(\mathbf{x}) = \frac{v_k}{nh_1 h_2 \ldots h_d}; \qquad \mathbf{x} \text{ in } B_k. \tag{9.14}$$

If we need an estimate of the probability density at $\mathbf{x}$, we first determine the bin that the observation falls into. The estimate of the probability density would be given by the number of observations falling into that same bin divided by the sample size and the bin widths of the partitions. The MATLAB code to create a bivariate histogram was given in Chapter 5. This could be easily extended to the general multivariate case.

For a density function that is sufficiently smooth [Scott, 2015], we can write the asymptotic MISE for a multivariate histogram as

$$\text{AMISE}_{Hist}(\mathbf{h}) = \frac{1}{nh_1 h_2 \ldots h_d} + \frac{1}{12} \sum_{j=1}^{d} h_j^2 R(f_j), \tag{9.15}$$

where $\mathbf{h} = (h_1, \ldots, h_d)$. As before, the first term indicates the asymptotic integrated variance and the second term provides the asymptotic integrated squared bias. This has the same general form as the 1D histogram and shows the same bias-variance trade-off. Minimizing Equation 9.15 with respect to h_i provides the following equation for optimal bin widths in the multivariate case

$$h_{i_{Hist}}^* = R(f_i)^{-1/2} \left(6 \prod_{j=1}^{d} R(f_j)^{1/2} \right)^{\frac{1}{2+d}} n^{\frac{-1}{2+d}}, \tag{9.16}$$

where

$$R(f_i) = \int_{\mathfrak{R}^d} \left(\frac{\partial}{\partial x_i} f(\mathbf{x}) \right)^2 d\mathbf{x}.$$

We can get a multivariate normal reference rule by looking at the special case where the data are distributed as multivariate normal with the covariance equal to a diagonal matrix with $\sigma_1^2, \ldots, \sigma_d^2$ along the diagonal. The normal reference rule in the multivariate case is given next [Scott, 2015].

NORMAL REFERENCE RULE – MULTIVARIATE HISTOGRAMS

$$h_{i_{Hist}}^* = 3.5\sigma_i n^{\frac{-1}{2+d}}; \qquad i = 1, \ldots, d.$$

Notice that this reduces to the same univariate normal reference rule when $d = 1$. As before, we can use a suitable estimate for σ_i.

Frequency Polygons

Another method for estimating probability density functions is to use a frequency polygon. A univariate frequency polygon approximates the density by linearly interpolating between the bin midpoints of a histogram with equal bin widths. Because of this, the frequency polygon extends beyond the histogram to empty bins at both ends.

The univariate probability density estimate using the frequency polygon is obtained from the following:

$$\hat{f}_{FP}(x) = \left(\frac{1}{2} - \frac{x}{h}\right)\hat{f}_k + \left(\frac{1}{2} + \frac{x}{h}\right)\hat{f}_{k+1}; \qquad \bar{B}_k \le x \le \bar{B}_{k+1}, \qquad (9.17)$$

where $\hat{f}_k$ and $\hat{f}_{k+1}$ are adjacent univariate histogram values and $\bar{B}_k$ is the center of bin B_k. An example of a *section* of a frequency polygon is shown in Figure 9.3.

As is the case with the univariate histogram, under certain assumptions, we can write the asymptotic MISE as [Scott, 2015, 1985]

$$\text{AMISE}_{FP}(h) = \frac{2}{3nh} + \frac{49}{2880}h^4 R(f''), \qquad (9.18)$$

where f'' is the second derivative of $f(x)$. The optimal bin width that minimizes the AMISE for the frequency polygon is given by

$$h_{FP}^* = 2\left(\frac{15}{49nR(f'')}\right)^{1/5}. \qquad (9.19)$$

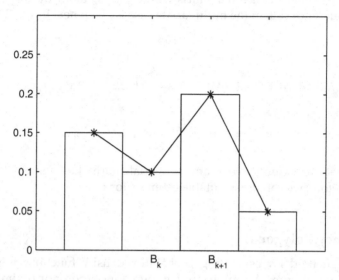

FIGURE 9.3
The frequency polygon is obtained by connecting the center of adjacent bins using straight lines. This figure illustrates a section of the frequency polygon.

If $f(x)$ is the probability density function for the standard normal, then $R(f'') = 3/(8\sqrt{\pi}\sigma^5)$. Substituting this in Equation 9.19, we obtain the following normal reference rule for a frequency polygon.

NORMAL REFERENCE RULE – FREQUENCY POLYGON

$$h_{FP}^* = 2.15\sigma n^{-1/5}.$$

We can use the sample standard deviation in this rule as an estimate of σ or choose a robust estimate based on the interquartile range. If we choose the IQR and use $\hat{\sigma} = IQR/1.348$, then we obtain a bin width of

$$\hat{h}_{FP}^* = 1.59 \times IQR \times n^{-1/5}.$$

As for the case of histograms, Scott [2015] provides a skewness factor for frequency polygons, given by

$$\text{skewness factor}_{FP} = \frac{12^{1/5}\sigma}{e^{7\sigma^2/4}(e^{\sigma^2}-1)^{1/2}(9\sigma^4 + 20\sigma^2 + 12)^{1/5}}. \tag{9.20}$$

If there is evidence that the data come from a skewed distribution, then the bin width should be multiplied by this factor. The kurtosis factor for frequency polygons can be found in Scott [2015].

Example 9.3

Here we show how to create a frequency polygon using the Old Faithful **geyser** data. We must first create the histogram from the data, where we use the frequency polygon normal reference rule to choose the smoothing parameter.

```
load geyser
n = length(geyser);
% Use normal reference rule for bin width
% of frequency polygon.
h = 2.15*sqrt(var(geyser))*n^(-1/5);
t0 = min(geyser)-1;
tm = max(geyser)+1;
bins = t0:h:tm;
vk = histc(geyser,bins);
vk(end) = [];
fhat = vk/(n*h);
```

We then use the MATLAB function called **interp1** to interpolate between the bin centers. This function takes three arguments (and an optional fourth argument). The first two arguments to **interp1** are the **xdata** and **ydata** vectors that contain the observed data. In our case, these are the bin centers and the bin heights from the density histogram. The third argument is a vector of **xinterp** values for which we would like to obtain interpolated **yinterp** values. There is an optional fourth argument that allows the user to select the type of interpolation (**linear, cubic, nearest**, and **spline**). The default is **linear**, which is what we need for the frequency polygon. The following code constructs the frequency polygon for the **geyser** data.

```
% For frequency polygon, get the bin centers,
% with empty bin center on each end.
bc2 = (t0-h/2):h:(tm+h/2);
binh = [0 fhat 0];
% Use linear interpolation between bin centers
% Get the interpolated values at x.
xinterp = linspace(min(bc2),max(bc2));
fp = interp1(bc2, binh, xinterp);
```

To see how this looks, we can plot the frequency polygon and underlying histogram, which is shown in Figure 9.4.

```
% To plot this, use bar with the bin centers
tm = max(bins);
bc = (t0+h/2):h:(tm-h/2);
```

```
bar(bc,fhat,1,'w')
hold on
plot(xinterp,fp)
hold off
axis([30 120 0 0.035])
xlabel('Waiting Time (minutes)')
ylabel('Probability Density Function')
title('Old Faithful-Waiting Times Between Eruptions')
```

To ensure that we have a valid probability density function, we can verify that the area under the curve is approximately one by using the **trapz** function.

```
area = trapz(xinterp,fp);
```

We get an approximate area under the curve of 0.9998, indicating that the frequency polygon is indeed a *bona fide* density estimate.
❑

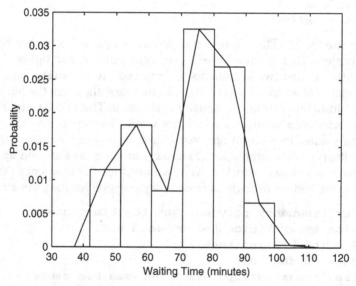

FIGURE 9.4
This shows a frequency polygon for the Old Faithful data (waiting time between eruptions).

The frequency polygon can be extended to the multivariate case. The interested reader is referred to Scott [1985, 2015] for more details on the multivariate frequency polygon. He proposes an approximate normal reference rule for the multivariate frequency polygon given by the following formula.

NORMAL REFERENCE RULE – FREQUENCY POLYGON (MULTIVARIATE)

$$h_i^* = 2\sigma_i n^{-1/(4+d)},$$

where a suitable estimate for σ_i can be used. This is derived using the assumption that the true probability density function is multivariate normal with covariance equal to the identity matrix. The following example illustrates the procedure for obtaining a bivariate frequency polygon.

Example 9.4

We first generate some random variables that are bivariate standard normal and then calculate the surface heights corresponding to the linear interpolation between the histogram density bin heights.

```
% First get the constants.
bin0 = [-4 -4];
n = 1000;
% Normal reference rule with sigma = 1.
h = 3*n^(-1/4)*ones(1,2);
% Generate bivariate standard normal variables.
x = randn(n,2);
% Find the number of bins.
nb1 = ceil((max(x(:,1))-bin0(1))/h(1));
nb2 = ceil((max(x(:,2))-bin0(2))/h(2));
% Find the mesh or bin edges.
t1 = bin0(1):h(1):(nb1*h(1)+bin0(1));
t2 = bin0(2):h(2):(nb2*h(2)+bin0(2));
[X,Y] = meshgrid(t1,t2);
```

Now that we have the random variables and the bin edges, the next step is to find the number of observations that fall into each bin. This is easily done with the MATLAB function **inpolygon**. This function can be used with any polygon (e.g., triangle or hexagon), and it returns the indices to the points that fall into that polygon.

```
% Find bin frequencies.
[nr,nc] = size(X);
vu = zeros(nr-1,nc-1);
for i = 1:(nr-1)
  for j = 1:(nc-1)
    xv = [X(i,j) X(i,j+1) X(i+1,j+1) X(i+1,j)];
    yv = [Y(i,j) Y(i,j+1) Y(i+1,j+1) Y(i+1,j)];
    in = inpolygon(x(:,1),x(:,2),xv,yv);
    vu(i,j) = sum(in(:));
  end
end
```

```
fhat = vu/(n*h(1)*h(2));
```

Now that we have the histogram density, we can use the MATLAB function `interp2` to linearly interpolate at points between the bin centers.

```
% Now get the bin centers for the frequency polygon.
% We add bins at the edges with zero height.
t1 = (bin0(1)-h(1)/2):h(1):(max(t1)+h(1)/2);
t2 = (bin0(2)-h(2)/2):h(2):(max(t2)+h(2)/2);
[bcx,bcy] = meshgrid(t1,t2);
[nr,nc] = size(fhat);
binh = zeros(nr+2,nc+2);   % add zero bin heights
binh(2:(1+nr),2:(1+nc))=fhat;
% Get points where we want to interpolate to get
% the frequency polygon.
[xint,yint]=meshgrid(linspace(min(t1),max(t1),30),...
    linspace(min(t2),max(t2),30));
fp = interp2(bcx,bcy,binh,xint,yint,'linear');
```

We can verify that this is a valid density by estimating the area under the curve.

```
df1 = xint(1,2)-xint(1,1);
df2 = yint(2,1)-yint(1,1);
area = sum(sum(fp))*df1*df2;
```

This yields an area of 0.9976. A surface plot of the frequency polygon is shown in Figure 9.5.
❑

Averaged Shifted Histograms

When we create a histogram or a frequency polygon, we need to specify a complete mesh determined by the bin width h and the starting point t_0. The reader should have noticed that the parameter t_0 did not appear in any of the asymptotic integrated squared bias or integrated variance expressions for the histograms or frequency polygons. The MISE is affected more by the choice of bin width than the choice of starting point t_0. The averaged shifted histogram (ASH) was developed to account for different choices of t_0, with the added benefit that it provides a *smoother* estimate of the probability density function.

The idea is to create many histograms with different bin origins t_0 (but with the same h) and average the histograms together. The histogram is a piecewise constant function, and the average of piecewise constant functions will also be the same type of function. Therefore, the ASH is also in the form of a histogram, and the following discussion treats it as such. The ASH is often implemented in conjunction with the frequency polygon, where the

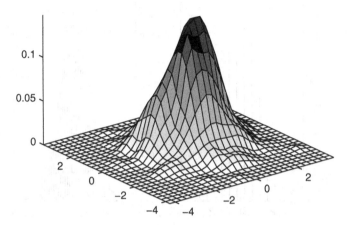

FIGURE 9.5
This is a frequency polygon of bivariate standard normal data.

latter is used to linearly interpolate between the smaller bin widths of the ASH.

To construct an ASH, we have a set of m histograms, $\hat{f}_1, \ldots, \hat{f}_m$ with constant bin width h. The origins are given by the sequence

$$t'_0 = t_0 + 0, t_0 + \frac{h}{m}, t_0 + \frac{2h}{m}, \ldots, t_0 + \frac{(m-1)h}{m}.$$

In the univariate case, the unweighted or naive ASH is given by

$$\hat{f}_{ASH}(x) = \frac{1}{m} \sum_{i=1}^{m} \hat{f}_i(x),$$ (9.21)

which is just the average of the histogram estimates at each point x. It should be clear that the $\hat{f}_{ASH}$ is a piecewise function over smaller bins, whose width is given by $\delta = (h/m)$. This is shown in Figure 9.6 where we have a single histogram $\hat{f}_i$ and the ASH estimate.

In what follows, we consider the ASH as a histogram over the narrower intervals given by $B'_k = [k\delta, (k+1)\delta)$, with $\delta = (h/m)$. As before we denote the bin counts for these bins by v_k. An alternative expression for the naive ASH can be written as

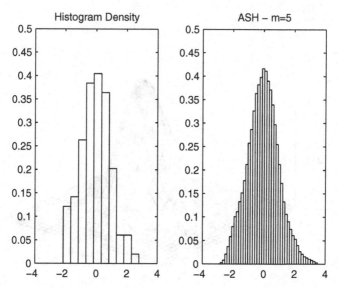

FIGURE 9.6
On the left is a histogram density based on 100 standard normal random variables, where we used the MATLAB default of 10 bins. On the right is an ASH estimate for the same data set, with m = 5.

$$\hat{f}_{ASH}(x) = \frac{1}{nh} \sum_{i=1-m}^{m-1} \left(1 - \frac{|i|}{m}\right) v_{k+i}; \qquad x \text{ in } B'_k. \tag{9.22}$$

To make this a little clearer, let's look at a simple example of the naive ASH, with $m = 3$. In this case, our estimate at a point x is

$$\hat{f}_{ASH}(x) = \frac{1}{nh}\left[\left(1 - \frac{2}{3}\right)v_{k-2} + \left(1 - \frac{1}{3}\right)v_{k-1} + \left(1 - \frac{0}{3}\right)v_{k-0} + \right.$$
$$\left.\left(1 - \frac{1}{3}\right)v_{k+1} + \left(1 - \frac{2}{3}\right)v_{k+2}\right]; \qquad x \text{ in } B'_k.$$

We can think of the factor $(1 - |i|/m)$ in Equation 9.22 as weights on the bin counts. We can use arbitrary weights instead, to obtain the general ASH.

GENERAL AVERAGED SHIFTED HISTOGRAM

$$\hat{f}_{ASH} = \frac{1}{nh} \sum_{|i| < m} w_m(i) v_{k+i}; \qquad x \text{ in } B'_k. \tag{9.23}$$

A general formula for the weights is given by

$$w_m(i) = m \times \frac{K(i/m)}{\sum\limits_{j=1-m}^{m-1} K(j/m)}; \qquad i = 1-m, \ldots, m-1, \qquad (9.24)$$

with K a continuous function over the interval $[-1, 1]$. This function K is sometimes chosen to be a probability density function. In Example 9.5, we use the biweight function:

$$K(t) = \frac{15}{16}(1 - t^2)^2 I_{[-1, 1]}(t) \qquad (9.25)$$

for our weights. Here $I_{[-1, 1]}$ is the indicator function over the interval $[-1, 1]$.

The algorithm for the general univariate ASH [Scott, 2015] is given here and is also illustrated in MATLAB in Example 9.5. This algorithm requires at least $m - 1$ empty bins on either end.

UNIVARIATE ASH – ALGORITHM:

1. Generate a mesh over the range $(t_0, nbin \times \delta + t_0)$ with bin widths of size δ, $\delta \ll h$ and $h = m\delta$. The quantity $nbin$ is the number of bins—see the comments below for more information on this number. Include at least $m - 1$ empty bins on either end of the range.
2. Compute the bin counts v_k.
3. Compute the weight vector $w_m(i)$ given in Equation 9.24.
4. Set all $\hat{f}_k = 0$.
5. Loop over $k = 1$ to $nbin$
   Loop over $i = max\{1, k - m + 1\}$ to $min\{nbin, k + m - 1\}$

$$\text{Calculate: } \hat{f}_i = \hat{f}_i + v_k w_m(i - k).$$

6. Divide all $\hat{f}_k$ by nh, these are the ASH heights.
7. Calculate the bin centers using $\bar{B}_k = t_0 + (k - 0.5)\delta$.

In practice, one usually chooses the m and h by setting the number of narrow (size δ) bins between 50 and 500 over the range of the sample. This is then extended to put some empty bins on either end of the range.

Example 9.5

In this example, we construct an ASH probability density estimate of the Buffalo **snowfall** data [Scott, 2015]. These data represent the annual snowfall in inches in Buffalo, New York over the years 1910 – 1972. First load the data and get the appropriate parameters.

```
load snowfall
n = length(snowfall);
m = 30;
h = 14.6;
delta = h/m;
```

The next step is to construct a mesh using the smaller bin widths of size δ over the desired range. Here we start the density estimate at zero.

```
% Get the mesh.
t0 = 0;
tf = max(snowfall)+20;
nbin = ceil((tf-t0)/delta);
binedge = t0:delta:(t0+delta*nbin);
```

We need to obtain the bin counts for these smaller bins, and we use the **histc** function since we want to use the bin edges rather than the bin centers.

```
% Get the bin counts for the smaller bin width delta.
vk = histc(snowfall,binedge);
% Put into a vector with m-1 zero bins on either end.
fhat = [zeros(1,m-1),vk,zeros(1,m-1)];
```

Next, we construct our weight vector according to Equation 9.24, where we use the biweight kernel given in Equation 9.25. Instead of writing the kernel as a separate function, we will use the MATLAB **inline** function to create a function object. We can then call that **inline** function just as we would an M-file function.

```
% Get the weight vector.
% Create an inline function for the kernel.
kern = inline('(15/16)*(1-x.^2).^2');
ind = (1-m):(m-1);
% Get the denominator.
den = sum(kern(ind/m));
% Create the weight vector.
wm = m*(kern(ind/m))/den;
```

The following section of code essentially implements steps 5 through 7 of the ASH algorithm.

```
% Get the bin heights over smaller bins.
fhatk = zeros(1,nbin);
```

```
for k = 1:nbin
    ind = k:(2*m+k-2);
    fhatk(k) = sum(wm.*fhat(ind));
end
fhatk = fhatk/(n*h);
bc = t0+((1:k)-0.5)*delta;
```

We use the following steps to obtain Figure 9.7, where we use a different type of MATLAB plot to show the ASH estimate. We use the bin edges with the **stairs** plot, so we must append an extra bin height at the end to ensure that the last bin is drawn and to make it dimensionally correct for plotting.

```
% To use the stairs plot, we need to use the
% bin edges.
stairs(binedge,[fhatk fhatk(end)])
axis square
title('ASH - Buffalo Snowfall Data')
xlabel('Snowfall (inches)')
```

❑

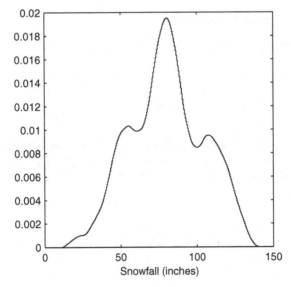

FIGURE 9.7
This is the ASH estimate for the Buffalo snowfall data. The parameters used to obtain this were h = 14.6 inches and m = 30. Notice that the ASH estimate reveals evidence of three modes.

The multivariate ASH is obtained by averaging shifted multivariate histograms. Each histogram has the same bin dimension $h_1 \times \ldots \times h_d$, and each is constructed using shifts along the coordinates given by multiples of δ_i / m_i, $i = 1, \ldots, d$. Scott [2015] provides a detailed algorithm for the bivariate ASH.

9.3 Kernel Density Estimation

Scott [2015] shows that as the number of histograms m approaches infinity, the ASH becomes a kernel estimate of the probability density function. The first published paper describing nonparametric probability density estimation was by Rosenblatt [1956], where he described the general kernel estimator. Many papers that expanded the theory followed soon after. A partial list includes Parzen [1962], Cencov [1962], and Cacoullos [1966]. Several references providing surveys and summaries of nonparametric density estimation are provided in Section 9.7. The following treatment of kernel density estimation follows that of Silverman [1986] and Scott [2015].

Univariate Kernel Estimators

The kernel estimator is given by

$$\hat{f}_{Ker}(x) = \frac{1}{nh} \sum_{i=1}^{n} K\left(\frac{x - X_i}{h}\right), \qquad (9.26)$$

where the function $K(t)$ is called a **kernel**. This must satisfy the condition that $\int K(t)dt = 1$ to ensure that our estimate in Equation 9.26 is a *bona fide* density estimate. If we define $K_h(t) = K(t/h)/h$, then we can also write the kernel estimate as

$$\hat{f}_{Ker}(x) = \frac{1}{n} \sum_{i=1}^{n} K_h(x - X_i). \qquad (9.27)$$

Usually, the kernel is a symmetric probability density function, and often a standard normal density is used. However, this does not have to be the case, and we will present other choices later in this chapter. From the definition of a kernel density estimate, we see that $\hat{f}_{Ker}(x)$ inherits all of the properties of the kernel function, such as continuity and differentiability.

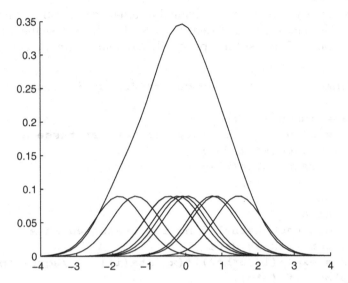

FIGURE 9.8
We obtain the above kernel density estimate for n = 10 random variables. A weighted kernel is centered at each data point, and the curves are averaged together to obtain the estimate.

From Equation 9.26, the estimated probability density function is obtained by placing a weighted kernel function, centered at each data point and then taking the average of them. See Figure 9.8 for an example of this procedure. Notice that the places where there are more curves or kernels yield "bumps" in the final estimate. An alternative implementation is discussed in the exercises.

PROCEDURE – UNIVARIATE KERNEL

1. Choose a kernel, a smoothing parameter h, and the domain (the set of x values) over which to evaluate $\hat{f}(x)$.
2. For each X_i, evaluate the following kernel at all x in the domain:

$$K_i = K\left(\frac{x - X_i}{h}\right); \qquad i = 1, \ldots, n.$$

The result from this is a set of n curves, one for each data point X_i.
3. Weight each curve by $1/h$.
4. For each x, take the average of the weighted curves.

Example 9.6

In this example, we show how to obtain the kernel density estimate for a data set, using the standard normal density as our kernel. We use the procedure outlined above. The resulting probability density estimate is shown in Figure 9.8.

```
% Generate standard normal random variables.
n = 10;
data = randn(1,n);
% We will get the density estimate at these x values.
x = linspace(-4,4,50);
fhat = zeros(size(x));
h = 1.06*n^(-1/5);
hold on
for i=1:n
    % get each kernel function evaluated at x
    % centered at data
    f = exp(-(1/(2*h^2))*(x-data(i)).^2)/sqrt(2*pi)/h;
    plot(x,f/(n*h));
    fhat = fhat+f/(n);
end
plot(x,fhat);
hold off
```

❑

As in the histogram, the parameter h determines the amount of smoothing we have in the estimate $\hat{f}_{Ker}(x)$. In kernel density estimation, the h is usually called the **window width**. A small value of h yields a rough curve, while a large value of h yields a smoother curve. This is illustrated in Figure 9.9, where we show kernel density estimates $\hat{f}_{Ker}(x)$ at various window widths. Notice that when the window width is small, we get a lot of noise or spurious structure in the estimate. When the window width is larger we get a smoother estimate, but there is the possibility that we might obscure bumps or other interesting structure in the estimate. In practice, it is recommended that the analyst examine kernel density estimates for different window widths to search for structures such as modes or bumps.

As with the other univariate probability density estimators, we are interested in determining appropriate values for the parameter h. These can be obtained by choosing values for h that minimize the asymptotic MISE. Scott [2015] shows that, under certain conditions, the AMISE for a nonnegative univariate kernel density estimator is

$$\text{AMISE}_{Ker}(h) = \frac{R(K)}{nh} + \frac{1}{4}\sigma_k^4 h^4 R(f''), \tag{9.28}$$

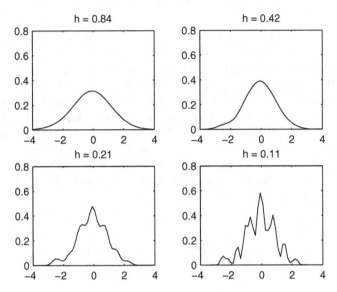

FIGURE 9.9
Four kernel density estimates using $n = 100$ standard normal random variables. Four different window widths are used. Note that as h gets smaller, the estimate gets rougher.

where the kernel K is a continuous probability density function with $\mu_K = 0$ and $0 < \sigma_K^2 < \infty$. The window width that minimizes this is given by

$$h^*_{Ker} = \left(\frac{R(K)}{n\sigma_k^4 R(f'')} \right)^{1/5}. \tag{9.29}$$

Parzen [1962] and Scott [2015] describe the conditions under which this holds. Notice in Equation 9.28 that we have the same bias-variance trade-off with h that we had in previous density estimates.

For a kernel that is equal to the normal density $R(f'') = 3/(8\sqrt{\pi}\sigma^5)$, we have the following normal reference rule for the window width h.

NORMAL REFERENCE RULE – KERNELS

$$h^*_{Ker} = \left(\frac{4}{3} \right)^{1/5} \sigma n^{-1/5} \approx 1.06\sigma n^{-1/5}.$$

We can use some suitable estimate for σ, such as the standard deviation, or $\hat{\sigma} = IQR/1.348$. The latter yields a window width of

$$\hat{h}^*_{Ker} = 0.786 \times IQR \times n^{-1/5}.$$

Silverman [1986] recommends that one use whichever is smaller, the sample standard deviation or $IQR/1.348$ as an estimate for σ.

We now turn our attention to the problem of what kernel to use in our estimate. It is known [Scott, 2015] that the choice of smoothing parameter h is more important than choosing the kernel. This arises from the fact that the effects from the choice of kernel (e.g., kernel tail behavior) are reduced by the averaging process. We discuss the efficiency of the kernels next, but what really drives the choice of a kernel are computational considerations or the amount of differentiability required in the estimate.

As for efficiency, the optimal kernel was shown to be the Epanechnikov kernel given by [Epanechnikov, 1969]

$$K(t) = \begin{cases} \dfrac{3}{4}(1 - t^2); & -1 \le t \le 1 \\ 0; & \text{otherwise.} \end{cases}$$

It is illustrated in Figure 9.10 along with some other kernels.

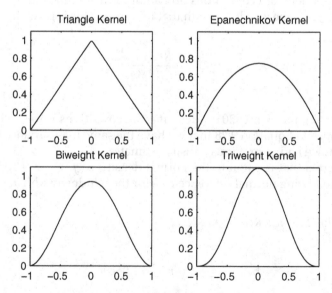

FIGURE 9.10
These illustrate four kernels that can be used in probability density estimation.

Several choices for kernels are given in Table 9.1. Silverman [1986] and Scott [2015] show that these kernels have efficiencies close to that of the Epanechnikov kernel, the least efficient being the normal kernel. Thus, it seems that efficiency should not be the major consideration in deciding what kernel to use. It is recommended that one choose the kernel based on other considerations as stated above.

TABLE 9.1

Examples of Kernels for Density Estimation

Kernel Name	Equation
Triangle	$K(t) = (1 - |t|)$ $\quad -1 \le t \le 1$
Epanechnikov	$K(t) = \frac{3}{4}(1 - t^2)$ $\quad -1 \le t \le 1$
Biweight	$K(t) = \frac{15}{16}(1 - t^2)^2$ $\quad -1 \le t \le 1$
Triweight	$K(t) = \frac{35}{32}(1 - t^2)^3$ $\quad -1 \le t \le 1$
Normal	$K(t) = \frac{1}{\sqrt{2\pi}} \exp\left\{\frac{-t^2}{2}\right\}$ $\quad -\infty < t < \infty$

Multivariate Kernel Estimators

Here we assume that we have a sample of size n, where each observation is a d-dimensional vector, $\mathbf{x}_i$, $i = 1, ..., n$. The simplest case for the multivariate kernel estimator is the product kernel. Descriptions of the general kernel density estimate can be found in Scott [2015] and in Silverman [1986]. The *product kernel* is

$$\hat{f}_{Ker}(\mathbf{x}) = \frac{1}{nh_1...h_d} \sum_{i=1}^{n} \left\{ \prod_{j=1}^{d} K\left(\frac{x_j - X_{ij}}{h_j}\right) \right\}, \tag{9.30}$$

where X_{ij} is the j-th component of the i-th observation. Note that this is the product of the same univariate kernel, with a (possibly) different window width in each dimension. Since the product kernel estimate is comprised of univariate kernels, we can use any of the kernels that were discussed previously.

Scott [2015] gives expressions for the asymptotic integrated squared bias and asymptotic integrated variance for the multivariate product kernel. If the normal kernel is used, then minimizing these yields a normal reference rule for the multivariate case, which is given next.

NORMAL REFERENCE RULE - KERNEL (MULTIVARIATE)

$$h^*_{j_{Ker}} = \left(\frac{4}{n(d+2)}\right)^{\frac{1}{d+4}} \sigma_j; \qquad j = 1, ..., d,$$

where a suitable estimate for σ_j can be used. If there is any skewness or kurtosis evident in the data, then the window widths should be narrower, as discussed previously. The skewness factor for the frequency polygon (Equation 9.20) can be used here.

Example 9.7

In this example, we construct the product kernel estimator for the **iris** data. To make it easier to visualize, we use only the first two variables (sepal length and sepal width) for each species. So, we first create a data matrix comprised of the first two columns for each species.

```
load iris
% Create bivariate data matrix with all three species.
data = [setosa(:,1:2)];
data(51:100,:) = versicolor(:,1:2);
data(101:150,:) = virginica(:,1:2);
```

Next we obtain the smoothing parameter using the normal reference rule.

```
% Get the window width using the Normal Ref Rule.
[n,p] = size(data);
s = sqrt(var(data));
hx = s(1)*n^(-1/6);
hy = s(2)*n^(-1/6);
```

The next step is to create a grid over which we will construct the estimate.

```
% Get the ranges for x and y & construct grid.
num_pts = 30;
minx = min(data(:,1));
maxx = max(data(:,1));
miny = min(data(:,2));
maxy = max(data(:,2));
gridx = ((maxx+2*hx)-(minx-2*hx))/num_pts
gridy = ((maxy+2*hy)-(miny-2*hy))/num_pts
[X,Y]=meshgrid((minx-2*hx):gridx:(maxx+2*hx),...
```

```
          (miny-2*hy):gridy:(maxy+2*hy));
x = X(:);     %put into col vectors
y = Y(:);
```

We are now ready to get the estimates. Note that in this example, we are changing the form of the loop. Instead of evaluating each weighted curve and then averaging, we will be looping over each point in the domain.

```
z = zeros(size(x));
for i=1:length(x)
    xloc = x(i)*ones(n,1);
    yloc = y(i)*ones(n,1);
    argx = ((xloc-data(:,1))/hx).^2;
    argy = ((yloc-data(:,2))/hy).^2;
    z(i) = (sum(exp(-.5*(argx+argy))))/(n*hx*hy*2*pi);
end
[mm,nn] = size(X);
Z = reshape(z,mm,nn);
```

We show the surface plot for this estimate in Figure 9.11. As before, we can verify that our estimate is a *bona fide* by estimating the area under the curve. In this example, we get an area of 0.9994.

```
          area = sum(sum(Z))*gridx*gridy;
```

❑

The Statistics toolbox has a function called **ksdensity** that will compute a univariate kernel density estimate of a random sample. There are several options for the type of kernel, the bandwidth, and more.

Before leaving this section, we present a summary of univariate probability density estimators and their corresponding normal reference rule for the smoothing parameter h. These are given in Table 9.2.

9.4 Finite Mixtures

So far, we have been discussing nonparametric density estimation methods that require a choice of smoothing parameter h. In the previous section, we showed that we can get different estimates of our probability density depending on our choice for h. It would be helpful if we could avoid choosing a smoothing parameter. In this section, we present a method called finite mixtures that does not require a smoothing parameter. However, as is often the case, when we eliminate one parameter we end up replacing it with another. In finite mixtures, we do not have to worry about the smoothing

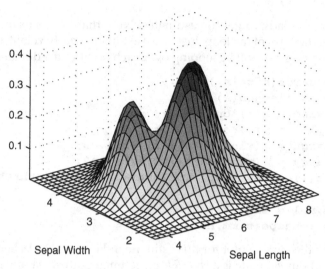

FIGURE 9.11
This is the product kernel density estimate for the sepal length and sepal width of the **iris** *data. These data contain all three species. The presence of peaks in the data indicate that two of the species might be distinguishable based on these variables.*

TABLE 9.2

Summary of Univariate Probability Density Estimators and the Normal Reference Rule for the Smoothing Parameter

Method	Estimator	Normal Reference Rule
Histogram	$\hat{f}_{Hist}(x) = \dfrac{v_k}{nh}$ x in B_k	$h^{*}_{Hist} = 3.5\sigma n^{-1/3}$
Frequency Polygon	$\hat{f}_{FP}(x) = \left(\dfrac{1}{2} - \dfrac{x}{h}\right)\hat{f}_k + \left(\dfrac{1}{2} + \dfrac{x}{h}\right)\hat{f}_{k+1}$ $\bar{B}_K \leq x \leq \bar{B}_{k+1}$	$h^{*}_{FP} = 2.15\sigma n^{-1/5}$
Kernel	$\hat{f}_{Ker}(x) = \dfrac{1}{nh}\sum_{i=1}^{n} K\left(\dfrac{x - X_i}{h}\right)$	$h^{*}_{Ker} = 1.06\sigma n^{-1/5}$; K is the normal kernel.

parameter. Instead, we have to determine the number of terms in the mixture.

Finite mixtures offer advantages in the area of the computational load put on the system. Two issues to consider with many probability density estimation methods are the computational burden in terms of the amount of information we have to store and the computational effort needed to obtain the probability density estimate at a point. We can illustrate these ideas using the kernel density estimation method. To evaluate the estimate at a point x (in the univariate case) we have to retain all of the data points because the estimate is a weighted sum of n kernels centered at each sample point. In addition, we must calculate the value of the kernel n times. The situation for histograms and frequency polygons is a little better. The amount of information we must store to provide an estimate of the probability density is essentially driven by the number of bins. Of course, the situation becomes worse when we move to multivariate kernel estimates, histograms, and frequency polygons. With the massive, high-dimensional data sets we often work with, the computational effort and the amount of information that must be stored to use the density estimates is an important consideration. Finite mixtures is a technique for estimating probability density functions that can require relatively little computer storage space or computations to evaluate the density estimates.

Univariate Finite Mixtures

The finite mixture method assumes the density $f(x)$ can be modeled as the sum of c weighted densities, with $c \ll n$. The most general case for the univariate finite mixture is

$$f(x) = \sum_{i=1}^{c} p_i g(x; \theta_i),$$

(9.31)

where p_i represents the *weight* or *mixing coefficient* for the i-th term, and $g(x; \theta_i)$ denotes a probability density, with parameters represented by the vector θ_i. To make sure that this is a *bona fide* density, we must impose the condition that $p_1 + \ldots + p_c = 1$ and $p_i > 0$. To evaluate $f(x)$, we take our point x, find the value of the component densities $g(x; \theta_i)$ at that point, and take the weighted sum of these values.

Example 9.8

The following example shows how to evaluate a finite mixture model at a given x. We construct the curve for a three-term finite mixture model, where the component densities are taken to be normal. The model is given by

$$f(x) = 0.3 \times \phi(x;-3, 1) + 0.3 \times \phi(x;0, 1) + 0.4 \times \phi(x;2, 0.5),$$

where $\phi(x;\mu, \sigma^2)$ represents the normal probability density function at x. We see from the model that we have three terms or component densities, centered at -3, 0, and 2. The mixing coefficient or weight for the first two terms are 0.3 leaving a weight of 0.4 for the last term. The following MATLAB code produces the curve for this model and is shown in Figure 9.12.

```
% Create a domain x for the mixture.
x = linspace(-6,5);
% Create the model - normal components used.
mix = [0.3 0.3 0.4];        % mixing coefficients
mus = [-3 0 2];             % term means
vars = [1 1 0.5];
nterm = 3;
% Use Statistics Toolbox function to evaluate
% normal pdf.
fhat = zeros(size(x));
for i = 1:nterm
    fhat = fhat+mix(i)*normpdf(x,mus(i),vars(i));
end
plot(x,fhat)
title('3 Term Finite Mixture')
```

❏

Hopefully, the reader can see the connection between finite mixtures and kernel density estimation. Recall that in the case of univariate kernel density estimators, we obtain these by evaluating a weighted kernel centered at each sample point, and adding these n terms. So, a kernel estimate can be considered a special case of a finite mixture where $c = n$.

The component densities of the finite mixture can be any probability density function, continuous or discrete. In this book, we confine our attention to the continuous case and use the normal density for the component function. Therefore, the estimate of a finite mixture would be written as

$$\hat{f}_{FM}(x) = \sum_{i=1}^{c} \hat{p}_i \phi(x;\hat{\mu}_i, \hat{\sigma}_i^2), \tag{9.32}$$

where $\phi(x;\hat{\mu}_i, \hat{\sigma}_i^2)$ denotes the normal probability density function with mean $\hat{\mu}_i$ and variance $\hat{\sigma}_i^2$. In this case, we have to estimate $c - 1$ independent mixing coefficients, as well as the c means and c variances using the data. Note that to evaluate the density estimate at a point x, we only need to retain these $3c - 1$ parameters. Since we typically have $c \ll n$, this can be a significant computational savings over evaluating density estimates using

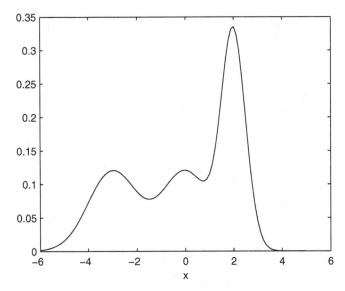

FIGURE 9.12
This shows the probability density function corresponding to the three-term finite mixture model from Example 9.8.

the kernel method. With finite mixtures much of the computational burden is shifted to the estimation part of the problem.

Visualizing Finite Mixtures

The methodology used to estimate the parameters for finite mixture models will be presented later on in this section. We first show a method for visualizing the underlying structure of finite mixtures with normal component densities [Priebe, et al. 1994] because it is used to help visualize and explain another approach to density estimation—a technique called adaptive mixtures. Here, structure refers to the number of terms in the mixture, along with the component means and variances. In essence, we are trying to visualize the high-dimensional parameter space (recall there are $3c - 1$ parameters for the univariate mixture of normals) in a 2D representation. This is called a *dF* *plot*, where each component is represented by a circle. The circles are centered at the mean μ_i and the mixing coefficient p_i. The size of the radius of the circle indicates the standard deviation. An example of a *dF* plot is given in Figure 9.13 and is discussed in the following example.

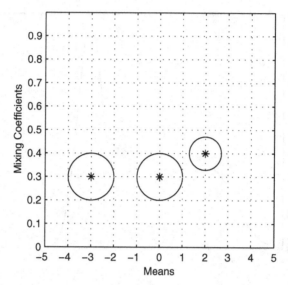

FIGURE 9.13
This shows the dF plot for the three-term finite mixture model of Figure 9.12.

Example 9.9

We construct a *dF* plot for the finite mixture model discussed in the previous example. Recall that the model is given by

$$f(x) = 0.3 \times \phi(x;-3, 1) + 0.3 \times \phi(x;0, 1) + 0.4 \times \phi(x;2, 0.5).$$

Our first step is to set up the model consisting of the number of terms, the component parameters, and the mixing coefficients.

```
% Recall the model - normal components used.
mix = [0.3 0.3 0.4];        % mixing coefficients
mus = [-3 0 2];             % term means
vars = [1 1 0.5];
nterm = 3;
```

Next we set up the figure for plotting. Note that we rescale the mixing coefficients for easier plotting on the vertical axis and then map the labels to the corresponding value.

```
t = 0:.05:2*pi+eps;  % values to create circle
% To get some scales right.
minx = -5;
maxx = 5;
```

```
scale = maxx-minx;
lim = [minx maxx minx maxx];
% Set up the axis limits.
figure
axis equal
axis(lim)
grid on
% Create and plot a circle for each term.
hold on
for i=1:nterm
    % rescale for plotting purposes
    ycord = scale*mix(i)+minx;
    xc = mus(i)+sqrt(vars(i))*cos(t);
    yc = ycord+sqrt(vars(i))*sin(t);
    plot(xc,yc,mus(i),ycord,'*')
end
hold off
% Relabel the axis to show the right coefficient.
tick = (maxx-minx)/10;
set(gca,'Ytick',minx:tick:maxx)
set(gca,'XTick',minx:tick:maxx)
set(gca,'YTickLabel',{'0','0.1','0.2','0.3',...
    '0.4','0.5','0.6','0.7','0.8','0.9','1'})
xlabel('Means'),ylabel('Mixing Coefficients')
title('dF Plot for Univariate Finite Mixture')
```

The first circle on the left corresponds to the component with $p_i = 0.3$ and $\mu_i = -3$. Similarly, the middle circle of Figure 9.13 represents the second term of the model. Note that this representation of the mixture makes it easier to see which terms carry more weight and where they are located in the domain.

❑

Multivariate Finite Mixtures

Finite mixtures are easily extended to the multivariate case. Here we define the multivariate finite mixture model as the weighted sum of multivariate component densities,

$$f(\mathbf{x}) = \sum_{i=1}^{c} p_i g(\mathbf{x}; \boldsymbol{\theta}_i).$$

As before, the mixing coefficients or weights must be nonnegative and sum to one, and the component density parameters are represented by $\boldsymbol{\theta}_i$. When we are estimating the function, we often use the multivariate normal as the component density. This gives the following equation for an estimate of a multivariate finite mixture:

$$\hat{f}_{FM}(\mathbf{x}) = \sum_{i=1}^{c} \hat{p}_i \phi(\mathbf{x}; \hat{\boldsymbol{\mu}}_i, \hat{\boldsymbol{\Sigma}}_i), \qquad (9.33)$$

where $\mathbf{x}$ is a d-dimensional vector, $\hat{\boldsymbol{\mu}}_i$ is a d-dimensional vector of means, and $\hat{\boldsymbol{\Sigma}}_i$ is a $d \times d$ covariance matrix. There are still $c - 1$ mixing coefficients to estimate. However, there are now $c \times d$ values that have to be estimated for the means and $(cd(c + 1))/2$ values for the component covariance matrices.

The dF representation has been extended [Solka, Poston, and Wegman, 1995] to show the structure of a multivariate finite mixture, when the data are 2D or 3D. In the 2D case, we represent each term by an ellipse centered at the mean of the component density $\hat{\boldsymbol{\mu}}_i$, with the eccentricity of the ellipse showing the covariance structure of the term. For example, a term with a covariance that is close to the identity matrix will appear as a circle. We label the center of each ellipse with text identifying the mixing coefficient. An example is illustrated in Figure 9.14.

A dF plot for a trivariate finite mixture can be fashioned by using color to represent the values of the mixing coefficients. In this case, we use the three dimensions in our plot to represent the means for each term. Instead of ellipses, we move to ellipsoids, with eccentricity determined by the covariance as before. See Figure 9.15 for an example of a trivariate dF plot. The dF plots are particularly useful when working with the adaptive mixtures density estimation method that will be discussed shortly. We provide a function called **csdfplot** that will implement the dF plots for univariate, bivariate, and trivariate data.

Example 9.10

In this example, we show how to implement the function called **csdfplot** and illustrate its use with bivariate and trivariate models. The bivariate case is the following three-component model:

$$p_1 = 0.5 \qquad p_2 = 0.3 \qquad p_3 = 0.2,$$

$$\boldsymbol{\mu}_1 = \begin{bmatrix} -1 \\ -1 \end{bmatrix} \qquad \boldsymbol{\mu}_2 = \begin{bmatrix} 1 \\ 1 \end{bmatrix} \qquad \boldsymbol{\mu}_3 = \begin{bmatrix} 5 \\ 6 \end{bmatrix},$$

$$\boldsymbol{\Sigma}_1 = \begin{bmatrix} 1 & 0 \\ 0 & 1 \end{bmatrix} \qquad \boldsymbol{\Sigma}_2 = \begin{bmatrix} 0.5 & 0 \\ 0 & 0.5 \end{bmatrix} \qquad \boldsymbol{\Sigma}_3 = \begin{bmatrix} 1 & 0.5 \\ 0.5 & 1 \end{bmatrix}.$$

```
% First create the model.
% The function expects a vector of weights;
% a matrix of means, where each column of the matrix
% corresponds to a d-D mean; a 3D array of
% covariances, where each page of the array is a
% covariance matrix.
pies = [0.5 0.3 0.2]; % mixing coefficients
mus = [-1 1 5; -1 1 6];
% Delete any previous variances in the workspace.
clear vars
vars(:,:,1) = eye(2);
vars(:,:,2) = eye(2)*.5
vars(:,:,3) = [1 0.5; 0.5 1];
figure
csdfplot(mus,vars,pies)
```

The resulting plot is shown in Figure 9.14. Note that the covariance of two of the component densities is represented by circles, with one larger than the other. These correspond to the first two terms of the model. The third component density has an elliptical covariance structure indicating non-zero off-diagonal elements in the covariance matrix. We now do the same thing for the trivariate case, where the model is

$$\mu_1 = \begin{bmatrix} -1 \\ -1 \\ -1 \end{bmatrix} \qquad \mu_2 = \begin{bmatrix} 1 \\ 1 \\ 1 \end{bmatrix} \qquad \mu_3 = \begin{bmatrix} 5 \\ 6 \\ 2 \end{bmatrix},$$

$$\Sigma_1 = \begin{bmatrix} 1 & 0 & 0 \\ 0 & 1 & 0 \\ 0 & 0 & 1 \end{bmatrix} \qquad \Sigma_2 = \begin{bmatrix} 0.5 & 0 & 0 \\ 0 & 0.5 & 0 \\ 0 & 0 & 0.5 \end{bmatrix} \qquad \Sigma_3 = \begin{bmatrix} 1 & 0.7 & 0.2 \\ 0.7 & 1 & 0.5 \\ 0.2 & 0.5 & 1 \end{bmatrix}.$$

The mixing coefficients are the same as before. We need only to adjust the means and the covariance accordingly.

```
mus(3,:) = [-1 1 2];
% Delete previous vars array or you will get an error.
clear vars
vars(:,:,1) = eye(3);
vars(:,:,2) = eye(3)*.5;
vars(:,:,3)=[1 0.7 0.2;
             0.7 1 0.5;
             0.2 0.5 1];
figure
csdfplot(mus,vars,pies)
```

```
% get a different viewpoint
view([-34,9])
```

The trivariate *dF* plot for this model is shown in Figure 9.15. Two terms (the first two) are shown as spheres and one as an ellipsoid.
❑

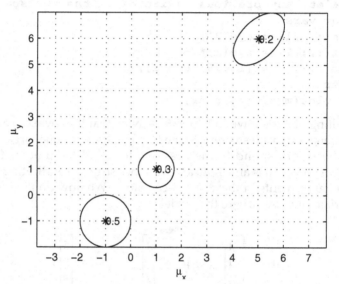

FIGURE 9.14
This is a bivariate dF plot for the three-term mixture model of Example 9.10.

EM Algorithm for Estimating the Parameters

The problem of estimating the parameters in a finite mixture has been studied extensively in the literature. The book by Everitt and Hand [1981] provides an excellent overview of this topic and offers several methods for parameter estimation. The technique we present here is called the *Expectation-Maximization* (EM) method. This is a general method for optimizing likelihood functions and is useful in situations where data might be missing or simpler optimization methods fail. The seminal paper on this topic is by Dempster, Laird, and Rubin [1977], where they formalize the EM algorithm and establish its properties. Redner and Walker [1984] apply it to mixture densities. The EM methodology is now a standard tool for statisticians and is used in many applications.

In this section, we discuss the EM algorithm as it can be applied to estimating the parameters of a finite mixture of normal or Gaussian densities.

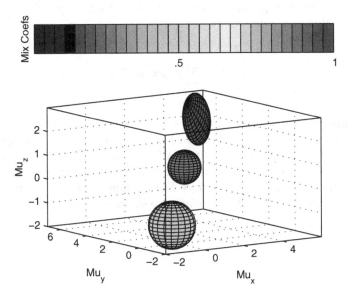

FIGURE 9.15
This is the trivariate dF plot for the three-term mixture model of Example 9.10.

We will revisit the EM algorithm in Chapter 11 in the context of model-based clustering, where we provide more details on the procedure.

To use the EM algorithm, we must have a value for the number of terms c in the mixture. This is usually obtained using prior knowledge of the application (the analyst thinks there are a certain number of groups), using graphical exploratory data analysis (looking for clusters or other group structure), or using some other method of estimating the number of terms. The approach called adaptive mixtures [Priebe, 1994] offers a way to address the problem of determining the number of component densities to use in the finite mixture model. This approach is discussed later in this chapter.

Besides the number of terms, we must also have an initial guess for the value of the component parameters. Once we have an initial estimate, we update the parameter estimates using the data and the equations given next. These are called the iterative EM update equations, and we provide the multivariate case as the most general one. The univariate case follows easily.

The first step is to determine the posterior probabilities given by

$$\hat{\tau}_{ij} = \frac{\hat{p}_i \phi(\mathbf{x}_j; \hat{\mu}_i, \hat{\Sigma}_i)}{\hat{f}(\mathbf{x}_j)}; \qquad i = 1, ..., c\,; j = 1, ..., n\,, \qquad (9.34)$$

where $\hat{\tau}_{ij}$ represents the estimated posterior probability that point x_j belongs to the i-th term, $\phi(x_j;\hat{\mu}_i, \hat{\Sigma}_i)$ is the multivariate normal density for the i-th term evaluated at x_j, and

$$\hat{f}(x_j) = \sum_{k=1}^{c} \hat{p}_k \phi(x_j;\hat{\mu}_i, \hat{\Sigma}_i) \tag{9.35}$$

is the finite mixture estimate at point x_j.

The posterior probability tells us the likelihood that a point belongs to each of the separate component densities. We can use this estimated posterior probability to obtain a weighted update of the parameters for each component. This yields the iterative EM update equations for the mixing coefficients, the means, and the covariance matrices. These are

$$\hat{p}_i = \frac{1}{n} \sum_{j=1}^{n} \hat{\tau}_{ij} \tag{9.36}$$

$$\hat{\mu}_i = \frac{1}{n} \sum_{j=1}^{n} \frac{\hat{\tau}_{ij} x_j}{\hat{p}_i} \tag{9.37}$$

$$\hat{\Sigma}_i = \frac{1}{n} \sum_{j=1}^{n} \frac{\hat{\tau}_{ij}(x_j - \hat{\mu}_i)(x_j - \hat{\mu}_i)^T}{\hat{p}_i} . \tag{9.38}$$

Note that if $d = 1$, then the update equation for the variance is

$$\hat{\sigma}_i^2 = \frac{1}{n} \sum_{j=1}^{n} \frac{\hat{\tau}_{ij}(x_j - \hat{\mu}_i)^2}{\hat{p}_i} . \tag{9.39}$$

The steps for the EM algorithm to estimate the parameters for a finite mixture with multivariate normal components are given here and are illustrated in Example 9.11.

FINITE MIXTURES – EM PROCEDURE

1. Determine the number of terms or component densities c in the mixture.

2. Determine an initial guess at the component parameters. These are the mixing coefficients, means, and covariance matrices for each normal density.

3. For each data point x_j, calculate the posterior probability using Equation 9.34.

4. Update the mixing coefficients, the means, and the covariance matrices for the individual components using Equations 9.36 through 9.38.

5. Repeat steps 3 through 4 until the estimates converge.

Typically, step 5 is implemented by continuing the iteration until the changes in the estimates at each iteration are less than some preset tolerance. Note that with the iterative EM algorithm, we need to use the entire data set to simultaneously update the parameter estimates. This imposes a high computational load when dealing with massive data sets.

Example 9.11

In this example, we provide the MATLAB code that implements the multivariate EM algorithm for estimating the parameters of a finite mixture probability density model. To illustrate this, we will generate a data set that is a mixture of two terms with equal mixing coefficients. One term is centered at the point $(-2, 2)$, and the other is centered at $(2, 0)$. The covariance of each component density is given by the identity matrix. Our first step is to generate 200 data points from this distribution.

```
% Create some artificial two-term mixture data.
n = 200;
data = zeros(n,2);
% Now generate 200 random variables. First find
% the number that come from each component.
r = rand(1,n);
% Find the number generated from component 1.
ind = length(find(r <= 0.5));
% Create some mixture data. Note that the
% component densities are multivariate normals.
% Generate the first term.
data(1:ind,1) = randn(ind,1) - 2;
data(1:ind,2) = randn(ind,1) + 2;
% Generate the second term.
data(ind+1:n,1) = randn(n-ind,1) + 2;
data(ind+1:n,2) = randn(n-ind,1);
```

We must then specify various parameters for the EM algorithm, such as the number of terms.

```
c = 2;     % number of terms
```

```
[n,d] = size(data);  % n=# pts, d=# dims
tol = 0.00001;    % set up criterion for stopping EM
max_it = 100;
totprob = zeros(n,1);
```

We also need an initial guess at the component density parameters.

```
% Get the initial parameters for the model to start EM
mu(:,1) = [-1 -1]';  % each column represents a mean
mu(:,2) = [1 1]';
mix_cof = [0.3 0.7];
var_mat(:,:,1) = eye(d);
var_mat(:,:,2) = eye(d);
varup = zeros(size(var_mat));
muup = zeros(size(mu));
% Just to get started.
num_it = 1;
deltol = tol+1;% to get started
```

The following code implements the EM updates found in Equations 9.34
through 9.38.

```
while num_it <= max_it & deltol > tol
   % get the posterior probabilities
   totprob = zeros(n,1);
   for i=1:c
     posterior(:,i) = mix_cof(i)*...
         csevalnorm(data,mu(:,i)',var_mat(:,:,i));
      totprob = totprob+posterior(:,i);
   end
   den = totprob*ones(1,c);
   posterior = posterior./den;
   % Update the mixing coefficients.
   mix_cofup = sum(posterior)/n;
   % Update the means.
   mut = data'*posterior;
   MIX = ones(d,1)*mix_cof;
   muup = mut./(MIX*n);
   % Update the means and the variances.
   for i=1:c
      cen_data = data-ones(n,1)*mu(:,i)';
      mat = cen_data'*...
         diag(posterior(:,i))*cen_data;
      varup(:,:,i)=mat./(mix_cof(i)*n);
   end
   % Get the tolerances.
   delvar = max(max(max(abs(varup-var_mat))));
   delmu = max(max(abs(muup-mu)));
```

```
      delpi = max(abs(mix_cof-mix_cofup));
      deltol = max([delvar,delmu,delpi]);
      % Reset parameters.
      num_it = num_it+1;
      mix_cof = mix_cofup;
      mu = muup;
      var_mat = varup;
  end   % while loop
```

For our data set, it took 37 iterations to converge to an answer. The convergence of the EM algorithm to a solution and the number of iterations depends on the tolerance, the initial parameters, the data set, etc. The estimated model returned by the EM algorithm is

$$\hat{p}_1 = 0.498 \qquad \hat{p}_2 = 0.502,$$

$$\hat{\mu}_1 = \begin{bmatrix} -2.08 \\ 2.03 \end{bmatrix} \qquad \hat{\mu}_2 = \begin{bmatrix} 1.83 \\ -0.03 \end{bmatrix}.$$

For brevity, we omit the estimated covariances, but we can see from these results that the model does match the data that we generated.

❑

We will return to the EM algorithm in Chapter 11, where we discuss the model-based clustering methodology. This clustering framework will offer another approach that addresses the issues of choosing the number of terms and calculating starting values of the parameters for the EM algorithm.

Adaptive Mixtures

The *adaptive mixtures* [Priebe, 1994] method for density estimation uses a data-driven approach for estimating the number of component densities in a mixture model. This technique uses the recursive EM update equations that are provided next. The basic idea behind adaptive mixtures is to take one point at a time and determine the distance from the observation to each component density in the model. If the distance to each component is larger than some threshold, then a new term is created. If the distance is less than the threshold for all terms, then the parameter estimates are updated based on the recursive EM equations.

We start our explanation of the adaptive mixtures approach with a description of the recursive EM algorithm for mixtures of multivariate normal densities. This method recursively updates the parameter estimates based on a new observation. As before, the first step is to determine the posterior probability that the new observation belongs to each term:

$$\hat{\tau}_i^{(n+1)} = \frac{\hat{p}_i^{(n)} \phi(\mathbf{x}^{(n+1)}; \hat{\mu}_i^{(n)}, \hat{\Sigma}_i^{(n)})}{\hat{f}^{(n)}(\mathbf{x}^{(n+1)})}; \qquad i = 1, \ldots, c, \qquad (9.40)$$

where $\hat{\tau}_i^{(n+1)}$ represents the estimated posterior probability that the new observation $\mathbf{x}^{(n+1)}$ belongs to the i-th term, and the superscript (n) denotes the estimated parameter values based on the previous n observations. The denominator is the finite mixture density estimate

$$\hat{f}^{(n)}(\mathbf{x}^{(n+1)}) = \sum_{i=1}^{c} \hat{p}_i \phi(\mathbf{x}^{(n+1)}; \hat{\mu}_i^{(n)}, \hat{\Sigma}_i^{(n)})$$

for the new observation using the mixture from the previous n points.

The remainder of the recursive EM update equations are given by Equations 9.41 through 9.43. Note that recursive equations are typically in the form of the old value for an estimate plus an update term using the new observation. The recursive update equations for mixtures of multivariate normals are

$$\hat{p}_i^{(n+1)} = \hat{p}_i^{(n)} + \frac{1}{n}(\hat{\tau}_i^{(n+1)} - \hat{p}_i^{(n)}) \qquad (9.41)$$

$$\hat{\mu}_i^{(n+1)} = \hat{\mu}_i^{(n)} + \frac{\hat{\tau}_i^{(n+1)}}{n\hat{p}_i^{(n)}}(\mathbf{x}^{(n+1)} - \hat{\mu}_i^{(n)}) \qquad (9.42)$$

$$\hat{\Sigma}_i^{(n+1)} = \hat{\Sigma}_i^{(n)} + \frac{\hat{\tau}_i^{(n+1)}}{n\hat{p}_i^{(n)}}\left[(\mathbf{x}^{(n+1)} - \hat{\mu}_i^{(n)})(\mathbf{x}^{(n+1)} - \hat{\mu}_i^{(n)})^T - \hat{\Sigma}_i^{(n)}\right]. \qquad (9.43)$$

This reduces to the 1D case in a straightforward manner, as was the case with the iterative EM update equations.

The adaptive mixtures approach updates our probability density estimate $\hat{f}(\mathbf{x})$ and also provides the opportunity to expand the parameter space (i.e., the model) if the data indicate that should be done. To accomplish this, we need a way to determine when a new component density should be added. This could be done in several ways, but the one we present here is based on the Mahalanobis distance. If this distance is too large for all of the terms (or alternatively if the minimum distance is larger than some threshold), then we can consider the new point too far away from the existing terms to update the current model. Therefore, we create a new term.

The squared Mahalanobis distance between the new observation $\mathbf{x}^{(n+1)}$ and the i-th term is given by

$$MD_i^2(\mathbf{x}^{(n+1)}) = (\mathbf{x}^{(n+1)} - \hat{\boldsymbol{\mu}}_i^{(n)})^T (\hat{\boldsymbol{\Sigma}}_i^{(n)})^{-1} (\mathbf{x}^{(n+1)} - \hat{\boldsymbol{\mu}}_i^{(n)}). \tag{9.44}$$

We create a new term if

$$\min_i \{ MD_i^2(\mathbf{x}^{(n+1)}) \} > t_C, \tag{9.45}$$

where t_C is a threshold to create a new term. The rule in Equation 9.45 states that if the smallest squared Mahalanobis distance is greater than the threshold, then we create a new term.

In the univariate case, if $t_C = 1$ is used, then a new term is created if a new observation is more than one standard deviation away from the mean of each term. For $t_C = 4$, a new term would be created for an observation that is at least two standard deviations away from the existing terms. For multivariate data, we would like to keep the same term creation rate as in the 1D case. Solka [1995] provides thresholds t_C based on the squared Mahalanobis distance for the univariate, bivariate, and trivariate cases. These are shown in Table 9.3.

TABLE 9.3

Recommended Thresholds for Adaptive Mixtures

Dimensionality	Create Threshold
1	1
2	2.34
3	3.54

When we create a new term, we can initialize the parameters using Equations 9.46 through 9.48. We denote the current number of terms in the model by N.

$$\hat{\boldsymbol{\mu}}_{N+1}^{(n+1)} = \mathbf{x}^{(n+1)}, \tag{9.46}$$

$$\hat{p}_{N+1}^{(n+1)} = \frac{1}{n+1}, \tag{9.47}$$

$$\hat{\boldsymbol{\Sigma}}_{N+1}^{(n+1)} = \Im(\hat{\boldsymbol{\Sigma}}_i), \tag{9.48}$$

where $\Im(\hat{\Sigma}_i)$ is a weighted average using the posterior probabilities. In practice, some other estimate or initial covariance can be used for the new term. To ensure that the mixing coefficients sum to one when a new term is added, the $\hat{p}_i^{(n+1)}$ must be rescaled using

$$\hat{p}_i^{(n+1)} = \frac{n\hat{p}_i^{(n)}}{n+1}; \qquad i = 1, \ldots, N.$$

We continue through the data set, one point at a time, adding new terms as necessary. Our density estimate is then given by

$$\hat{f}_{AM}(\mathbf{x}) = \sum_{i=1}^{N} \hat{p}_i \phi(\mathbf{x}; \hat{\mu}_i, \hat{\Sigma}_i). \qquad (9.49)$$

This allows for a variable number of terms N, where usually $N \ll n$. The adaptive mixtures technique is captured in the procedure given here, and a function called **csadpmix** is provided with the Computational Statistics Toolbox. Its use in the univariate case is illustrated in Example 9.12.

ADAPTIVE MIXTURES PROCEDURE

1. Initialize the adaptive mixtures procedure using the first data point $\mathbf{x}^{(1)}$, as follows:

$$\hat{\mu}_1^{(1)} = \mathbf{x}^{(1)}, \hat{p}_1^{(1)} = 1, \text{ and } \hat{\Sigma}_1^{(1)} = \mathbf{I},$$

where $\mathbf{I}$ denotes the identity matrix. In the univariate case, the variance of the initial term is one.

2. For a new data point $\mathbf{x}^{(n+1)}$, calculate the squared Mahalanobis distance as in Equation 9.44.

3. If the minimum squared distance is greater than t_C, then create a new term using Equations 9.46 through 9.48. Increase the number of terms N by one.

4. If the minimum squared distance is less than the create threshold t_C, then update the existing terms using Equations 9.41 through 9.43.

5. Continue steps 2 through 4 using all data points.

In practice, the adaptive mixtures method is used to get initial values for the parameters, as well as an estimate of the number of terms needed to model the density. One would then use these as a starting point and apply the iterative EM algorithm to refine the estimates.

Example 9.12

In this example, we illustrate the MATLAB code that implements the univariate adaptive mixtures density estimation procedure. We generate random variables using the three-term mixture model that was discussed in Example 9.9. Recall that the model is given by

$$f(x) = 0.3 \times \phi(x;-3, 1) + 0.3 \times \phi(x;0, 1) + 0.4 \times \phi(x;2, 0.5).$$

```
% Get the true model to generate data.
pi_tru = [0.3 0.3 0.4];
n = 100;
x = zeros(n,1);
% Now generate 100 random variables. First find
% the number that fall in each one.
r = rand(1,100);
% Find the number generated from each component.
ind1 = length(find(r <= 0.3));
ind2 = length(find(r > 0.3 & r <= 0.6));
ind3 = length(find(r > 0.6));
% create some artificial 3 term mixture data
x(1:ind1) = randn(ind1,1) - 3;
x(ind1+1:ind2+ind1)=randn(ind2,1);
x(ind1+ind2+1:n) = randn(ind3,1)*sqrt(0.5)+2;
```

We now call the adaptive mixtures function **csadpmix** to estimate the model.

```
% Now call the adaptive mixtures function.
maxterms = 25;
[pihat,muhat,varhat] = csadpmix(x,maxterms);
```

The following MATLAB commands provide the plots shown in Figure 9.16.

```
% Get the plots.
csdfplot(muhat,varhat,pihat,min(x),max(x));
axis equal
nterms = length(pihat);
figure
csplotuni(pihat,muhat,varhat,...
min(x)-5,max(x)+5,100)
```

We reorder the observations and repeat the process to get the plots in Figure 9.17.

```
% Now re-order the points and repeat
% the adaptive mixtures process.
ind = randperm(n);
x = x(ind);
```

```
[pihat,muhat,varhat] = csadpmix(x,maxterms);
```
❑

Our example above demonstrates some interesting things to consider with adaptive mixtures. First, the model complexity or the number of terms is sometimes greater than is needed. For example, in Figure 9.16, we show a *dF* plot for the three-term mixture model in Example 9.12. Note that the adaptive mixture approach yields more than three terms. This is a problem with mixture models in general. Different models (number of terms and estimated component parameters) can produce essentially the same function estimate (or curve) for $\hat{f}(x)$. This is illustrated in Figures 9.16 and 9.17, where we see that similar curves are obtained from two different models for the same data set. These results are straight from the adaptive mixtures density estimation approach. In other words, we did not use this estimate as an initial starting point for the EM approach. If we had applied the iterative EM to these estimated models, then the curves should be the same.

The other issue that must be considered when using the adaptive mixtures approach is that the resulting model or estimated probability density function depends on the order in which the data are presented to the algorithm. This is also illustrated in Figures 9.16 and 9.17, where the second estimated model is obtained after reordering the data. These issues were addressed by Solka [1995].

9.5 Generating Random Variables

In the introduction, we discussed several uses of probability density estimates, and it is our hope that the reader will discover many more. One of the applications of density estimation is in the area of modeling and simulation. Recall that a key aspect of modeling and simulation is the collection of data generated according to some underlying random process and the desire to generate more random variables from the same process for simulation purposes. One option is to use one of the density estimation techniques discussed in this chapter and randomly sample from that distribution. In this section, we provide the methodology for generating random variables from finite or adaptive mixtures density estimates.

We have already seen an example of this process in Example 9.11 and Example 9.12. The procedure is to first choose the class membership of generated observations based on uniform (0,1) random variables. The number of random variables generated from each component density is given by the corresponding proportion of these uniform variables that are in the required range. The steps are outlined here.

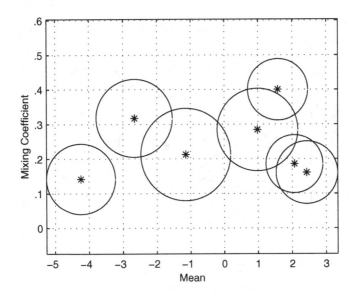

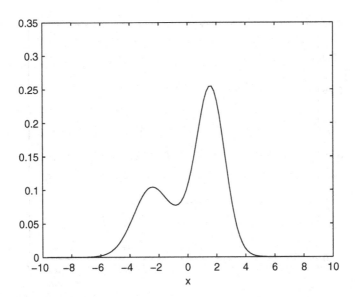

FIGURE 9.16

The upper plot shows the dF representation for Example 9.12. Compare this with Figure 9.17 for the same data. Note that the curves are essentially the same, but the number of terms and associated parameters are different. Thus, we can get different models for the same data.

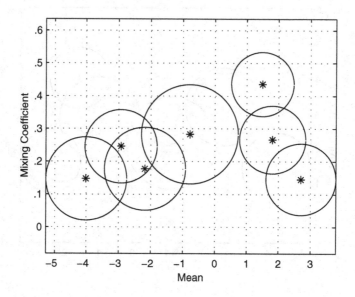

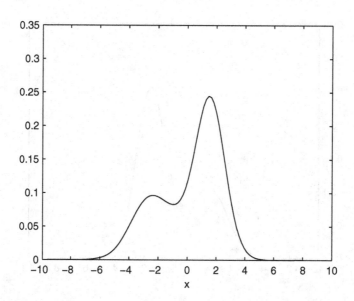

FIGURE 9.17
This is the second estimated model using adaptive mixtures for the data generated in Example 9.12. This second model was obtained by reordering the data set and then implementing the adaptive mixtures technique. This shows the dependence of the technique on the order in which the data are presented to the method.

PROCEDURE – GENERATING RANDOM VARIABLES (FINITE MIXTURE)

1. We are given a finite mixture model $(p_i,\ g_i(\mathbf{x};\boldsymbol{\theta}_i))$ with c components, and we want to generate n random variables from that distribution.

2. First determine the component membership of each of the n random variables. We do this by generating n uniform $(0,1)$ random variables (U_i). Component membership is determined as follows:

   If $0 \le U_i < p_1$, then X_i is from component density 1.

   If $p_1 \le U_i < p_1 + p_2$, then X_i is from component density 2.

   $\cdots$

   If $\displaystyle\sum_{j=1}^{c-1} p_j \le U_i \le 1$, then X_i is from component density c.

3. Generate the X_i from the corresponding $g_i(\mathbf{x};\boldsymbol{\theta}_i)$ using the component membership found in step 2.

Note that with this procedure, one could generate random variables from a mixture of any component densities. For instance, the model could be a mixture of exponentials, betas, etc.

Example 9.13

Generate a random sample of size n from a finite mixture estimate of the Old Faithful Geyser data (**geyser**). First we have to load up the data and build a finite mixture model.

```
load geyser
% Expects rows to be observations.
data = geyser';
% Get the finite mixture.
% Use a two term model.
% Set initial model to means at 50 and 80.
muin = [50, 80];
% Set mixing coefficients equal.
piesin = [0.5, 0.5];
% Set initial variances to 1.
varin = [1, 1];
max_it = 100;
tol = 0.001;
% Call the finite mixtures.
[pies,mus,vars]=...
    csfinmix(data,muin,varin,piesin,max_it,tol);
```

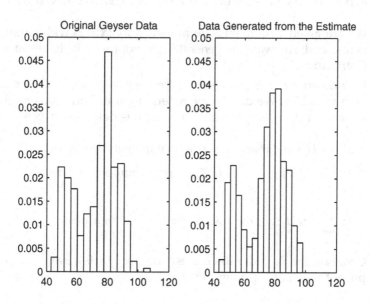

FIGURE 9.18
Histogram density estimates of the Old Faithful geyser data. The one on the right shows the estimate from the data that was sampled from the finite mixture density estimate of the original data.

Now generate some random variables according to this estimated model.

```
% Now generate some random variables from this model.
% Get the true model to generate data from this.
n = 300;
x = zeros(n,1);
% Now generate 300 random variables. First find
% the number that fall in each one.
r = rand(1,n);
% Find the number generated from component 1.
ind = length(find(r <= pies(1)));
% Create some mixture data. Note that the
% component densities are normals.
x(1:ind) = randn(ind,1)*sqrt(vars(1)) + mus(1);
x(ind+1:n) = randn(n-ind,1)*sqrt(vars(2)) + mus(2);
```

We can plot density histograms to compare the two data sets. These are shown in Figure 9.18. Not surprisingly, they have a similar distribution shape. The user is asked to explore this further in the exercises.
❑

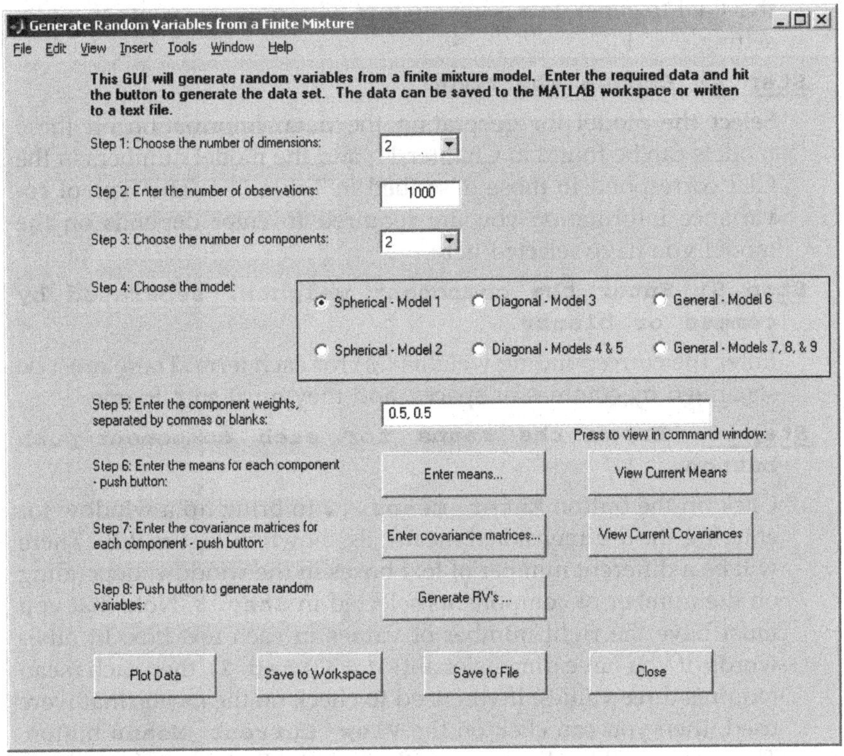

FIGURE 9.19

This shows the **genmix** *GUI. The steps that must be taken to enter the parameter information and generate the data are shown on the left side of the window.*

We also provide a GUI called **genmix** that will generate multivariate random variables from a finite mixture distribution [Martinez et al., 2010]. A screen shot of the GUI is given in Figure 9.19. One can see the steps for entering the required information on the left-side of the GUI window. Now, we give a brief description of how they work and the information that must be entered into the GUI.

Step 1: Choose the number of dimensions.

This is a pop-up menu. Simply select the number of dimensions for the data.

Step 2: Enter the number of observations.

Type the total number of points n in the data set.

Step 3: Choose the number of components.

This is the number of terms or component densities in the mixture, which is the value for c in Equation 9.31. Note that one can use this GUI to generate a finite mixture with only one component by setting $c = 1$.

Step 4: Choose the model.

Select the model for generating the data. Information on these models can be found in Chapter 11, and the model numbers in the GUI correspond to those described in Table 11.1. The type of co-variance information you are required to enter depends on the model you have selected here.

Step 5: Enter the component weights, separated by commas or blanks.

Enter the corresponding weights (π_k) for each term. These must be separated by commas or spaces, and they must sum to one.

Step 6: Enter the means for each component-push button.

Click on the button **Enter means...** to bring up a window for entering the d-dimensional means, as shown in Figure 9.20. There will be a different number of text boxes in the window, depending on the number of components selected in **Step 3**. Note that you must have the right number of values in each text box. In other words, if you have dimensionality $d = 3$ (**Step 1**), then each mean requires three values. If you need to check on the means that were used, then you can click on the **View Current Means** button. The means will be displayed in the MATLAB command window.

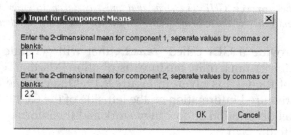

FIGURE 9.20
This shows the pop-up window for entering 2D means and for the two components or terms in the finite mixture.

Step 7: Enter the covariance matrices for each component - push button.

Click on the button **Enter covariance matrices...** to activate a pop-up window. You will get a different window, depending on the chosen model (**Step 4**). See Figure 9.21 for examples of the three types of covariance matrix input windows. As with the

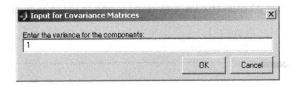

(a) This shows the pop-up window(s) for the spherical family of models. The only value that must be entered for each covariance matrix is the volume λ (see Table 11.1).

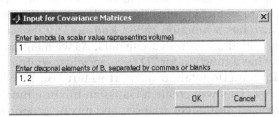

(b) This shows the pop-up window(s) for the diagonal family of models. One needs to enter the volume λ and the diagonal elements of the matrix **B** *(see Table 11.1).*

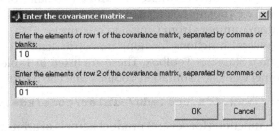

(c) When the general family is selected, one of these pop-up windows will appear for each unique covariance matrix. Each text box corresponds to a row of the covariance matrix.

FIGURE 9.21
Here we have examples of the covariance matrix inputs for the three families of models.

means, you can push the **View Current Covariances** button to view the covariance matrices in the MATLAB command window.

Step 8: Push the button to generate random variables.

After all of the variables have been entered, push the button labeled **Generate RVs...** to generate the data set.

Once the variables have been generated, you have several options. The data can be saved to the workspace using the button **Save to Workspace**. When this is activated, a pop-up window appears, and you can enter a variable name in the text box. The data can also be saved to a text file for use in other software applications by clicking on the button **Save to File**. This brings up the usual window for saving files. An added feature to verify the output is the **Plot Data** button. The data are displayed in a scatterplot matrix (using the **plotmatrix** function) when this is pushed.

9.6 MATLAB® Code

The MATLAB Statistics Toolbox has many options for probability density estimation. It has functions for estimating distribution parameters (e.g., **mle**, **normfit, expfit, betafit**, etc.) that can be used for *parametric* density estimation. It also has a function called **ksdensity** that implements the kernel smoothing method for univariate data. As we mentioned previously, the toolbox now has a function for 2D frequency histograms called **hist3**.

A constructor function called **gmdistribution** is available in the Statistics Toolbox. This will create a Gaussian mixture model (or mixture object) with specified means, covariances, and mixing proportions. There are several methods that can be used with this type of object: **cdf** (cumulative distribution function for a Gaussian mixture), **pdf** (probability density function), **fit** (parameter estimates), **mahal** (Mahalanobis distance to means), **posterior** (posterior probabilities of components), and **random** (generates random numbers from Gaussian mixture).

We provide several functions for nonparametric density estimation with the Computational Statistics Toolbox. These are listed in Table 9.4.

There is a Kernel Density Estimation Toolbox that can be downloaded from

http://www.ics.uci.edu/~ihler/code/kde.html

This toolbox has functions for several product kernels of arbitrary dimension, including the Gaussian, and it has different bandwidth selection methods.

There are also many user-contributed functions that have been posted to MATLAB Central. Go to the following site:

http://www.mathworks.com/matlabcentral/fileexchange

and search for "kernel density estimation" to find some useful tools.

9.7 Further Reading

The discussion of histograms, frequency polygons, and averaged shifted histograms presented in this book follows that of Scott [2015]. Scott's book is an excellent resource for univariate and multivariate density estimation, and it describes many applications of the techniques. It includes a comprehensive treatment of the underlying theory on selecting smoothing parameters,

TABLE 9.4

List of Functions from Chapter 9 Included in the
Computational Statistics Toolbox

Purpose	MATLAB Function
These provide a bivariate histogram.	`cshist2d` `cshistden`
This returns a frequency polygon density estimate.	`csfreqpoly`
This function returns the Averaged Shifted Histogram.	`csash`
These functions perform kernel density estimation.	`cskernnd` `cskern2d`
Create plots	`csdfplot` `csplotuni`
Functions for finite and adaptive mixtures	`csfinmix` `csadpmix`
Generate random variables for a multivariate finite mixture	`genmix`

analyzing the performance of density estimates in terms of the asymptotic mean integrated squared error, and also addresses high dimensional data.

The summary book by Silverman [1986] provides a relatively nontheoretical treatment of density estimation. He includes a discussion of histograms, kernel methods, and others. This book is readily accessible to most statisticians, data analysts, or engineers. It contains applications and computational details, making the subject easier to understand.

Other books on density estimation include Tapia and Thompson [1978], Devroye and Gyorfi [1985], Wand and Jones [1995], and Simonoff [1996]. The Tapia and Thompson book offers a theoretical foundation for density estimation and includes a discussion of Monte Carlo simulations. The Devroye and Gyorfi text describes the theory of density estimation using the L_1 (absolute error) viewpoint instead of L_2 (squared error). The books by Wand and Jones and Simonoff look at using kernel methods for smoothing and exploratory data analysis.

A paper by Izenman [1991] provides a comprehensive review of many methods in univariate and multivariate density estimation and includes an extensive bibliography. Besides histograms and kernel methods, he discusses projection pursuit density estimation [Friedman, Stuetzle, and Schroeder, 1984], maximum penalized likelihood estimators, sieve estimators, and orthogonal estimators. Jones, Marron, and Sheather [1996] provide a survey of bandwidth selection methods for probability density estimation. The article is easy to read and provides a good introduction for those not familiar with the area.

For the reader who would like more information on finite mixtures, we recommend Everitt and Hand [1981] for a general discussion of this topic. The book provides a summary of the techniques for obtaining mixture models (estimating the parameters) and illustrates them using applications. That text also discusses ways to handle the problem of determining the number of terms in the mixture and other methods for estimating the parameters. It is appropriate for someone with a general statistics or engineering background. For readers who would like more information on the theoretical details of finite mixtures, we refer them to McLachlan and Basford [1988] or Titterington, Smith, and Makov [1985]. The book by McLachlan and Peel [2000] provides many examples of finite mixtures, linking them to machine learning, data mining, and pattern recognition.

We discuss model-based clustering in Chapter 11, but we mention its use here because it is can be used to estimate a probability density function based on the finite mixture model. An excellent paper that describes the uses of the model-based clustering framework for density estimation is by Fraley and Raftery [2002].

The EM algorithm is described in the text by McLachlan and Krishnan [1997]. This offers a unified treatment of the subject and provides numerous applications of the EM algorithm to regression, factor analysis, medical imaging, experimental design, finite mixtures, and others.

For a theoretical discussion of the adaptive mixtures approach, the reader is referred to Priebe [1993, 1994]. These examine the error in the adaptive mixtures density estimates and its convergence properties. A paper by Priebe and Marchette [2000] describes a data-driven method for obtaining parsimonious mixture model estimates. This methodology addresses some problems with the adaptive/finite mixtures approach: (1) adaptive mixtures are not designed to yield a parsimonious model, and (2) how many terms or component densities should be used in a finite mixture model.

Solka, Poston, and Wegman [1995] extend the static *dF* plot to a dynamic one. References to MATLAB code are provided in this paper describing a dynamic view of the adaptive mixtures and finite mixtures estimation process in time (i.e., iterations of the EM algorithm).

Exercises

9.1.   Create a MATLAB function that will return the value of the histogram estimate for the probability density function. Do this for the 1D case.

9.2. Generate a random sample of data from a standard normal. Construct a kernel density estimate of the probability density function and verify that the area under the curve is approximately 1 using **trapz**.

9.3. Generate 100 univariate normals and construct a histogram. Calculate the MSE at a point x_0 using Monte Carlo simulation. Do this for varying bin widths. What is the better bin width? Does the sample size make a difference? Does it matter whether x_0 is in the tails or closer to the mean? Repeat this experiment using the absolute error. Are your conclusions similar?

9.4. Generate univariate normal random variables. Using the normal reference rules for h, construct a histogram, a frequency polygon, and a kernel estimate of the data. Estimate the MSE at a point x_0 using Monte Carlo simulation.

9.5. Generate a random sample from the exponential distribution. Construct a histogram using the normal reference rule. Using Monte Carlo simulation, estimate the MISE. Use the skewness factor to adjust h and reestimate the MISE. Which window width is better?

9.6. Use the **snowfall** data and create a MATLAB **movie** that shows how 1D histograms change with bin width. See **help** on **movie** for information on how to do this. Also make a **movie** showing how changing the bin origin affects the histogram.

9.7. Repeat Example 9.2 for bin widths given by the Freedman-Diaconis Rule. Is there a difference in the results? What does the histogram look like if you use Sturge's Rule?

9.8. Write a MATLAB function that will return the value of a bivariate histogram at a point, given the bin counts, the sample size, and the window widths.

9.9. Write a MATLAB function that will evaluate the cumulative distribution function for a univariate frequency polygon. You can use the **trapz**, **quad**, or **quadl** functions.

9.10. Load the **iris** data. Create a 150×2 matrix by concatenating the first two columns of each species. Construct and plot a frequency polygon of these data. Do the same thing for all possible pairs of columns. You might also look at a **contour** plot of the frequency polygons. Is there evidence of groups in the plots?

9.11. In this chapter, we showed how you could construct a kernel density estimate by placing a weighted kernel at each data point, evaluating the kernels over the domain, and then averaging the n curves. In that implementation, we are looping over all of the data points. An alternative implementation is to loop over all points in the domain where you want to get the value of the estimate, evaluate a weighted kernel at each point, and take the average. The following code shows you how to do this. Implement this using the Buffalo **snowfall** data. Verify that this is a valid density by estimating the area under the curve.

```
load snowfall
x = 0:140;
n = length(snowfall);
h = 1.06*sqrt(var(snowfall))*n^(-1/5);
fhat = zeros(size(x));
% Loop over all values of x in the domain
% to get the kernel evaluated at that point.
for i = 1:length(x)
xloc = x(i)*ones(1,n);
% Take each value of x and evaluate it at
% n weighted kernels -
% each one centered at a data point,
% then add them up.
arg = ((xloc-snowfall)/h).^2;
fhat(i) = (sum(exp(-.5*(arg)))/(n*h*sqrt(2*pi)));
end
```

9.12. Write a MATLAB function that will construct a kernel density estimate for the multivariate case.

9.13. Write a MATLAB function that will provide the finite mixture density estimate at a point in d dimensions.

9.14. Implement the univariate adaptive mixtures density estimation procedure on the Buffalo **snowfall** data. Once you have your initial model, use the EM algorithm to refine the estimate.

9.15. In Example 9.13, we generate a random sample from the kernel estimate of the Old Faithful **geyser** data. Repeat this example to obtain a new random sample of **geyser** data from the estimated model and construct a new density estimate from the second sample. Find the integrated squared error between the two density estimates. Does the error between the curves indicate that the second random sample generates a similar density curve?

9.16. Say we have a kernel density estimate where the kernel used is a normal density. If we put this in the context of finite mixtures, then what are the values for the component parameters (p_i, μ_i, σ_i^2) in the corresponding finite mixture?

9.17. Repeat Example 9.12. Plot the curves from the estimated models. What is the ISE between the two estimates? Use the iterative EM algorithm on both models to refine the estimates. What is the ISE after you do this? What can you say about the two different models? Are your conclusions different if you use the IAE?

9.18. Write a MATLAB function that will generate random variables (univariate or multivariate) from a finite mixture of normals.

9.19. Using the method for generating random variables from a finite mixture that was discussed in this chapter, develop and implement an algorithm for generating random variables based on a kernel density estimate.

9.20. Write a function that will estimate the MISE between two functions. Convert it to also estimate the MIAE between two functions.

9.21. Apply some of the univariate density estimation techniques from this chapter to the **forearm** data.

9.22. The **elderly** data set contains the height measurements (in centimeters) of 351 elderly females [Hand, et al., 1994]. Use some of the univariate density estimation techniques from this chapter to explore the data. Is there evidence of bumps and modes?

9.23. Apply the multivariate techniques of this chapter to the **nfl** data [Csorgo and Welsh, 1989; Hand, et al., 1994]. These data contain bivariate measurements of the game time to the first points scored by kicking the ball between the end posts and the game time to the first points scored by moving the ball into the end zone. The times are in minutes and seconds. Plot your results.

9.24. Using the **genmix** GUI, generate 2D data ($n = 500$) for each of the major finite mixture models (see Table 11.1 for more information on the models). Plot the data in a scatterplot and compare.

Chapter 10

Supervised Learning

10.1 Introduction

Statistical pattern recognition is an area of computational statistics that uses many of the concepts we have covered so far, such as probability density estimation and cross-validation. Examples where statistical pattern recognition techniques can be used are numerous and arise in disciplines such as medicine, computer vision, robotics, manufacturing, finance, and many others. Some of these include the following:

- A doctor diagnoses a patient's illness based on the symptoms and test results.
- A military analyst classifies regions of an image as natural or man-made for use in targeting systems.
- A loan manager at a bank must decide whether a customer is a good credit risk based on their income, past credit history, and other variables.

In all of these applications, the human is often assisted by computational and statistical pattern recognition techniques.

Pattern recognition methods for *supervised learning* situations are covered in this chapter. With supervised learning, we have a data set where we know the classification of each observation. Thus, each case or data point has a class label associated with it. We construct a classifier based on this data set, which is then used to classify future observations.

Sometimes we are in a situation where we do not know the class membership for our observations, or we do not know how many classes are represented by the data. In this case, we are in the *unsupervised learning* mode. We cover techniques for unsupervised learning in the next chapter.

In this section, we first provide a brief introduction to the goals of pattern recognition and a broad overview of the main steps of building classifiers. In Section 10.2, we present a discussion of Bayes classifiers and pattern recognition in a hypothesis testing framework. Section 10.3 contains techniques for evaluating the performance of the classifier. In Section 10.4, we illustrate how to construct classification trees. In Section 10.5, we briefly describe several supervised learning methods for combining classifiers, such as bagging and boosting. We cover the nearest neighbor classifier in Section 10.6 and conclude the chapter with a discussion of support vector machines in Section 10.7.

Figure 10.1 illustrates the major steps of statistical pattern recognition. The first step in pattern recognition is to select *features* that will be used to distinguish between the classes. As the reader might suspect, the choice of features is perhaps the most important part of the process. Building accurate classifiers is much easier with features that allow one to readily distinguish between classes.

Once features are selected, we obtain a sample of these features for the different classes. This means that we find objects that belong to the classes of interest and then measure the features. Thus, each observation (sometimes also called a *case* or *pattern*) has a class label attached to it. Now that we have data that are known to belong to the different classes, we can use this information to create a methodology that will take as input a set of feature measurements and output the estimated class membership. How these classifiers are created will be the topic of this chapter.

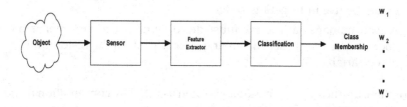

FIGURE 10.1
This shows a schematic diagram of the major steps for statistical pattern recognition [Duda and Hart, 1973].

One of the main data sets we will use to illustrate these ideas is the `iris` data set that we encountered in a previous chapter. Recall that we have three species of iris: *Iris setosa*, *Iris versicolor*, and *Iris virginica*. The data were originally used by Fisher [1936] to develop a classifier that would take measurements from a new iris and determine its species based on the features [Hand, et al., 1994].

We now use this example to describe the steps of the supervised pattern recognition process. First, we need to choose some features we hope will distinguish one type of iris from another. In this case, we have four features: sepal length, sepal width, petal length, and petal width. The next step in the process is to find many flowers from each species and measure the sepal length, sepal width, petal length, and petal width for each one. The measured features for each flower yield one observation or 4D data point. We also attach a class label to each of the observations that indicates which species it belongs to. Finally, we build a classifier using these data and one of the techniques that are described in this chapter. Next, we would measure the four features for an iris of unknown species and use the classifier to assign the species membership.

10.2 Bayes Decision Theory

The Bayes approach to pattern classification is a fundamental technique, and we recommend it as the starting point for most pattern recognition applications. If this method is not adequate, then more complicated techniques may be used (e.g., neural networks, classification trees, support vector machines). Bayes decision theory poses the classification problem in terms of probabilities. Therefore, all of the probabilities must be known or estimated from the data.

We have already seen an application of Bayes decision theory in Chapter 2. There we wanted to know the probability that a piston ring came from a particular manufacturer given that it failed. It makes sense to make the decision that the part came from the manufacturer that has the highest posterior probability. To put this in the pattern recognition context, we could think of the part failing as the feature. The resulting classification would be the manufacturer (M_A or M_B) that sold us the part. In the following, we will see that Bayes decision theory is an application of Bayes' theorem, where we will classify observations using the posterior probabilities.

We start off by fixing some notation. Let the class membership be represented by $\omega_j, j = 1, ..., J$ for a total of J classes. For example, with the **iris** data, we have $J = 3$ classes:

$\omega_1 = $ *Iris setosa*

$\omega_2 = $ *Iris versicolor*

$\omega_3 = $ *Iris virginica.*

The features we are using for classification are denoted by the d-dimensional vector $\mathbf{x}, d = 1, 2, ...$ With the **iris** data, we have four measurements, so $d = 4$. In the supervised learning situation, each of the observed feature vectors will also have a class label attached to it.

Our goal is to use the data to create a decision rule or classifier that will take a feature vector **x** whose class membership is unknown and return the class it most likely belongs to. A logical way to achieve this is to assign the class label to this feature vector using the class corresponding to the highest *posterior probability*. This probability is given by

$$P(\omega_j|\mathbf{x}); \qquad j = 1, ..., J. \tag{10.1}$$

Equation 10.1 represents the probability that the case belongs to the j-th class given the observed feature vector $\mathbf{x}$. To use this rule, we would evaluate all of the J posterior probabilities, and the class corresponding to the highest probability would be the one we choose.

We can find the posterior probabilities using Bayes' theorem:

$$P(\omega_j|\mathbf{x}) = \frac{P(\omega_j)P(\mathbf{x}|\omega_j)}{P(\mathbf{x})}, \tag{10.2}$$

where

$$P(\mathbf{x}) = \sum_{j=1}^{J} P(\omega_j)P(\mathbf{x}|\omega_j). \tag{10.3}$$

We see from Equation 10.2 that we must know the *prior probability* that it would be in class j given by

$$P(\omega_j); \qquad j = 1, ..., J, \tag{10.4}$$

and the *class-conditional probability*

$$P(\mathbf{x}|\omega_j); \qquad j = 1, ..., J. \tag{10.5}$$

The class-conditional probability in Equation 10.5 represents the probability distribution of the features for each class. The prior probability in Equation 10.4 represents our initial degree of belief that an observed set of features is a case from the j-th class. The process of estimating these probabilities is how we build the classifier.

The prior probabilities can either be inferred from prior knowledge of the application, estimated from the data, or assumed to be equal. In the piston ring example, we know how many parts we buy from each manufacturer. So, the prior probability that the part came from a certain manufacturer would be based on the percentage of parts obtained from that manufacturer. In other applications, we might know the prevalence of some class in our population. This might be the case in medical diagnosis, where we have some idea of the percentage of the population who are likely to have a certain disease or medical condition.

Using our **iris** data as an example, we could estimate the prior probabilities using the proportion of each class in our sample. We had 150 observed feature vectors, with 50 coming from each class. Therefore, our estimated prior probabilities would be

$$\hat{P}(\omega_j) = \frac{n_j}{n} = \frac{50}{150} \approx 0.33; \qquad j = 1, 2, 3.$$

Finally, we might use equal priors when we believe each class is equally likely.

Now that we have our prior probabilities, $\hat{P}(\omega_j)$, we turn our attention to the class-conditional probabilities $P(\mathbf{x}|\omega_j)$. We can use the density estimation techniques covered in Chapter 9 to obtain these probabilities. In essence, we take all of the observed feature vectors that are known to come from class ω_j and estimate the density using only those cases. We will cover two approaches for doing this: parametric and nonparametric.

Estimating Class-Conditional Probabilities: Parametric Method

In parametric density estimation, we assume a distribution for the class-conditional probability densities and estimate them by estimating the corresponding distribution parameters. For example, we might assume the features come from a multivariate normal distribution. To estimate the density, we find estimates of μ_j and Σ_j for each class. This procedure is illustrated in Example 10.1 for the **iris** data.

Example 10.1

We now estimate our class-conditional probability density functions using the **iris** data as an example. We assume that the required probabilities are multivariate normal for each class. The following MATLAB® code shows how to get the class-conditional probabilities for each species of iris.

```
load iris
% This loads up three matrices:
% setosa, virginica, and versicolor
% We will assume each class is multivariate normal.
% To get the class-conditional probabilities, we
% get estimates for the parameters for each class.
muset = mean(setosa);
covset = cov(setosa,1); % Divide by n rather than n-1.
muvir = mean(virginica);
covvir = cov(virginica,1);
muver = mean(versicolor);
covver = cov(versicolor,1);
```

Note that we are using the maximum likelihood estimate for the covariance matrix, where we divide by n rather than $n - 1$. The estimated multivariate normal densities for each class are then used in Bayes decision rule (to be covered shortly) to classify observations.
❑

The method shown in Example 10.1 yields what is commonly known as a *quadratic classifier*. When we construct this type of classifier, we assume that the class-conditional probability density functions are Gaussian, where each one has a different covariance matrix. We can simplify this somewhat by making the assumption that the classes *share* the same covariance matrix, yielding what is known as the *linear classifier*.

We saw how to estimate the class-conditional probability density functions for the quadratic classifier in the previous example. We did this by using the data known to belong to the j-th class to estimate the mean and covariance matrix for that class. We used the maximum likelihood estimate of the covariance matrix [Webb, 2011] given by

$$\hat{\Sigma}_j = \frac{1}{n_j} \sum_{x_i \in \omega_j} (x_i - \mu_j)(x_i - \mu_j)^T .$$

Now, let's see how we estimate parameters for the linear classifier. The d-dimensional means for each class are found separately, as in Example 10.1. We then find the common covariance matrix as a weighted average of the individual covariances for each class. The maximum likelihood estimate of the *pooled covariance matrix* is given by

$$S_W = \sum_{j=1}^{J} \frac{n_j}{n} \hat{\Sigma}_j .$$

This is sometimes also known as the *within-class scatter matrix*. An unbiased estimate of the pooled covariance matrix is

$$\frac{n}{n-J} S_W .$$

Since we use *covariance* in our definition of these classifiers, we are making the assumption that we have multivariate data. However, these classifiers are also defined for univariate observations. If we have univariate data, then a linear classifier has equal variances. Similarly, a quadratic classifier has (possibly) unequal variances among the classes. Thus, the methods described above for finding the class-conditional densities in the multivariate case are easily adapted to the univariate case by using variances rather than covariances.

Naive Bayes Classifiers

To construct a *naive Bayes classifier*, we first assume that the individual features are independent, given the class. This makes it easier to estimate the multivariate class-conditional probabilities. The probability density function for the class-conditional probability can be written as

$$P(\mathbf{x}|\omega_m) = P(x_1|\omega_m) \times \ldots \times P(x_p|\omega_m).$$

To estimate the class-conditional probabilities, we first estimate the univariate density for each feature or dimension using the data from each known class. We then multiply them together to get the joint density. This can save on computations because fewer parameters need to be estimated. Another benefit of fewer parameters to estimate is that we typically require fewer data points for training in these situations, which can be important when we have many features or dimensions to deal with.

MATLAB has functions for constructing a naive Bayes classifier. One is **fitcnb** and the other is **fitNaiveBayes**. These are object-oriented and they produce a special type of data object (see Appendix A). There are five distributions that can be specified for the class-conditional probabilities. These are the normal distribution, kernel density estimates, multinomials, and strings that name other probability distributions in the Statistics Toolbox (see Chapter 2).

Estimating Class-Conditional Probabilities: Nonparametric

If it is not appropriate to assume the features for a class follow a known distribution, then we can use the nonparametric density estimation techniques from Chapter 9. These include the averaged shifted histogram, the frequency polygon, kernel densities, finite mixtures, and adaptive mixtures. To obtain the class-conditional probabilities, we take the set of measured features from each class and estimate the density using one of these methods. This is illustrated in Example 10.2, where we use the product kernel to estimate the probability densities for the **iris** data.

Example 10.2
We can estimate the class-conditional probability densities for the **iris** data using the product kernel. We illustrate the use of two functions for estimating the product kernel. One is called **cskern2d** that can only be used for bivariate data. The output arguments from this function are matrices for use in the MATLAB plotting functions **surf** and **mesh**. The **cskern2d** function should be used when the analyst wants to plot the resulting probability density. We use it on the first two dimensions of the **iris** data and plot the surface for *Iris virginica* in Figure 10.2.

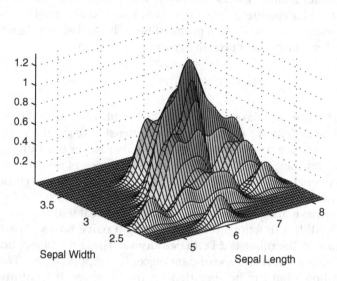

FIGURE 10.2
Using only the first two features of the data for Iris virginica, we construct an estimate of the corresponding class-conditional probability density using the product kernel. This is the output from the function **cskern2d**.

```
load iris
% This loads up three matrices:
% setosa, virginica, and versicolor
% We will use the product kernel to estimate densities.
% To try this, get the kernel estimate for the first
% two features and plot.
% The arguments of 0.1 indicate the grid size in
% each dimension. This creates the domain over
% which we will estimate the density.
[xset,yset,pset]=cskern2d(setosa(:,1:2),0.1,0.1);
[xvir,yvir,pvir]=cskern2d(virginica(:,1:2),0.1,0.1);
[xver,yver,pver]=cskern2d(versicolor(:,1:2),0.1,0.1);
mesh(xvir,yvir,pvir)
colormap(gray(256))
```

A more useful function for statistical pattern recognition is **cskernmd**, which returns the value of the probability density $\hat{f}(\mathbf{x})$ for a given d-dimensional vector $\mathbf{x}$. We can find all of the class-conditional probabilities for the first observation in the *Iris setosa* group using the following.

```
% If one needs the value of the probability surface,
% then use this function. For example...
ps = cskernmd(setosa(1,1:2),setosa(:,1:2));
```

```
pver = cskernmd(setosa(1,1:2),versicolor(:,1:2));
pvir = cskernmd(setosa(1,1:2),virginica(:,1:2));
```

The function call above uses the optimal bandwidth using the normal reference rule.
❏

Bayes Decision Rule

Now that we know how to get the prior probabilities and the class-conditional probabilities, we can use Bayes' theorem to obtain the posterior probabilities. Bayes decision rule is based on these posterior probabilities.

BAYES DECISION RULE:
Given a feature vector **x***, assign it to class* ω_j *if*

$$P(\omega_j|\mathbf{x}) > P(\omega_i|\mathbf{x}); \qquad i = 1, ..., J; \ i \neq j. \tag{10.6}$$

This states that we will classify an observation **x** as belonging to the class that has the highest posterior probability. It is known [Duda and Hart, 1973] that the decision rule given by Equation 10.6 yields a classifier with the minimum probability of error.

We can use an equivalent rule by recognizing that the denominator of the posterior probability (see Equation 10.2) is simply a normalization factor and is the same for all classes. So, we can use the following alternative decision rule:

$$P(\mathbf{x}|\omega_j)P(\omega_j) > P(\mathbf{x}|\omega_i)P(\omega_i); \qquad i = 1, ..., J; \ i \neq j. \tag{10.7}$$

Equation 10.7 is Bayes decision rule in terms of the class-conditional and prior probabilities. If we have equal priors for each class, then our decision is based only on the class-conditional probabilities.

The decision rule partitions the feature space into J decision regions $\Omega_1, \Omega_2, ..., \Omega_J$. If **x** is in region Ω_j, then we will say it belongs to class ω_j.

We now turn our attention to the error we have in our classifier when we use Bayes decision rule. An error is made when we classify an observation as class ω_i when it is really in the j-th class. We denote the complement of region Ω_i as Ω_i^c, which represents every region except Ω_i. To get the probability of error, we calculate the following integral over all values of **x** [Duda and Hart, 1973; Webb, 2011]:

$$P(\text{error}) = \sum_{i=1}^{J} \int_{\Omega_i^c} P(\mathbf{x}|\omega_i)P(\omega_i)d\mathbf{x}. \tag{10.8}$$

Thus, to find the probability of making an error (i.e., assigning the wrong class to an observation), we find the probability of error for each class and add the probabilities together. In the following example, we make this clearer by looking at a two-class case and calculating the probability of error.

Example 10.3

We will look at a univariate classification problem with equal priors and two classes. The class-conditionals are given by $\phi(x; \mu, \sigma^2)$:

$$P(x|\omega_1) = \phi(x; -1, 1)$$
$$P(x|\omega_2) = \phi(x; 1, 1).$$

The priors are

$$P(\omega_1) = 0.6$$
$$P(\omega_2) = 0.4.$$

The following MATLAB code creates the required curves for the decision rule of Equation 10.7.

```
% This illustrates the 1D case for two classes.
% We will shade in the area where there can be
% misclassified observations.
% Get the domain for the densities.
dom = -6:.1:8;
dom = dom';
% Note that we could use csnormp or normpdf.
pxg1 = csevalnorm(dom,-1,1);
pxg2 = csevalnorm(dom,1,1);
plot(dom,pxg1,dom,pxg2)
% Find decision regions - multiply by priors
ppxg1 = pxg1*0.6;
ppxg2 = pxg2*0.4;
plot(dom,ppxg1,'k',dom,ppxg2,'k')
xlabel('x')
```

The resulting plot is given in Figure 10.3, where we see that the decision regions given by Equation 10.7 are obtained by finding where the two curves intersect. For example, if we observe a value of a feature given by $x = -2$, then we would classify that object as belonging to class ω_1. If we observe $x = 4$, then we would classify that object as belonging to class ω_2. Let's see what happens when $x = -0.75$. We can find the probabilities using

```
x = -0.75;
% Evaluate each non-normalized posterior.
```

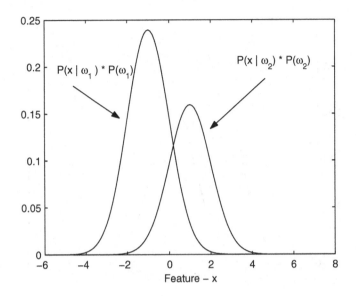

FIGURE 10.3
Here we show the univariate, two-class case from Example 10.3. Note that each curve represents the probabilities in Equation 10.7. The point where the two curves intersect partitions the domain into one where we would classify observations as class 1 (ω_1) and another where we would classify observations as class 2 (ω_2).

```
po1 = csevalnorm(x,-1,1)*0.6;
po2 = csevalnorm(x,1,1)*0.4;
```

$$P(-0.75|\omega_1)P(\omega_1) = 0.23$$
$$P(-0.75|\omega_2)P(\omega_2) = 0.04.$$

These are shown in Figure 10.4. Note that we would label this observation as belonging to class ω_1, but there is non-zero probability that the case corresponding to $x = -0.75$ could belong to class 2. We now turn our attention to how we can estimate this error.

```
% To get estimates of the error, we can
% estimate the integral as follows
% Note that 0.1 is the step size, and we
% are approximating the integral using a sum.
% The decision boundary is where the two curves meet.
ind1 = find(ppxg1 >= ppxg2);
% Now find the other part.
ind2 = find(ppxg1<ppxg2);
pmis1 = sum(ppxg1(ind2))*.1;
```

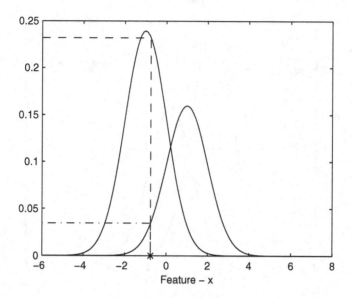

FIGURE 10.4
The vertical dotted line represents $x = -0.75$. The probabilities needed for the decision rule of Equation 10.7 are represented by the horizontal dotted lines. We would classify this case as belonging to class 1 (ω_1), but there is a possibility that it could belong to class 2 (ω_2).

```
pmis2 = sum(ppxg2(ind1))*.1;
errorhat = pmis1 + pmis2;
```

From the above, we estimate the probability of error as 0.15. To get this probability, we find the shaded area under the curves (see Figure 10.5). ❑

It is sometimes better (in terms of computations and implementation) to devise a *discriminant function* $g_j(\mathbf{x})$, $j = 1, ..., J$, and use it to classify our observations. For example, we would assign a feature $\mathbf{x}$ to class ω_j if

$$g_j(\mathbf{x}) > g_i(\mathbf{x}) ; \qquad \text{for all } i \neq j .$$

This means that we would compute J discriminant functions and estimate the class membership as the one corresponding to the largest value of the discriminant function $g_j(\mathbf{x})$. The classifier we get using Bayes decision rule can be represented in this way, where we have

$$g_j(\mathbf{x}) = P(\omega_j|\mathbf{x}) .$$

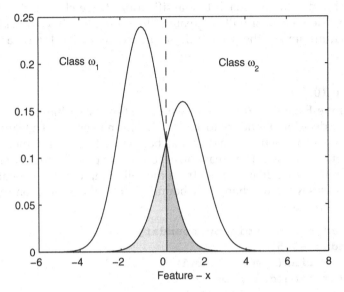

FIGURE 10.5
The shaded regions show the probability of misclassifying an object. The lighter region shows the probability of classifying as class 1 when it is really class 2. The darker region shows the probability of classifying as class 2, when it belongs to class 1.

The discriminant function we choose is not unique [Duda and Hart, 1973]. In general, if f is a monotonically increasing function, then we can use $f(g_j(\mathbf{x}))$ as our discriminant function instead. This sometimes makes the decision rules easier to understand and to implement. Any of the following discriminant functions can be used for minimum-error-rate classification:

$$g_j(\mathbf{x}) = P(\omega_j|\mathbf{x})$$
$$g_j(\mathbf{x}) = P(\mathbf{x}|\omega_j)P(\omega_j)$$
$$g_j(\mathbf{x}) = \log P(\mathbf{x}|\omega_j) + \log P(\omega_j).$$

The discriminant functions are really just another way to express our decision rule. We saw in Example 10.3 that the decision rule divided our feature space into two decision regions. In general, decision rules will separate the space into J regions separated by **decision boundaries**. These boundaries are places where ties occur between the discriminant functions with the highest value. For the two-class case in the previous example, this can be found as the x such that the following equation is satisfied:

$$P(x|\omega_1)P(\omega_1) = P(x|\omega_2)P(\omega_2).$$

This decision region can be changed, as we will see shortly, when we discuss the likelihood ratio approach to classification. If we change the decision boundary, then the error will be greater, illustrating that Bayes decision rule is one that minimizes the probability of misclassification [Duda and Hart, 1973].

Example 10.4

We continue Example 10.3, where we now show what happens when we change the decision boundary to $x = -0.5$. This means that if a feature has a value of $x < -0.5$, then we classify it as belonging to class 1. Otherwise, we say it belongs to class 2. The areas corresponding to possible misclassification are shown in Figure 10.6. We see from the following that the probability of error increases when we change the boundary from the optimal one given by Bayes decision rule.

```
% Change the decision boundary.
bound = -0.5;
ind1 = find(dom <= bound);
ind2 = find(dom > bound);
pmis1 = sum(ppxg1(ind2))*.1;
pmis2 = sum(ppxg2(ind1))*.1;
errorhat = pmis1 + pmis2;
```

This yields an estimated error of 0.20.
❑

Bayes decision theory can address more general situations where there might be a variable cost or risk associated with classifying something incorrectly or allowing actions in addition to classifying the observation. For example, we might want to penalize the error of classifying some section of tissue in an image as cancerous when it is not, or we might want to include the action of not making a classification if our uncertainty is too great. We will provide references at the end of the chapter for those readers who require a more general treatment of statistical pattern recognition.

Likelihood Ratio Approach

The likelihood ratio technique addresses the issue of variable misclassification costs in a hypothesis testing framework. This methodology does not assign an explicit cost to making an error as in the Bayes approach, but it enables us to set the amount of error we will tolerate for misclassifying one of the classes.

Recall from Chapter 7 that in hypothesis testing we have two types of errors. One type of error is when we wrongly reject the null hypothesis when it is really true. This is the Type I error. The other way we can make a wrong

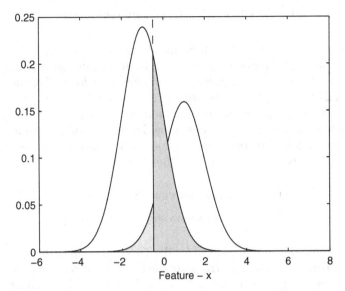

FIGURE 10.6
If we move the decision boundary to $x = -0.5$, then the probability of error is given by the shaded areas. Not surprisingly, the error increases when we change from the boundary given by Bayes decision rule.

decision is to not reject the null hypothesis when we should. Typically, we try to control the probability of Type I error by setting a desired significance level α, and we use this level to determine our decision boundary. We can fit our pattern recognition process into the same framework.

In the rest of this section, we consider only two classes, ω_1 and ω_2. First, we have to determine what class corresponds to the null hypothesis and call this the nontarget class. The other class is denoted as the target class. In this book, we use ω_1 to represent the target class and ω_2 to represent the nontarget class. The following examples should clarify these concepts.

- We are building a classifier for a military command and control system that will take features from images of objects and classify them as targets or nontargets. If an object is classified as a target, then we will destroy it. Target objects might be tanks or military trucks. Nontarget objects are such things as school buses or automobiles. We would want to make sure that when we build a classifier we do not classify an object as a tank when it is really a school bus. So, we will control the amount of acceptable error in wrongly saying it (a school bus or automobile) is in the target class. This is the same as our Type I error, if we write our hypotheses as

H_0 Object is a school bus, automobile, etc.

H_1 Object is a tank, military vehicle, etc.

- Another example, where this situation arises is in medical diagnosis. Say that the doctor needs to determine whether a patient has cancer by looking at radiographic images. The doctor does not want to classify a region in the image as cancer when it is not. So, we might want to control the probability of wrongly deciding that there is cancer when there is none. However, failing to identify a cancer when it is really there is more important to control. Therefore, in this situation, the hypotheses are

H_0 X-ray shows cancerous tissue

H_1 X-ray shows only healthy tissue

The terminology sometimes used for the Type I error in pattern recognition is *false alarms* or *false positives*. A false alarm is wrongly classifying something as a target (ω_1), when it should be classified as nontarget (ω_2). The probability of making a false alarm (or the probability of making a Type I error) is denoted as

$$P(FA) = \alpha.$$

This probability is represented as the shaded area in Figure 10.7.

Recall that Bayes decision rule gives a rule that yields the minimum probability of incorrectly classifying observed patterns. We can change this boundary to obtain the desired probability of false alarm α. Of course, if we do this, then we must accept a higher probability of misclassification as we illustrated in Example 10.4.

In the two-class case, we can put our Bayes decision rule in a different form. Starting from Equation 10.7, we have our decision as

$$P(\mathbf{x}|\omega_1)P(\omega_1) > P(\mathbf{x}|\omega_2)P(\omega_2) \Rightarrow \mathbf{x} \text{ is in } \omega_1, \tag{10.9}$$

or else we classify $\mathbf{x}$ as belonging to ω_2. Rearranging this inequality yields the following decision rule:

$$L_R(\mathbf{x}) = \frac{P(\mathbf{x}|\omega_1)}{P(\mathbf{x}|\omega_2)} > \frac{P(\omega_2)}{P(\omega_1)} = \tau_C \Rightarrow \mathbf{x} \text{ is in } \omega_1. \tag{10.10}$$

The ratio on the left of Equation 10.10 is called the *likelihood ratio,* and the quantity on the right is the threshold. If $L_R > \tau_C$, then we decide that the case belongs to class ω_1. If $L_R < \tau_C$, then we group the observation with class ω_2.

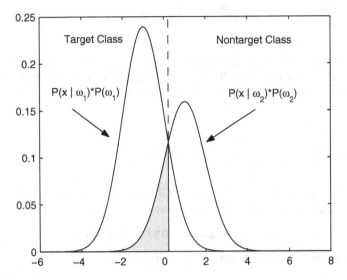

FIGURE 10.7
The shaded region shows the probability of false alarm or the probability of wrongly classifying as belonging to the target class ω_1 when it really belongs to the nontarget class ω_2.

If we have equal priors, then the threshold is one ($\tau_C = 1$). Thus, when $L_R > 1$, we assign the observation or pattern to ω_1, and if $L_R < 1$, then we classify the observation as belonging to ω_2. We can also adjust this threshold to obtain a desired probability of false alarm, as we show in Example 10.5.

Example 10.5

We use the class-conditional and prior probabilities of Example 10.3 to show how we can adjust the decision boundary to achieve the desired probability of false alarm. Looking at Figure 10.7, we see that

$$P(FA) = \int_{-\infty}^{C} P(x|\omega_2)P(\omega_2)dx,$$

where C represents the value of x that corresponds to the decision boundary. We can factor out the prior, so

$$P(FA) = P(\omega_2)\int_{-\infty}^{C} P(x|\omega_2)dx.$$

We then have to find the value for C such that

$$\int_{-\infty}^{C} P(x|\omega_2)dx = \frac{P(FA)}{P(\omega_2)}.$$

From Chapter 3, we recognize that C is a quantile. Using the probabilities in Example 10.3, we know that $P(\omega_2) = 0.4$ and $P(x|\omega_2)$ are normal with mean 1 and variance of 1. If our desired $P(FA) = 0.05$, then

$$\int_{-\infty}^{C} P(x|\omega_2)dx = \frac{0.05}{0.40} = 0.125.$$

We can find the value for C using the inverse cumulative distribution function for the normal distribution. In MATLAB, this is

```
c = norminv(0.05/0.4,1,1);
```

This yields a decision boundary of $x = -0.15$.
❑

10.3 Evaluating the Classifier

Once we have our classifier, we need to evaluate its usefulness by measuring the percentage of observations that we correctly classify. This yields an estimate of the probability of correctly classifying cases. It is also important to report the probability of false alarms, when the application requires it (e.g., when there is a target class). We will discuss two methods for estimating the probability of correctly classifying cases and the probability of false alarm: the use of an independent test sample and cross-validation.

Independent Test Sample

If our sample is large, we can divide it into a training set and a testing set. We use the training set to build our classifier and then we classify observations in the test set using our classification rule. The proportion of correctly classified observations is the *estimated classification rate*. Note that the classifier has not seen the patterns in the test set, so the classification rate estimated in this way is not biased. Of course, we could collect more data to be used as the independent test set, but that is often impossible or impractical.

By biased we mean that the estimated probability of correctly classifying a pattern is not overly optimistic. A common mistake that some researchers make is to build a classifier using their sample and then use the same sample to determine the proportion of observations that are correctly classified. That procedure typically yields much higher classification success rates because the classifier has already *seen* the patterns. It does not provide an accurate idea of how the classifier recognizes patterns it has not seen before. However, for a thorough discussion on these issues, see Ripley [1996]. The steps for evaluating the classifier using an independent test set are outlined next.

PROBABILITY OF CORRECT CLASSIFICATION – INDEPENDENT TEST SAMPLE

1. Randomly separate the sample into two sets of size n_{TEST} and n_{TRAIN}, where $n_{TRAIN} + n_{TEST} = n$. One is for building the classifier (the training set), and one is used for testing the classifier (the testing set).

2. Build the classifier (e.g., Bayes decision rule, classification tree, etc.) using the training set.

3. Present each pattern from the test set to the classifier and obtain a class label for it. Since we know the correct class for these observations, we can count the number we have successfully classified. Denote this quantity as N_{CC}.

4. The rate at which we correctly classified observations is

$$P(CC) = \frac{N_{CC}}{n_{TEST}}.$$

The higher this proportion, the better the classifier. We illustrate this procedure in Example 10.6.

Example 10.6

We first load the data and then divide the data into two sets, one for building the classifier and one for testing it. We use the two species of **iris** that are hard to separate: *Iris versicolor* and *Iris virginica*.

```
load iris
% This loads up three matrices:
% setosa, versicolor, and virginica.
% We will use the versicolor and virginica.
% To make it interesting, we will use only the
% first two features.
% Get the data for the training and testing set. We
% will just pick every other one for the testing set.
indtrain = 1:2:50;
indtest = 2:2:50;
versitest = versicolor(indtest,1:2);
```

```
versitrain = versicolor(indtrain,1:2);
virgitest = virginica(indtest,1:2);
virgitrain = virginica(indtrain,1:2);
```

We now build the classifier by estimating the class-conditional probabilities. We use the parametric approach, making the assumption that the class-conditional densities are multivariate normal. In this case, the estimated priors are equal.

```
% Get the classifier. We will assume a multivariate
% normal model for these data.
muver = mean(versitrain);
covver = cov(versitrain);
muvir = mean(virgitrain);
covvir = cov(virgitrain);
```

Note that the classifier is obtained using the training set only. We use the testing set to estimate the probability of correctly classifying observations.

```
% Present each test case to the classifier. Note that
% we are using equal priors, so the decision is based
% only on the class-conditional probabilities.
% Put all of the test data into one matrix.
X = [versitest;virgitest];
% These are the probability of x given versicolor.
pxgver = csevalnorm(X,muver,covver);
% These are the probability of x given virginica.
pxgvir = csevalnorm(X,muvir,covvir);
% Check which are correctly classified.
% In the first 25, pxgver > pxgvir are correct.
ind = find(pxgver(1:25)>pxgvir(1:25));
ncc = length(ind);
% In the last 25, pxgvir > pxgver are correct.
ind = find(pxgvir(26:50) > pxgver(26:50));
ncc = ncc + length(ind);
pcc = ncc/50;
```

Using this type of classifier and this partition of the learning sample, we estimate the probability of correct classification to be 0.74.
❑

Cross-Validation

When the sample is too small to partition the data into a training set and an independent test set, then we can use the cross-validation approach. Recall from Chapter 8 that with cross-validation, we systematically partition the data into testing sets of size k. The $n - k$ observations are used to build the classifier, and the remaining k patterns are used to test it. The following is the procedure for calculating the probability of correct classification using cross-validation with $k = 1$.

PROBABILITY OF CORRECT CLASSIFICATION – CROSS-VALIDATION

1. Set the number of correctly classified patterns to 0, $N_{CC} = 0$.
2. Keep out one observation, call it x_i.
3. Build the classifier using the remaining $n - 1$ observations.
4. Present the observation x_i to the classifier and obtain a class label using the classifier from the previous step.
5. If the class label is correct, then increment the number correctly classified using

$$N_{CC} = N_{CC} + 1.$$

6. Repeat steps 2 through 5 for each pattern in the sample.
7. The probability of correctly classifying an observation is given by

$$P(CC) = \frac{N_{CC}}{n}.$$

Example 10.7
We return to the **iris** data of Example 10.6, and we estimate the probability of correct classification using cross-validation with $k = 1$. We first set up some preliminary variables and load the data.

```
load iris
% This loads up three matrices:
% setosa, versicolor, and virginica.
% We will use the versicolor and virginica.
% Note that the priors are equal, so the decision is
% based on the class-conditional probabilities.
ncc = 0;
% We will use only the first two features of
% the iris data for our classification.
% This should make it more difficult to
% separate the classes.
% Delete 3rd and 4th features.
virginica(:,3:4) = [];
versicolor(:,3:4) = [];
[nver,d] = size(versicolor);
[nvir,d] = size(virginica);
```

```
n = nvir + nver;
```

First, we will loop through all of the **versicolor** observations. We build a classifier, leaving out one pattern at a time for testing purposes. Throughout this loop, the class-conditional probability for **virginica** remains the same, so we find that first.

```
% Loop first through all of the patterns corresponding
% to versicolor. Here correct classification
% is obtained if pxgver > pxgvir;
muvir = mean(virginica);
covvir = cov(virginica);
% These will be the same for this part.
for i = 1:nver
    % Get the test point and the training set
    versitrain = versicolor;
    % This is the testing point.
    x = versitrain(i,:);
    % Delete from training set.
    % The result is the training set.
    versitrain(i,:)=[];
    muver = mean(versitrain);
    covver = cov(versitrain);
    pxgver = csevalnorm(x,muver,covver);
    pxgvir = csevalnorm(x,muvir,covvir);
    if pxgver > pxgvir
    % then we correctly classified it
        ncc = ncc+1;
    end
end
```

We repeat the same procedure leaving out each **virginica** observation as the test pattern.

```
% Loop through all of the patterns of virginica notes.
% Here correct classification is obtained when
% pxgvir > pxxgver
muver = mean(versicolor);
covver = cov(versicolor);
% Those remain the same for the following.
for i = 1:nvir
    % Get the test point and training set.
    virtrain = virginica;
    x = virtrain(i,:);
    virtrain(i,:)=[];
    muvir = mean(virtrain);
    covvir = cov(virtrain);
    pxgver = csevalnorm(x,muver,covver);
```

```
pxgvir = csevalnorm(x,muvir,covvir);
if pxgvir > pxgver
  % then we correctly classified it
  ncc = ncc+1;
end
end
```

Finally, the probability of correct classification is estimated using

```
pcc = ncc/n;
```

The estimated probability of correct classification for the **iris** data using cross-validation is 0.68.

❏

Receiver Operating Characteristic (ROC) Curve

We now turn our attention to how we can use cross-validation to evaluate a classifier that uses the likelihood approach with varying decision thresholds τ_C. It would be useful to understand how the classifier performs for various thresholds (corresponding to the probability of false alarm) of the likelihood ratio. This will tell us what performance degradation we have (in terms of correctly classifying the target class) if we limit the probability of false alarm to some level.

We start by dividing the sample into two sets: one with all of the target observations and one with the nontarget patterns. Denote the observations as follows

$$\mathbf{x}_i^{(1)} \Rightarrow \text{Target pattern } (\omega_1)$$

$$\mathbf{x}_i^{(2)} \Rightarrow \text{Nontarget pattern } (\omega_2).$$

Let n_1 represent the number of target observations (class ω_1) and n_2 denote the number of nontarget (class ω_2) patterns. We work first with the nontarget observations to determine the threshold we need to get a desired probability of false alarm. Once we have the threshold, we can determine the probability of correctly classifying the observations belonging to the target class.

Before we go on to describe the receiver operating characteristic (ROC) curve, we first describe some terminology. For any boundary we might set for the decision regions, we are likely to make mistakes in classifying cases. There will be some target patterns that we correctly classify as targets, and some we misclassify as nontargets. Similarly, there will be nontarget patterns that are correctly classified as nontargets and some that are misclassified as targets. This is summarized as follows:

- <u>True Positives – TP</u>: This is the fraction of patterns correctly classified as target cases.
- <u>False Positives – FP</u>: This is the fraction of nontarget patterns incorrectly classified as target cases.
- <u>True Negatives –TN</u>: This is the fraction of nontarget cases correctly classified as nontarget.
- <u>False Negatives – FN</u>: This is the fraction of target cases incorrectly classified as nontarget.

In our previous terminology, the false positives (FP) correspond to the false alarms. Figure 10.8 shows these areas for a given decision boundary.

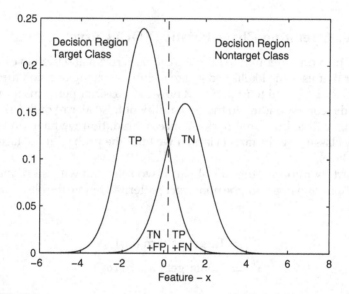

FIGURE 10.8
In this figure, we see the decision regions for deciding whether a feature corresponds to the target class or the nontarget class.

A **ROC** *curve* is a plot of the true positive rate against the false positive rate. ROC curves are used primarily in signal detection and medical diagnosis [Egan, 1975; Lusted, 1971; McNeil, et. al., 1975; Hanley and McNeil, 1983; Hanley and Hajian-Tilaki, 1997]. In their terminology, the true positive rate is also called the sensitivity. **Sensitivity** is the probability that a classifier will classify a pattern as a target when it really is a target. **Specificity** is the probability that a classifier will correctly classify the true nontarget cases. Therefore, we see that a ROC curve is also a plot of sensitivity against 1 minus specificity.

One of the purposes of a ROC curve is to measure the discriminating power of the classifier. It is used in the medical community to evaluate the diagnostic power of tests for diseases. By looking at a ROC curve, we can understand the following about a classifier:

- It shows the trade-off between the probability of correctly classifying the target class (sensitivity) and the false alarm rate (1 − specificity).
- The area under the ROC curve can be used to compare the performance of classifiers.

We now show in more detail how to construct a ROC curve. Recall that the likelihood ratio is given by

$$L_R(\mathbf{x}) = \frac{P(\mathbf{x}|\omega_1)}{P(\mathbf{x}|\omega_2)}.$$

We start off by forming the likelihood ratios using the nontarget (ω_2) observations and cross-validation to get the distribution of the likelihood ratios when the class membership is truly ω_2. We use these likelihood ratios to set the threshold that will give us a specific probability of false alarm.

Once we have the thresholds, the next step is to determine the rate at which we correctly classify the target cases. We first form the likelihood ratio for each target observation using cross-validation, yielding a distribution of likelihood ratios for the target class. For each given threshold, we can determine the number of target observations that would be correctly classified by counting the number of L_R that are greater than that threshold. These steps are described in detail in the following procedure.

CROSS-VALIDATION FOR SPECIFIED FALSE ALARM RATE

1. Given observations with class labels ω_1 (target) and ω_2 (nontarget), set desired probabilities of false alarm and a value for k.
2. Leave k points out of the nontarget class to form a set of test cases denoted by *TEST*. We denote cases belonging to class ω_2 as $\mathbf{x}_i^{(2)}$.
3. Estimate the class-conditional probabilities using the remaining $n_2 - k$ nontarget cases and the n_1 target cases.
4. For each of those k observations, form the likelihood ratios

$$L_R(\mathbf{x}_i^{(2)}) = \frac{P(\mathbf{x}_i^{(2)}|\omega_1)}{P(\mathbf{x}_i^{(2)}|\omega_2)}; \qquad \mathbf{x}_i^{(2)} \text{ in } TEST.$$

5. Repeat steps 2 through 4 using all of the nontarget cases.

6. Order the likelihood ratios for the nontarget class.

7. For each probability of false alarm, find the threshold that yields that value. For example, if the $P(FA) = 0.1$, then the threshold is given by the quantile $\hat{q}_{0.9}$ of the likelihood ratios. Note that higher values of the likelihood ratios indicate the target class. We now have an array of thresholds corresponding to each probability of false alarm.

8. Leave k points out of the target class to form a set of test cases denoted by *TEST*. We denote cases belonging to ω_1 by $x_i^{(1)}$.

9. Estimate the class-conditional probabilities using the remaining $n_1 - k$ target cases and the n_2 nontarget cases.

10. For each of those k observations, form the likelihood ratios

$$L_R(\mathbf{x}_i^{(1)}) = \frac{P(\mathbf{x}_i^{(1)}|\omega_1)}{P(\mathbf{x}_i^{(1)}|\omega_2)}; \qquad \mathbf{x}_i^1 \text{ in } TEST.$$

11. Repeat steps 8 through 10 using all of the target cases.

12. Order the likelihood ratios for the target class.

13. For each threshold and probability of false alarm, find the proportion of target cases that are correctly classified to obtain the $P(CC_{Target})$. If the likelihood ratios $L_R(\mathbf{x}_i^{(1)})$ are sorted, then this would be the number of cases that are greater than the threshold.

This procedure yields the rate at which the target class is correctly classified for a given probability of false alarm. We show in Example 10.8 how to implement this procedure in MATLAB and plot the results in a ROC curve.

Example 10.8
In this example, we illustrate the cross-validation procedure and ROC curve using the univariate model of Example 10.3. We first use MATLAB to generate some data.

```
% Generate some data, use the model in Example 10.3.
% p(x|w1) ~ N(-1,1), p(w1) = 0.6
% p(x|w2) ~ N(1,1),p(w2) = 0.4;
% Generate the random variables.
n = 1000;
u = rand(1,n);% find out what class they are from
n1 = length(find(u <= 0.6));% # in target class
n2 = n-n1;
x1 = randn(1,n1) - 1;
x2 = randn(1,n2) + 1;
```

We set up some arrays to store the likelihood ratios and estimated probabilities. We also specify the values for the $P(FA)$. For each $P(FA)$, we will be estimating the probability of correctly classifying objects from the target class.

```
% Set up some arrays to store things.
lr1 = zeros(1,n1);
lr2 = zeros(1,n2);
pfa = 0.01:.01:0.99;
pcc = zeros(size(pfa));
```

We now implement steps 2 through 7 of the cross-validation procedure. This is the part where we find the thresholds that provide the desired probability of false alarm.

```
% First find the threshold corresponding
% to each false alarm rate.
% Build classifier using target data.
mu1 = mean(x1);
var1 = cov(x1);
% Do cross-validation on non-target class.
for i = 1:n2
    train = x2;
    test = x2(i);
    train(i) = [];
    mu2 = mean(train);
    var2 = cov(train);
    lr2(i) = csevalnorm(test,mu1,var1)./...
            csevalnorm(test,mu2,var2);
end
% sort the likelihood ratios for the non-target class
lr2 = sort(lr2);
% Get the thresholds.
thresh = zeros(size(pfa));
for i = 1:length(pfa)
    thresh(i) = csquantiles(lr2,1-pfa(i));
end
```

For the given thresholds, we now find the probability of correctly classifying the target cases. This corresponds to steps 8 through 13.

```
% Now find the probability of correctly
% classifying targets.
mu2 = mean(x2);
var2 = cov(x2);
% Do cross-validation on target class.
for i = 1:n1
    train = x1;
```

```
      test = x1(i);
      train(i) = [];
      mu1 = mean(train);
      var1 = cov(train);
      lr1(i) = csevalnorm(test,mu1,var1)./...
         csevalnorm(test,mu2,var2);
   end
   % Find the actual pcc.
   for i = 1:length(pfa)
      pcc(i) = length(find(lr1 >= thresh(i)));
   end
   pcc = pcc/n1;
```

The ROC curve is given in Figure 10.9. We estimate the area under the curve as 0.91, using

```
   area = sum(pcc)*.01;
```

❑

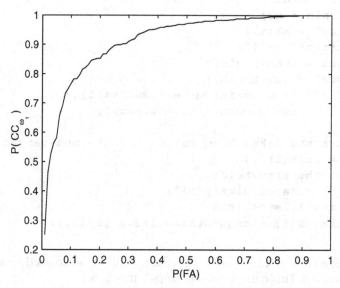

FIGURE 10.9
This shows the ROC curve for Example 10.8.

10.4 Classification Trees

While Bayes decision theory yields a classification rule that is intuitively appealing, it does not provide insights about the structure or the nature of the classification rule or help us determine what features are important. *Classification trees* can yield complex decision boundaries, and they are appropriate for ordered data, categorical data, or a mixture of the two types. In this book, we will be concerned only with the case where all features are continuous random variables. The interested reader is referred to Breiman, et al. [1984], Webb [2011], and Duda, Hart, and Stork [2001] for more information on the other cases.

A decision or classification tree represents a multi-stage decision process, where a binary decision is made at each stage. The tree is made up of *nodes* and *branches*, with nodes being designated as an internal or a terminal node. *Internal nodes* are ones that split into two children, while *terminal nodes* do not have any children. A terminal node has a class label associated with it, such that observations that fall into the particular terminal node are assigned to that class.

To use a classification tree, a feature vector is presented to the tree. If the value for a feature is less than some number, then the decision is to move to the left child. If the answer to that question is no, then we move to the right child. We continue in that manner until we reach one of the terminal nodes, and the class label that corresponds to the terminal node is the one that is assigned to the pattern. We illustrate this with a simple example.

Example 10.9

We show a simple classification tree in Figure 10.10, where we are concerned with only two features. Note that all internal nodes have two children and a splitting rule. The split can occur on either variable, with observations that are less than that value being assigned to the left child and the rest going to the right child. Thus, at node 1, any observation where the first feature is less than 5 would go to the left child. When an observation stops at one of the terminal nodes, it is assigned to the corresponding class for that node. We illustrate these concepts with several cases. Say that we have a feature vector given by $\mathbf{x} = (4, 6)$, then passing this down the tree, we get

$$\text{node } 1 \rightarrow \text{node } 2 \Rightarrow \omega_1 .$$

If our feature vector is $\mathbf{x} = (6, 6)$, then we travel the tree as follows:

$$\text{node } 1 \rightarrow \text{node } 3 \rightarrow \text{node } 4 \rightarrow \text{node } 6 \Rightarrow \omega_2 .$$

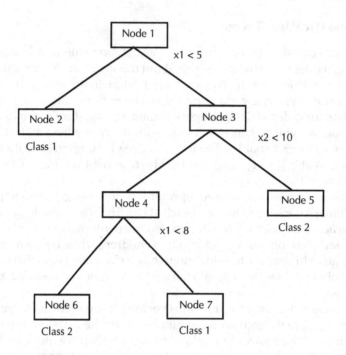

FIGURE 10.10
This simple classification tree for two classes is used in Example 10.9. Here we make decisions based on two features, x_1 and x_2.

For a feature vector given by $\mathbf{x} = (10, 12)$, we have

$$\text{node } 1 \rightarrow \text{node } 3 \rightarrow \text{node } 5 \Rightarrow \omega_2.$$

❏

We give a brief overview of the steps needed to create a tree classifier and then explain each one in detail. To start the process, we must grow an overly large tree using a criterion that will give us optimal splits for the tree. It turns out that these large trees fit the training data set very well. However, they do not generalize well, so the rate at which we correctly classify new patterns is low. The proposed solution [Breiman, et al., 1984] to this problem is to continually prune the large tree using a minimal cost complexity criterion to get a sequence of sub-trees. The final step is to choose a tree that is the "right size" using cross-validation or an independent test sample. These three main procedures are described in the remainder of this section. However, to make things easier for the reader, we first provide the notation that will be used to describe classification trees.

CLASSIFICATION TREES – NOTATION

- **L** denotes a learning set made up of n observed d-dimensional feature vectors and their class label.
- T is a classification tree, and t represents a node in the tree.
- T_t is a branch of tree T starting at node t.
- $\widehat{T}$ is the set of terminal nodes in the tree, and $|\widehat{T}|$ is the number of terminal nodes in tree T.
- n_j is the number of observations in the learning set that belong to the j-th class ω_j, $j = 1, ..., J$.
- $n(t)$ is the number of observations that fall into node t.
- $n_j(t)$ is the number of observations at node t that belong to class ω_j.
- $P(\omega_j, t)$ represents the joint probability that an observation will be in node t, and it will belong to class ω_j. It is calculated using the prior probability an observation belongs to the j-th class, $P(\omega_j)$:

$$P(\omega_j, t) = \frac{P(\omega_j)n_j(t)}{n_j}.$$

- $P(t)$ is the probability that an observation falls into node t and is given by

$$P(t) = \sum_{j=1}^{J} P(\omega_j, t).$$

- $P(\omega_j|t)$ denotes the probability that an observation is in class ω_j given it is in node t. This is calculated from

$$P(\omega_j|t) = \frac{P(\omega_j, t)}{P(t)}.$$

- $r(t)$ represents the resubstitution estimate of the probability of misclassification for node t and a given classification into class ω_j. This is found by subtracting the maximum conditional probability $P(\omega_j|t)$ for the node from 1:

$$r(t) = 1 - \max_{j} \{P(\omega_j|t)\}. \tag{10.11}$$

- $R(t)$ is the resubstitution estimate of risk for node t. This is

$$R(t) = r(t)P(t).$$

- $R(T)$ denotes a resubstitution estimate of the overall misclassification rate for a tree T. This can be calculated using every terminal node in the tree as follows:

$$R(T) = \sum_{t \in \hat{T}} r(t)P(t) = \sum_{t \in \hat{T}} R(t).$$

Growing the Tree

The idea behind binary classification trees is to split the d-dimensional space into smaller and smaller partitions, such that the partitions become purer in terms of the class membership. In other words, we are seeking partitions where the majority of the members in each partition belong to one class. To illustrate these ideas, we use a simple example where we have patterns from two classes, each one containing two features, x_1 and x_2. How we obtain these data are discussed in the following example.

Example 10.10
We use synthetic data to illustrate the concepts of classification trees. There are two classes in the data set, and we generate 50 points from each class. This means that our prior probabilities are equal. Each class is generated with the following bivariate normal distributions:

$$\mu_1 = [1, 1]^T \qquad \mu_2 = [-1, -1]^T$$

$$\Sigma_1 = \begin{bmatrix} 2 & 0 \\ 0 & 2 \end{bmatrix} \qquad \Sigma_2 = \begin{bmatrix} 2 & 0 \\ 0 & 2 \end{bmatrix}$$

$$P(\omega_1) = 0.5 \qquad P(\omega_2) = 0.5.$$

The following MATLAB code will generate the random sample.

```
% Generate some observations that are
% bivariate normal.
mu1 = [1;1];
mu2 = [-1;1];
sig1 = 2*eye(2);
sig2 = 2*eye(2);
% We will use equal class sizes,
% so the priors are 0.5.
```

```
data1 = mvnrnd(mu1,sig1,50);
data2 = mvnrnd(mu2,sig2,50);
X = [data1;data2];
clab = [ones(50,1);2*ones(50,1)];
```

A scatterplot of these data is given in Figure 10.11. One class is depicted by the "*" and the other is represented by the "o." It appears that we have some regions of overlap between the classes, making it harder to build a classifier. ❑

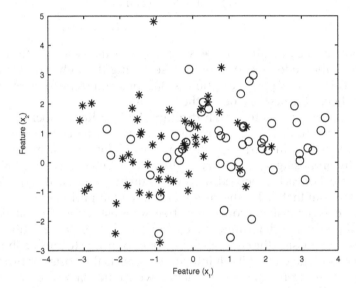

FIGURE 10.11
This shows a scatterplot of the data that will be used in some of our classification tree examples. Data that belong to class 1 are shown by the "o," and those that belong to class 2 are denoted by an "."*

To grow a tree, we need to have some criterion to help us decide how to split the nodes. We also need a rule that will tell us when to stop splitting the nodes, at which point we are finished growing the tree. The stopping rule can be quite simple, since we want to first grow an overly large tree. One possible choice is to continue splitting terminal nodes until each one contains observations from the same class, in which case some nodes might have only one observation in the node. Another option is to continue splitting nodes until there are too few observations left in a node or the terminal node is pure (all observations belong to one class).

We now discuss the splitting rule in more detail. When we split a node, our goal is to find a split that reduces the impurity in some manner. So, we need a measure of impurity $i(t)$ *for a node t.* Breiman, et al. [1984] discuss several possibilities, one of which is called the ***Gini diversity index***. This is the one we will use in our implementation of classification trees. The Gini index is given by

$$i(t) = \sum_{i \neq j} P(\omega_i | t) P(\omega_j | t),$$

which can also be written as

$$i(t) = 1 - \sum_{j=1}^{J} P^2(\omega_j | t). \tag{10.12}$$

To grow a tree at a given node, we search for the best split (in terms of decreasing the node impurity) by first searching through each variable or feature. This yields d possible optimal splits for a node (one for each feature), and we choose the best one out of these d splits.

The problem now is to search through the infinite number of possible splits. We can limit our search by using the following convention. For all feature vectors in our learning sample, we search for the best split at the k-th feature by proposing splits that are halfway between consecutive values for that feature. For each proposed split, we evaluate the impurity criterion and choose the split that yields the largest decrease in impurity.

Once we have finished growing our tree, we must assign class labels to the terminal nodes and determine the corresponding misclassification rate. It makes sense to assign the class label to a node according to the likelihood that it is in class ω_j given that it fell into node t. This is the posterior probability $P(\omega_j | t)$. So, using Bayes decision theory, we would classify an observation at node t with the class ω_j that has the highest posterior probability. The error in our classification is then given by $r(t)$ (Equation 10.11).

We summarize the steps for growing a classification tree in the following procedure. In the learning set, each observation will be a row in the matrix **X**. The measured value of the k-th feature for the i-th observation is denoted by x_{ik}.

PROCEDURE – GROWING A TREE

1. Determine the maximum number of observations n_{max} that will be allowed in a terminal node.

2. Determine the prior probabilities of class membership. These can be estimated from the data, or they can be based on prior knowledge of the application.

3. If a terminal node in the current tree contains more than the maximum allowed observations and contains observations from more than one class, then search for the best split.

a. Put the x_{ik} in ascending order to give the ordered values $x_{(i)k}$, $k = 1, ..., J$.

b. Determine all splits $s_{(i)k}$ in the k-th feature using

$$s_{(i)k} = x_{(i)k} + (x_{(i)k} - x_{(i+1)k})/2.$$

c. For each proposed split, evaluate the impurity function $i(t)$ and the goodness of the split.

d. Pick the best, which is the one that yields the largest decrease in impurity $\Delta i(s, t)$, which is given by

$$\Delta i(s, t) = i(t) - p_R i(t_R) - p_L i(t_L),$$

where p_L and p_R are the proportion of data sent to the left and right child nodes by the split s.

4. Using the information in step 3, split the node on the variable that yields the best overall split.

5. For that split found in step 4, determine the observations that go to the left child and those that go to the right child.

6. Repeat steps 3 through 5 until each terminal node satisfies the stopping rule (has observations from only one class or does not have enough observations left to split the node).

Example 10.11

In this example, we grow the initial large tree on the data set given in the previous example. We stop growing the tree when each terminal node has fewer than five observations or the node is pure. Version 6 of the MATLAB Statistics Toolbox contains many useful functions for constructing, using, and displaying classification trees. This new version implements a tree constructor to produce an object-oriented *tree* class. However, we do not need to know the details of what this means to use the functions, as we show next. We first load the data that we generated in the previous example.

```
% Use the data from the previous example.
% Construct a tree using classregtree.
% The 'splitmin' parameter tells MATLAB that
% there has to be at least 5 observations in
% a node to split it.
t = classregtree(X,clab,'splitmin',5,...
    'method','classification')
```

The first argument is the observed features, and the second argument to the function contains the class labels. The parameter **'splitmin'** specifies the smallest number of observations that can be in a node to split it further; the default is ten. Because we did not have the semi-colon in the call to **classregtree**, the following is displayed in the command window:

```
Decision tree for classification
  1   if x1<0.421191 then node 2 else node 3
  2   if x1<-1.18865 then node 4 else node 5
  3   if x2<3.10127 then node 6 else node 7
  4   if x2<0.198493 then node 8 else node 9
  5   if x1<0.285778 then node 10 else node 11
  6   if x1<1.50404 then node 12 else node 13
  7   class = 2
  8   class = 2
  9   if x2<1.29411 then node 14 else node 15
 10   if x1<0.227184 then node 16 else node 17
 11   class = 2
 12   ...
```

This is useful because it can be used to code the decision rules in another programming language. The amount of information can be overwhelming if the tree is too large, so it is also a good idea to graphically view the tree. The following MATLAB command will plot the tree in a special GUI viewer:

view(t)

The tree is shown in Figure 10.12, where we see that the tree has partitioned the feature space into many decision regions. The **Click to display** menu allows one to view different information when clicking on the nodes, such as classification and probabilities. The **Pruning level** control shows the sequence of optimally pruned subtrees by clicking on the arrow. We describe the pruning procedure next.

❏

Pruning the Tree

Recall that the classification error for a node is given by Equation 10.11. If we grow a tree until each terminal node contains observations from only one class, then the error rate will be zero. Therefore, if we use the classification error as a stopping criterion or as a measure of when we have a good tree, then we would grow the tree until there are pure nodes. However, as we mentioned before, this procedure over fits the data and the classification tree will not generalize well to new patterns. The suggestion made in Breiman, et al. [1984] is to grow an overly large tree, denoted by T_{max}, and then to find a

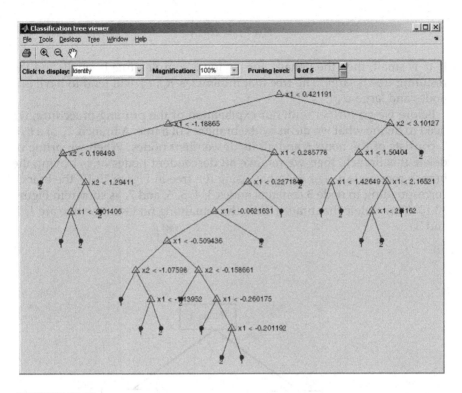

FIGURE 10.12
This is the classification tree for the data shown in Figure 10.11. This tree partitions the feature space into many decision regions.

nested sequence of subtrees by successively pruning branches of the tree. The best tree from this sequence is chosen based on the misclassification rate estimated by cross-validation or an independent test sample. We describe the two approaches after we discuss how to prune the tree.

The pruning procedure uses the misclassification rates along with a cost for the complexity of the tree. The complexity of the tree is based on the number of terminal nodes in a subtree or branch. The cost complexity measure is defined as

$$R_\alpha(T) = R(T) + \alpha |\widehat{T}|; \qquad \alpha \geq 0. \qquad (10.13)$$

We look for a tree that minimizes the cost complexity given by Equation 10.13. The α is a parameter that represents the complexity cost per terminal node. If we have a large tree and every terminal node contains observations from only one class, then $R(T)$ will be zero. However, there will be a penalty paid because of the complexity, and the cost complexity measure becomes

$$R_\alpha(T) = \alpha|\widehat{T}|.$$

If α is small, then the penalty for having a complex tree is small, and the resulting tree is large. The tree that minimizes $R_\alpha(T)$ will tend to have few nodes and large α.

Before we go further with our explanation of the pruning procedure, we need to define what we mean by the branches of a tree. A branch T_t of a tree T consists of the node t and all its descendent nodes. When we prune or delete this branch, then we remove all descendent nodes of t, leaving the branch root node t. For example, using the tree in Figure 10.10, the branch corresponding to node 3 contains nodes 3, 4, 5, 6, and 7, as shown in Figure 10.13. If we delete that branch, then the remaining nodes in the tree are 1, 2, and 3.

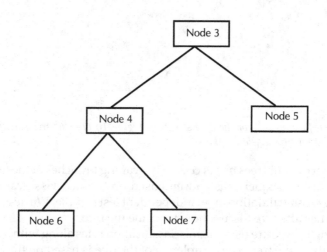

FIGURE 10.13
These are the nodes that comprise the branch corresponding to node 3.

Minimal complexity pruning searches for the branches that have the weakest link, which we then delete from the tree. The pruning process produces a sequence of subtrees with fewer terminal nodes and decreasing complexity.

We start with our overly large tree and denote this tree as T_{max}. We are searching for a finite sequence of subtrees such that

$$T_{max} > T_1 > T_2 > \dots > T_K = \{t_1\}.$$

There is a continuum of values for the complexity parameter α, but if a tree $T(\alpha)$ is a tree that minimizes $R_\alpha(T)$ for a given α, then it will continue to minimize it until a jump point for α is reached. Thus, we will be looking for a sequence of complexity values α and the trees that minimize the cost complexity measure for each level. Once we have our tree T_1, we start pruning off the branches that have the weakest link. To find the weakest link, we first define a function on a tree as follows:

$$g_k(t) = \frac{R(t) - R(T_{kt})}{|\widehat{T}_{kt}| - 1} \qquad t \text{ is an internal node,} \qquad (10.14)$$

where T_{kt} is the branch T_t corresponding to the internal node t of subtree T_k. From Equation 10.14, for every internal node in tree T_k, we determine the value for $g_k(t)$. We define the weakest link t_k^* in tree T_k as the internal node t that minimizes Equation 10.14,

$$g_k(t_k^*) = min_t\{g_k(t)\}.$$

Once we have the weakest link, we prune the branch defined by that node. The new tree in the sequence is obtained by

$$T_{k+1} = T_k - T_{t_k^*}, \qquad (10.15)$$

where the subtraction in Equation 10.15 indicates the pruning process. We set the value of the complexity parameter to

$$\alpha_{k+1} = g_k(t_k^*).$$

The result of this pruning process will be a decreasing sequence of trees,

$$T_{max} > T_1 > T_2 > \dots > T_K = \{t_1\},$$

along with an increasing sequence of values for the complexity parameter

$$0 = \alpha_1 < \dots < \alpha_k < \alpha_{k+1} < \dots < \alpha_K.$$

We need the following key fact when we describe the procedure for choosing the best tree from the sequence of subtrees:

For $k \geq 1$, the tree T_k is the minimal cost complexity tree for the interval $\alpha_k \leq \alpha < \alpha_{k+1}$, and

$$T(\alpha) = T(\alpha_k) = T_k.$$

PROCEDURE – PRUNING THE TREE

1. Start with a large tree T_{max}.
2. For all internal nodes in the current tree, calculate $g_k(t)$ as given in Equation 10.14.
3. The weakest link is the node that has the smallest value for $g_k(t)$.
4. Prune off the branch that has the weakest link.
5. Repeat steps 2 through 4 until only the root node is left.

By default, the MATLAB tree constructor function **classregtree** returns a tree that contains the necessary information for pruning the tree, so we do not have to implement the pruning procedure. We can look at the sequence of subtrees by clicking on different pruning levels in the tree viewer. We illustrate this in Figure 10.14 using the tree constructed in the previous example. This pruning information is also used by MATLAB with the function **test** to pick the optimal tree.

Choosing the Best Tree

In the previous section, we discussed the importance of using independent test data to evaluate the performance of our classifier. We now use the same procedures to help us choose the right size tree. It makes sense to choose a tree that yields the smallest true misclassification cost, but we need a way to estimate this.

The values for misclassification rates that we get when constructing a tree are really estimates using the learning sample. We would like to get less biased estimates of the true misclassification costs, so we can use these values to choose the tree that has the smallest estimated misclassification rate. We can get these estimates using either an independent test sample or cross-validation. In this text, we cover the situation where there is a unit cost for misclassification, and the priors are estimated from the data. For a general treatment of the procedure, the reader is referred to Breiman, et al. [1984].

Selecting the Best Tree Using an Independent Test Sample

We first describe the independent test sample case because it is easier to understand. The notation that we use is summarized next.

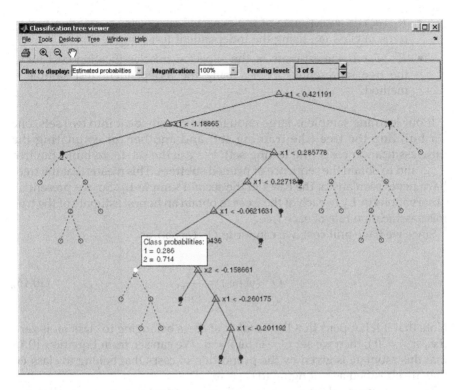

FIGURE 10.14

*This is the classification tree we constructed in Example 10.11. Here we are showing pruning level 3 in the optimal pruning sequence calculated with the **classregtree** function. The branches that have been pruned are shown with gray dotted lines. In MATLAB, a pruning level of zero corresponds to the full tree. Note that we have **Click to display** set to **Estimated probabilities**. The classification probabilities are displayed when we click on a node, as we see in this figure.*

NOTATION – INDEPENDENT TEST SAMPLE METHOD

- L_1 is the subset of the learning sample L that will be used for building the tree (i.e., the training set).
- L_2 is the subset of the learning sample L that will be used for testing the tree and choosing the best subtree.
- $n^{(2)}$ is the number of cases in L_2.
- $n_j^{(2)}$ is the number of observations in L_2 that belong to class ω_j.
- $n_{ij}^{(2)}$ is the number of observations in L_2 that belong to class ω_j that were classified as belonging to class ω_i.
- $\hat{Q}^{TS}(\omega_i | \omega_j)$ represents the estimate of the probability that a case belonging to class ω_j is classified as belonging to class ω_i, using the independent test sample method.

- $\hat{R}^{TS}(\omega_j)$ is an estimate of the expected cost of misclassifying patterns in class ω_j, using the independent test sample.
- $\hat{R}^{TS}(T_k)$ is the estimate of the expected misclassification cost for the tree represented by T_k using the independent test sample method.

If our learning sample is large enough, we can divide it into two sets, one for building the tree (the training set) and another for estimating the misclassification costs (the testing set). We use the set L_1 to build the tree T_{max} and to obtain the sequence of pruned subtrees. This means that the trees have never seen any of the cases in the second sample L_2. So, we present all observations in L_2 to each of the trees to obtain an honest estimate of the true misclassification rate of each tree.

Since we have unit cost, we can write $\hat{Q}^{TS}(\omega_i|\omega_j)$ as

$$\hat{Q}^{TS}(\omega_i|\omega_j) = \frac{n_{ij}^{(2)}}{n_j^{(2)}}. \tag{10.16}$$

Note that if it happens that the number of cases belonging to class ω_j is zero (i.e., $n_j^{(2)} = 0$), then we set $\hat{Q}^{TS}(\omega_i|\omega_j) = 0$. We can see from Equation 10.16 that this estimate is given by the proportion of cases that belong to class ω_j that are classified as belonging to class ω_i.

The total proportion of observations belonging to class ω_j that are misclassified is given by

$$\hat{R}^{TS}(\omega_j) = \sum_{i \neq j} \hat{Q}^{TS}(\omega_i|\omega_j).$$

This is our estimate of the expected misclassification cost for class ω_j. Finally, we use the total proportion of test cases misclassified by tree T as our estimate of the misclassification cost for the tree classifier. This can be calculated using

$$\hat{R}^{TS}(T_k) = \frac{1}{n^{(2)}} \sum_{i,j} n_{ij}^{(2)}. \tag{10.17}$$

Equation 10.17 is easily calculated by simply counting the number of misclassified observations from L_2 and dividing by the total number of cases in the test sample.

The rule for picking the best subtree requires one more quantity. This is the standard error of our estimate of the misclassification cost for the trees. In our case, the prior probabilities are estimated from the data, and we have unit cost for misclassification. Thus, the standard error is estimated by

$$\hat{SE}\left(\hat{R}^{TS}(T_k)\right) = \left\{\hat{R}^{TS}(T_k)\left(1 - \hat{R}^{TS}(T_k)\right)/n^{(2)}\right\}^{1/2}, \qquad (10.18)$$

where $n^{(2)}$ is the number of cases in the independent test sample.

To choose the right size subtree, Breiman, et al. [1984] recommend the following. First, find the smallest value of the estimated misclassification error. Then, add the standard error given by Equation 10.18 to the minimum misclassification error. Finally, find the smallest tree (the tree with the largest subscript k) such that its misclassification cost is less than the minimum misclassification plus its standard error. In essence, we are choosing the least complex tree whose accuracy is comparable to the tree yielding the minimum misclassification rate.

PROCEDURE – CHOOSING THE BEST SUBTREE – TEST SAMPLE METHOD

1. Randomly partition the learning set into two parts, $\mathbf{L}_1$ and $\mathbf{L}_2$ or obtain an independent test set by randomly sampling from the population.
2. Grow a large tree T_{max} using $\mathbf{L}_1$.
3. Prune T_{max} to get the sequence of subtrees T_k.
4. For each tree in the sequence, take the cases in $\mathbf{L}_2$ and present them to the tree.
5. Count the number of cases that are misclassified.
6. Calculate the estimate for $\hat{R}^{TS}(T_k)$ using Equation 10.17.
7. Repeat steps 4 through 6 for each tree in the sequence.
8. Find the minimum error

$$\hat{R}^{TS}_{min} = \min_k \left\{\hat{R}^{TS}(T_k)\right\}.$$

9. Calculate the standard error in the estimate of $\hat{R}^{TS}_{min}$ using Equation 10.18.
10. Add the standard error to $\hat{R}^{TS}_{min}$ to get

$$\hat{R}^{TS}_{min} + \hat{SE}(\hat{R}^{TS}_{min}).$$

11. Find the tree with the fewest number of nodes (or equivalently, the largest k) such that its misclassification error is less than the amount found in step 10.

Example 10.12

The MATLAB Statistics Toolbox has a function called **test** that will return the error rate of a tree that was constructed with **classregtree**. The function will estimate the error using resubstitution, an independent test sample, or cross-validation. We generated our own data set in Example 10.10 and used it to construct a classification tree in the previous example. Because this is simulated data, we can generate an independent test set using the same distribution.

```
% Find the best tree from the previous example.
% First create an independent test sample.
n = 100;
u = rand(1,n);
% Find the number in class 1.
n1 = length(find(u<=0.5));
% The rest are in class 2.
n2 = n - n1;
% Generate the data for the classes.
testdata1 = mvnrnd(mu1,sig1,n1);
testdata2 = mvnrnd(mu2,sig2,n2);
% Put them into one matrix.
Xtest = [testdata1;testdata2];
% Get the class labels.
clabtest = [ones(n1,1);2*ones(n2,1)];
```

The MATLAB test function will return a vector containing the estimated error for each subtree in the pruning sequence, their associated standard errors, the number of terminal nodes in the subtree, and the best level of the tree.

```
% Now find the error in the tree using
% the independent test set.
[erri,sei,nti,besti] = test(t,'test',Xtest,clabtest);
```

The best level returned by this function is the one that corresponds to the smallest tree that is within one standard error of the minimum-cost subtree. The best level returned for our tree using the independent test sample method is **besti** = **3**. The tree corresponding to this pruning level is the one given in Figure 10.14. We show a plot of the errors as a function of the tree size (number of terminal nodes) in Figure 10.15.

❑

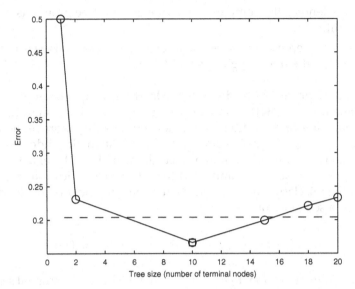

FIGURE 10.15
This solid line represents the estimated error for each tree size. The dotted line is one standard error above the minimum cost (error) tree. Note that the optimal tree is the one with ten terminal nodes (Figure 10.14).

Selecting the Best Tree Using Cross-Validation

We now turn our attention to the case where we use cross-validation to estimate our misclassification error for the trees. In cross-validation, we divide our learning sample into several training and testing sets. We use the training sets to build sequences of trees and then use the test sets to estimate the misclassification error.

In previous examples of cross-validation, our testing sets contained only one observation. In other words, the learning sample was sequentially partitioned into *n* test sets. As we discuss shortly, it is recommended that far fewer than *n* partitions be used when estimating the misclassification error for trees using cross-validation. We first provide the notation that will be used in describing the cross-validation method for choosing the right size tree.

NOTATION – CROSS-VALIDATION METHOD

- $\mathbf{L}_v$ denotes a partition of the learning sample $\mathbf{L}$, such that

$$\mathbf{L}^{(v)} = \mathbf{L} - \mathbf{L}_v; \qquad v = 1, ..., V.$$

- $T_k^{(v)}$ is a tree grown using the partition $\mathbf{L}^{(v)}$.
- $\alpha_k^{(v)}$ denotes the complexity parameter for a tree grown using the partition $\mathbf{L}^{(v)}$.
- $\hat{R}^{CV}(T)$ represents the estimate of the expected misclassification cost for the tree using cross-validation.

We start the procedure by dividing the learning sample $\mathbf{L}$ into V partitions $\mathbf{L}_v$. Breiman, et al. [1984] recommend a value of $V = 10$ and show that cross-validation using finer partitions does not significantly improve the results. For better results, it is also recommended that systematic random sampling be used to ensure a fixed fraction of each class will be in $\mathbf{L}_v$ and $\mathbf{L}^{(v)}$. These partitions $\mathbf{L}_v$ are set aside and used to test our classification tree and to estimate the misclassification error. We use the remainder of the learning set $\mathbf{L}^{(v)}$ to get a sequence of trees

$$T_{max}^{(v)} > T_1^{(v)} > \ldots > T_k^{(v)} > T_{k+1}^{(v)} > \ldots > T_K^{(v)} = \{t_1\},$$

for each training partition. Keep in mind that we have our original sequence of trees that were created using the entire learning sample $\mathbf{L}$, and that we are going to use these sequences of trees $T_k^{(v)}$ to evaluate the classification performance of each tree in the original sequence T_k. Each one of these sequences will also have an associated sequence of complexity parameters

$$0 = \alpha_1^{(v)} < \ldots < \alpha_k^{(v)} < \alpha_{k+1}^{(v)} < \ldots < \alpha_K^{(v)}.$$

At this point, we have $V + 1$ sequences of subtrees and complexity parameters.

We use the test samples $\mathbf{L}_v$ along with the trees $T_k^{(v)}$ to determine the classification error of the subtrees T_k. To accomplish this, we have to find trees that have equivalent complexity to T_k in the sequence of trees $T_k^{(v)}$.

Recall that a tree T_k is the minimal cost complexity tree over the range $\alpha_k \leq \alpha < \alpha_{k+1}$. We define a representative complexity parameter for that interval using the geometric mean

$$\alpha'_k = \sqrt{\alpha_k \alpha_{k+1}}. \tag{10.19}$$

The complexity for a tree T_k is given by this quantity. We then estimate the misclassification error using

$$\hat{R}^{CV}(T_k) = \hat{R}^{CV}(T(\alpha'_k)), \tag{10.20}$$

where the right hand side of Equation 10.20 is the proportion of test cases that are misclassified, using the trees $T_k^{(v)}$ that correspond to the complexity parameter α'_k.

To choose the best subtree, we need an expression for the standard error of the misclassification error $\hat{R}^{CV}(T_k)$. When we present our test cases from the partition $\mathbf{L}_v$, we record a zero or a one, denoting a correct classification and an incorrect classification, respectively. We see then that the estimate in Equation 10.20 is the mean of the ones and zeros. We estimate the standard error of this from

$$\hat{SE}\left(\hat{R}^{CV}(T_k)\right) = \sqrt{\frac{s^2}{n}}, \qquad (10.21)$$

where s^2 is $(n-1)/n$ times the sample variance of the ones and zeros.

The cross-validation procedure for estimating the misclassification error when we have unit cost and the priors are estimated from the data is outlined next.

PROCEDURE – CHOOSING THE BEST SUBTREE (CROSS-VALIDATION)

1. Obtain a sequence of subtrees T_k that are grown using the learning sample $\mathbf{L}$.

2. Determine the cost complexity parameter α'_k for each T_k using Equation 10.19.

3. Partition the learning sample into V partitions, $\mathbf{L}_v$. These will be used to test the trees.

4. For each $\mathbf{L}_v$, build the sequence of subtrees using $\mathbf{L}^{(v)}$. We should now have $V+1$ sequences of trees.

5. Now find the estimated misclassification error $\hat{R}^{CV}(T_k)$. For α'_k corresponding to T_k, find all equivalent trees $T_k^{(v)}$, $v = 1, ..., V$. We do this by choosing the tree $T_k^{(v)}$ such that

$$\alpha'_k \in [\alpha_k^{(v)}, \alpha_{k+1}^{(v)}).$$

6. Take the test cases in each $\mathbf{L}_v$ and present them to the tree $T_k^{(v)}$ found in step 5. Record a one if the test case is misclassified and a zero if it is classified correctly. These are the classification costs.

7. Calculate $\hat{R}^{CV}(T_k)$ as the proportion of test cases that are misclassified (or the mean of the array of ones and zeros found in step 6).

8. Calculate the standard error as given by Equation 10.21.

9. Continue steps 5 through 8 to find the misclassification cost for each subtree T_k.

10. Find the minimum error

$$\hat{R}_{min}^{CV} = \min_{k} \left\{ \hat{R}^{CV}(T_k) \right\}.$$

11. Add the estimated standard error to it to get

$$\hat{R}_{min}^{CV} + \hat{SE}(\hat{R}_{min}^{CV}).$$

12. Find the tree with the largest k or fewest number of nodes such that its misclassification error is less than the amount found in step 11.

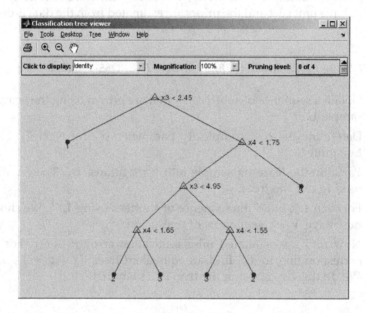

FIGURE 10.16
This is the classification tree for the **iris** *data.*

Example 10.13

For this example, we return to the **iris** data that we described at the beginning of this chapter. We saw in the previous example that the **test** function will return estimates of the error for a tree that was constructed with **classregtree**. So, we first get a classification tree for the **iris** data.

```
load iris
% Put into data matrix.
```

```
X = [setosa; versicolor; virginica];
% Create class labels.
clab = [ones(50,1);2*ones(50,1);3*ones(50,1)];
% Create the tree.
t = classregtree(X,clab,'splitmin',5,...
    'method','classification')
```

The resulting tree is shown in Figure 10.16, and we see from the **Pruning level** control that there are five subtrees in all, with level zero corresponding to the full tree and level four corresponding to the root node. We will use the **test** function to obtain the cross-validation estimate of the error with five cross-validation subsamples.

```
% First find the error in the tree using
% the cross-validation method.
% The default number of subsamples is ten.
[errCV,seCV,ntCV,bestCV]=test(t,'crossvalidate',...
    X,clab,...
    'nsamples',5);
```

Next, we will use the resubstitution method to estimate the error for this tree, so we can compare it to what we get with cross-validation. With this method, the error is found using the training data to test the tree. Because of this, it underestimates the likelihood of misclassification when using the tree to classify new data.

```
% Now find the error in the tree using
% the resubstitution method.
[errR,seR,ntR,bestR] = test(t,'resubstitution');
```

The best level or subtree based on cross-validation and resubstitution is

```
bestCV = 2
bestR = 0
```

So, we would prune to the second level of the tree based on cross-validation, and we would not prune the tree at all based on the resubstitution estimation of error. We plot the two types of errors as a function of tree size in Figure 10.17, along with the tree pruned to the best level.
❑

Other Tree Methods

There are other methods for splitting nodes, terminating the node splitting, and pruning trees. We briefly describe some of them here.

The **classregtree** function in the Statistics Toolbox will create trees based on three rules: the Gini diversity index, the twoing index, and the maximum reduction in deviance. A splitting rule based on the Gini index

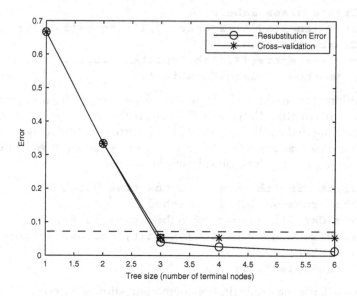

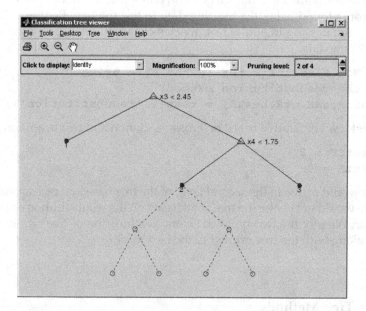

FIGURE 10.17

In the top portion of this figure, we have the estimated error using cross-validation and resubstitution. Note that the resubstitution method underestimates the error because it uses the training set to test the tree and estimate the error. Thus, it provides an overly optimistic view of the performance of the classifier. The dotted line is one standard error above the minimum error (via cross-validation) tree. The optimal tree with three terminal nodes is shown in the bottom part of the figure.

tries to find the largest homogeneous class within the dataset and isolate it from the rest of the data. The tree construction process continues to split nodes in the same way until nodes cannot be divided any more.

The *twoing index* tries to divide the data more evenly than the Gini rule would do by separating whole classes of data and finding the largest group first. The twoing index is given by

$$i_{two}(t) = \frac{p_L p_R}{4} \left[\sum_j (|P(\omega_j|t_L) - P(\omega_j|t_R)|) \right]^2,$$

where $P(\omega_j|t)$ is the relative frequency of class ω_j at node t.

The maximum *reduction in deviance* splitting rule for a classification tree uses the following definition of deviance for the tree:

$$i_{dev}(t) = -2 \sum_j n_j(t) \log \frac{n_j(t)}{n(t)}.$$

In general, the deviance of a statistical model m measures how far the model is from some target model.

These are not the only rules that have been proposed in the literature. Another popular one is the *entropy impurity* or *information impurity* that measures the homogeneity of a node. This is defined as

$$i_{ent}(t) = -\sum_j \frac{n_j(t)}{n(t)} \log_2 \frac{n_j(t)}{n(t)}.$$

It has been discussed in the literature that the choice of the impurity function or splitting rule does not seem to affect the accuracy of the final classifier [Mingers, 1989a; Duda, Hart, and Stork, 2001]. However, Zambon, et al. [2006] showed that the Gini index worked better when classification trees were used to process remotely sensed image data, while the entropy and twoing splitting rules performed poorly.

Mingers [1989b] describes several pruning rules, including the *error complexity pruning* of Breiman, et al. [1984]. Another pruning method is called *critical value pruning* that uses an estimate of the importance or strength of a node. This estimate is based on the value of the node impurity. This pruning method specifies a critical value and prunes the nodes that do not reach this value, unless a node lower in the tree does reach it.

The *reduced error pruning* [Quinlan, 1987] is a method that uses the test data to prune trees. One starts with a complete tree and uses it to process the test data, keeping track of the value of $n_j(t)$ at each node t. For each of the internal nodes, one counts the number of errors if the subtree is kept and the number of errors if it is pruned. The pruned node will sometimes make fewer errors than the subtree makes. One prunes the node with the largest decrease in misclassification (errors). This process continues until further pruning increases the misclassification rate.

Mingers [1989b] describes two more pruning schemes, and he shows that the three methods we just mentioned—error complexity pruning, critical value pruning, and reduced error pruning—perform well. He also shows that the performance of the resulting classification tree does not depend significantly on interactions between the splitting rule and the pruning method.

Other tree-based classification methods have been proposed, but most of them incorporate the basic steps we described previously. Duda, Hart, and Stork [2001] summarize two popular tree classifiers: ID3 and C4.5. The ID3 tree classifier was the third in a series of *interactive dichotomizer* procedures, hence its name. The method can only be used for unordered (nominal) data. If the data include real-valued variables, then they can be binned and each bin would be treated as nominal. The C4.5 method is a refinement and successor of the ID3 tree-based classification technique. Real-valued variables are processed the same way as the classification tree method we described in this chapter. The C4.5 method uses heuristics for pruning that are based on the statistical significance of node splits.

10.5 Combining Classifiers

We now present several methods for pattern recognition that primarily came out of the machine learning or data mining communities. These are bagging, boosting, arcing, and random forests. All of these methods are based on various ways of combining ensembles of classifiers.

Bagging

Bagging is a procedure for combining different classifiers that have been constructed using the same data set. The term *bagging* is an acronym for *bootstrap aggregating*. It is not surprising then that the method involves the use of the bootstrap methodology.

The motivation for methods that combine classifiers is to improve an unstable classifier. An *unstable classifier* is one where a small change in the learning set or the classification parameters produces a large change in the classifier [Breiman, 1996a]. Instability was studied by Breiman [1996b], and he concluded that neural network classifiers and classification trees were unstable classifiers. Studies have shown that bagging works well for unstable classifiers [Breiman, 1996a]. On the other hand, if the classifier is stable, then bagging will yield little improvement.

In the bagging procedure, we find B bootstrap replicates of our data set. We then construct our chosen classifier using each bootstrap sample, resulting in B classifiers. To classify a new observation **x**, we find the class predicted by each of the B bootstrap classifiers and assign the class label that is predicted most often among the bootstrap classifiers. Thus, the resulting classification is done by simple majority vote among the B classifiers. We outline the procedure next.

PROCEDURE – BAGGING

1. Assume we have a set of n observations with class labels, ω_j.
2. Generate a bootstrap sample of size n by sampling with replacement from the original data set.
3. Construct a classifier using the bootstrap sample in step 2.
4. Repeat steps 2 and 3, B times. This yields B classifiers.
5. To classify a new observation **x**, assign a class label to the observation using each of the B classifiers. This yields B class labels for **x**. Assign the class label to **x** that occurs with the highest frequency among the B class labels (majority vote). Ties are broken arbitrarily.

Note that we assume that the individual classifiers (step 3) yield a class label in step 5 of the bagging procedure. As we know, we could have classifiers that produce estimates of the posterior probabilities (Equation 10.2), such as linear and quadratic classifiers. In this case, we could use Bayes' decision rule to produce a class label and then use the majority vote procedure outlined above. Another option is to average the estimated posterior probabilities over all of the bootstrap replications and then use Bayes' decision rule on this bagged estimate of the posterior probabilities. Brieman [1996a] shows that the misclassification rate of these two approaches is almost identical when applied to several data sets from the machine learning community.

Webb [2011] makes the point that if the size of the training sample is small, then bagging can improve the classification error, even when the classifier is stable. Also, if the sample size is very large, then combining classifiers is not useful because the individual bootstrap classifiers tend to be similar.

Example 10.14
In this example, we use the **insect** data set to illustrate the bagging procedure. This data set contains thirty observations pertaining to three characteristics measured on three classes of insects with ten observations in each class. We will use the linear classifier on our bootstrap replicates in this example. We know that this classifier is stable, so we do not expect much improvement, if any, with bagging. First, we load up the data set and then

create a training and test set. The test set consists of one observation from each of the classes.

```
load insect
% Let's keep out one observation from each
% class for testing purposes.
X = insect;
X([1,11,21],:) = [];
Xtest = insect([1,11,21],:);
[n,d] = size(X);
% We also need group labels.
clabs = [ones(9,1);2*ones(9,1);3*ones(9,1)];
```

Next, we get indices to the observations for our bootstrap samples. We use the **unidrnd** function in the Statistics Toolbox to get the indices.

```
% Find 50 bootstrap replicates and classifiers.
B = 50;
boot = unidrnd(n,B,n);
```

Now, we use the **classify** function in the Statistics Toolbox to get the bootstrap classifiers.

```
% Classify the three test cases using
% each bootstrap sample.
results = zeros(3,B);
for i = 1:B
    % Get the indices to the bootstrap sample.
    ind = boot(i,:);
    % Classify the test observations using the
    % classifier based on the bootstrap sample.
    class = classify(Xtest,X(ind,:),clabs(ind),...
        'linear');
    % Store in the results array.
    results(:,i) = class(:);
end
```

Looking at the matrix **results**, we see that the test observations for class one and class three were correctly classified in every bootstrap classifier. The test observation belonging to class two was classified as belonging to either class one or two. We can find the majority vote for this case as follows.

```
% Find the majority class for the second test case.
ind1 = find(results(2,:)==1);
ind2 = find(results(2,:)==2);
length(ind1)
length(ind2)
```

There were 15 votes for class one and 35 votes for class two. So, this data point will be classified as belonging to the second class, since class two occurs with higher frequency than class one. The reader is asked in the exercises to find out whether or not bagging was an improvement over using just one classifier.

❑

In bagging classifiers, one can take advantage of the fact that we are using bootstrap samples from the training set to obtain a combined predictor. Breiman [1996c] showed that each bootstrap sample leaves out about 37% of the observations. The observations that are left out of each bootstrap sample can serve as an independent test set for the component classifiers, yielding a better estimate of generalization errors for bagged predictors. Breiman calls this *out-of-bag* estimation.

Boosting

Boosting is a procedure for combining weak or unstable classifiers in order to *boost* or improve their performance. A *weak classifier* is one where the performance is just slightly better than random [Bishop, 2006]. Boosting is like bagging in that it builds many classifiers and combines them to predict the class. Also, like bagging, it works best on unstable and weak classifiers. However, unlike bagging, the classifiers must be constructed sequentially.

The main idea of boosting is to first assign equal weights to each observation in the data set. A classifier is constructed using the weighted points, and the classification errors are assessed with the *training* set. At the next iteration, misclassified data points are given higher weights, increasing their importance. The ones that are correctly classified receive less weight at the next iteration. Thus, misclassified observations have more influence on the learning method, and the next classifier is forced to focus on those points.

From this description, we see that boosting requires a classifier that can handle weights on the observations. This is called *boosting by weighting*. If weights cannot be incorporated into the classifier, then we can get a random sample from the training data proportional to their weights at each iteration of the boosting procedure. This is called *boosting by sampling* [Dietterich, 2000].

Schapire was the first to come up with a polynomial-time boosting algorithm [1990], which was followed by a more efficient method from Freund [1995]. Then, Freund and Schapire [1997] joined together to develop *AdaBoost* (*Adaptive Boosting*), which now serves as the basic boosting procedure. Several slightly different versions of the AdaBoost algorithm can be found in the literature (see some of the references in the last section), but they all have the same basic steps.

The AdaBoost method we outline next is taken from Webb [2011]. It is applicable to the two-class case, where we denote the binary classifier as $g(\mathbf{x})$. Thus, we have

$$g(\mathbf{x}) = 1 \Rightarrow \mathbf{x} \in \omega_1$$
$$g(\mathbf{x}) = -1 \Rightarrow \mathbf{x} \in \omega_2$$

The notation above means the following. If the output of our classifier is 1, then this implies that the pattern $\mathbf{x}$ belongs to class one. If the output of our classifier is -1, then we say that the pattern belongs to class two.

PROCEDURE – ADABOOST

1. Start with the following inputs:
   a. A data set with n observations, where each one is labeled with class label y_i, $y_i \in \{-1, 1\}$, $i = 1, ..., n$.
   b. A maximum number of iterations, T.
   c. A weak classifier $g(\mathbf{x})$.
2. Initialize the weights to $w_i^{(1)} = 1/n$, for $i = 1, ..., n$. The superscript represents the iteration number (t) of the boosting procedure.
3. Do the following for $t = 1, ..., T$:
   a. Apply the weights to the training data or obtain a new data set of size n by sampling with replacement and with probability proportional to their weights $w_i^{(t)}$.
   b. Construct a classifier $g_t(\mathbf{x})$ using the points from step 3a.
   c. Classify all observations using $g_t(\mathbf{x})$.
   d. Calculate the error e_t for the current classifier by adding up all of the weights $w_i^{(t)}$ that correspond to misclassified observations.
   e. If $e_t > 0.5$ or $e_t = 0$, then terminate the procedure. If this is not true, we set the weights for the *misclassified* patterns to

$$w_i^{(t+1)} = \frac{w_i^{(t)}(1 - e_t)}{e_t}.$$

   f. Normalize the weights $w_i^{(t+1)}$ so they add up to one.
   g. Calculate

$$\alpha_t = \log\left[\frac{1 - e_t}{e_t}\right]. \tag{10.22}$$

4. Return the coefficients α_t and the component classifiers $g_t(\mathbf{x})$, for $t = 1, ..., T$.

The final classifier is a weighted sum of the classifiers $g_t(\mathbf{x})$, and the class membership is determined using

$$G(\mathbf{x}) = \sum_{t=1}^{T} \alpha_t g_t(\mathbf{x}),$$ (10.23)

where we assign $\mathbf{x}$ to class ω_1 if $G(\mathbf{x}) > 0$. Thus, the output from the boosted classifiers is a weighted majority vote of the base classifiers, with α_t representing the importance (or weight) assigned to the t-th classifier.

Duda, Hart, and Stork [2001] note that if the component classifiers are weak learners, and the value for T is sufficiently high, then the training error can be made very low. However, this does not guarantee that this type of classifier generalizes well when used with observations that are not in the training set.

We would also like to point out that to use the classifier in Equation 10.23 on new observations, we need to store the coefficients α_t and the classifiers $g_t(\mathbf{x})$. So, this could be expensive in terms of storage and computations if the classifiers are not something simple. It appears from the literature that boosting is used most often with classification trees that are called decision *stumps*. These are trees that contain just one node and two terminal leaves, so they are a simple threshold on one of the variables.

Example 10.15

In this example, we will use boosting code called **adabstdemo**, some **stumps** classification functionality, and a data set, all of which were downloaded, with permission, from:

http://www.work.caltech.edu/~htlin/

First, we generate some bivariate, two-class data. The boundary between these two classes is given by a sine function, so simpler classifiers (e.g., linear or quadratic) would not do a good job of separating the classes.

```
% The data are generated according to:
x = rand(400, 2) * 2 - 1;
y = (x(:, 2) > sin(pi * x(:, 1))) * 2 - 1;
```

We first run AdaBoost for ten iterations using the following command.

```
% Run for 10 iterations.
[w, g] = adabstdemo(x, y, 10, ...
    @stumplearn, @stumpfunc, 1, 0, 100);
```

This yields the decision boundaries shown in the top of Figure 10.18. The dotted line shows the decision boundary for the current classifier, and the solid line separates the space into discriminant regions based on the combined classifier. Next, we get the classifier after 100 iterations to see if it will improve.

```
% Run for 100 iterations.
[w, g] =  adabstdemo(x, y, 100, ...
    @stumplearn, @stumpfunc, 1, 0, 100);
```

The results are illustrated in the plot at the bottom of Figure 10.18. We see that the discriminant boundary based on the combined classifier is doing a better job of separating the classes as the number of classifiers increases. Note that the function **adabstdemo** returns the classifier weights (**w**) and the classifier (**g**).
❑

Arcing Classifiers

Breiman [1998] introduced the notion of *arcing classifiers*. This stands for *adaptively resample and combine*, and he uses it to describe a family of algorithms that do just that: resample adaptively and then combine the classifiers. He notes that AdaBoost belongs to this family of algorithms. Breiman [1998] compares this with another family of methods called *perturb and combine* (P&C). An example of this type of classification approach is bagging.

One of the main arcing algorithms is called Arc-x4. This algorithm is comparable to AdaBoost, and Breiman uses it to make the point that the strength of AdaBoost comes from the adaptive reweighting of the observations, not from the final combination. It is like AdaBoost in that it sequentially builds T classifiers. However, the weighting scheme is simpler. The weight of a data point is proportional to the number of mistakes previous classifiers made to the fourth power, plus one. The final classification of observations is done using a straight majority vote with ties broken arbitrarily. This part differs from AdaBoost because each classifier has equal vote in Arc-x4. The Arch-x4 method is taken from Breiman [1998] and is outlined next.

PROCEDURE – ARC-X4

1. Start with the following inputs:
   a. A data set with n observations, where each one is labeled with class label y_i, $y_i \in \{-1, 1\}$, $i = 1, ..., n$.
   b. A maximum number of iterations, T.
   c. A weak classifier $g(\mathbf{x})$.
2. Initialize the weights to $w_i^{(1)} = 1/n$, for $i = 1, ..., n$. The superscript represents the iteration number (t) of the procedure.

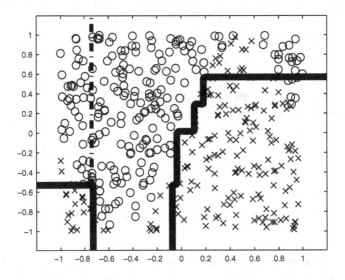

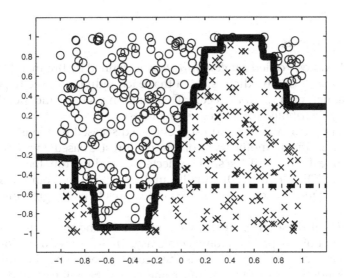

FIGURE 10.18
Here we have scatterplots of the data generated in Example 10.15, where observations are shown using different plotting symbols for each class. These plots show the evolution of the AdaBoost classification procedure. The dotted lines indicate the decision boundary of the current classifier, and the solid lines correspond to the boundary based on the weighted combination of all classifiers. Note that the ensemble classifier does a better job separating the classes as more classifiers are added. The top figure shows what happens after 10 iterations of AdaBoost, and the bottom plot is the classifier after 100 iterations.

3. Do the following for $t = 1, ..., T$:

   a. Obtain a new data set by sampling with replacement and with probability proportional to their weights $w_i^{(t)}$.

   b. Construct a classifier $g_t(\mathbf{x})$.

   c. Classify all observations using $g_t(\mathbf{x})$.

   d. Let $m(i)$ be the number of misclassifications of the i-th observation by all classifiers $g_1(\mathbf{x}), ..., g_t(\mathbf{x})$.

   e. Update the weights to step $t + 1$ using

$$w_i^{(t+1)} = \frac{[1 + m(i)^4]}{\sum (1 + m(i)^4)}.$$

4. Combine classifiers $g_1(\mathbf{x}), ..., g_T(\mathbf{x})$ using unweighted majority vote.

To classify new observations using Arc-x4, we only need to keep the individual component classifiers $g_t(\mathbf{x})$, since the classifiers are combined using a simple majority vote. As with bagging, ties are broken arbitrarily.

Random Forests

The methodology called ***random forests*** was defined by Breiman [2001] as a combination of tree predictors, such that each individual tree depends, in some manner, on a random vector. This random vector is independent and identically distributed for all trees in the forest.

One example of a random forest is one we discussed previously: bagging with tree classifiers. Another type of random forest is based on random splits at each node. In this case, the split is selected at random from the K best splits at the node [Dietterich, 2000]. Finally, another type of random forest can be grown by randomly selecting a subset of features to use in growing the tree [Ho, 1998].

For each of these procedures, we generate a random vector Θ that is independent of any past random vectors used in the procedure. However, it has the same distribution. A tree is then constructed using the training set and the random vector Θ. The nature and size of Θ will vary and depends on the type of tree construction. For example, in bagging tree classifiers, the random vector would be the indices for the bootstrap sample. In the case of random splits at the tree nodes, the random vector would be selecting an integer between 1 and K. Once a large number of trees are generated, the final classification is given using a majority vote using all trees.

10.6 Nearest Neighbor Classifier

The MATLAB Statistics Toolbox has several nonparametric approaches for classification or supervised learning. These include support vector machines, classification trees, ensemble methods, and nearest neighbors [Bishop, 2006]. We just discussed classification trees and ensemble methods. We describe K-nearest neighbors in this section and support vector machines in the next.

The *K-nearest neighbor classifier* uses a rule that is based on what class occurs most often in the set of K-nearest neighbors of a feature vector that we want to classify. We can also cast it in the framework of Bayes decision theory from Section 10.2, where we classify an observation based on the highest posterior probability $P(\omega_m | \mathbf{x})$, $m = 1, ..., M$.

The K-nearest neighbor classifier has some nice properties. It is suitable for problems where we have a large number of features or when we have more features than observations ($p > n$) because it depends only on the interpoint distances. It is also appropriate for multi-class applications.

Suppose we want to classify a feature vector $\mathbf{x}$, which is not necessarily one of our training points. We construct a sphere centered at this data point $\mathbf{x}$, and the size of the sphere is such that exactly K points in our training data set are within that sphere. It does not matter what class they belong to—we just want K-labeled points from our training data in the sphere.

We let K_m represent the number of points in the sphere that are from the m-th class. An estimate of the density associated with each class in the sphere is given by

$$P(\mathbf{x} | \omega_m) = \frac{K_m}{n_m V},$$

where V is the volume of the sphere and n_m is the number of observations in class m. The unconditional density is found using

$$P(\mathbf{x}) = \frac{K}{nV},$$

and the priors are estimated from the training data as

$$P(\omega_m) = \frac{n_m}{n}.$$

Putting these together using Bayes' theorem, we get the following posterior probability:

$$P(\omega_m|\mathbf{x}) = \frac{P(\omega_m)P(\mathbf{x}|\omega_m)}{P(\mathbf{x})} = \frac{K_m}{K}.$$

We would assign the class label m corresponding to the highest posterior probability as given above.

So, to classify a new observation, we first have to specify a value of K. We then find the K closest points to this observation that are in our training data set. We assign the class label that has the highest number of cases in this nearest neighbor set.

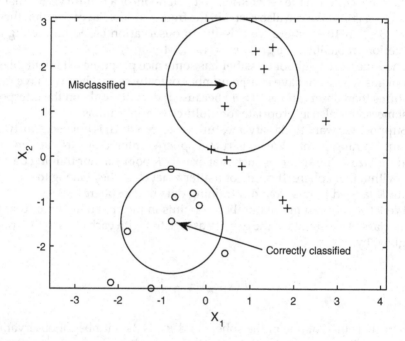

FIGURE 10.19

This shows the concepts of the K-nearest neighbor classifier using k = 5. We have two classes shown by different symbols—"o" and "+." We would classify the "o" observation in the center of the top circle as class "+," which is incorrect. The "o" point in the bottom circle would be classified correctly.

This is illustrated in Figure 10.19, where we have two classes plotted with different symbols. We show what happens when we use $K = 5$. The data point with the "o" symbol at the top is misclassified, while the point at the bottom is classified correctly.

MATLAB's function for constructing a nearest neighbor classifier is called **fitcknn**. The general syntax for this function is shown next.

```
mdl = fitcknn(X,y,'name',value)
```

As in our previous examples, the name-value pairs are optional and allow one to specify many aspects of the classification model, including the number of neighbors K. The default is $K = 1$.

Example 10.16

We are not going to use the MATLAB function **fitcknn** in this example because we want to show the details of the K-nearest neighbor classifier to illustrate the concepts. We are going to use just two of the species in the **iris** data set and include two of the features—petal length and width. First, we load the data and put them into a data matrix **X**.

```
load iris
% Put the two species into a data matrix.
% Just use two dimensions (3 and 4) for
% ease of visualization.
X = [versicolor(:,3:4);virginica(:,3:4)];
```

Next, we add a third column that contains the class labels.

```
% Add another column that has the class labels.
X(:,3) = [zeros(50,1);ones(50,1)];
```

We pick a new observation to classify and add 99 as the class label to indicate a missing value.

```
% Designate a point in the space to classify.
% Make the third element '99' to indicate
% missing class label.
x = [5.4107, 1.6363, 99];
```

We create a new matrix with the unlabeled observation as the first row. We can use this with the **pdist** function to find the distance with this unlabeled data point and all training observations.

```
% Create a matrix with the new observation
% as the first row and the remaining ones as the
% observed data.
Xn = [x ; X];
```

We can convert the pairwise distances from **pdist** to a square matrix using **squareform**, as shown here. Then we sort the distances of either the first row or the first column.

```
% First, get the pairwise distances.
D = squareform(pdist(Xn(:,1:2)));
% Then, find the ones closest to the new point.
% Do this by looking in the first row or column.
[pd, ind ] = sort(D(:,1));
```

The output variable **ind** contains the sort order, and we can use the indices to find the labels of the closest training observations to our new data point. Note that the first element of **ind** is the index of the new observation, so we need to start after that one. Using $k = 10$, the indexes of the observations that fall in that neighborhood are in elements 2 through 11 of **ind**. We use the **tabulate** function to get the frequencies of the classes in the neighborhood, as shown here.

```
% Find the frequency of classes within the 10
% nearest neighbors.
tabulate(Xn(ind(2:11),3))
```

The results show that 9 of the 10 nearest neighbors to our new point belong to the *Virginica* class (class 1), so we would label it as belonging to *Virginica*.

```
% Following shows that 9 of the 10 nearest neighbors
% are class 1 or Virginica. So, that is how we
% would classify it.
```

Value	Count	Percent
0	1	10.00%
1	9	90.00%

Figure 10.20 shows the results of applying this classification to our new data point.

❑

10.7 Support Vector Machines

A support vector machine is a general approach for supervised learning that has been around since the 1990s and has grown in popularity. It is now one of the main tools for classification used in the machine learning community. Support vector machines encompasses three different classifiers, each is a generalization of a linear decision boundary (see Section 10.2). These are known as the maximal margin classifier, the support vector classifier, and the support vector machine. We use the notation of support vector machines given in James, et al. [2013] in what follows.

Maximal Margin Classifier

Recall that with supervised learning, we have n d-dimensional observations providing measurements of the features that hopefully allow us to discriminate between classes. Furthermore, we have two classes with class labels denoted by -1 and 1. This means our data are given by

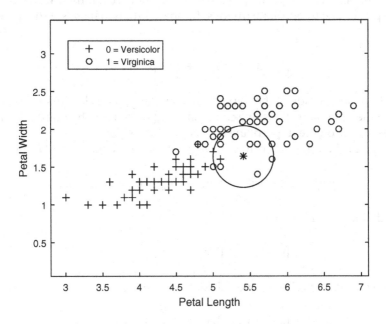

FIGURE 10.20
This plot illustrates the K-nearest neighbor classifier (K = 10) applied to a new observation (note the "*"). We would classify it as species Virginica.

$$\mathbf{X} = \begin{bmatrix} x_{11} & \dots & x_{1d} \\ \dots & \dots & \dots \\ x_{n1} & \dots & x_{nd} \end{bmatrix} \text{ and } y_1, y_2, \dots, y_n, \tag{10.24}$$

where $y_i \in \{-1, 1\}$. As with all supervised learning applications, we want to construct a classifier that allows us to assign a label to a new observation that we will denote by

$$\mathbf{x}^{new} = (x_1^{new}, \dots x_d^{new})^T.$$

The *maximal margin classifier* seeks to find the hyperplane that separates the two classes in some optimal sense. A *hyperplane* is a flat subspace that has dimension $d - 1$.. In our situation, it is not constrained to pass through the origin. Thus, if we have a space with $d = 2$, then the hyperplane is a line. If $d = 3$, then our hyperplane is a flat two-dimensional space, which is a plane.

The hyperplane is given by the following equation

$$\beta_0 + \beta_1 X_1 + \beta_2 X_2 + \dots + \beta_d X_d = 0. \tag{10.25}$$

If a point $X = (X_1, X_2, ..., X_d)^T$ satisfies Equation 10.25, then it lies on the hyperplane. The hyperplane divides the space into two halves. If the point X yields

$$\beta_0 + \beta_1 X_1 + \beta_2 X_2 + ... + \beta_d X_d > 0,$$

then it lies on one side of the hyperplane. Similarly, if it is less than zero, then it is on the other half. Thus, we can easily find out where the point lies in the space—either on the hyperplane or on one of the sides.

Using our data (Equation 10.24), we have the following formulation

$$y_i(\beta_0 + \beta_1 x_{i1} + \beta_2 x_{i2} + ... + \beta_d x_{id}) > 0, \quad i = 1, ...n.$$

Note that if $y_i = 1$, then

$$\beta_0 + \beta_1 x_{i1} + \beta_2 x_{i2} + ... + \beta_d x_{id} > 0,$$

and the observation lies on one side of the hyperplane. If $y_i = -1$, then

$$\beta_0 + \beta_1 x_{i1} + \beta_2 x_{i2} + ... + \beta_d x_{id} < 0,$$

and it lies on the other side of the hyperplane.

If we can find a hyperplane that separates our two-class data, then we can easily classify a new observation by looking at the sign of

$$f(x^{new}) = \beta_0 + \beta_1 x_1^{new} + \beta_2 x_2^{new} + ... + \beta_d x_d^{new}. \tag{10.26}$$

If 10.26 is negative, then it is in class -1, and if it is positive, then it is in class 1. Furthermore, the magnitude of $f(x^{new})$ provides an indication of how far it is from the separating hyperplane. If it is close to zero, then it is near to the hyperplane, and we might have less confidence in the classification [James, 2013].

Let's assume that the data are linearly separable. In other words, there exists a hyperplane that can separate the observed data into the two classes without error. If this is the case, then there are an infinite number of possible hyperplanes because we can shift them slightly and still separate the data. So, we need to find a hyperplane that is optimal, where optimality is determined by the location of the hyperplane in relation to the labeled observations.

To do this, we find the perpendicular distance from each data point to a proposed hyperplane. The **margin** is the smallest one of these distances. It makes sense to look for the hyperplane that corresponds to the one with the largest margin or the largest minimum distance. If we use this hyperplane to classify observations (Equation 10.26), then we have the **maximal margin classifier**.

We find the largest margin by solving the following optimization problem. Given a set of n p-dimensional observations with associated class labels y, as in 10.24, then we want to find the coefficients of the separating hyperplane that maximizes M, as shown here:

$$\max_\beta M$$

$$\text{subject to } \sum_{j=1}^{d} \beta_j^2 = 1 \,, \tag{10.27}$$

$$y_i(\beta_0 + \beta_1 x_{i1} + \beta_2 x_{i2} + \dots + \beta_d x_{id}) \geq M$$

for all $i = 1, \dots, n$.

This formulation says we want to find the largest margin M, such that each observation is on the correct side of the hyperplane with a positive cushion M. It can be shown that the perpendicular distance from $\mathbf{x}_i$ to the hyperplane is given by $y_i(\beta_0 + \beta_1 x_{i1} + \beta_2 x_{i2} + \dots + \beta_d x_{id})$ [James, 2013]. Therefore, the requirements in our optimization problem dictate that the data points will be a distance M or more from the hyperplane and will also be on the correct side.

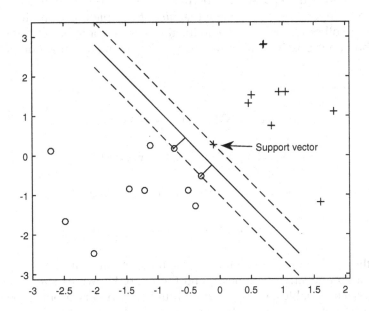

FIGURE 10.21

Here we show a small two-class data set. The separating hyperplane is given by the solid line, and the margin edges are shown as dotted lines. The support vectors are the points on the dotted lines and they define the margins. The margin is given by the perpendicular distance between the support vectors and the hyperplane.

An example of a small 2D data set is shown in Figure 10.21, where the two classes are linearly separable. The separating hyperplane is shown as a solid line. The perpendicular distance from the dashed line to the separating hyperplane is the margin. A certain number of data points will lie on the margin, and these are called the *support vectors*.

It is clear from the plot that moving or adding training observations outside of the support vectors (i.e., away from the hyperplane) would not affect the classifier. This is different from the linear and quadratic classifiers we looked at previously. It also gives a fairly robust classifier because adding outlying observations far from the hyperplane will not affect the boundary.

Support Vector Classifier

It is likely that in most real-world applications our data will not be separable. So, a separating hyperplane would not exist, and we cannot use the technique in the previous section. A generalization of the maximal margin classifier has been developed for this situation. It is known as the *support vector classifier*.

The support vector classifier is also found using an optimization problem. Because there is no separating hyperplane in the non-separable case, we now look for a separating hyperplane that *almost* completely separates the two classes. In other words, we have to tolerate some error in our classifier.

To build our classifier, we now solve the following optimization problem:

$$\max_{\beta, \, \varepsilon} M$$

$$\text{subject to } \sum_{j=1}^{d} \beta_j^2 = 1 , \ \sum_{i=1}^{n} \varepsilon_i \leq C , \ \varepsilon_i \geq 0 \tag{10.28}$$

$$y_i(\beta_0 + \beta_1 x_{i1} + \beta_2 x_{i2} + \ldots + \beta_d x_{id}) \geq M(1 - \varepsilon_i)$$

As before, we classify a new observation using the sign of

$$f(x^{new}) = \beta_0 + \beta_1 x_1^{new} + \beta_2 x_2^{new} + \ldots + \beta_d x_d^{new},$$

with the coefficients β_j chosen using the training data with Equation 10.28.

C is a non-negative parameter, and the ε_i are known as *slack variables*. These slack variables allow a certain number of observations to be on the wrong side of the margin or the wrong side of the separating hyperplane. If $\varepsilon_i = 0$, then the data point is on the right side of the margin (and the correct side of the hyperplane, of course!). An observation with $\varepsilon_i > 0$ is on the wrong side of the margin. If an observation has $\varepsilon_i > 1$, then it is on the incorrect side of the hyperplane [James et al., 2013].

The parameter C can be used to control the number and the degree of the errors we will allow in the classifier. If $C = 0$, then we are requiring zero violations to the margin or hyperplane. This also means all slack variables are zero, and we have the maximal margin optimization problem of the previous section that can only be solved if the classes are separable. As C increases, we allow more violations, and the margin will widen. Conversely, as C decreases, the margin will become smaller.

This situation is illustrated in Figure 10.22, where we use two species of the Fisher's **iris** data. It is known that these are not linearly separable. We show the support vector classifier (solid line) and the margins (dotted lines). Note that we have some violations to the margins and the hyperplane.

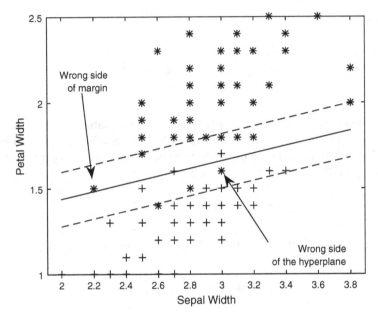

FIGURE 10.22
This shows the situation where we can not perfectly separate the classes using a hyperplane.

Support Vector Machines

The small data sets we used so far to illustrate the concepts of support vectors and separating hyperplanes have been simple, and where the discriminant boundary appears to be linear. However, we often encounter data where non-linear boundaries are more suitable. For instance, we show an example in Figure 10.23, where we have two-class data that have a circular boundary separating them. It is clear that a linear discriminant would not separate the classes.

James et al. [2013] make the point that this situation is analogous to regression when the relationship between a single predictor and the response is not given by a straight line. For example, a quadratic or cubic model might be more appropriate depending on the data, so we can add non-linear functions (e.g., squares or cubes) of the predictor to the model. This has the effect of expanding the feature space because there are now more coefficients or terms in the model.

Keep in mind that the separating hyperplane (Equation 10.25) is also a model, as we have in the regression setting. So, we can do something similar when a non-linear decision boundary is better. For example, we might use

$$y_i[\beta_0 + \beta_1 x_{i1} + \dots + \beta_d x_{id} + \beta_{d+1}(x_{i1}^2 + \dots + x_{id}^2)] \tag{10.29}$$

or the following:

$$y_i\left[\beta_0 + \sum_{j=1}^{d} \beta_j x_{ij} + \sum_{j=d+1}^{2d} \beta_j x_{ij}^2\right].$$

To illustrate this concept, we apply the model in Equation 10.29 to the data set shown in Figure 10.23 and plot them in the enlarged 3D space (see Figure 10.24). Note that the data are now linearly separable in this larger space. For this example, the hyperplane (3D) is given by

$$\beta_0 + \beta_1 X_1 + \beta_2 X_2 + \beta_3(X_1^2 + X_2^2) = 0.$$

Rearranging this a little, we have the familiar equation for a circle

$$X_1^2 + X_2^2 + \frac{\beta_1}{\beta_3}X_1 + \frac{\beta_2}{\beta_3}X_2 + \frac{\beta_0}{\beta_3} = 0.$$

Thus, we see that the 3D linear hyperplane boundary maps into a circular discriminant boundary in the 2D space, as expected.

There are many ways to map our data into a higher-dimensional space. The *support vector machines* expand the number of features through the use of a particular type of kernel. To describe the kernel approach used in support vector machines, we first express the discriminant function (Equation 10.26) in terms of the inner products between a new observation $\mathbf{x}^{new}$ and the n observations:

$$f(\mathbf{x}^{new}) = \beta_0 + \sum_{i=1}^{n} \alpha_i \langle \mathbf{x}^{new}, \mathbf{x}_i \rangle, \tag{10.30}$$

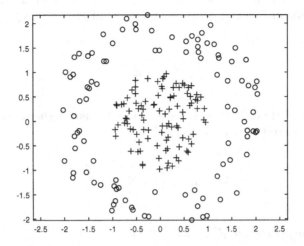

FIGURE 10.23

Here we have two classes that are separable, but not by a hyperplane. The decision boundary is nonlinear or a circle in this case.

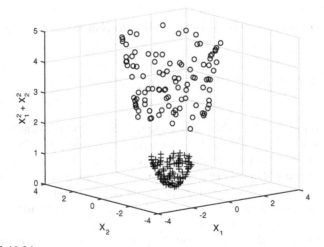

FIGURE 10.24

We use the model in Equation 10.29 to enlarge the feature space and show the data in a 3D scatterplot. Note that the data can now be separated with a hyperplane.

where $\langle \mathbf{a}, \mathbf{b} \rangle$ denotes the inner product between two vectors $\mathbf{a}$ and $\mathbf{b}$. We now have to estimate the parameters α_i and β_0 to obtain our classifier. However, it turns out that to do this, we only need to calculate all pair-wise inner products of the n observations—$\langle \mathbf{x}_i, \mathbf{x}_k \rangle$, $i \neq k$.

This simple inner product can be replaced by a generalized function called a ***reproducing kernel***. It provides a measure of similarity between two vectors or observations. The kernel function will be denoted as $K(\mathbf{x}_i, \mathbf{x}_k)$.

There are many options for kernels, and we list the common ones here. We can use the *linear kernel* function for the i-th and j-th vectors given by

$$K(\mathbf{x}_i, \mathbf{x}_k) = \sum_{j=1}^{d} x_{ij} x_{ik}$$

to get the support vector classifier we had previously. Other popular kernels are the *polynomial kernel* of degree D, $D > 0$:

$$K(\mathbf{x}_i, \mathbf{x}_k) = \left(1 + \sum_{j=1}^{d} x_{ij} x_{ik}\right)^D,$$

the *radial basis function (Gaussian) kernel*:

$$K(\mathbf{x}_i, \mathbf{x}_k) = \exp\left[-\frac{1}{2\sigma^2} \sum_{j=1}^{d} (x_{ij} - x_{kj})^2\right],$$

and the *sigmoid kernel* (also known as the *neural network kernel*):

$$K(\mathbf{x}_i, \mathbf{x}_k) = \tanh\left(\kappa \sum_{j=1}^{d} x_{ij} x_{ik} + \Theta\right),$$

where $\kappa > 0$ and $\Theta > 0$. In this kernel, the gain is κ, and the offset is Θ. Only some values for these constants yield valid reproducing or inner product kernels [Vapnik, 1998; Cristianini and Shawe-Taylor, 2000].

There is no general best kernel to use in building a support vector machine classifier. Therefore, we have to try different options and then use the tools for assessing classification performance (e.g., cross-validation) to select a good kernel and tuning parameters.

We might suspect that our computational burden would increase because we are enlarging our feature space. However, the inner product kernels actually enable us to do this efficiently because they only require the pairwise inner products between the observations. Even better, we can do this implicitly without actually moving to the new larger feature space and doing the calculations there.

Recall that we have to solve an optimization problem to find the support vector machine classifier (Equation 10.30). The formulation to construct a support vector machine is given by the following [Cristianini and Shawe-Taylor, 2000]:

$$\max_\alpha \left\{ \sum_i \alpha_i - \frac{1}{2} \sum_i \sum_j \alpha_i \alpha_j y_i y_j \langle \mathbf{x}_i, \mathbf{x}_j \rangle \right\}$$

$$\text{(10.31)}$$

$$\text{subject to} \qquad \sum_i y_i \alpha_i = 0 \ , \quad 0 \le \alpha_i \le C$$

The value C is called the **box constraint** in the MATLAB function for building support vector machines, and it serves the same purpose as before; i.e., it is a tuning parameter that can be chosen using cross-validation.

The parameters α_i provide a measure of how each training data point influences the resulting classifier. Those that do not affect the classifier have a value of 0. This has implications for the "size" of the classifier. We just have to keep the observations $\mathbf{x}_i$ such that $\alpha_i \ne 0$ to classify new observations, and this is usually much less than n.

We skip the details on how to solve the optimization problem because it does not add any insight to our understanding of the classifier. However, it is useful to note that the objective function in Equation 10.31 can be solved with existing quadratic programming techniques because it depends on the α_i quadratically, and the α_i are linear in the constraints. This means that there are efficient methods for finding the support vector machine.

Example 10.17

We are going to use the bank note authentication data set for this example. These data were downloaded from the UCI Machine Learning Repository. Images were taken of genuine and forged bank notes, and statistics based on wavelet transforms were calculated. These include the variance, skewness, kurtosis, and the entropy. The data are saved in the file **bankdata**.

```
% Loads bank authentication data and class labels.
% These are called BankDat and Class.
load bankdata
```

We will use just two of the dimensions for our classifier to make it easier to visualize the results. We create a data matrix **X** using the features of skewness and entropy.

```
% Extract the two dimensions to use in classification.
X = BankDat(:,[2,4]);
% Change the class labels to -1 and 1.
ind = find(Class == 0);
Class(ind) = Class(ind) - 1;
```

The MATLAB Statistics Toolbox has a function for fitting support vector machines called **fitcsvm**. It has the linear, Gaussian or polynomial kernels as options. We will use the radial basis function or Gaussian kernel.

```
% Fit the Gaussian SVM
cl_svm = fitcsvm(X,Class,'KernelFunction','rbf',...
    'BoxConstraint',10,'ClassNames',[-1,1]);
```

The following commands create a set of data points in the domain and then predict the class membership based on the support vector machine.

```
% Set up a grid and use to find classification regions.
d = 0.02;
[x1G,x2G] = meshgrid(min(X(:,1)):d:max(X(:,1)),...
    min(X(:,2)):d:max(X(:,2)));
xGrid = [x1G(:),x2G(:)];
% The function predict will apply the classifier
% to the specified grid points.
[~,scores] = predict(cl_svm,xGrid);
```

We now display the scatterplot of points, along with the decision boundary in Figure 10.25. The code to construct this plot is shown next.

```
% Display points in scatterplot.
% Add the decision boundary - single contour.
gscatter(X(:,1),X(:,2),Class,'rg','o+',3);
xlabel('Skewness')
ylabel('Entropy')
title('Regions from SVM Classifier')
hold on
% Display a single contour line at z = 0.
Z = reshape(scores(:,2),size(x1G));
contour(x1G,x2G,Z,[0 0],'k');
hold off
```

It might be a little difficult to see in the black-and-white figure in the book, but there are some regions where the two classes overlap. So, we know the support vector machine classifier is not going to separate the classes perfectly. We can estimate the error using cross-validation, as shown here.

```
% We have some regions where the classes overlap.
% so, there will be some error.
% Use cross-validation to estimate the error.
cv_svm = crossval(cl_svm);
kfoldLoss(cv_svm)
```

The error using the default of 10-fold cross-validation is 0.1013.
❏

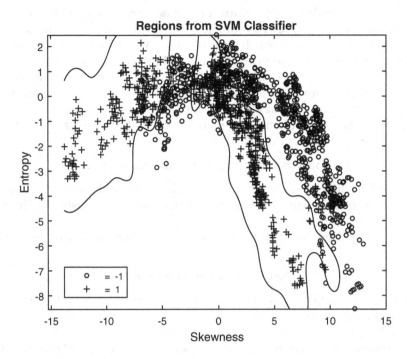

FIGURE 10.25

This shows the decision boundary from the support vector machine applied to the bank data (**bankdat**). *The radial basis function or Gaussian kernel was used.*

10.8 MATLAB® Code

There are many MATLAB functions that the analyst can use to develop classifiers using Bayes decision theory. For the parametric approach to probability density estimation, one can use functions in the Statistics Toolbox (e.g., **normfit**, **expfit**, **gamfit**, **unifit**, **betafit**, **weibfit**, etc.) to estimate a probability density function. These functions return the appropriate distribution parameters estimated from the sample. For the nonparametric approach to density estimation, one can use techniques from Chapter 9, such as histograms, frequency polygons, kernel methods, finite mixtures, or adaptive mixtures. As we saw, there is also a function in the Statistics Toolbox called **classify**. This performs linear and quadratic discriminant analysis.

The Statistics Toolbox has added capabilities to create and work with classification trees. The main function for finding a tree, along with an optimal pruning sequence, is the **classregtree** function that creates a tree object. Prior versions of the Statistics Toolbox had a function called **treefit**.

This worked similarly to the **classregtree** function, except that it was not object-oriented. The **treefit** function, along with its companion functions, have been kept for backwards compatibility. We summarize the functions that can be used for supervised learning in Tables 10.1 and 10.2. It should be noted that some of the functions for fitting and working with classification trees will be removed in a future release of MATLAB. They recommend using the function **fitctree** and related functions.

A set of M-files implementing many of the methods described in Ripley [1996] are available for download at

www.mathworks.com/matlabcentral

Search for "discrim" to locate the Discriminant Analysis Toolbox. There are functions for Bayesian classifiers, neural net classifiers, logistic discriminant analysis, and others.

Another pattern recognition toolbox that is available for free download can be found at

http://www.prtools.org/

This is the basic MATLAB software used in the text by van der Heijden, et al. [2004]. This toolbox includes many of the common classifiers, as well as neural network classifiers and support vector machines.

There is also a computer manual [Stork and Yom-Tov, 2004] that describes and includes a downloadable toolbox for the text *Pattern Classification* by Duda, Hart, and Stork [2001]. Access to the toolbox is given when purchasing the manual. This has functions and GUIs for many classification methods. Some free functions, including ones for boosting, C4.5 decision trees, and support vector machines can be found at

http://www.yom-tov.info/computer_manual.html

A free Support Vector Machine Toolbox can be found at

http://www.isis.ecs.soton.ac.uk/resources/svminfo/

This website also has links to a tutorial and other useful papers on support vector machines [Burges, 1998].

Another website that has MATLAB code for pattern recognition (Bayesian classifiers, clustering, and support vector machines) can be found at

http://cmp.felk.cvut.cz/cmp/software/stprtool/

Finally, there is an AdaBoost Toolbox available for download at

http://graphics.cs.msu.ru/en/science/research/...

machinelearning/adaboosttoolbox

This has functions for AdaBoost, gentle AdaBoost [Friedman, et al., 2000], and modest AdaBoost [Vezhnevets and Vezhnevets, 2005]. Classification and regression trees are implemented as weak learners.

TABLE 10.1

Statistics Toolbox Functions[a]

Purpose	MATLAB Function
Tree-class constructor and related functions	`classregtree`
	`prune`
	`eval`
	`test`
	`view`
Tree fitting and related functions	`treefit`
	`treeprune`
	`treeval`
	`treetest`
	`treedisp`
	`fitctree`
Helper functions for tree objects	`nodesize`
	`nodeprob`
	`nodeerr`
	`children`
	`parent`
	`numnodes`
	`risk`
	`isbranch`
	`cutvar`
	`cuttype`
Linear or quadratic classifier	`classify`
	`fitcdiscr`
Naive Bayes classifier	`fitcnb`
	`fitNaiveBayes`
K nearest neighbor	`fitcknn`
Support vector machines	`fitcsvm`

[a] Future releases of MATLAB will remove some of the functions for working with classification trees. They recommend using the function fitctree.

10.9 Further Reading

There are many excellent books on statistical pattern recognition that can be used by students at the graduate level or researchers with a basic knowledge of calculus and linear algebra. The text by Duda and Hart [1973] is a classic book on pattern recognition and includes the foundational theory behind Bayes decision theory, classification, and discriminant analysis. It has

TABLE 10.2

Computational Statistics Toolbox Functions

Purpose	MATLAB Function
Statistical pattern recognition using Bayes decision theory	`csrocgen`
	`cskernmd`
	`cskern2d`
Boosting with decision stumps	`csboost`

recently been revised and updated [Duda, Hart, and Stork, 2001]. This second edition contains many new topics, examples, and pseudo-code. As mentioned before, they also have a separate computer manual to accompany the text [Stork and Yom-Tov, 2004].

Fukunaga [1990] is at the same level and includes similar subjects; however, it goes into more detail on the feature extraction aspects of pattern recognition. Devroye, Gyorfi, and Lugosi [1996] contains an extensive treatment of the probabilistic theory behind pattern recognition. Ripley [1996] covers pattern recognition from a neural network perspective. This book is recommended for both students and researchers as a standard reference. An excellent book that discusses all aspects of statistical pattern recognition is the text by Webb [2011]. This is suitable for advanced undergraduate students and professionals. The author explains the techniques in a way that is understandable, and he provides enough theory to explain the methodology, but does not overwhelm the reader with it.

The definitive book on classification trees is the one by Breiman, et al. [1984]. This text provides algorithms for building classification trees using ordered or categorical data, mixtures of data types, and splitting nodes using more than one variable. They also provide the methodology for using trees in regression. A paper by Safavian and Landgrebe [1991] provides a review of methodologies for building and using classification trees. A description of classification trees can also be found in Webb [2011] and Duda, Hart, and Stork [2001]. Loh [2014] has an excellent overview paper on the development of classification and regression trees over the past fifty years.

Two excellent books on pattern recognition from a machine learning and data mining perspective are *The Elements of Statistical Learning: Data Mining, Inference, and Prediction* by Hastie, Tibshirani, and Friedman [2009] and *Pattern Recognition and Machine Learning* by Bishop [2006]. These texts are easy to read, and they include pseudo-code for many supervised (and unsupervised) learning methods. A good tutorial introduction to boosting is by Meir and Rätsch [2003]. They provide both theory and practical aspects of ensemble learning.

Many papers are available that provide experimental comparison and assessment of ensemble classifiers such as bagging, boosting, and arcing. The first is by Quinlan [1996] where he compares bagging and boosting using C4.5 (a type of decision tree classification methodology, [Quinlan, 1993]). This is followed by a comparison of bagging, boosting, and variants, including Arc-x4 [Bauer and Kohavi, 1999]. Among other things, they compare voting methods to nonvoting methods and show that voting methods lead to large reductions in the mean-square errors of the classifiers.

A paper that focuses on comparing ensembles of decision trees is by Dieterich [2000]. He compares bagging, boosting, and randomization (a type of random forest) using the C4.5 decision-tree algorithm. His experiments show that randomization is competitive with bagging, but not as good as boosting, when there is little or no classification noise. If there is a lot of classification noise, then bagging is better than boosting.

A good study of bagging that provides insights about the performance of bagging with linear classifiers is by Skurichina and Duin [1998]. Friedman, Hastie, and Tibshirani [2000] describe a statistical view of boosting, and Schapire, et al. [1998] make a connection between support vector machines and boosting.

There are also some tutorial articles on pattern recognition and machine learning. A special invited article by Friedman [2006] provides an introduction to support vector machines and boosted decision trees. A nice tutorial on support vector machines was written by Burges [1998], and a gentle introduction on the topic can be found in Bennett and Campbell [2000]. Other overviews of statistical learning theory can be found in Vapnik [1999] and Jain, Duin, and Mao [2000]. An accessible book on support vector machines is *An Introduction to Support Vector Machines and Other Kernel-Based Learning Methods* by Cristianini and Shawe-Taylor [2000].

Exercises

10.1. Load the **insect** data [Hand, et al., 1994; Lindsey, et al., 1987]. These are three variables measured on each of ten insects from three species. Using the parametric approach and assuming that these data are multivariate normal with different covariances, construct a Bayes classifier. Use the classifier to classify the following vectors as species I, II, or III:

10.2. Apply the cross-validation procedure and ROC curve analysis of Example 10.8 to the **tibetan** data. Designate Type A skulls as the target class and Type B skulls as the nontarget class.

10.3. Use the **bank** data along with the independent test sample approach to estimate the probability of correctly classifying patterns (see Example 10.7). The file contains two matrices, one corresponding to features taken from 100 forged Swiss bank notes and

X_1	X_2	X_3
190	143	52
174	131	50
218	126	49
130	131	51
138	127	52
211	129	49

the other comprising features from 100 genuine Swiss bank notes [Flury and Riedwyl, 1988]. There are six features: length of the bill, left width of the bill, right width of the bill, width of the bottom margin, width of the top margin, and length of the image diagonal. Compare classifiers obtained from: (1) the parametric approach, assuming the class-conditionals are multivariate normal with different covariances, and (2) the nonparametric approach, estimating the class-conditional probabilities using the product kernel. Which classifier performs better based on the estimated probability of correct classification?

10.4. Apply the cross-validation procedure and ROC curve analysis of Example 10.8 to the **bank** data. The target class corresponds to the forged bills. Obtain ROC curves for a classifier built using: (1) the parametric approach, assuming the class-conditionals are multivariate normal with different covariances, and (2) the nonparametric approach, estimating the class-conditional probabilities using the product kernel. Which classifier performs better, based on the ROC curve analysis?

10.5. For the **bank** data, obtain a classification tree using half of the data set. Use the independent test sample approach with the remaining data to pick a final pruned tree.

10.6. Do a Monte Carlo study of the probability of misclassification. Generate n random variables using the class-conditional probabilities and the priors from Example 10.3. Estimate the probability of misclassification based on the data. Note that you will have to do some probability density estimation here. Record the probability of error for this trial. Repeat for M Monte Carlo trials. Plot a histogram of the errors. What can you say about the probability of error based on this Monte Carlo experiment?

10.7. The **flea** data set [Hand, et al., 1994; Lubischew, 1962] contains measurements on three species of flea beetle: *Chaetocnema concinna*, *Chaetocnema heikertingeri*, and *Chaetocnema heptapotamica*. The features for classification are the maximal width of aedeagus in the forepart (microns) and the front angle of the aedeagus (units are 7.5 degrees). Build a classifier for these data using a Bayes classifier.

For the Bayes classifier, experiment with different methods of estimating the class-conditional probability densities. Construct ROC curves and use them to compare the classifiers.

10.8. Build a classification tree using the **flea** data. Based on a three-term multivariate normal finite mixture model for these data, obtain an estimate of the model. Using the estimated model, generate an independent test sample. Finally, use the generated data to pick the best tree in the sequence of subtrees.

10.9. The *k-nearest neighbor rule* assigns patterns x to the class that is the most common amongst its k nearest neighbors. To fix the notation, let k_m represent the number of cases belonging to class ω_m out of the k nearest neighbors to x. We classify x as belonging to class ω_m, if $k_m \geq k_i$, for $i = 1, ..., J$. Write a MATLAB function that implements this classifier.

10.10. Repeat Example 10.7 using all of the features for **versicolor** and **virginica**. What is your estimated probability of correct classification?

10.11. Apply the method of Example 10.7 to the **virginica** and **setosa** classes. Discuss your results.

10.12. We applied bagging to the **insect** data using linear classifiers on each of the bootstrap samples. Use the training set in Example 10.14 and a linear classifier to classify the three test cases of the same example. Did bagging improve the classification results?

10.13. Generate two-class bivariate data using the model in Example 10.10. Use the **classify** function to determine the decision regions when you use a linear classifier and a quadratic classifier. Plot the regions on a scatterplot of the data. Use different symbols for each class. Compare the two decision regions.

10.14. Plot three univariate class-conditional probabilities, all with variance of one and means given by $-1.5, 0, 1.5$. Assuming equal priors, what is the decision boundary using Bayes rule? What is the probability of misclassification?

10.15. Repeat the previous problem, but use different variances. Compare the decision boundaries of this classifier (linear) with what you had in the previous problem (quadratic).

10.16. Derive the discriminant function for the linear and quadratic classifier using

$$g_j(\mathbf{x}) = \log P(\mathbf{x}|\omega_j) + \log P(\omega_j).$$

10.17. Run the **demo1**, **demo2**, and **demo3** example files. These run the **adabstdemo** for various data sets. Describe your results.

10.18. Apply the naive Bayes classifier to the **iris** data and explore your results. Assuming that the data have been merged into one data matrix **X**, and the class labels are in the object **clab**, then the command for the default of Gaussian marginals is

NBk = fitcnb(X,clab)

10.19. Use the support vector machines approach to build classifiers for the **bank**, **insect**, and **tibetan** data sets. Compare the classification results with classifiers constructed in previous problems.

Chapter 11

Unsupervised Learning

11.1 Introduction

In the last chapter, we described the *supervised learning* setting where we had observations with given class labels. In these cases, we know that the data come from different groups or classes, and we know how many groups are represented by the data. However, in many situations, we do not know the class membership for our observations at the start of our analysis. These are called **unsupervised learning** or **clustering** applications.

Clustering methodology is used to explore a data set where the goal is to separate the observations into groups or to provide understanding about the underlying structure of the data. The results from clustering methods can also be used to prototype supervised classifiers and to generate hypotheses.

Keep in mind that clustering is an exploratory data analysis approach that looks for a particular type of structure in the data—groups. There are different methods for locating specific types of clusters or groups, such as spherical clusters, elongated clusters, and so forth. As we will see, some methods will find the specified number of groups, regardless of whether or not the groups are really there. This means that the methods used for clustering impose a structure on the data, and this structure might not be valid. Thus, analysts should use different methods of finding groups in data and assess the results to make sure the clusters are reasonable.

In this chapter, we discuss three main methods for clustering. These are hierarchical clustering, k-means, and model-based clustering. The chapter concludes with a discussion of methods for assessing the validity of the clusters. However before we delve into the methods, we first need to cover the notion of distance because it is foundational to some of the clustering techniques discussed later.

11.2 Measures of Distance

The goal of clustering is to partition our data into groups such that the observations that are in one group are similar to members of the same group and are dissimilar to those in other groups. We need to have some way of measuring the dissimilarity, and there are several measures that have been developed in the literature. We describe a small subset of these next.

The first measure of dissimilarity is the *Euclidean distance* given by

$$d_{rs} = \sqrt{(\mathbf{x}_r - \mathbf{x}_s)^T (\mathbf{x}_r - \mathbf{x}_s)}, \tag{11.1}$$

where $\mathbf{x}_r$ is a column vector representing one observation. We could also use the *Mahalanobis distance* defined as

$$d_{rs} = \sqrt{(\mathbf{x}_r - \mathbf{x}_s)^T \Sigma^{-1} (\mathbf{x}_r - \mathbf{x}_s)}, \tag{11.2}$$

where Σ^{-1} denotes the inverse covariance matrix. The *city block distance* is found using absolute values, and it is calculated using

$$d_{rs} = \sum_{j=1}^{d} |x_{rj} - x_{sj}|. \tag{11.3}$$

In Equation 11.3, we take the absolute value of the difference between the observations $\mathbf{x}_r$ and $\mathbf{x}_s$ componentwise and then add up the values. The final distance that we present covers the more general case of the Euclidean distance or the city block distance. This is called the *Minkowski distance*, and it is found using

$$d_{rs} = \left\{ \sum_{j=1}^{d} |x_{rj} - x_{sj}|^p \right\}^{1/p}. \tag{11.4}$$

If $p = 1$, then we have the city block distance, and if $p = 2$ we have the Euclidean distance.

Researchers should be aware that distances might be affected by differing scales or magnitude among the variables. For example, suppose our data measured two variables: age and annual income in dollars. Because of its magnitude, the income variable could influence the distances between observations, and we would end up clustering mostly on the incomes. Thus, in some situations, we might want to standardize the observations to have zero mean and unit standard deviation. The MATLAB® Statistics Toolbox contains a function called **zscore** that will perform this standardization.

The Statistics Toolbox also has a function that calculates distances. It is called **pdist** and takes as its argument a matrix X that is dimension $n \times d$. As always, each row represents an observation in our data set. The **pdist** function returns a vector containing the distance information. The default distance is Euclidean, but the user can specify other distances. See the MATLAB **help** on **pdist** for a list of dissimilarity measures that one can obtain. We illustrate the use of this function in the following example.

Example 11.1

We use a small data set to illustrate the various distances available in the MATLAB Statistics Toolbox. We have only five data points. The following commands set up the matrix of values and plots the points in Figure 11.1.

```
% Let's make up a data set - 2D.
x = [1 1; 1 2; 2 1; -1 -1; -1 -2];
plot(x(:,1),x(:,2),'kx') % plots the points.
axis([-3 3 -3 3])
text(x(:,1)+.1,x(:,2)+.1,...
    {'1','2','3','4','5'});
```

We first find the Euclidean distance between the points using the **pdist** function. We also illustrate the use of the function **squareform** that puts the distances in a more familiar matrix form, where the *ij*-th element corresponds to the distance between the *i*-th and *j*-th observation.

```
% Find the Euclidean distance using pdist.
% Convert to matrix form for easier reading.
ye = pdist(x,'euclid');
ye_mat = squareform(ye);
```

The matrix we get from this is

```
ye_mat =
```

0	1.0000	1.0000	2.8284	3.6056
1.0000	0	1.4142	3.6056	4.4721
1.0000	1.4142	0	3.6056	4.2426
2.8284	3.6056	3.6056	0	1.0000
3.6056	4.4721	4.2426	1.0000	0

We contrast this with the city block distance.

```
% Contrast with city block metric.
ycb = pdist(x,'cityblock');
ycb_mat = squareform(ycb);
```

The result we get from this is

```
ycb_mat =
        0      1      1      4      5
        1      0      2      5      6
        1      2      0      5      6
        4      5      5      0      1
        5      6      6      1      0
```

Note that both of these interpoint distance matrices are square and symmetric, as expected.
❑

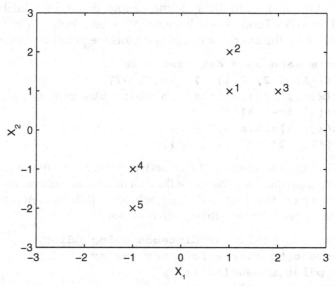

FIGURE 11.1
These are the observations used in Example 11.1. Two clusters are clearly seen.

11.3 Hierarchical Clustering

There are two types of hierarchical clustering methods: agglomerative and divisive. *Divisive* methods start with one large group and successively split the groups until there are n groups with one observation per group. In general, methods for this type of hierarchical clustering are computationally inefficient [Webb, 2011], so we do not discuss them further. *Agglomerative* methods are just the opposite; we start with n groups (one observation per group) and successively merge the two most similar groups until we are left with only one group.

There are five commonly employed methods for merging clusters in agglomerative clustering. These are single linkage, complete linkage, average linkage, centroid linkage, and Ward's method. The MATLAB Statistics Toolbox has a function called **linkage** that will perform agglomerative clustering using any of these methods. Its use is illustrated in the next example, but first we briefly describe each of the methods [Hair, et al., 1995].

The *single linkage* method uses minimum distance, where the distance between clusters is defined as the smallest distance between pairs having one point in each cluster. The first cluster is formed by merging the two groups with the shortest distance. Then the next smallest distance is found between all of the clusters. The two clusters corresponding to the smallest distance are then merged. The process continues in this manner until there is one group. In some cases, single linkage can lead to chaining of the observations, where those on the ends of the chain might be very dissimilar.

The process for the *complete linkage* method is similar to single linkage, but the clustering criterion is different. The distance between groups is defined as the most distant pair of observations, with one coming from each group. The logic behind using this type of similarity criterion is that the maximum distance between observations in each cluster represents the smallest sphere that can enclose all of the objects in both clusters. Thus, the closest of these cluster pairs should be grouped together. The complete linkage method does not have the chaining problem that single linkage has.

The *average linkage* method for clustering starts out the same way as single and complete linkage. In this case, the cluster criterion is the average distance between all pairs, where one member of the pair comes from each cluster. Thus, we find all pairwise distances between observations in each cluster and take the average. This linkage method tends to combine clusters with small variances and to produce clusters with approximately equal variance.

Centroid linkage calculates the distance between two clusters as the distance between the centroids. The *centroid* of a cluster is defined as the d-dimensional sample mean for those observations that belong to the cluster. Whenever we merge clusters together or add an observation to a cluster, the centroid is recalculated.

The distance between two clusters using *Ward's linkage* method is defined as the incremental sum of the squares between two clusters. To merge clusters, the within-group sum-of-squares is minimized over all possible partitions obtained by combining two clusters. The within-group sum-of-squares is defined as the sum of the squared distances between all observations in a cluster and its centroid. This method tends to produce clusters with approximately the same number of observations in each one.

Example 11.2

We illustrate the **linkage** function using the data and distances from the previous example. We look only at single linkage and complete linkage using the Euclidean distances. We can view the results of the clustering in dendrograms given in Figures 11.2 and 11.3.

```
% Get the cluster output from the linkage function.
zsingle = linkage(ye,'single');
zcomplete = linkage(ye,'complete');
% Get the dendrogram.
dendrogram(zsingle)
title('Clustering - Single Linkage')
dendrogram(zcomplete)
title('Clustering - Complete Linkage')
```

A *dendrogram* shows the links between objects as inverted U-shaped lines, where the height of the U represents the distance between the objects. The cases are listed along the horizontal axis. Cutting the tree at various y values of the dendrogram yields different clusters. For example, cutting the complete linkage tree in Figure 11.3 at $y = 1.2$ would yield three clusters with the following members:

Group 1: x_4, x_5

Group 2: x_1, x_3

Group 3: x_2

Alternatively, if we cut the tree at $y = 3$, then we have two clusters. One cluster has observations x_1, x_2, x_3, and the other one containing x_4, x_5. Note that if we choose to create two clusters in the single and complete linkage dendrograms, then the two methods give the same cluster memberships (compare dendrograms in Figures 11.2 and 11.3).
❏

Now that we have our cases clustered, we would like to measure the validity of the clustering. One way to do this would be to compare the distances between all observations with the links in the dendrogram. If the clustering is a valid one, then there should be a strong correlation between them. We can measure this using the *cophenetic correlation coefficient*. A cophenetic matrix is defined using the results of the linkage procedure. The ij-th entry of the cophenetic matrix is the fusion level (the height of the link) at which the i-th and j-th objects appear together in the same cluster for the first time. The correlation coefficient between the distances and the corresponding cophenetic entries is the cophenetic correlation coefficient. Large values indicate that the linkage provides a reasonable clustering of the data. The MATLAB Statistics Toolbox provides a function that will calculate the cophenetic correlation coefficient. Its use is illustrated in the following example.

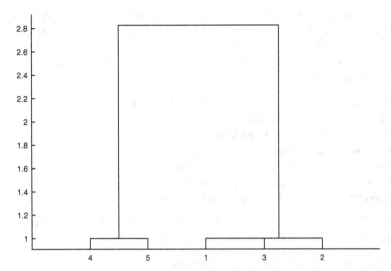

FIGURE 11.2
This is the dendrogram using Euclidean distances and single linkage.

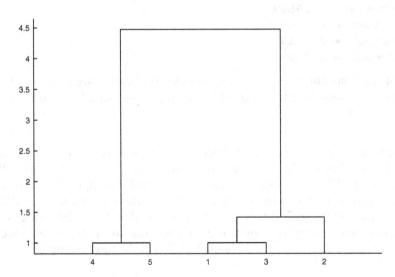

FIGURE 11.3
This is the dendrogram using Euclidean distances and complete linkage.

Example 11.3

In this example, we show how to obtain the cophenetic correlation coefficient in MATLAB. We use the same small data set from the previous examples and calculate the cophenetic correlation coefficient when we have clusters based on different distances and linkages. First, we get the clusters using the following commands.

```
x = [1 1; 1 2; 2 1; -1 -1; -1 -2];
ye = pdist(x,'euclid');
ycb = pdist(x,'cityblock');
zsineu = linkage(ye,'single');
zcompeu = linkage(ye,'complete');
zsincb = linkage(ycb,'single');
zcomcb = linkage(ycb,'complete');
```

We now have four different cluster hierarchies. Their cophenetic correlation coefficients can be found from the following:

```
ccompeu = cophenet(zcompeu,ye)
csineu = cophenet(zsineu,ye)
csincb = cophenet(zsincb,ycb)
ccomcb = cophenet(zcomcb,ycb)
```

From this we get:

```
ccompeu = 0.9503
csineu = 0.9495
csincb = 0.9546
ccomcb = 0.9561
```

All of the resulting cophenetic correlation coefficients are large, with the largest corresponding to the complete linkage clustering based on the city block distance.
❑

We now illustrate hierarchical clustering using a real data set that we have seen before. These data comprise measurements of a type of insect called *Chaetocnema* [Lindsey, Herzberg, and Watts, 1987; Hand, et al., 1994]. The variables measure the width of the first joint of the first tarsus, the width of the first joint of the second tarsus, and the maximal width of the aedegus. All measurements are in microns. We suspect that there are three species represented by these data, and we explore this in the next example.

Example 11.4

The data we need are contained in the file called **insect**. For this example, we will use only the first two features of the data for our hierarchical agglomerative clustering. The following MATLAB code loads the data and constructs the scatterplot shown in Figure 11.4. For this data set, we happen to know the true class membership. The first ten observations are one class of

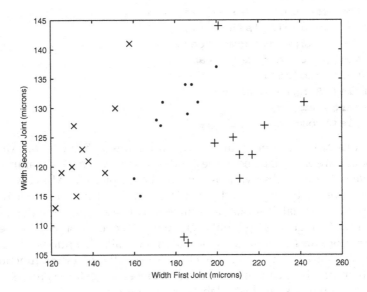

FIGURE 11.4

This is a scatterplot of the first two features of the **insect** *data. The classes are shown with different plotting symbols. Of course, in a truly unsupervised learning case, we would not have this information.*

insect, observations eleven through twenty are another class, and the remainder are from a third class. We use this information to plot the data points using a different plotting symbol for each class.

```
% First load the data.
load insect
% We will look only at the first
% two dimensions in this example.
X = insect(:,1:2);
plot(X(1:10,1),X(1:10,2),'.')
hold on
plot(X(11:20,1),X(11:20,2),'+')
plot(X(21:30,1),X(21:30,2),'x')
hold off
xlabel('Width First Joint (microns)')
ylabel('Width Second Joint (microns)')
```

Next, we get the pairwise distances using the default Euclidean distance for **pdist**. Then we get the hierarchical clustering using two types of linkage: single and complete.

```
% First get the pairwise distances.
% We will use Euclidean distance.
```

```
d = pdist(X);
% Now get two types of linkage.
zs = linkage(d,'single');
zc = linkage(d,'complete');
% Now plot in dendrograms.
dendrogram(zs);
title('Single Linkage')
dendrogram(zc);
title('Complete Linkage')
```

The results of the clustering are shown in Figures 11.5 and 11.6. All of the data points are easily labeled in the dendrogram leaves in this example, since we have only thirty observations. To avoid over-plotting, MATLAB plots a maximum of thirty leaves in the dendrogram. So, if $n \le 30$, the leaf labels correspond to actual observation numbers. This number is the default, and the user can change it using input arguments to the function. See the **help** on the **dendrogram** function for information on available options. We can get the membership for any number of clusters using the **cluster** function. We know there are three clusters, so we use **cluster** to get three groups and the **find** function to assign cluster labels to the observations.

```
% Get the membership for 3 clusters.
Ts = cluster(zs,'maxclust',3);
Tc = cluster(zc,'maxclust',3);
% Now find out what ones are in each cluster
% for the complete linkage case.
ind1 = find(Tc == 1);
ind2 = find(Tc == 2);
ind3 = find(Tc == 3);
```

Choosing three groups in the complete linkage agglomerative clustering yields the following partition:

```
1, 2, 4, 5, 7, 8, 9, 11, 15
3, 12, 13, 14, 16, 17, 18, 19, 20
6, 10, 21, 22, 23, 24, 25, 26, 27, 28, 29, 30
```

Since we know the true classes or groups in the data, we can see that this is a reasonable result. We can get the cophenetic coefficient to see which type of linkage correlates better with the interpoint distances.

```
% Now look at the cophenetic coefficient to
% compare single and complete linkage.
cophs = cophenet(zs,d);
cophc = cophenet(zc,d);
```

The cophenetic coefficient for single linkage is 0.5049, and complete linkage has a coefficient of 0.7267. Thus, complete linkage is better.

❑

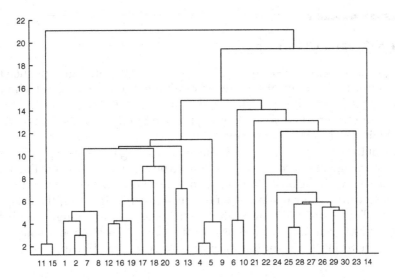

FIGURE 11.5

This is the dendrogram from using single linkage for the **insect** *data. If we cut the tree to give us three groups (say at y = 18), then one group would have observations 11 and 15, and another would contain observation 14. The rest would be in the third group.*

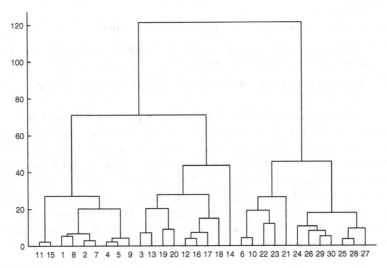

FIGURE 11.6

This is the dendrogram from complete linkage agglomerative clustering of the **insect** *data. Looking at the labels at the bottom of the tree and choosing three clusters (cutting the tree at y = 60), we see that we have a more reasonable grouping, as compared to the single linkage case.*

11.4 *K*-Means Clustering

The goal of **k-means clustering** is to partition the data into k groups such that the within-group sum-of-squares is minimized. One way this technique differs from hierarchical clustering is that we must specify the number of groups or clusters that we are looking for. We briefly describe two methods for obtaining clusters via k-means.

One of the basic algorithms for k-means clustering is a two-step procedure. First, we assign observations to its closest group, usually using the Euclidean distance between the observation and the cluster centroid. The second step of the procedure is to calculate the new cluster centroid using the assigned objects. These steps are alternated until there are no changes in cluster membership or until the centroids do not change. This method is sometimes referred to as HMEANS or the basic ISODATA method [Spath, 1980].

PROCEDURE – HMEANS METHOD

1. Specify the number of clusters k.
2. Determine initial cluster centroids. These can be randomly chosen or the user can specify them.
3. Calculate the distance between each observation and each cluster centroid.
4. Assign every observation to the closest cluster.
5. Calculate the centroid (i.e., the d-dimensional mean) of every cluster using the observations that were just grouped there.
6. Repeat steps 3 through 5 until no more changes are made.

There are two problems with the HMEANS method. The first one is that this method could lead to empty clusters. As the centroid is recalculated and observations are reassigned to groups, some clusters could become empty. The second issue concerns the optimality of the partitions. With k-means, we are actually searching for partitions where the within-group sum-of-squares is minimum. It can be shown [Webb, 2011] that in some cases, the final k-means cluster assignment is not optimal, in the sense that moving a single point from one cluster to another may reduce the sum of squared errors. The following procedure helps address the second problem.

PROCEDURE – K-MEANS

1. Obtain a partition of k groups, possibly from the HMEANS method.

2. Take each data point $\mathbf{x}_i$ and calculate the Euclidean distance between it and every cluster centroid.

3. Here $\mathbf{x}_i$ is in the r-th cluster, n_r is the number of points in the r-th cluster, and d_{ir}^2 is the Euclidean distance between $\mathbf{x}_i$ and the centroid of cluster r. If there is a group s such that

$$\frac{n_r}{n_r - 1} d_{ir}^2 > \frac{n_s}{n_s + 1} d_{is}^2 ,$$

then move $\mathbf{x}_i$ to cluster s.

4. If there are several clusters that satisfy the above inequality, then move the $\mathbf{x}_i$ to the group that has the smallest value for

$$\frac{n_s}{n_s + 1} d_{is}^2 .$$

5. Repeat steps 2 through 4 until no more changes are made.

We note that there are many algorithms for k-means clustering described in the literature. We provide some references to these in the last section of this chapter.

Example 11.5

We show how to implement HMEANS in MATLAB, using the **insect** data of the previous example. We first obtain the cluster centers by randomly picking observations from the data set. Note that initial cluster centers do not have to be actual observations.

```
% First load the data.
load insect
% We will look only at the first
% two dimensions in this example.
X = insect(:,1:2);
[n,d] = size(X);
% Choose the number of clusters
k = 3;

% Pick some observations to be the cluster centers.
ind = randperm(n);
ind = ind(1:k);
nc = X(ind,:);
% Set up storage.
% Integers 1,...,k indicating cluster membership
cid = zeros(1,n);
```

```
% Make this different to get the loop started.
oldcid = ones(1,n);
% The number in each cluster.
nr = zeros(1,k);
% Set up maximum number of iterations.
maxiter = 100;
iter = 1;

while ~isequal(cid,oldcid) && iter < maxiter
    oldcid = cid;
    % Implement the hmeans method.
    % For each point, find the distance
    % to all cluster centers.
    for i = 1:n
        dist = sum((repmat(X(i,:),k,1)-nc).^2,2);
         % assign it to this cluster
        [m,ind] = min(dist);
        cid(i) = ind;
    end
    % Find the new cluster centers.
    for i = 1:k
        % Find all points in this cluster.
        ind = find(cid==i);
        % Find the centroid.
        nc(i,:) = mean(X(ind,:));
        % Find the number in each cluster;
        nr(i) = length(ind);
    end
    iter = iter + 1
end
```

To check these results, we show a scatterplot of the group membership found using *k*-means clustering, where the three classes are represented by different plotting symbols. This is given in Figure 11.7.
❏

The MATLAB Statistics Toolbox now has a function for *k*-means clustering called **kmeans**. We will use this function later on when we illustrate some techniques for assessing cluster results and for estimating the number of clusters represented by our data. In the next section, we present another approach that incorporates a way of estimating the number of groups as part of the clustering process.

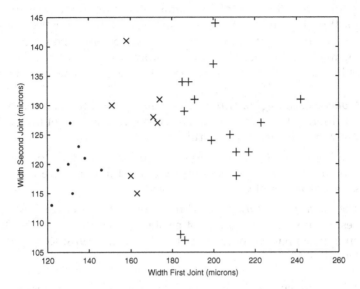

FIGURE 11.7
This is a scatterplot of the **insect** *data. The three groups found using k-means are shown with different plotting symbols. Compare this with the known groups in Figure 11.4.*

11.5 Model-Based Clustering

In this section, we describe a method for finding groups based on probability density estimation via finite mixtures called ***model-based clustering*** [Fraley, 1998; Fraley and Raftery, 1998]. As we will see, it combines several ideas: (1) a form of agglomerative clustering, (2) Gaussian finite mixtures, (3) the EM algorithm, and (4) an approximation to Bayes factors to choose the best model. Of course, the use of probability density models for clustering is not a new idea, and we provide several references at the end of this chapter.

Recall that a finite mixture probability density function models the density function as a weighted sum of component densities. When we use finite mixtures for cluster analysis, we assume that each component density function corresponds to a cluster. We can then assign observations to a group using the probability that a data point belongs to the cluster, just as we do with Bayesian classifiers. So, the problem of finding the groups or clusters becomes one of estimating a finite mixture.

We learned in a previous chapter that the EM (Expectation-Maximization) algorithm can be used to estimate a finite mixture, and we saw that several issues must be addressed. First, we need to specify the number of component densities in our finite mixture. Since we are in a clustering context, this would be the same as assigning a value for k in k-means. Second, we need a good starting point (initial values of the parameters of the finite mixture) for the EM algorithm. Finally, we have to assume some form for the component densities, such as Gaussian, the t distribution, etc. The model-based clustering methodology addresses these problems.

Model-based clustering has three main components as described next.

1. *Model-based agglomerative clustering* is used to get a reasonable partition of the data. Each partition is then used to get initial starting values for the EM algorithm.

2. The *EM algorithm* is used to estimate the parameters of the finite mixture. Each component density of the finite mixture will represent a group or cluster.

3. The *Bayesian Information Criterion* (BIC) is used to choose the best grouping and number of clusters, given the data. The BIC is an approximation to Bayes factors [Schwarz, 1978; Kass and Raftery, 1995].

We will discuss each of these in more detail and then put the pieces together to present the overall model-based clustering methodology.

Finite Mixture Models and the EM Algorithm

Recall from Chapter 9 that a *multivariate finite mixture* probability density is given by

$$f(\mathbf{x}; p, \boldsymbol{\theta}) = \sum_{k=1}^{c} p_k \, g_k(\mathbf{x}; \boldsymbol{\theta}_k). \tag{11.5}$$

The *component density* is denoted by $g_k(\mathbf{x}; \boldsymbol{\theta}_k)$ with parameters represented by $\boldsymbol{\theta}_k$. Note that $\boldsymbol{\theta}_k$ is used to denote any type and number of parameters. The *weights* are given by p_k, with the constraint that they are nonnegative and sum to one. These weights are also called the *mixing proportions* or *mixing coefficients*.

The component densities can be any *bona fide* probability density, but one of the most commonly used ones is the multivariate normal. This yields the following equation for a multivariate Gaussian finite mixture

$$f(\mathbf{x}; p, \boldsymbol{\mu}, \boldsymbol{\Sigma}) = \sum_{k=1}^{c} p_k \phi\,(\mathbf{x}; \boldsymbol{\mu}_k, \boldsymbol{\Sigma}_k), \tag{11.6}$$

where ϕ represents a multivariate normal probability density function given by

$$\phi(\mathbf{x}_i; \boldsymbol{\mu}_k, \boldsymbol{\Sigma}_k) = \frac{\exp\left\{-\frac{1}{2}(\mathbf{x}_i - \boldsymbol{\mu}_k)^T \boldsymbol{\Sigma}_k^{-1}(\mathbf{x}_i - \boldsymbol{\mu}_k)\right\}}{(2\pi)^{d/2}\sqrt{|\boldsymbol{\Sigma}_k|}}.$$

Thus, we have the parameters $\boldsymbol{\mu}_k$, where each one is a d-dimensional vector of means, and $\boldsymbol{\Sigma}_k$, where each is a $d \times d$ covariance matrix. To cluster the data we need to estimate the weights p_k, the d-dimensional means for each term, and the covariance matrices.

The term *model* in model-based clustering has to do, in part, with the idea of constraining the covariance matrices of the component densities in various ways. This yields different models for the finite mixture. Celeux and Govaert [1995] provide the following decomposition of the k-th covariance matrix $\boldsymbol{\Sigma}_k$ in Equation 11.6:

$$\boldsymbol{\Sigma}_k = \lambda_k \mathbf{D}_k \mathbf{A}_k \mathbf{D}_k^T. \tag{11.7}$$

The factor λ_k in Equation 11.7 governs the volume of the k-th component density (or cluster). The factor $\mathbf{D}_k$ is a matrix with columns corresponding to the eigenvectors of $\boldsymbol{\Sigma}_k$, and it determines the orientation of the cluster. The matrix $\mathbf{A}_k$ is a diagonal matrix that contains the normalized eigenvalues of $\boldsymbol{\Sigma}_k$ along the diagonal. This matrix is associated with the shape of the cluster.

Celeux and Govaert [1995] show that the eigenvalue decomposition given in Equation 11.7 produces 14 models obtained by adjusting or restricting the form of the component covariance matrices $\boldsymbol{\Sigma}_k$. They provide covariance matrix update equations based on these models for use in the EM algorithm. Some of these have a closed form update formula, and others must be solved in an iterative manner. In this text, we work only with the nine models that have a closed form update formula for the covariance matrix. Thus, what we present is a subset of the available models. See Table 11.1 for a summary of the nine models pertaining to the decomposition given in Equation 11.7. For more information on these models, including examples with MATLAB, see Martinez et al. [2010].

We discussed the EM algorithm in Chapter 9, and we repeat some of the concepts here for ease of reading. Recall that we wish to estimate the parameters

$$\boldsymbol{\theta} = p_1, ..., p_{c-1}, \boldsymbol{\mu}_1, ..., \boldsymbol{\mu}_c, \boldsymbol{\Sigma}_1, ..., \boldsymbol{\Sigma}_c,$$

by maximizing the log-likelihood given by

TABLE 11.1 MULTIVARIATE NORMAL MIXTURE MODELS

Closed Form Solution to Covariance Matrix Update Equation in EM Algorithm[a] [Martinez et al., 2010]

Mode l #	Covariance	Distribution	Description
1	$\Sigma_k = \lambda I$	Family: Spherical Volume: Fixed Shape: Fixed Orientation: NA	•Diagonal covariance matrices •Diagonal elements are equal •Covariance matrices are equal •I is a $p \times p$ identity matrix
2	$\Sigma_k = \lambda_k I$	Family: Spherical Volume: Variable Shape: Fixed Orientation: NA	•Diagonal covariance matrices •Diagonal elements are equal •Covariance matrices may vary •I is a $p \times p$ identity matrix
3	$\Sigma_k = \lambda B$	Family: Diagonal Volume: Fixed Shape: Fixed Orientation: Axes	•Diagonal covariance matrices •Diagonal elements may be unequal •Covariance matrices are equal •B is a diagonal matrix
4	$\Sigma_k = \lambda B_k$	Family: Diagonal Volume: Fixed Shape: Variable Orientation: Axes	•Diagonal covariance matrices •Diagonal elements may be unequal •Covariance matrices may vary among components
5	$\Sigma_k = \lambda_k B_k$	Family: Diagonal Volume: Variable Shape: Variable Orientation: Axes	•Diagonal covariance matrices •Diagonal elements may be unequal •Covariance matrices may vary among components
6	$\Sigma_k = \lambda DAD^T$	Family: General Volume: Fixed Shape: Fixed Orientation: Fixed	•Covariance matrices can have nonzero off-diagonal elements •Covariance matrices are equal

7	$\Sigma_k = \lambda D_k A D_k^T$	Family: General
		Volume: Fixed
		Shape: Fixed
		Orientation: Variable
			•Covariance matrices can have
			  nonzero off-diagonal elements
			•Covariance matrices may vary
			  among components

8	$\Sigma_k = \lambda D_k A_k D_k^T$	Family: General
		Volume: Fixed
		Shape: Variable
		Orientation: Variable
			•Covariance matrices can have
			  nonzero off-diagonal elements
			•Covariance matrices may vary
			  among components

9	$\Sigma_k = \lambda_k D_k A_k D_k^T$	Family: General
		Volume: Variable
		Shape: Variable
		Orientation: Variable
			•Covariance matrices can have
			  nonzero off-diagonal elements
			•Covariance matrices may vary
			  among components

a.   This is a subset of the available models.

$$\ln[L(\theta|\mathbf{x}_1, \dots, \mathbf{x}_n)] = \sum_{i=1}^{n} \ln\left[\sum_{k=1}^{c} p_k \phi(\mathbf{x}_i; \mu_k, \Sigma_k)\right]. \tag{11.8}$$

Note that we do not have to estimate all c of the weights because

$$p_c = 1 - \sum_{i=1}^{c-1} p_i.$$

We assume that the components exist in a fixed proportion in the mixture, given by the p_k. It is this component membership (or cluster membership) that is unknown and is the reason why we need to use something like the EM algorithm to maximize Equation 11.8. We can write the *posterior probability* that an observation $\mathbf{x}_i$ belongs to component k as

$$\hat{\tau}_{ik}(\mathbf{x}_i) = \frac{\hat{p}_k \phi(\mathbf{x}_i; \hat{\mu}_k, \hat{\Sigma}_k)}{\hat{f}(\mathbf{x}_i; \hat{p}, \hat{\mu}, \hat{\Sigma})}; \qquad k = 1, \dots, c; \; i = 1, \dots, n, \tag{11.9}$$

where

$$\hat{f}(\mathbf{x}_i; \hat{p}, \hat{\mu}, \hat{\Sigma}) = \sum_{j=1}^{c} \hat{p}_j \phi(\mathbf{x}_i; \hat{\mu}_j, \hat{\Sigma}_j).$$

In the clustering context, we use the posterior probability in Equation 11.9 to estimate the probability that an observation belongs to the k-th cluster.

Recall from calculus, that to maximize the function given in Equation 11.8, we must find the first partial derivatives with respect to the parameters and then set them equal to zero. The solutions [Everitt and Hand, 1981] to the likelihood equations are

$$\hat{p}_k = \frac{1}{n} \sum_{i=1}^{n} \hat{\tau}_{ik} \tag{11.10}$$

$$\hat{\mu}_k = \frac{1}{n} \sum_{i=1}^{n} \frac{\hat{\tau}_{ik}\mathbf{x}_i}{\hat{p}_k} \tag{11.11}$$

$$\hat{\Sigma}_k = \frac{1}{n} \sum_{i=1}^{n} \frac{\hat{\tau}_{ik}(\mathbf{x}_i - \hat{\mu}_k)(\mathbf{x}_i - \hat{\mu}_k)^T}{\hat{p}_k}. \tag{11.12}$$

The update for the covariance matrix given in Equation 11.12 is what we use for the unconstrained case described in the previous section (model nine in Table 11.1). We do not provide the update equations for the other models in this text, but they are implemented in the model-based clustering functions that are included in the Computational Statistics Toolbox. Please see Celeux and Govaert [1995] for a complete description of the update equations for all models.

The *EM algorithm* is a two step process, consisting of an E-step and an M-step, as outlined next. These two steps are repeated until the estimated values converge.

E-Step

We calculate the posterior probability that the i-th observation belongs to the k-th component, given the current values of the parameters. This is given by Equation 11.9.

M-Step

Update the parameter estimates using the posterior probability from the E-step and Equations 11.10 and 11.11. The covariance matrix is updated using the equation for the desired model from Table 11.1.

Note that the E-step involves estimating the posterior probability using the current parameter values. Thus, we need to start with initial estimates of the mixture parameters. We also need to specify how many terms or components we have in our finite mixture. It is known that the likelihood surface typically has many modes or many locally optimal solutions. Additionally, the EM algorithm may even diverge, depending on the starting point. However, the EM can provide improved estimates of our parameters if the starting point is a good one [Fraley and Raftery, 2002]. The idea that partitions based on model-based agglomerative clustering provide a good starting point for the EM algorithm was first discussed in Dasgupta and Raftery [1998].

Model-Based Agglomerative Clustering

Model-based agglomerative clustering uses the same general ideas as the agglomerative clustering we described earlier. In particular, all observations start out in their own singleton group, and two groups are merged at each step until we have just one group. It also provides a partition of the data for any given number of clusters between one and n. The difference between the two methods has to do with the criterion used to decide which two groups to merge at each step. With regular agglomerative clustering, the two closest groups (in terms of the selected distance and type of linkage) are merged. In the case of model-based agglomerative clustering, we do not use a distance. Instead, we use the *classification likelihood* as our objective function, which is given by

$$L_{CL}(\theta_k, \gamma_i; \mathbf{x}_i) = \prod_{i=1}^{n} f_{\gamma_i}(\mathbf{x}_i; \theta_{\gamma_i}), \tag{11.13}$$

where γ_i is a label indicating a classification for the i-th observation. We have $\gamma_i = k$, if $\mathbf{x}_i$ belongs to the k-th component. In the mixture approach, the number of observations in each component has a multinomial distribution with sample size of n and probability parameters given by $p_1, ..., p_c$. The exact form of the classification likelihood (Equation 11.13) will change depending on the model chosen for our finite mixture.

In model-based agglomerative clustering we seek to find partitions that maximize (approximately) the classification likelihood. To accomplish this, the two clusters producing the largest increase in the classification likelihood are merged at each step. This process continues until all observations are in one group. Note that the form of the objective function can be adjusted as in Fraley [1998] to handle singleton clusters.

We could use any of the nine models in Table 11.1 for our finite mixture, but previous research indicates that the unconstrained model (number 9) with *agglomerative* model-based clustering yields reasonable initial partitions for the EM algorithm, regardless of the final model used for the component densities. Thus, our implementation of *agglomerative* model-based clustering includes the unconstrained model only. Fraley [1998] provides efficient algorithms for the four basic models (1, 2, 6, 9) and shows how the techniques developed for these can be extended to the other models.

Model-based agglomerative clustering was developed as a component of the model-based clustering methodology. However, it can also be used as a stand-alone agglomerative clustering procedure, as we demonstrate in the next example.

Example 11.6

We return to the **insect** data of previous examples to show how to use a function called **agmbclust**. This is included in the Computational Statistics Toolbox.

```
% First load the data.
load insect
% Again, we will look only at the first
% two dimensions in this example.
X = insect(:,1:2);
% Now call the model-based agglomerative
```

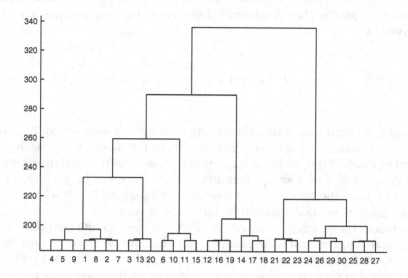

FIGURE 11.8
This is the dendrogram from model-based agglomerative clustering for the **insect** *data.*

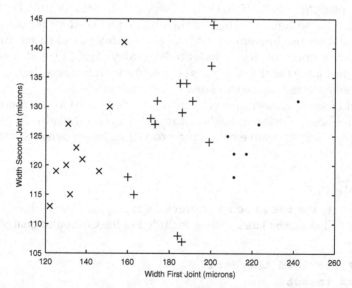

FIGURE 11.9
This is a scatterplot of the **insect** *data. The data were clustered using model-based agglomerative clustering. Three groups were selected and cluster labels were assigned. Different plotting symbols are used for each group. The reader should compare this to the true class membership in Figure 11.4.*

```
% clustering function.
z = agmbclust(X);
```

The output from this function is the same *structure* that MATLAB outputs for the **linkage** function. This means that we can use the same functions that we used with **linkage**, such as **dendrogram**.

```
% We can plot in a dendrogram.
dendrogram(z);
title('Model-Based Agglomerative Clustering')
```

This is shown in Figure 11.8, where the scale on the vertical axis no longer represents distance. Rather, it is a quantity that is related to the classification likelihood. If we use **cluster** to obtain three clusters, then we can show a scatterplot of the groups (Figure 11.9).

❏

Bayesian Information Criterion

We now turn our attention to the third major component in model-based clustering, which is choosing the model that best fits our data according to some criterion. The Bayesian approach to model selection started with Jeffreys [1935, 1961]. He developed a framework for calculating the evidence in favor of a null hypothesis (or model) using the Bayes factor, which is the posterior odds of one hypothesis when the prior probabilities of the two hypotheses are equal [Kass and Raftery, 1995].

The *Bayesian Information Criterion* (BIC) is given by

$$BIC = 2\log[L_M(\mathbf{X}, \hat{\boldsymbol{\theta}})] - m\log(n), \tag{11.14}$$

where L_M is the likelihood given the data, the model M, and the estimated parameters $\hat{\boldsymbol{\theta}}$. The number of independent parameters to be estimated for model M is given by m. It is well known that finite mixture models fail to satisfy the regularity conditions to make Equation 11.14 valid, but previous research shows that the use of the BIC in model-based clustering produces reasonable results [Fraley and Raftery, 1998; Dasgupta and Raftery, 1998].

Model-Based Clustering Procedure

Now that we have the pieces of model-based clustering, we can provide a procedure that (hopefully) puts it all together. We first show a diagram of the steps in Figure 11.10. From this we see that one estimates many finite mixtures of Gaussians using different values for the number of terms (c) and different models (Table 11.1). This produces many groupings of our data, and the "best" one is chosen using the BIC.

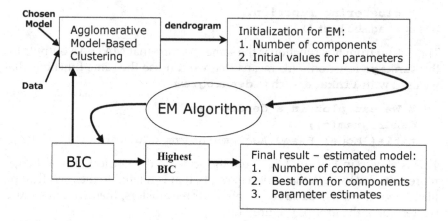

FIGURE 11.10

This diagram shows the flow of the steps in the model-based clustering procedure. [Martinez et al., 2010]

PROCEDURE – MODEL-BASED CLUSTERING

1. Using the unconstrained model (model 9), apply the model-based agglomerative clustering procedure to the data. This provides a partition of the data for any given number of clusters.
2. Choose a model, M (see Table 11.1).
3. Choose a number of clusters or component densities, c.
4. Find a partition with c groups using the results of the agglomerative model-based clustering (step 1).
5. Using this partition, find the mixing coefficients, means, and covariances for each cluster. The covariances are constrained according to the model chosen in step 2.
6. Using the chosen c (step 3) and the initial values (step 5), apply the EM algorithm to obtain the final estimates.
7. Calculate the value of the BIC for this value of c and M.
8. Go to step 3 to choose another value of c.
9. Go to step 2 to choose another model M.
10. Choose the configuration (number of clusters c and form for the covariance matrices) that corresponds to the highest BIC.

The model-based clustering methodology is essentially an exhaustive search where we vary the number of groups (c) that we look for, as well as the types of clusters (models in Table 11.1). Thus, if we wanted to look for 1 to C groups using all nine of the models, then we would have to estimate

$C \times 9$ finite mixtures. This can be computationally intensive, so one should keep C to a reasonable number.

Example 11.7

We provide a function with the text that implements the entire model-based clustering procedure as outlined above. This function is called **mbclust**. The inputs to the function are the $n \times d$ data matrix and the maximum number of clusters to look for (C). Several variables are returned by **mbclust**:

- **bics** is a matrix containing the BIC values for each model and number of clusters. Each row corresponds to one of the nine models, and each column corresponds to the number of clusters. So, **bics** is a $9 \times C$ matrix.
- **bestmodel** returns the "best" model. This is the model that yields the highest BIC value.
- **allmodel** is a structure that includes all of the estimated models (weights, means, and covariance matrices). To access the i-th model for j clusters, use the syntax: **allmodels(i).clus(j)**. We will explore this further in the next example.
- **z** is the agglomerative clustering linkage matrix.
- **clabs** contains the clustering labels using the grouping from the **bestmodel**.

We return to the **insect** data to demonstrate the **mbclust** function.

```
% First load the data.
load insect
% Again, we will look only at the first
% two dimensions in this example.
X = insect(:,1:2);
% Call the model-based clustering function
% using a maximum of 6 clusters.
[bics,bestmodel,allmodel,Z,clabs] = ...
    mbclust(X,6);
```

The current status of the model-based clustering methodology is shown in the MATLAB command window while the procedure is running, as is the model yielding the highest BIC, once the procedure finishes. We can plot the BIC values using

plotbic(bics)

This produces the plot shown in Figure 11.11, where we see that the model corresponding to the highest BIC value is model 6 at 3 groups. Hopefully, the highest BIC value corresponds to a nice elbow in the curve, as it does in this case. However, it often shows that other models might be reasonable

alternatives, since they have similar BIC values. This is why we return all
models, so users can explore them. We found the individual cluster labels
from the **clabs** output variable, and this produced the following groups for
the best model:

```
1, 2, 3, 4, 5, 6, 7, 8, 9, 10, 13
11, 12, 14, 15, 16, 17, 18, 19, 20
21, 22, 23, 24, 25, 26, 27, 28, 29, 30
```

These are excellent results with only one observation (13) that is grouped
incorrectly.
❑

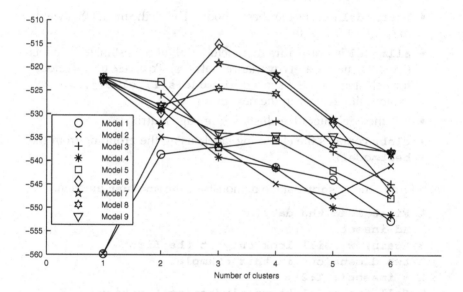

FIGURE 11.11
*This shows the BIC values for the **insect** data. Note that we have a nice upside down
elbow in the curve corresponding to the highest BIC value. Based on this, we would choose
model 6 and 3 clusters.*

In the previous example, the cluster labels for the 'best' model in the **clabs**
array were returned by **mbclust**. We noted earlier, that each component of
the finite mixture corresponds to a cluster. We can use any of the models
(number of terms and form of covariance matrix) to group data based on
their posterior probability. We label an observation according to its highest
posterior probability given by the cluster index j:

$$\{ j | \hat{\tau}_{ij}^{*} = \max_k \hat{\tau}_{ik} \}.$$

In other words, we find the posterior probability that the i-th observation belongs to each of the component densities. We then say that it belongs to the cluster with the highest posterior probability. One benefit of using this approach is that we can use the quantity $1 - \max_k \hat{\tau}_{ik}$ as a measure of the uncertainty in the classification [Bensmail, et al. 1997].

Example 11.8

We provide a function called **mixclass** that returns group labels given the data and the model (weights, means, and covariance matrices). So, we can use this to get cluster labels for any finite mixture model. First, we have to get the model of interest. Recall that the structure **allmodel** contains all models that were estimated in the model-based clustering process. It is a structure with one field called **clus**. The index to **allmodel** indicates the model, and the index to **clus** corresponds to the number of clusters. The field **clus** is itself a structure with **maxclus** records and three fields: **pies, mus,** and **vars**. It appears from Figure 11.11 that another competing grouping would be given by model 7 and 3 groups. We can extract the information (weights, means, and covariance matrices) for **mixclass** as follows.

```
% We extract the estimated parameter information:
pies = allmodel(7).clus(3).pies;
mus = allmodel(7).clus(3).mus;
vars = allmodel(7).clus(3).vars;
```

Next, we find group labels using the **mixclass** function. This function returns a measure of the uncertainty in the classification, as well as the group labels.

```
% Find the group labels.
[clabs7,unc] = mixclass(X,pies,mus,vars);
```

We can find the observations in each group using

```
% Find the observations in each group.
ind1 = find(clabs7 == 1);
ind2 = find(clabs7 == 2);
ind3 = find(clabs7 == 3);
```

Looking at the observations in each of these groups (i.e., **ind1, ind2,** and **ind3**) reveals no difference from the groups we had with the best model (model 6, 3 groups) returned by **mbclust**.

❑

11.6 Assessing Cluster Results

Clustering is an exploratory data analysis approach that can help us understand the structure represented by our data. As with all types of modeling and data analysis, we need to go back and assess the results that we get. For clustering, there are two main questions that need to be addressed:

1. Is the cluster structure we found an adequate representation of the true groupings?
2. What is the correct number of clusters or groups?

In the previous section, we described model-based clustering as one way to address these questions. Prior to that, we saw how to compare the results of hierarchical agglomerative clustering with the interpoint distances using the cophenetic coefficient. Recall that this coefficient is a measure of how well the clustering (i.e., the dendrogram) represents the dissimilarities.

We now look at other ways to assess the quality of the cluster results and to estimate the "correct" number of groups in our data. We first present a method due to Mojena [1977] for determining the number of groups in hierarchical clustering. Next, we cover the silhouette statistic and plot of Kaufman and Rousseeuw [1990]. We conclude with a brief summary of other methods that have appeared in the literature.

Mojena – Upper Tail Rule

The *upper tail rule* was developed by Mojena [1977] as a way of determining the number of clusters when hierarchical clustering is used to group the data. It uses the relative sizes of the different fusion levels in the hierarchy.

We let the fusion levels $\alpha_0, \alpha_1, \alpha_2, ..., \alpha_{n-1}$ correspond to the stages in the hierarchy with $n, n-1, ..., 1$ clusters. We denote the average and standard deviation of the j previous fusion levels by $\overline{\alpha}_j$ and $s_{\overline{\alpha}(j)}$. To apply this rule, we estimate the number of groups as the first level at which we have

$$\alpha_{j+1} > \overline{\alpha}_j + c s_{\overline{\alpha}(j)}, \tag{11.15}$$

where c is a constant. Mojena says that a value of c between 2.75 and 3.50 is good, but Milligan and Cooper [1985] suggest a value of 1.25 instead, based on their study of simulated data sets.

One could also look at a plot of the values

$$\frac{(\alpha_{j+1} - \overline{\alpha}_j)}{s_{\overline{\alpha}(j)}} \tag{11.16}$$

versus the number of clusters j. An elbow in the plot is an indication of the number of clusters.

Example 11.9

In our version of the Mojena plot shown here, an elbow in the curve indicates the number of clusters. We need the output from the **linkage** function, so let's return to the hierarchical clustering in Example 11.4.

```
load insect
% We will look only at the first
% two dimensions in this example.
X = insect(:,1:2);
d = pdist(X);
% Get the linkage.
zc = linkage(d,'complete');
```

We will actually be plotting the raw fusion levels instead of the standardized version given in Equation 11.16. It is our experience that plots of the raw levels are easier to understand.

```
% Plot the raw fusion levels, rather
% than the standardized version of
% Equation 11.16.
nc = 10;
Zf = flipud(zc);
plot(1:10,Zf(1:10,3),'o-')
xlabel('Number of Clusters')
ylabel('Raw Fusion Levels')
```

This plot is shown in Figure 11.12, where we see an elbow at three clusters. We provide a function called **mojenaplot** that implements the Mojena graphical approach given by Equation 11.16.
❑

Silhouette Statistic

Kaufman and Rousseeuw [1990] present the *silhouette statistic* as a way of estimating the number of groups in a data set. Given observation i, we denote the average distance to all other points in its own cluster as a_i. For any other cluster C, we let $\overline{d}(i, C)$ represent the average distance of i to all objects in cluster C. Finally, we let b_i denote the minimum of the average distances $\overline{d}(i, C)$, where the minimum is taken as C ranges over all clusters except the observation's own cluster.

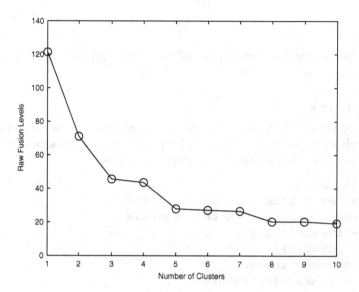

FIGURE 11.12
This is a plot of the raw fusion levels from complete linkage of the **insect** *data. The elbow in the curve shows that three is a reasonable estimate of the number of groups.*

The *silhouette width* for the i-th observation is

$$sw_i = \frac{(b_i - a_i)}{\max(a_i, b_i)},$$

(11.17)

and it ranges from –1 to 1. If an observation has a value close to 1, then the data point is closer to its own cluster than a neighboring one. If it has a silhouette width close to –1, then it is not very well clustered. A silhouette width close to zero indicates that the observation could just as well belong to its current cluster or another one that is near to it.

The number of clusters in the data set is estimated by using the partition with two or more clusters that yields the largest average silhouette width. We can find the *average silhouette width* by averaging over all observations:

$$\overline{sw} = \frac{1}{n} \sum_{i=1}^{n} sw_i.$$

Kaufman and Rousseeuw state that an average silhouette width greater than 0.5 indicates a reasonable partition of the data, and a value of less than 0.2 indicates the data do not exhibit cluster structure.

There is also a nice graphical display called a *silhouette plot* in the Statistics Toolbox, which is illustrated in the next example. This type of plot displays the silhouette values for each cluster separately, allowing the analyst to rapidly visualize and assess the cluster structure.

Example 11.10

We will use the Statistics Toolbox *k*-means function to illustrate the silhouette statistic and silhouette plot. The **kmeans** function has many options. One of them is to run the *k*-means procedure many times from different random starting points.

```
load insect
X = insect(:,1:2);
% Get a k-means clustering using 3 clusters,
% and 5 trials or replicates. We also ask
% MATLAB to display the final results for
% each replicate.
km3 = kmeans(X,3,'replicates',5,'display','final');
```

Something similar to the following should appear in the command window:

```
3 iterations, total sum of distances = 5748.72
4 iterations, total sum of distances = 6236
5 iterations, total sum of distances = 5748.72
3 iterations, total sum of distances = 5748.72
4 iterations, total sum of distances = 5748.72
```

This shows that locally optimal solutions do exist because we have different values for the total sum of distances. Cluster labels corresponding to the best one will be returned by the **kmeans** function. Now, let's use *k*-means with a different value of *k*. Looking back at the plot in Figure 11.12, we see another elbow at five clusters, so we now try $k = 5$.

```
% Get k-means clusters using k = 5.
km5 = kmeans(X,5,'replicates',5,'display','final');
```

Next, we use the **silhouette** function and the results of **kmeans** to get the silhouette values and plots.

```
% Now get the silhouette values and plots.
[sil3,h3] = silhouette(X,km3);
[sil5,h5] = silhouette(X,km5);
```

These plots are shown in Figure 11.13. Since this data set is small, we get a good picture of the cluster membership. For example, with the five cluster grouping, we have one cluster with only two observations, and they both have high silhouette values. The second cluster in both plots has mostly high values, indicating that observations in this group are similar. The others have some observations with low silhouette values. We can find the average silhouette values, as follows:

```
% Find the mean silhouette values for each grouping.
mean(sil3)
mean(sil5)
```

The average silhouette value for three clusters is 0.6581, while the average value for five clusters is 0.6884. This is an indication that there is some cluster structure in the data, with five clusters being slightly better than three.
❑

It is important to note that the silhouette width as defined in Equation 11.17 requires a distance or dissimilarity. The MATLAB function **silhouette** has many options for distances, with squared Euclidean being the default. See the **help** on this function for more information. Also, while the example shown above was for group labels from k-means clustering, one could use it with anything that returns cluster labels, such as **cluster** or the model-based clustering functions.

Other Methods for Evaluating Clusters

We now briefly describe other methods for evaluating clusters. These include methods for assessing the quality of the clusters and choosing an appropriate number of groups.

Sugar and James [2003] described a method for identifying the number of groups based on distortion, which measures the average Mahalanobis distance, per dimension, between each data point and its closest cluster. In practice, one estimates the *distortion* by applying k-means clustering to the observed data using different values for k and finding the mean-squared error. A transformation is applied to the distortions, and one plots the transformed distortions versus the number of clusters. The number of clusters is estimated as the value of k associated with the largest jump.

Tibshirani, et al. [2001] developed a technique called the *gap statistic method* for estimating the number of clusters in a data set. The gap statistic method compares the within-cluster dispersion with what one might expect given a reference null distribution with no clusters. The gap statistic can be used with any clustering technique. Please see Martinez et al. [2010] for more information on the method and MATLAB code for implementing the procedure.

Calinski and Harabasz [1974] defined an index for determining the number of clusters as follows

$$ch_k = \frac{tr(\mathbf{S}_{B_k})/(k-1)}{tr(\mathbf{S}_{W_k})/(n-k)},$$

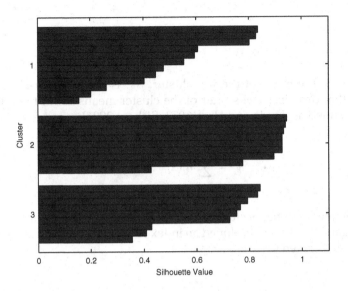

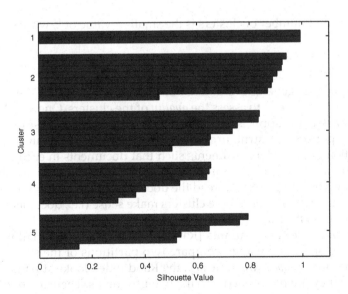

FIGURE 11.13

*The top panel shows the silhouette plot for three clusters obtained using k-means clustering on the **insect** data. The bottom panel shows the silhouette plot for five clusters. Data points with values close to one indicate they are closer to their own cluster than a neighboring one. There were no negative silhouette values for these clusters, indicating reasonable groupings of the data. Note that there is one bar for each observation.*

where $\mathbf{S}_{W_k}$ is the *within-class scatter matrix* for k clusters, given by

$$\mathbf{S}_{W_k} = \sum_{\mathbf{x} \in c_j}^{k} (\mathbf{x} - \bar{\mathbf{x}}_j)(\mathbf{x} - \bar{\mathbf{x}}_j)^T,$$

where $\bar{\mathbf{x}}_j$ is the mean of the j-th cluster. $\mathbf{S}_{B_k}$ is the *between-class scatter matrix* that describes the scatter of the cluster means about the total mean and is defined as [Duda and Hart, 1973; Bishop, 2006]

$$\mathbf{S}_{B_k} = \sum_{j=1}^{k} n_j(\bar{\mathbf{x}}_j - \bar{\mathbf{x}})(\bar{\mathbf{x}}_j - \bar{\mathbf{x}})^T.$$

The estimated number of clusters is given by the largest value of ch_k for $k \geq 2$.

Hartigan [1985] also developed an index to estimate the number of clusters. This index is given by

$$hart_k = \left[\frac{tr(\mathbf{S}_{W_k})}{tr(\mathbf{S}_{W_{k+1}})} - 1 \right](n - k - 1).$$

The estimated number of clusters is the smallest value of k, where

$$1 \leq k \leq 10.$$

What we just gave above is a list of methods for finding the *number* of clusters. We also need to assess the *quality* of the clusters. One way to do this is to go back to the application and see if the groups make sense or if one can discover patterns or structure given the grouping. For example, say the application is to cluster documents such that documents in each cluster are on the same topic or have similar meaning. To make sure the results of the clustering are valid, we could read the documents that are grouped together to determine whether or not the clusters make sense (i.e., documents in each cluster discuss the same topic).

The **Rand index** and variants [Rand, 1971; Fowlkes and Mallows, 1983] were developed as a way to compare two partitions of the same data set. Since the cluster *labels* are arbitrary, the Rand index looks at pairs of points and how they are partitioned into the two groupings. There are two ways in which the i-th and j-th point could be grouped similarly, in which case the groupings agree with each other. They agree if the i-th and j-th points are in the *same* cluster in both groupings. The groupings are also in agreement if the i-th and j-th observations are in *different* clusters in both groupings.

The Rand index calculates the proportion of the total of n choose 2 objects that agree between the two groupings. It is a measure of similarity between the groupings. It is bounded below by zero when the groupings are not similar and have an upper bound of one when the groupings are exactly the same. Note that one could use the Rand index to compare the grouping to known class labels, if available.

It has been shown that the Rand index increases as the number of clusters increases, which can be a problem. Also, the possible range of values for the Rand index can be very narrow, in spite of the bounds given above. Hubert and Arabie [1985] developed an *adjusted Rand index* that addresses these issues. The adjusted Rand index provides a standardized measure such that its expected value is zero when the partitions are selected at random and one when the partitions match completely. We provide functions for calculating both of these measures.

Previously, we saw that the silhouette plot is a graphical way to assess cluster quality. Other graphical methods could be used such as the dendrogram, scatterplots, the rectangle cluster plot, and others. See Martinez et al. [2010] for more information on these methods and other graphical approaches.

11.7 MATLAB® Code

We provide a function called **cshmeans** that implements the HMEANS method given above. We also have a function called **cskmeans** that checks to see if moving individual observations changes the error. With both of these functions, the user can specify the initial centers as an input argument. However, if that argument is omitted, then the function will randomly pick the initial cluster centers.

We provide functions for the Rand index and the adjusted Rand index that were taken from the EDA Toolbox. These are called **randind** and **adjrand**, respectively. The **mojenaplot** function will construct a plot using the standardized fusion levels to help choose the number of clusters. We also include several functions that pertain to model-based clustering, such as **agmclust** for model-based agglomerative clustering, **mbcfinmix** that runs the model-based EM algorithm to find the mixture parameters, and **mbclust** that implements the entire model-based clustering methodology. See Table 11.2 for a list of relevant functions in the Computational Statistics Toolbox.

The MATLAB Statistics Toolbox has several functions for clustering. In this chapter, we showed how to use **pdist**, **linkage**, **cluster**, and **cophenet** functions. There are other clustering capabilities that the data analyst might find useful. One pertains to the function **cluster**, which is used to divide the output of linkage into clusters. It can do this in one of two ways: (1) by finding the natural divisions or (2) by the user specifying a number of clusters (as we did in this chapter). The function **clusterdata** combines the capabilities of the three functions: **pdist, linkage,** and **cluster** into one function.

The function **inconsistent** helps the user find natural divisions in the data set by comparing the length of the links in a cluster tree with the lengths of neighboring links. If the link is approximately the same as its neighbors, then it exhibits a high level of consistency. If not, then they are considered to be inconsistent. Inconsistent links might indicate a division of the data.

A **gmdistribution** constructor is available in the Statistics Toolbox. This will create Gaussian mixture models with specified means, covariances, and mixing proportions. There are several methods that can be used with this type of object, including the function **cluster**. See Table 11.3 for a list of relevant functions in the Statistics Toolbox.

TABLE 11.2

Functions in the Computational Statistics Toolbox

Purpose	MATLAB Function
Clustering	`cshmeans`
	`cskmeans`
	`agmclust`
	`mbcfinmix`
	`mbclust`
Assessing clusters	`mojenaplot`
	`adjrand`
	`randind`

TABLE 11.3

Statistics Toolbox Functions for Clustering

Purpose	MATLAB Function
Clustering	`kmeans`
	`linkage`
	`cluster`
	`dendrogram`
	`clusterdata`
	`gmdistribution`
Helper functions	`pdist`
	`squareform`
	`cophenet`
	`inconsistent`
	`silhouette`

11.8 Further Reading

Many books are available that describe clustering techniques, and we mention a few of them here. The books by Hartigan [1975], Spath [1980], Anderberg [1973], Kaufman and Rousseeuw [1990], and Jain and Dubes [1988] provide treatments of the subject at the graduate level. Most of the books mentioned in the text on statistical pattern recognition discuss clustering also. For example, see Duda and Hart [1973], Duda, Hart, and Stork [2001], Ripley [1996], or Webb [2011]. For two books that are appropriate at the undergraduate level, we refer the reader to Everitt [1993] and Gordon [1999]. The text by Martinez et al. [2010] has more detail on many of the approaches described in this chapter.

See Bock [1996] for a survey of cluster analysis in a probabilistic and inferential framework. Finite mixture models have been proposed for cluster analysis by Edwards and Cavalli-Sforza [1965], Day [1969], Wolfe [1970], Scott and Symons [1971], and Binder [1978]. Later researchers recognized that this approach can be used to address some of the issues in cluster analysis we just discussed [Fraley and Raftery, 2002; Everitt, Landau, and Leese, 2001; McLachlan and Peel, 2000; McLachlan and Basford, 1988; Banfield and Raftery, 1993].

Survey papers on clustering are also available. A recent one that provides an overview of clustering from the machine learning viewpoint is Jain, Murty, and Flynn [1999]. A discussion of clustering from an exploratory data analysis point of view is by Dubes and Jain [1980]. An early review paper on grouping is by Cormack [1971], where he provides a summary of distances, clustering techniques, and their various limitations. An interesting position paper on clustering algorithms from a data mining point of view is Estivill-Castro [2002].

Milligan and Cooper [1985] provide an excellent survey and analysis of procedures for estimating the number of clusters. One of the successful ones in their study uses a statistic based on the within-group sum-of-squares, which was developed by Krzanowski and Lai [1988]. Tibshirani, et al. [2001] develop a new method called *prediction strength* for validating clusters and assessing the number of groups, based on cross-validation. Bailey and Dubes [1982] present a method called *cluster validity profiles* that quantify the interaction between a cluster and its environment in terms of its compactness and isolation, thus providing more information than a single index would. Finally, Gordon [1994] proposes a U-statistic based on sets of pairwise dissimilarities for assessing the validity of clusters.

Exercises

11.1.   The **household** [Hand, et al., 1994; Aitchison, 1986] data set contains the expenditures for housing, food, other goods, and services (four expenditures) for households comprised of single people. Apply the clustering methods of this chapter to see if there are two groups in the data, one for single women and one for single men. To check your results, the first 20 cases correspond to single men, and the last 20 cases are for single women.

11.2.   The **measure** [Hand, et. al., 1994] data set contains 20 measurements of the chest, waist, and hip. Half of the measured individuals are women, and half are men. Use the techniques in this chapter to see if there is evidence of two groups.

11.3.   Use the online **help** to find out more about the MATLAB Statistics Toolbox functions **cluster** and **inconsistent**. Use these to find clusters in the **iris** data set. Assess your results.

11.4.   Apply k-means clustering to the complete **bank** data, without class labels. Apply the hierarchical clustering methods to the data. Is there significant evidence of two groups?

11.5.   Compare the two models in Example 11.7 and 11.8 using the cluster membership uncertainty. You will need to use the **mixclass** function to get the uncertainty in the cluster membership for the best model (model six, three groups).

11.6.   Using the **genmix** GUI, generate 2D data ($n = 500$) for each of the major finite mixture models (see Table 11.1). Use the model-based clustering methodology to group the data and assess your results.

11.7.   Cluster the **iris** data using hierarchical clustering with various types of linkage and distances. Use the silhouette method to estimate the number of groups in the data. Is there evidence of chaining with single linkage?

11.8.   Repeat Example 11.3 using other distances and types of linkage. Compare the results.

11.9.   The Mojena plot of Example 11.9 shows another elbow at five clusters for the **insect** data. Get the five clusters (see Example 11.4) and show the different groups in a scatterplot. Are five clusters reasonable?

11.10. Get the silhouette plot and the average silhouette value using the cluster labels for three groups (Example 11.4) and five groups (Problem 11.9) of the **insect** data. Discuss the results.

11.11. Invoke the **help** feature on the functions for Rand indexes (**adjrand** and **randind**) **to find out how they work.** Use these to assess all of the groupings we had for the **insect** data by comparing them with the true groups.

11.12. Write a MATLAB function that implements the Calinski and Harabasz index for determining the number of clusters and apply it to the **insect** data.

11.13. Write a MATLAB function that implements the Hartigan index for determining the number of clusters and apply it to the **insect** data.

11.14. Apply the clustering methods of this chapter to find groups (if any) in the following data sets: **flea, household,** and **posse**.

11.15. Agglomerative clustering is sometimes used to obtain a starting point for *k*-means clustering. Write a MATLAB function that will run agglomerative clustering, find *k* clusters, and estimate starting points for each of the *k* clusters. Apply this to the data sets mentioned in the previous problem.

Chapter 12

Parametric Models

12.1 Introduction

Chapter 8 briefly introduced the concepts of linear regression and showed how cross-validation can be used to find a model that provides a good fit to the data. We return to linear regression in this section to help introduce other parametric models for estimating relationships between variables. We first revisit classical linear regression, providing more information on how to analyze and visualize the results of the model. We will also examine more of the capabilities available in MATLAB® for this type of analysis. In Section 12.2, we present spline regression models. Logistic regression is discussed in Section 12.3. Logistic regression is a special case of generalized linear models, which are presented in Section 12.4. We discuss model selection and regularization methods in Section 12.5 and conclude the chapter with a discussion of partial least squares regression in Section 12.6.

Recall from Chapter 8 that one model for linear regression is

$$Y = \beta_0 + \beta_1 X + \beta_2 X^2 + \ldots + \beta_D X^D + \varepsilon, \tag{12.1}$$

where D represents the degree of the polynomial. This is sometimes known as *simple regression* because we have one predictor Y and one response variable X. We follow the terminology of Draper and Smith [1981], where the word *linear* refers to the fact that the model is linear with respect to the coefficients, β_j. This does not mean that we are restricted to fitting only straight lines to the data.

The model given in Equation 12.1 can be expanded to include multiple predictors $X_j, j = 1, \ldots, d$. An example of this type of model is

$$Y = \beta_0 + \beta_1 X_1 + \ldots + \beta_d X_d + \varepsilon. \tag{12.2}$$

The model in Equation 12.2 is often referred to as *multiple regression*, and it reflects the simplest case for this class of models. For example, we could include interaction terms such as $\beta_j X_k X_m$ in the model. For the most part, we will be focusing on the simple regression case for the remainder of this chapter, but it should be clear from the context what model we are referring to.

In parametric linear regression, we can model the relationship using any combination of predictor variables, order (or degree) of the variables, etc. and use the least squares approach to estimate the parameters. Note that it is called *parametric* because we are assuming an explicit model for the relationship between the predictors and the response. Several *nonparametric* methods will be presented in the next chapter.

To make our notation consistent, we will use the matrix formulation of linear regression for the model in Equation 12.1. Let $\mathbf{Y}$ be an $n \times 1$ vector of observed values for the response variable. The matrix $\mathbf{X}$ is called the *design matrix*, where each row of $\mathbf{X}$ corresponds to one observation evaluated using the specified model (e.g., Equation 12.1). $\mathbf{X}$ has dimension $n \times (D + 1)$, where we have $D + 1$ columns to accommodate a constant term in the model. Thus, the first column of $\mathbf{X}$ is a column of ones. In general, the number of columns in $\mathbf{X}$ depends on the chosen parametric model (e.g., the number of predictor variables, cross terms, and degree) that is used.

We can write the model in matrix form as

$$\mathbf{Y} = \mathbf{X}\boldsymbol{\beta} + \boldsymbol{\varepsilon}, \tag{12.3}$$

where $\boldsymbol{\beta}$ is a vector of parameters to be estimated, and $\boldsymbol{\varepsilon}$ is an $n \times 1$ vector of errors, such that

$$E[\boldsymbol{\varepsilon}] = \mathbf{0}$$

$$V(\boldsymbol{\varepsilon}) = \sigma^2 \mathbf{I}.$$

The least squares solution for the parameters can be found by solving the so-called *normal equations*. This solution, denoted by $\hat{\boldsymbol{\beta}}$, is given by

$$\hat{\boldsymbol{\beta}} = (\mathbf{X}^T \mathbf{X})^{-1} \mathbf{X}^T \mathbf{Y}, \tag{12.4}$$

assuming the matrix inverse $(\mathbf{X}^T \mathbf{X})^{-1}$ exists. Please see any multivariate analysis book, such as Draper and Smith [1981], for a derivation of least squares regression. We also discuss this briefly in Chapter 8.

The matrix formulation given above can represent either simple regression or multiple regression. If we are using a simple polynomial regression model, then we have one predictor variable X (e.g., time), and the columns of $\mathbf{X}$ might correspond to X (time), X^2 (time-squared), X^3 (time-cubed), along with the constant (a column of ones). With multiple regression, we have different predictor variables, e.g., time, price, and age. In this case, the columns of **X** might represent X_1 (time), X_2 (price), X_3 (age), interaction terms, and the constant.

The solution formed by the parameter estimate $\hat{\beta}$ obtained using Equation 12.4 is valid in that it is the solution that minimizes the error sum-of-squares regardless of the distribution of the errors. However, normality assumptions (for the errors) must be satisfied if one is conducting hypothesis testing or constructing confidence intervals that depend on these estimates. Generalized linear models enables us to relax some of these assumptions, as we will see later in this chapter.

Example 12.1

In this example, we explore two ways to perform least squares regression in MATLAB. The first way is to use Equation 12.4 to explicitly calculate the inverse. The data in this example were used by Longley [1967] to verify the computer calculations from a least squares fit to data. They can be downloaded from

http://www.itl.nist.gov/div898/strd/

The data set contains six predictor variables, so the model follows the one in Equation 12.2:

$$y = \beta_0 + \beta_1 x_1 + \beta_2 x_2 + \beta_3 x_3 + \beta_4 x_4 + \beta_5 x_5 + \beta_6 x_6 + \varepsilon.$$

We added a column of ones to the original data to allow for a constant term in the model. The following sequence of MATLAB code obtains the parameter estimates using Equation 12.4:

```
load longley
bhat = inv(X'*X)*X'*Y;
```

The results are

-3482258.65, 15.06, -0.04, -2.02, -1.03, -0.05, 1829.15

A more efficient way to get the estimates is to use MATLAB's backslash operator "\". Not only is the backslash more efficient, it is less prone to numerical problems. When we try it on the **longley** data, we see that the parameter estimates match. The command

```
bhat = X\Y;
```

yields the same parameter estimates.
❏

Recall that the purpose of regression is to model the relationship between the independent (or predictor) variable and the dependent (or response) variable. We obtain the model by finding the values of the parameters that minimize the sum of the squared errors. Once we have such a model, we can use it to predict a value of y for a given x.

After we build our model, it is important to assess the model's fit to the data and to see if any of the assumptions are violated. For example, the least squares method assumes that the errors are normally distributed with the same variance. To determine whether these assumptions are reasonable, we can look at the difference between the observed Y_i and the predicted value $\hat{Y}_i$ that we obtain from the fitted model. These differences are called the *residuals* and are defined as

$$\hat{\varepsilon}_i = Y_i - \hat{Y}_i; \qquad i = 1, ..., n, \tag{12.5}$$

where Y_i is the observed response at X_i, and $\hat{Y}_i$ is the corresponding prediction at X_i using the model. The residuals can be thought of as the observed errors.

We can use the visualization techniques of Chapter 5 to make plots of the residuals to see if the assumptions are violated. For example, we can check the assumption of normality by plotting the residuals against the quantiles of a normal distribution in a q-q plot. If the points fall (roughly) on a straight line, then the normality assumption seems reasonable. Other ways to assess normality include a histogram (if n is large), box plots, etc.

Another and more common method of examining the residuals using graphics is to construct a scatterplot of the residuals against the fitted values. Here the vertical axis units are given by the residuals $\hat{\varepsilon}_i$, and the fitted values $\hat{Y}_i$ are shown on the horizontal axis. If the assumptions are correct for the model, then we would expect a horizontal band of points with no patterns or trends. We do not plot the residuals versus the observed values Y_i because they are correlated [Draper and Smith, 1981], while the $\hat{\varepsilon}_i$ and $\hat{Y}_i$ are not. We can also plot the residuals against the X_i, which is called a *residual dependence plot* [Cleveland, 1993]. If this scatterplot shows a relationship between the residuals (the remaining variation not explained by the model) and the predictor variable, then the model is inadequate and adding additional columns in the X matrix is indicated. These ideas are explored further in the exercises.

Example 12.2

The purpose of this example is to illustrate another method in MATLAB for fitting polynomials to data, as well as to show what happens when the model is not adequate. We use the function **polyfit** to fit polynomials of various degrees to data where we have one predictor and one response. Recall that the function **polyfit** takes three arguments: a vector of measured values of the predictor, a vector of response measurements, and the degree of the polynomial. One of the outputs from the function is a vector of estimated parameters. Note that MATLAB reports the coefficients in descending powers: $\hat{\beta}_D, ..., \hat{\beta}_0$. We use the **filip** data in this example, which can be downloaded from **http://www.itl.nist.gov/div898/strd/**. Like the **longley** data, this data set is used as a standard to verify the results of least squares regression. The model we use for these data is

$$y = \beta_0 + \beta_1 x + \beta_2 x^2 + ... + \beta_{10} x^{10} + \varepsilon.$$

We first load up the data and then naively fit a straight line. We suspect that this model will not be a good representation of the relationship between x and y.

```
load filip
% This loads up two vectors: x and y.
[p1,s] = polyfit(x,y,1);
% Get the curve from this fit.
yhat1 = polyval(p1,x);
plot(x,y,'k.',x,yhat1,'k')
```

By looking at **p1** we see that the estimates for the parameters are a y-intercept of 1.06 and a slope of 0.03. A scatterplot of the data points, along with the estimated line are shown in Figure 12.1. Not surprisingly, we see that the model is not adequate. Next, we try a polynomial of degree $d = 10$.

```
[p10,s] = polyfit(x,y,10);
% Get the curve from this fit.
yhat10 = polyval(p10,x);
plot(x,y,'k.',x,yhat10,'k')
```

The curve obtained from this model is shown in Figure 12.2, and we see that it is a much better fit. The reader will be asked to explore these data further in the exercises.

❑

The standard MATLAB program has an interface that can be used to fit curves. It is only available for 2D data (i.e., fitting Y as a function of one predictor variable X). It enables the user to perform many of the tasks of curve fitting and interpolation (e.g., choosing the degree of the polynomial, plotting the residuals, annotating the graph, etc.) through one graphical interface. The **Basic Fitting** interface is enabled through the **Figure** window **Tools** menu. To activate this graphical interface, plot a 2D curve using the **plot** command (or something equivalent) and click on **Basic Fitting** from the **Figure** window **Tools** menu.

The MATLAB Statistics Toolbox has an interactive graphical tool called **polytool** that allows the user to see what happens when the degree of the polynomial that is used to fit the data is changed.

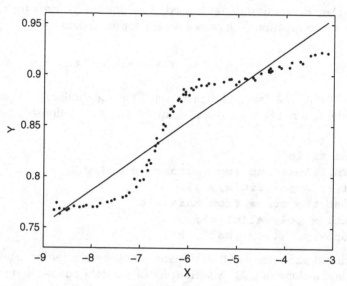

FIGURE 12.1
This shows a scatterplot of the **filip** *data, along with the resulting line obtained using a polynomial of degree one as the model. It is obvious that this model does not result in an adequate fit.*

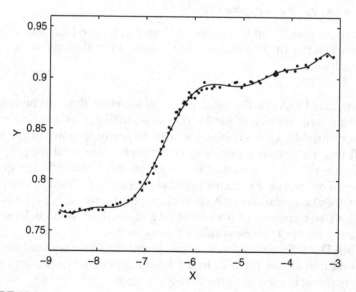

FIGURE 12.2
In this figure, we show the scatterplot for the **filip** *data along with a curve using a polynomial of degree ten as the model.*

12.2 Spline Regression Models

In the last example, we showed how to fit a polynomial regression model, and we saw that a higher degree polynomial improved our estimated model. One issue with this approach is that collinearity between the terms can appear as more of them are added to the polynomial. *Collinearity* means that there is a perfect or nearly perfect linear relationship between two or more of the monomials (terms) in the polynomial. In multiple regression, this is called *multicollinearity*, and it refers to a linear relationship between two or more predictors. *Spline regression* models are one way to achieve the desired flexibility that enables us to adequately model the relationship between the predictor and the response while avoiding collinearity. Some good references for the following discussion on splines are Marsh and Cormier [2002], Wahba [1990], Hastie and Tibshirani [1990], and de Boor [2001].

Regression splines use a piecewise polynomial model to find the fit between the predictor and the response variable. The regions over which the polynomials are defined are given by a sequence of points called *knots*. The set of *interior knots* will be denoted by t_i, where $t_1 < \ldots < t_K$. To this set of knots, we add the *boundary knots* denoted by t_0 and t_{K+1}.

As we will see shortly, depending on the model, different conditions are satisfied at the knot points. For example, in the piecewise linear case, the function is continuous at the knot points, but the first (and higher) derivative is discontinuous. In what follows, we assume that the number of knots and their locations are known. If the knot locations are unknown, then the spline regression model is nonlinear in its parameters. In this case, one would have to use a nonlinear estimation method, such as nonlinear least squares. If the number and location of knots must be estimated, then one could use a method such as stepwise regression. We do not address these situations in this text and refer the reader to Marsh and Cormier [2002] or Lee [2002] for details on these approaches and others.

The spline model is given by

$$f(X) = \beta_0 + \sum_{j=1}^{D} \beta_j X^j + \sum_{j=1}^{K} \beta_{j+D} \delta(t_j)(X - t_j)^D, \qquad (12.6)$$

where

$$\delta(t_j) = \begin{cases} 0; & X \le t_j \\ 1; & X > t_j. \end{cases}$$

The quantity $\delta(t_j)$ is sometimes known as a ***dummy variable*** often used in regression analysis to distinguish different groups. They allow us to switch off the parameters in our model, yielding the desired piecewise polynomial. It should be clear from Equation 12.6 that our function $f(X)$ is a linear combination of monomials.

Another way to represent this model is to use a truncated power basis of degree D:

$$f(X) = \beta_0 + \beta_1 X + \dots + \beta_D X^D + \sum_{j=1}^{K} \beta_{j+D}(X - t_j)_+^D, \qquad (12.7)$$

where $(a)_+$ is the positive part of a, which has the same effect as our dummy variable in Equation 12.6 [Lee, 2002]. Other basis functions such as B-splines or radial basis functions can be used [Green and Silverman, 1994], but the forms shown in Equations 12.6 and 12.7 are easier to understand. We show in Example 12.3 how to use Equation 12.6 for the piecewise linear case.

Example 12.3

We will use simulated data to illustrate spline regression when our true function is piecewise linear [Marsh and Cormier, 2002]:

$$f(x) = \begin{cases} 55 - 1.4x; & 0 \le x \le 12 \\ 15 + 1.8x; & 13 \le x \le 24 \\ 101 - 1.7x; & 25 \le x \le 36 \\ -24 + 1.7x; & 37 \le x \le 48. \end{cases} \qquad (12.8)$$

Given these equations, we would have three internal knots (approximately) at 12, 24, and 36. We first generate some noisy data according to these equations.

```
x1 = linspace(0,12,15);
y1 = 55 - 1.4*x1 + randn(size(x1));
x2 = linspace(13,24,20);
y2 = 15 + 1.8*x2 + randn(size(x2));
x3 = linspace(25,36,20);
y3 = 101 - 1.7*x3 + randn(size(x3));
x4 = linspace(37,48,15);
y4 = -24 + 1.7*x4 + randn(size(x4));
x = [x1, x2, x3, x4];
y = [y1, y2, y3, y4];
```

Next, we find the values of the dummy variables, recalling that dummy variables are zero when x is less than or equal to the corresponding knot. We implement this in MATLAB as follows:

```
% Get the values of the dummy variables.
D1 = ones(size(x));
D1(x <= 12) = 0;
D2 = ones(size(x));
D2(x <= 24) = 0;
D3 = ones(size(x));
D3(x <= 36) = 0;
```

The next step is to obtain the matrix **X**. This will contain a column of ones to represent the constant term, a column of the x observations to represent $\beta_1 x$, and three columns corresponding to the dummy variable terms.

```
% Next get the X matrix.
coL1 = D1.*(x-12);
coL2 = D2.*(x-24);
coL3 = D3.*(x-36);
n = length(x);
X = [ones(n,1),x(:),coL1(:),coL2(:),coL3(:)];
```

We can now solve for the coefficients β_j using

```
beta = X\y(:);
```

This yields a vector with five elements corresponding to $\hat{\beta}_0, \hat{\beta}_1, ..., \hat{\beta}_4$, the last three of which are the estimated coefficients for the dummy variable terms. The estimates for the coefficients are

$$54.66, \ -1.45, \ 3.30, \ -3.56, \ 3.14.$$

We now show how to obtain the polynomial for each interval. All of the dummy variables are equal to zero before the first knot at $t_1 = 12$, so our estimated polynomial is

$$\hat{f}(x \mid 0 \leq x \leq 12) = \hat{\beta}_0 + \hat{\beta}_1 x \tag{12.9}$$

In the second interval between t_1 and t_2, the first dummy variable is non-zero, but the other are still zero. We only need to add this first dummy term, so our estimated polynomial on this second interval is

$$\hat{f}(x \mid 12 < x \leq 24) = \hat{\beta}_0 + \hat{\beta}_1 x + \hat{\beta}_2(x - t_1)$$

$$= \hat{\beta}_0 + \hat{\beta}_1 x + \hat{\beta}_2 x - \hat{\beta}_2 t_1 \tag{12.10}$$

$$= (\hat{\beta}_0 - \hat{\beta}_2 t_1) + (\hat{\beta}_1 + \hat{\beta}_2)x.$$

Similarly, we add another dummy variable for the next interval between t_2 and t_3 to get:

$$\hat{f}(x|24 < x \le 36) = \hat{\beta}_0 + \hat{\beta}_1 x + \hat{\beta}_2(x - t_1) + \hat{\beta}_3(x - t_2)$$

$$= (\hat{\beta}_0 - \hat{\beta}_2 t_1 - \hat{\beta}_3 t_2) + (\hat{\beta}_1 + \hat{\beta}_2 + \hat{\beta}_3)x. \qquad (12.11)$$

The final interval is obtained in the same way. The following MATLAB code finds the polynomial for each of the intervals. Note that in MATLAB we represent these as row vectors where the elements are the coefficients of the monomials in descending powers. Thus, the coefficient for the constant term is the last element of the vector.

```
% Get the lines over the intervals.
% Polynomial for interval 1.
p1 = [beta(2), beta(1)];
% Polynomial for interval 2.
p2 = [beta(2) + beta(3), beta(1)-12*beta(3)];
% Polynomial for interval 3.
p3 = [beta(2)+beta(3)+beta(4), ...
      beta(1)-12*beta(3)-24*beta(4)];
% Polynomial for interval 4.
p4 = [beta(2)+beta(3)+beta(4)+beta(5), ...
      beta(1)-12*beta(3)-24*beta(4)-36*beta(5)];
```

These give us our estimated polynomials as

```
p1 = -1.45 54.66
p2 = 1.85 15.082
p3 = -1.71 100.42
p4 = 1.43 -12.78,
```

which seem reasonable when we compare them with Equation 12.8. We are now ready to evaluate the polynomials using the **polyval** function and to produce the plot in Figure 12.3.

```
% Get x, y values using the estimated polynomials
% to plot construct the plot.
y1hat = polyval(p1,x1);
y2hat = polyval(p2,x2);
y3hat = polyval(p3,x3);
y4hat = polyval(p4,x4);
yhat = [y1hat(:);y2hat(:);y3hat(:);y4hat(:)];
plot(x,y,'.',x,yhat)
xlabel('X'),ylabel('Y')
title('Simulated Data with Spline Regression Fit')
```

As expected, the fit is reasonable, except maybe on the last interval. While we know what the true piecewise polynomial is (Equation 12.8), it might be more visually appealing to use higher degree spline regression models. The reader is asked to explore these possibilities in the exercise.
❑

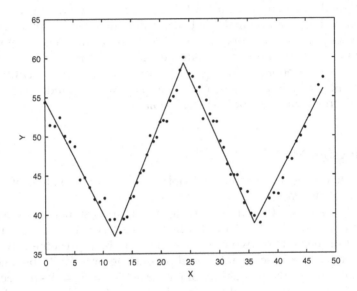

FIGURE 12.3
This is the spline regression fit and scatterplot for the simulated data of Example 12.3. We used a piecewise linear fit over each of the intervals.

We used piecewise *linear* regression in the previous example, and we can see from Figure 12.3 that the function is continuous at the knot points, but the first derivative is not continuous. This means that the slopes are different over each interval. If we use higher degree spline regression, such as quadratic or cubic, then we will have smoother functions because additional constraints are placed on higher derivatives at the knot points. In the quadratic case, we have a continuous function with continuous slopes (first derivative) at the knots. For cubic splines, we have a continuous function with continuous first and second derivatives. So, spline regression of degree $D \geq 1$ will produce continuous functions that have continuous derivatives up to order $D - 1$ at the knots. Of course, we can also have the piecewise constant case, where $D = 0$. Here the function and all derivatives are discontinuous at the knots.

12.3 Logistic Regression

In the previous discussions on linear regression, we have been focusing on situations where the response variable is continuous. However, there are many processes that produce outcomes better described by a dichotomous (two-valued or binary) variable. For example, in medical sciences, the probability that an individual will become ill with a disease can be assessed by detecting the presence of one or more risk factors. In the field of finance, it is critical to determine if a customer is a credit risk or not based on predetermined economic red flags. In general, we can classify entities as to their membership in a class or not. A popular way to model and analyze binary phenomena is via *logistic regression*.

Creating the Model

We denote our binary response variable as Y, where the values of Y might represent a customer's financial profile (not a credit risk = 0, credit risk = 1), a patient's diagnosis (no cancer = 0, cancer = 1), a consumer's preference for a product (brand A = 0, a brand different from A = 1), etc. That is, each observation has one of two outcomes that can be represented by $Y = 0$ (failure) or $Y = 1$ (success). We will denote the probability of success as $P(Y = 1) = p$ and the probability of failure as $P(Y = 0) = 1 - p$. Recall that such a binary response would be called a Bernoulli random variable. If we have n independent observations of such a random variable, then the number of successes has the binomial distribution with parameters n and p.

We seek to describe the relationship between p and our explanatory variable x. As in our previous approaches, we could define our regression model for a binary response as

$$P(Y = 1) = p(x) = \beta_0 + \beta_1 x. \tag{12.12}$$

To simplify the discussion, only one explanatory variable is used in the following explanation. However, keep in mind that we could have multiple predictor variables [Agresti, 2002].

Although similar to the linear regression equations from previous sections, the model in Equation 12.12 presents some difficulties in its use. For example, $p(x)$ is a probability and can only take on values between zero and one. However, the linear relationship shown in Equation 12.12 does not have such constraints and can assume values greater than one and less than zero. Additionally, assumptions used with linear regression, such as normality and constant variance, do not apply to binary responses. As we have already discussed, a binomial distribution is more appropriate in this case.

The relationship between the predictor x and the response $p(x)$ is usually nonlinear because a fixed change in x has a different impact as p approaches zero or one [Pampel, 2000]. Agresti [1996, 2002] states that, in practice, the relationship is often monotonically increasing ($\beta_1 > 0$) or decreasing ($\beta_1 < 0$) with S-shaped curves yielding a reasonable shape for the relationship. The following logistic regression function provides this S-shaped form:

$$p(x) = \frac{\exp(\beta_0 + \beta_1 x)}{1 + \exp(\beta_0 + \beta_1 x)}. \tag{12.13}$$

The shape of the curve for the logistic regression function shows that the rate of change of $p(x)$ is low when the values of x are close to zero and to one. An example of such a curve is given in Figure 12.4. The fastest rate of change happens when x is close to the mid-range. We should also note that Equation 12.13 implies that $p(x)$ can only take on values between zero and one, which is what we want.

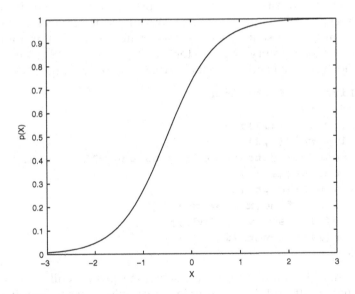

FIGURE 12.4
This shows an example of an S-shaped curve found using Equation 12.13 with $\beta_1 > 0$.

Equation 12.13 has the success probability p as a response, but the right-hand side is nonlinear. We can put this into a more familiar linear form by using the *logit* of the success probability:

$$\text{logit}[p(x)] = \log\left\{ \frac{p(x)}{1 - p(x)} \right\} = \beta_0 + \beta_1 x. \tag{12.14}$$

The logit in the above model is also known as the log of the odds function, where the *odds* is the success probability divided by the probability of failure.

The two models (Equations 12.13 and 12.14) provide modeling approaches with different interpretations of the parameters. We will address this issue in more detail after we see how logistic regression can be accomplished using functions in the Statistics Toolbox.

It is also important to note that the right-hand side of Equation 12.14 can be expanded in the same manner as regular linear regression. Thus, we can include higher order monomials (x^2, x^3, etc.), more than one predictor, and interaction terms.

Example 12.4

We will use a data set from Dobson [2002] to demonstrate logistic regression. These data comprise a sample of older people who were tested using a subset of the Wechsler Adult Intelligent Scale (WAIS). The testers recorded whether or not symptoms of senility were present (success, $y = 1$) or not (failure, $y = 0$). The explanatory or predictor variable is the WAIS score. First, we have to load the data and group them according to the unique WAIS scores.

```
% First load the data.
load wais
xall = wais(:,1);
yall = wais(:,2);
% Group the data according to the WAIS score.
x = unique(xall);
for i = 1:length(x)
    ind = find(x(i)== xall);
    y(i) = sum(yall(ind));
    n(i) = length(ind);
end
```

We now have the number of people showing signs of senility (success) for each of the 17 different WAIS scores in this sample. As we will see in the next section, we could use generalized linear models to find the logistic regression fit. However, we can also use the multinomial regression model. This model assumes that there are m categories for the response variable, and it reduces to the binomial case (success or failure) when $m = 2$. The Statistics Toolbox has a function called **mnrfit** that will fit nominal or ordinal multinomial regression. *Nominal* variables represent classifications where ordering does not make sense (bike/car/bus or democrat/republican/independent), while *ordinal* categories have some order to them (low/medium/high). The nominal model is the default for **mnrfit**, but you can specify the ordinal model (see the **help** documentation on the function for more information). One can also specify a link function for ordinal models. We will discuss link functions in the next section on generalized linear models. The following syntax will return the estimated coefficients (**B**), the deviance statistic (**dev**), and a structure with several useful quantities (**stats**).

```
% We can use the multinomial fit with two
% groups to get our logistic regression fit.
[B,dev,stats] = mnrfit(x(:),[y(:), n(:)-y(:)]);
```

Note that the inputs to the **mnrfit** function are the explanatory values and the response. The response is a matrix with m columns, where each column contains the number of successes in each of the m categories and for each of the unique x values. The estimated coefficients are

$$B = 2.4040 \quad -0.3235$$

This shows that $\hat{\beta}_0 = 2.4040$ and $\hat{\beta}_1 = -0.3235$. We can use the function **mnrval** to get estimates of the success probabilities using the logistic regression fit represented by our vector **B**.

```
% Now find the observed success probabilities for
% each WAIS score.
prop = y./n;
% Find estimates of the success probability.
propfit = mnrval(B,x(:));
```

We show these results in Table 12.1. We can also show this graphically, as the following code illustrates.

```
% Find fitted values for a range of x values and plot
% the curve with the observations.
xfit = linspace(min(x), max(x), 30);
propfitp = mnrval(B,xfit(:));
plot(x,prop,'o',xfit(:),propfitp(:,1));
axis([3 21 -.1 1.1])
xlabel('WAIS Score')
ylabel('Proportion')
title('Proportion with Symptoms of Senility')
```

The plot is shown in Figure 12.5, where we see that the curve obtained from the fitted model seems to be reasonable. Other diagnostic tools are available, and we will present some of these in the next section on generalized linear models and in the exercises.
❑

We do not address the issue of how one obtains parameter estimates for the coefficients in the logistic regression model in Equation 12.14, except to note that maximum likelihood is used rather than least squares. Most statistical

TABLE 12.1

Number of elderly with symptoms of senility (y) and their WAIS score (x)

Explanatory Variable (x)	Response Variable (y)	Observed Relative Frequencies (p)	Estimated Success Probability ($\hat{p}$)
4	1	0.5	0.752
5	1	1.0	0.687
6	1	0.5	0.614
7	2	0.67	0.535
8	2	1.0	0.454
9	2	0.33	0.376
10	1	0.17	0.303
11	1	0.17	0.240
12	0	0	0.186
13	1	0.17	0.142
14	2	0.29	0.107
15	0	0	0.080
16	0	0	0.059
17	0	0	0.043
18	0	0	0.032
19	0	0	0.023
20	0	0	0.017

packages use a technique called iterative weighted least squares [Nelder and Wedderburn, 1972]. This is a root-finding method that finds the estimated coefficients for any generalized linear model (logistic regression is part of this family).

Interpreting the Model Parameters

Interpreting the coefficients in Equation 12.14 can be difficult. As in linear regression, the slope β_1 represents the change in the left-hand side quantity for a one-unit change in the variable on the right-hand side. In the case of the logistic regression model, the units of the dependent variable represent the log odds, which is not very meaningful for researchers. This section explores the various ways in which one can interpret the coefficients in these models [Pampel, 2000; Agresti, 2002]: log odds, odds, and probabilities.

The first way we can interpret the logistic regression coefficients is directly using Equation 12.14. As in linear regression, the effects of the explanatory variable on the log odds (right side of Equation 12.14) are additive and linear. So, the coefficient β_1 represents the change in the predicted logit for a one-unit change in x (e.g., at $x + 1$). In the case of the WAIS score data in Example 12.4, the estimated coefficient is $\hat{\beta}_1 = -0.3235$. This means that an increase of

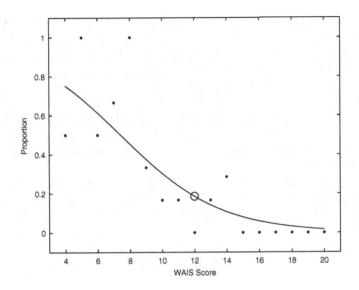

FIGURE 12.5
This shows the proportion of elderly with symptoms of senility as a function of their WAIS score. The observed relative frequencies are displayed as points. The smooth curve was obtained using the fitted logistic regression. The circle on the curve shows the estimated probability at the mean WAIS score (x).

one on the WAIS score decreases the logged odds of showing symptoms of senility by approximately 0.32.

From Equation 12.14, we can exponentiate both sides of the equation to get an expression relating the coefficients to the odds:

$$\frac{p(x)}{1-p(x)} = \exp\{\beta_0 + \beta_1 x\}. \qquad (12.15)$$

Thus, the odds are an exponentiated function of x, where a unit change in x changes the odds by a factor of e^{β_1}. In other words, the odds at $x + 1$ are the odds at x multiplied by e^{β_1}:

$$e^{\beta_0} e^{\beta_1(x+1)} = e^{\beta_0} e^{\beta_1 x} e^{\beta_1} = e^{\{\beta_0 + \beta_1 x\}} e^{\beta_1}.$$

This shows that the effects of the explanatory variable on the odds are multiplicative rather than additive, as we had with the log odds. Because of this, an exponentiated coefficient equal to one corresponds to a slope of zero in the additive case:

$$e^{\beta_1} = e^0 = 1.$$

An exponentiated coefficient (e^{β_1}) greater than one increases the odds, while an exponentiated coefficient less than one decreases the odds. For the WAIS senility data, we have $e^{\beta_1} = 0.72$, so the odds of having symptoms of senility are decreased by a factor of 0.72 when the WAIS score increases by one.

We can also interpret the logistic regression coefficients in terms of the probabilities by using the model in Equation 12.13. In this case, the relationship between the explanatory variable and the probabilities is nonlinear and nonadditive, as we see in Figure 12.4. Thus, the effect of the regression coefficients on probabilities depends on specific values of x or where we are on the curve.

One way to understand the effect of a continuous explanatory variable on the probability is to determine the slope of a line that is tangent to the curve (Equation 12.13 and Figure 12.4). For a logistic parameter β_1, this slope is given by

$$\beta_1 p(x)[1 - p(x)]. \qquad (12.16)$$

Because of the S shape of the curve, the slope defined by Equation 12.16 tends towards zero as the probability approaches zero or one. It reaches its highest value when $p(x) = 0.5$.

Returning to Example 12.4, the average WAIS score is 12, and the estimated probability for symptoms of senility at this value is $\hat{p}(x = 12) = 0.19$. The incremental rate of change in the estimated probability at $x = 12$ is

$$\hat{\beta}_1 \hat{p}(x)[1 - \hat{p}(x)] = (-0.32)(0.19)(0.81) = -0.049.$$

The estimated probability at $x = 12$ is shown as a circle on the curve in Figure 12.5.

We did not address the issue of assessing the model's goodness of fit using summary statistics, hypothesis testing, or residual visualization. We leave the discussion of these areas for the next section because the model approach described there encompasses logistic regression as a special case.

12.4 Generalized Linear Models

The logistic regression example of the previous section illustrated a situation where the normal-theory linear model should not be expected to provide satisfactory results because it is an inappropriate model for the data. Another happens when the response variable is a count of the number of occurrences of an event, as in Poisson regression. Nelder and Wedderburn [1972] provided the theory for a unified modeling framework that will handle the normal case, as well as many common nonnormal error distribution and mass functions. This methodology is called *generalized linear models* (GLMs).

In this section, we will follow the notation and treatment of GLMs found in Gill [2001] and Dobson [2002]. First, we provide some background theory on the exponential family of distributions, which is key to unifying the various error functions and models. Then, we will give examples of the exponential family form for the binomial, the Poisson, and the normal. This is followed by a discussion of the generalized linear model and information on MATLAB code for finding the fit in a Poisson regression example. Finally, we present several methods for assessing the goodness-of-fit and exploring the fitted model via the residuals.

Exponential Family Form

The key to generalized linear models is the *exponential family form*, which provides a common notation and enables a unified methodology for many regression modeling applications. Most common probability distribution functions and probability mass functions can be written in this form. They include the binomial, the Poisson, the negative binomial, the normal, and the gamma probability distribution. Recall that the gamma includes the chi-square and the exponential distributions as special cases.

Let us say we have a one-parameter (ζ) probability density (or mass) function for a random variable X that can be written in the form

$$f(x|\zeta) = r(x)s(\zeta)\exp[t(x)u(\zeta)], \qquad (12.17)$$

where r and t are real-valued functions that do not depend on the parameter ζ, and the real-valued functions s and u do not depend on x. Additionally, we have $r(x) > 0$ and $s(\zeta) > 0$.

Note that we can move the first two factors of Equation 12.17 into the exponent as follows

$$\begin{aligned} f(x|\zeta) &= r(x)s(\zeta)\exp[t(x)u(\zeta)] \\ &= \exp[\log\{r(x)\}] \times \exp[\log\{s(\zeta)\}] \times \exp[t(x)u(\zeta)] \quad (12.18) \\ &= \exp[\log\{r(x)\} + \log\{s(\zeta)\} + t(x)u(\zeta)], \end{aligned}$$

where $\log(*)$ is the natural logarithm. The first two terms in the exponent of Equation 12.18, $\log\{r(x)\} + \log\{s(\zeta)\}$ are known as the *additive component* because these are additive with respect to the random variable and the parameter. The remaining term $t(x)u(\zeta)$ is called the *interaction component*, as this is reminiscent of interaction terms in the usual linear regression context.

If $t(x) = x$, then the probability function is in canonical form for the random variable. Similarly, when $u(\zeta) = \zeta$, then it is in canonical form for the parameter. We can convert to the canonical form by transformations:

$$y = t(x)$$
$$\theta = u(\zeta).$$

As we will see, it is usually not necessary to apply transformations because the canonical form exists for many commonly-used probability density (or mass) functions. Thus, we will use the following canonical or standard form in the remainder of the section:

$$f(y|\theta) = \exp[y\theta - b(\theta) + c(y)]. \tag{12.19}$$

The term $b(\theta)$ in Equation 12.19 is called the **normalizing constant**, and it is a function of the parameter only, so it can be used to ensure that it is a valid probability function. McCullagh and Nelder [1989] call it the **cumulant function**. It is also known as the **natural parameter** of the distribution. Another important component of this equation is the form of θ, as this gives us the **canonical link**, which is a key ingredient of generalized linear models. In fact, it allows us to connect the linear or systematic component of our regression model to the nonnormal response variable.

Sometimes the distribution of interest has two parameters: one for location and one for scale. In these cases, the exponential form for Equation 12.19 can be written as

$$f(y|\theta) = \exp\left[\frac{y\theta - b(\theta)}{a(\psi)} + c(y, \psi)\right].$$

If the probability density (mass) function does not have this parameter, then $a(\psi) = 1$. The quantity $a(\psi)$ is also known as the **dispersion parameter**.

Gill [2001] shows that the expected value of Y for the exponential family of probability functions (Equation 12.19) can be found using

$$E[Y] = \mu = \frac{\partial}{\partial\theta}b(\theta). \tag{12.20}$$

He also shows that the variance of Y depends on $b(\theta)$ as follows

$$Var[Y] = a(\psi)\frac{\partial^2}{\partial\theta^2}b(\theta). \tag{12.21}$$

Equations 12.20 and 12.21 illustrate how the exponential family form unifies the theory and treatment for many probability density (mass) functions because they can be used to find the familiar expected value and variance for distributions in this family.

We give examples of some common probability functions that belong to the exponential family: the binomial, the Poisson, and the normal distribution.

Binomial

Suppose our random variable Y represents the number of successes in n trials. Then Y has the binomial distribution with parameters n and p, where p is the probability of success. This is given by

$$f(y|n, p) = \binom{n}{y} p^y (1 - p)^{n-y}, \qquad y = 0, 1, 2, \dots, n.$$

We can put this in the form we want by first exponentiating the factors and rearranging as follows:

$$
\begin{aligned}
f(y|n, p) &= \binom{n}{y} p^y (1 - p)^{n-y} \\
&= \exp\left[\log\binom{n}{y}\right] \times \exp[\log(p^y)] \times \exp[\log(1 - p)^{n-y}] \\
&= \exp\left[\log\binom{n}{y}\right] \times \exp[y\log(p)] \times \exp[(n - y)\log(1 - p)] \\
&= \exp\left[\log\binom{n}{y} + y\log(p) + n\log(1 - p) - y\log(1 - p)\right] \\
&= \exp\left[y\log\left(\frac{p}{1 - p}\right) - [-n\log(1 - p)] + \log\binom{n}{y}\right].
\end{aligned}
$$

From this we have the canonical link given by

$$\theta = \log\left(\frac{p}{1 - p}\right). \tag{12.22}$$

The natural parameter corresponds to the term

$$b(\theta) = -n\log(1 - p). \tag{12.23}$$

Note that here and what follows, we are treating n as a nuisance parameter.

The expression for the natural parameter in Equation 12.23 is not in terms of the parameter θ on the right-hand side, so we need to use some algebra to get $b(\theta)$ in the desired form. Using properties of logarithms, we can write Equation 12.23 as

$$b(\theta) = n\log\left(\frac{1}{1-p}\right).$$

If we add and subtract p to the numerator of the argument in the logarithm, we have

$$b(\theta) = n\log\left(\frac{1-p+p}{1-p}\right)$$

$$= n\log\left(\frac{1-p}{1-p} + \frac{p}{1-p}\right) = n\log\left(1 + \frac{p}{1-p}\right).$$

We convert the second term of the logarithm to the form of θ, as follows:

$$b(\theta) = n\log\left(1 + \frac{p}{1-p}\right).$$

$$= n\log\left(1 + \exp\left[\log\left(\frac{p}{1-p}\right)\right]\right)$$

$$= n\log(1 + \exp[\theta]).$$

Thus, we have the natural parameter given by $b(\theta) = n\log(1 + \exp[\theta])$.

Poisson

The Poisson distribution is used when the random variable Y is the number of occurrences of some event (or counts). This distribution is given by

$$f(y|\mu) = \frac{\mu^y e^{-\mu}}{y!}, \qquad y = 0, 1, 2, \ldots,$$

where the parameter μ is the mean number of occurrences.

Using properties of logarithms and some algebra, we can convert this to the exponential form, as follows:

$$f(y|\mu) = \frac{\mu^y e^{-\mu}}{y!}$$

$$= \exp(-\mu) \times \exp(\log[\mu^y]) \times \exp(-\log[y!])$$

$$= \exp[y\log(\mu) - \mu - \log(y!)].$$

Comparing this with Equation 12.19, we see that the canonical link is

$$\theta = \log(\mu), \qquad (12.24)$$

with the natural parameter given by

$$b(\theta) = \mu. \qquad (12.25)$$

Once again, this is not in the standard form as a function of θ. However, we can solve for μ in Equation 12.24 to get $\mu = \exp(\theta)$. Substituting this into Equation 12.25 yields

$$b(\theta) = \exp(\theta).$$

Normal Distribution

Our examples so far have been single-parameter probability distributions. The standard expression for the exponential family form can be expanded to include multiparameter models, as follows:

$$f(y|\Theta) = \exp\left[\sum_{j=1}^{k} y\theta_j - b(\theta_j) + c(y) \right],$$

where $\Theta = (\theta_1, ..., \theta_k)^T$ is a vector of parameters. The univariate normal is an example of such a distribution, where

$$\Theta = (\mu, \sigma^2)^T.$$

We will treat the scale parameter of the normal distribution as a nuisance parameter and focus on the location parameter μ. In this case, the exponential form for the normal distribution is derived as follows:

$$f(y|\mu, \sigma^2) = \frac{1}{\sqrt{2\pi\sigma^2}} \exp\left[-\frac{1}{2\sigma^2}(y-\mu)^2\right]$$

$$= \exp\left[\log\left\{\frac{1}{\sqrt{2\pi\sigma^2}}\right\} - \frac{1}{2\sigma^2}(y-\mu)^2\right].$$

We use properties of logarithms to adjust the first term in the exponent and also expand the second term to get

$$f(y|\mu, \sigma^2) = \exp\left[-\frac{1}{2}\log(2\pi\sigma^2) - \frac{1}{2\sigma^2}(y^2 - 2y\mu + \mu^2)\right].$$

Finally, we rearrange terms to match up with the canonical form, which gives us

$$f(y|\mu, \sigma^2) = \exp\left[\frac{y\mu - \mu^2/2}{\sigma^2} - \frac{1}{2}\left\{\frac{y^2}{\sigma^2} + \log(2\pi\sigma^2)\right\}\right]. \tag{12.26}$$

From Equation 12.26, we have the canonical link given by $\theta = \mu$ (the identity function), with the natural parameter given as $b(\theta) = \mu^2/2$. Substituting for μ yields

$$b(\theta) = \frac{\theta^2}{2}.$$

See Gill [2001] for a derivation of the exponential family form when the scale parameter σ^2 is the one of interest. Gill also provides derivations of the exponential form for the gamma and the negative binomial distributions.

Generalized Linear Model

We are now ready to discuss generalized linear models. Because they are an extension of the standard linear model, we once again begin our explanation with that model. Recall that, using matrix notation, the observational model can be written as

$$\mathbf{Y} = \mathbf{X}\beta + \varepsilon, \tag{12.27}$$

where $\mathbf{Y}$ is a vector of the independent and identically distributed responses, and ε is the error or the random component. The term $\mathbf{X}\beta$ is sometimes known as the *linear structure*. Alternatively, we can write the expectation model as

$$E[\mathbf{Y}] = \theta = \mathbf{X}\beta. \tag{12.28}$$

Our familiar assumptions regarding the standard linear model are that the Y_i are normally distributed with mean μ_i and variance σ^2. Furthermore, this implies that the errors ε_i are normally distributed with mean zero and variance σ^2.

The assumption of normality for the response variable is too restrictive, as we saw in the case of logistic regression. So, we generalize the standard linear model with a new predictor that uses the mean of the response variable. This generalization is given by

$$g(\mu) = \theta = \mathbf{X}\beta, \tag{12.29}$$

where g is a smooth, invertible function of the mean vector μ called the *link function*. In generalized linear models, the linear predictor $\mathbf{X}\beta$ is connected to the mean of the response variable and not directly to the outcome variable itself, as in Equation 12.27.

The generalized linear model consists of three components: the stochastic component, the systematic component, and the link function. Definitions for these are given next [Gill, 2001].

Stochastic Component

This is sometimes known as the *random component* or the *error structure*, and it corresponds to the response variable Y. The Y_i are still independent and identically distributed, but they are assumed to be distributed according to a specific exponential family distribution (e.g., normal, binomial, Poisson) with mean μ. Thus, generalized linear models include the case where the stochastic component is normally distributed (the standard linear model).

This adds another layer of complexity to our data analysis because we now have to choose a distribution for the response variables. Our knowledge of some exponential family forms can help us in our decision. For example, if our response variable is binary (e.g., 0 or 1, success or failure), then we might assume a binomial distribution for the random component. This was the case in the previous section on logistic regression. If the Y_i are continuous, then the assumption of normality might be appropriate. Or, it could be Poisson when we have counts as our response variable, as illustrated in Example 12.5.

Systematic Component

The systematic component $\mathbf{X}\beta$ of a generalized linear model is given by

$$\theta = X\beta,$$

and it specifies the linear predictor θ. The systematic component $X\beta$ identifies the variables used on the right-hand side of Equation 12.29. Note that they affect the responses Y_i only through the form of g.

Link Function

This function g is what connects the stochastic component and the systematic component (Equation 12.29). In most cases, it is given by the canonical link function in the exponential family form. We saw examples of these for the normal, the binomial, and the Poisson. The link function is important because it connects the possibly nonnormal response variable to standard linear model theory. This means that the response variable from the exponential family is affected by the predictor variables only through the link function, as follows

$$g^{-1}(\theta) = g^{-1}(X\beta) = \mu = E[Y].$$

We summarize the link functions for some exponential family distributions in Table 12.2.

Gill [2001] makes the point that most texts describe generalized linear models in terms of these three components. However, he states that there is also a fourth component that should be considered—the residuals. As in standard linear model theory, the residuals are a key component used to evaluate the quality and fit of the resulting generalized linear model.

To fit the models, we need to find estimates of the coefficients β_i. We do not provide details on how to do this in generalized linear models because most statistics packages (including the MATLAB Statistics Toolbox) have functions that will fit the model for many common exponential family forms. The method of maximum likelihood is used to estimate the parameters, but closed form expressions for these estimates do not exist. Thus, some sort of iterative algorithm for solving the nonlinear maximum likelihood equations has to be used. For more information on estimation for generalized linear models, see Nelder and Wedderburn [1972] or Gill [2001].

Example 12.5

For this example, we use a data set that was analyzed in Agresti [2002] to illustrate generalized linear models when the response variable consists of counts. These data come from a study of nesting horseshoe crabs (*Limulus polyphemus*) [Brockmann, 1996]. Each of the female crabs had a male crab attached to her in the nest. Some had other male crabs surrounding the nesting pair, which are called *satellites*. Brockmann found that the satellite males were fertilizing a large number of the eggs showing that they were in competition with the attached male. They also found that female crabs who

TABLE 12.2

Generalized Linear Models – Example Distributions

Distribution	Canonical Link $\theta = g(\mu)$	Inverse Link $\mu = g^{-1}(\theta)$ $= g^{-1}(X\beta)$	Natural Parameter $b(\theta)$
Binomial (logit)	$\log\left(\frac{p}{1-p}\right)$	$\frac{\exp(\theta)}{1+\exp(\theta)}$	$b(\theta) = n\log[1 + \exp(\theta)]$
Normal	μ	θ	$b(\theta) = \theta^2/2$
Poisson	$\log(\mu)$	$\exp(\theta)$	$b(\theta) = \exp(\theta)$

were in better condition were more likely to attract more satellites. Characteristics such as color, spine condition, weight, carapace width, and number of satellites were observed for female crabs. The first can serve as the explanatory variables, while the number of satellites represents the response. As in Agresti, we will look at a univariate case using the carapace width as the only explanatory variable. First, we load the data and show a plot of the width of the female crab and the number of satellites in Figure 12.6.

```
load crab
% The width of the female crab is in
% the first column of X.
plot(X(:,1),satellites,'.')
axis([20 34 -0.5 16])
xlabel('Carapace Width (cm)')
ylabel('Number of Satellites')
```

It makes sense to use the Poisson distribution for the response variable because they correspond to counts. If μ denotes the expected value for a Poisson random variable Y, then we have the *Poisson loglinear model*:

$$\log(\mu) = \beta_0 + \beta_1 x.$$

We find the fit using the **glmfit** function in the MATLAB Statistics Toolbox.

```
% Find the fit using a function from
% the MATLAB Statistics Toolbox.
beta = glmfit(X(:,1),satellites,'poisson');
```

The estimated coefficients yield the following model:

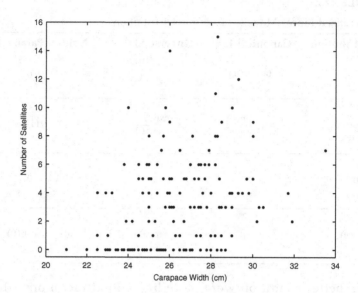

FIGURE 12.6
This is a scatterplot of the carapace width (cm) of female crabs and the number of satellites.

$$\log(\hat{\mu}) = \hat{\beta}_0 + \hat{\beta}_1 x = -3.305 + 0.164x.$$

This indicates that the estimated effect of carapace width on the number of satellites is positive. We can use the model fit to find the estimated mean number of satellites for any carapace width using

$$\hat{\mu} = \exp(\hat{\beta}_0 + \hat{\beta}_1 x) = \exp(-3.305 + 0.164x).$$

Note that the expression given above is the inverse of the Poisson link function. The **glmval** function will return the desired estimate for a given x using the coefficient estimates from **glmfit**. The function requires the user to specify the link function. Since this is Poisson regression, the default link function in **glmfit** is the log link.

```
% Use the glmval function in the Statistics
% Toolbox to find the estimated mean number of
% satellites for various values of x.
x0 = linspace(min(X(:,1)),max(X(:,1)),50);
yhat = glmval(beta,x0,'log');
```

We plot these results in Figure 12.7, where we see the estimated positive effect of width on the mean number of satellites.

❏

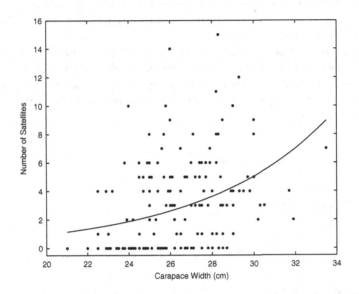

FIGURE 12.7
The curve shows the estimated mean number of satellites from the generalized linear model fit using a Poisson distribution for the response variable.

Model Checking

The variety of response variable distributions available in generalized linear models makes the assessment of model fit more difficult. Several approaches have been developed, but one has to be careful how they are applied and interpreted because conclusions can depend on the distribution type, sample size, observed counts, and more.

We will give a brief overview of the main approaches, and we encourage the reader to consult McCullagh and Nelder [1989], Gill [2001], and Dobson [2002] for more information on the derivations and the underlying theory. In what follows, we will discuss some statistics that can be used to assess the model's goodness of fit. This is followed by a short discussion of a Wald confidence interval for the estimated parameters. Finally, we will also provide information on several types of residuals, such as Pearson residuals, the Anscombe residuals, and the deviance residuals. As with standard linear models, these can be explored graphically.

Goodness of Fit

There are several goodness of fit statistics for generalized linear models. We first focus on one called the *deviance* and briefly mention another based on the Pearson residuals.

A common approach that can be used to assess our model fit is to compare it to another model that uses the same probability distribution and link function, but includes a different number of parameters or systematic component. For example, a generalized linear model for a binomial response variable might be

$$\log\left(\frac{p}{1-p}\right) = \beta_0, \tag{12.30}$$

and another could be

$$\log\left(\frac{p}{1-p}\right) = \beta_0 + \beta_1 x.$$

The model shown in Equation 12.30 is sometimes known as the *minimal model* [Dobson, 2002].

Another baseline model used in generalized linear models is known as the *saturated* or *maximal model*. Here, we specify a different parameter for each of the unique n observed predictors. This means that there could be up to n different parameters. We can then compare our model with one of these baseline models using a test statistic and its associated sampling distribution.

We will use N to represent the maximum number of parameters that can be estimated in our specified generalized linear model. The estimated parameter vector obtained via maximum likelihood for the saturated model will be denoted by $\hat{\beta}_{max}$. Then the likelihood function for the saturated model evaluated at $\hat{\beta}_{max}$ is represented by $L(\hat{\beta}_{max}; \mathbf{y})$. We can write the likelihood ratio λ as

$$\lambda = \frac{L(\hat{\beta}_{max}; \mathbf{y})}{L(\hat{\beta}; \mathbf{y})},$$

where the denominator is the likelihood function evaluated at the maximum likelihood estimate for our specified model of interest.

The logarithm of the likelihood ratio λ is more useful. So, using properties of logarithms, this produces the difference between the log likelihood functions:

$$\log(\lambda) = \log[L(\hat{\beta}_{max}; \mathbf{y})] - \log[L(\hat{\beta}; \mathbf{y})]. \tag{12.31}$$

If $\log(\lambda)$ is large, then this implies that the specified model is not an adequate description or representation of the data, relative to the saturated model.

In order to use Equation 12.31 as a goodness of fit statistic, we need to know its sampling distribution. It can be shown that $2 \cdot \log(\lambda)$ is distributed as a chi-squared random variable [Dobson, 2002]. In general, this statistic can be used to assess and compare the fit of two models. Nelder and Wedderburn [1972] named this statistic (based on Equation 12.31) the *deviance* for the model of interest. Thus, the deviance is given by

$$D = 2\{\log[L(\hat{\beta}_{max};y)] - \log[L(\hat{\beta};y)]\}.$$

Given a large sample size, the sampling distribution for the deviance is (approximately):

$$D \sim \chi^2(N - k), \tag{12.32}$$

where k is the number of parameters in our specified model of interest. If the response variable is normally distributed, then the distribution for the deviance in Equation 12.32 is exact. Large values of the deviance reside in the tails of the distribution. Thus, we might conclude that models with large deviances fit the data poorly when we use the deviance as a measure of the goodness of fit. A rule of thumb [Gill, 2001] states that the deviance should not be substantially larger than the degrees of freedom $N - k$.

Another statistic that can be used to assess the model fit is the *Pearson statistic*, which is the sum of the squared Pearson residuals:

$$X^2 = \sum \mathbf{R}^2_{Pearson}.$$

We will define and describe the residuals in more detail later in this section. This statistic is also distributed as chi-squared with $N - k$ degrees of freedom. Like the deviance, the distribution is exact for response variables that are normally distributed. McCullagh and Nelder [1989] and Gill [2001] urge users to be careful using both of these statistics when the response variable is nonnormal and the sample size is small.

Wald Confidence Interval

As we saw in Example 12.5, we can get maximum likelihood estimates of the model parameters, but it is often more informative to construct confidence intervals for them. Several types of confidence intervals are available for generalized linear models. However, we will discuss the simplest of these, which is known as called the *Wald confidence interval*.

The Wald $100 \times (1 - \alpha)$ % confidence interval [Agresti, 2002] is given by

$$\hat{\beta} \pm z^{(1-\alpha/2)} \times SE(\hat{\beta}),$$

where $\hat{\beta}$ is the parameter estimate, and $SE(\hat{\beta})$ is its standard error. (Recall that for a 95% confidence interval we have $z^{(1-\alpha/2)} = 1.96$.) The **glmfit** function will return the standard errors in a structure. We show how to construct confidence interval estimates in the next example.

Other confidence intervals in generalized linear models include one that is based on the likelihood ratio. However, this one is not as easy as the Wald interval to calculate. For more information on this type of interval, and its use in logistic and Poisson regression, see Agresti [2002].

Residuals

We have already learned about the importance of residuals with the standard linear model. We would like to use residuals from generalized linear models in a similar way. However, since these new models apply to a large class of distributions for the response variable, the residuals in generalized linear models are often not normally distributed.

Some of the residuals that have been developed for generalized linear models were constructed in the hopes of producing residuals that are close to being normally distributed with mean zero. There are [Gill, 2001] five major types of residuals for generalized linear models: response, Pearson, Anscombe, deviance, and working. We briefly define the first four types of residuals because these are the ones returned by **glmfit**.

The *response residual* is similar to the familiar residuals in standard linear regression. We calculate it, as follows

$$\mathbf{R}_{\text{Response}} = \mathbf{Y} - g^{-1}\left(\mathbf{X}\hat{\beta}\right) = \mathbf{Y} - \hat{\mu}.$$

This residual is often not very useful with generalized linear models because it is more likely to produce residuals that deviate quite a bit from those for a normal variate with zero mean. Thus, response residuals are often not informative.

Another type of residual is the *Pearson residual*. Recall that this was used to calculate the Pearson goodness of fit statistic. The Pearson residual is the raw response residual divided by the standard deviation of the prediction:

$$\mathbf{R}_{\text{Pearson}} = \frac{\mathbf{Y} - \hat{\mu}}{\sqrt{\text{Var}(\hat{\mu})}}.$$

Pearson residuals can sometimes have the desired normal distribution. However, they often have a very skewed distribution, so using these to check the resulting model fit can be misleading.

The next type of residual we will discuss is the **Anscombe residual**. The basic notion behind this type of residual is to transform the residuals so the skewness is lessened. Thus, the result will be approximately symmetric and unimodal. The Anscombe residuals are different for the various exponential family forms, so we do not provide details here. See Gill [2001] for a table of Anscombe residuals for the common exponential family distributions.

Finally, we have the **deviance residuals**, which are perhaps the ones used most often. The deviance residual is given by

$$\mathbf{R}_{\text{Deviance}} = \frac{(y_i - \hat{\mu}_i)}{\left| y_i - \hat{\mu}_i \right|} \sqrt{d_i} \, ,$$

where the first factor $(y_i - \hat{\mu}_i) / \left| y_i - \hat{\mu}_i \right|$ preserves the sign. The quantity d_i is the contribution of the i-th point to the deviance. As with the Anscombe residuals, there is a different closed-form deviance function for common distributions of the exponential family. Gill [2001] gives a table of deviance functions for some of the popular distributions, and he shows how they depend on the cumulant function, $b(\theta)$.

Pierce and Schafer [1986] provide a detailed study of the performance of residuals for generalized linear models, and they recommend the use of the deviance residuals. Gill [2001] likes both the deviance and the Anscombe residuals, as they generally produce residuals that are normally distributed with mean zero.

Several types of model-checking plots can be constructed. These include plotting the residuals versus the response, the residuals versus explanatory variables (X_i), a normal probability plot of the residuals, residuals against the fitted values, fitted values versus their quantile equivalent values, and residuals against transformed fitted values. However, it is important to keep in mind that residuals from generalized linear models do not *have* to be normally distributed with zero mean. Instead, we should be looking for systematic patterns in the distribution, as these could indicate a poor choice of link function, scale problems, or model misspecification.

In our list of plots given above, we mentioned one that had residuals plotted against transformed fitted values. McCullagh and Nelder [1989] and Nelder [1990] make the recommendation that one first transform the fitted values to the **constant-information scale** of the distribution and then plot the transformed values against the residuals. The goal of these transformations is to lessen the degree of curvature in the residual plots. The transformations for the normal, binomial, and Poisson distributions are shown in Table 12.3.

TABLE 12.3

Transformations to Constant-Information Scale

Distribution	Transformation
Binomial	$2 \times \operatorname{asin} \sqrt{\hat{\mu}}$
Normal	$\hat{\mu}$
Poisson	$2 \times \sqrt{\hat{\mu}}$

TABLE 12.4

Beetle Mortality Data

Dose ($\log_{10}$), x_i	Number of Beetles, n_i	Number Killed, y_i
1.6907	59	6
1.7242	60	13
1.7552	62	18
1.7842	56	28
1.8113	63	52
1.8369	59	53
1.8610	62	61
1.8839	60	60

Example 12.6

We will use a data set from Dobson [2002] to illustrate some of the techniques we just discussed for checking the fit of generalized linear models. These data represent the number of dead beetles after they have been exposed for 5 hours to different concentrations of gaseous carbon disulphide [Bliss, 1935]. These data are shown in Table 12.4 and are in the file **beetle.mat**. First, we load the data and show a plot of the proportions in Figure 12.8.

```
load beetle
% First plot the proportions.
plot(x,y./n,'.')
xlabel('Dose (log_10)')
ylabel('Proportion Killed')
```

The logistic model (binomial distribution for the error) is appropriate for these data, so we can fit the model in Equation 12.14. This corresponds to the logit link in generalized linear models. This is the default link for the **glmfit**

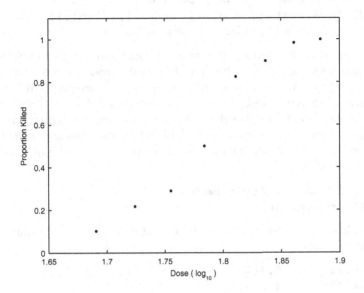

FIGURE 12.8
This is a plot of the proportion of beetles killed as a function of the quantity (log$_{10}$) of carbon disulphide.

function when we have the binomial distribution. Two other links can also be used with binomial response variables. One is called the *probit link*, which is one of the original models used for bioassay data [Dobson, 2002]. The probit model is based on the cumulative probability function for the standard normal distribution and is given by

$$\Phi^{-1}(p) = \beta_0 + \beta_1 x.$$

The other model we can use has a link function called the *complementary log log function*. This model is

$$\log[-\log(1-p)] = \beta_0 + \beta_1 x.$$

The **glmfit** function will estimate the parameters for these three links. We now show an alternative output syntax that will return several quantities that are useful for model checking. Note that the binomial **glmfit** requires a two-column array for the response. The first column is the count y_i, and the second column is the number of trials n_i.

```
% Get the model fits for the three binomial links.
[B1,dev1,stats1]=glmfit(x(:),[y(:) n(:)],'binomial');
[B2,dev2,stats2] = glmfit(x(:),[y(:) n(:)],...
```

```
      'binomial','link','probit');
[B3,dev3,stats3] = glmfit(x(:),[y(:) n(:)],...
      'binomial','link','comploglog');
```

The output array **B1** contains the parameter estimates in ascending order of the parameters, e.g., β_0, β_1. The output quantity **dev1** is the deviance of the fit. The output variable **stats1** is a structure with several useful statistics, such as residuals (response, Pearson, Anscombe, and deviance), the standard errors, and the degrees of freedom. See the **help** documentation on **glmfit** for information on the other fields and the field names that are in the **stats1** structure. One way to see the parameter estimates and their standard errors for the models is:

```
% This is the logit model:
[B1, stats1.se]
```

This yields the following parameters (first column) and standard errors (second column):

```
-60.7175    5.1807
 34.2703    2.9121
```

The estimated parameters, the standard errors, and the deviances for each model are shown in Table 12.5. We see from the deviance values that the complementary log log link has the lowest deviance, indicating a better fit than the others. We could plot the residuals against the fitted values, which are found using the **glmval** function.

TABLE 12.5

Estimated Binomial Models and Standard Errors (in parentheses) for the **beetle** Data

Quantity	Logit	Probit	CompLogLog
$\hat{\beta}_0$	-60.72 (5.18)	-34.94 (2.65)	-39.57 (3.24)
$\hat{\beta}_1$	34.27 (2.91)	19.73 (1.49)	22.04 (1.79)
Deviance	11.23	10.12	3.45

```
% Next we get the estimated fits for
% the three models.
fit1 = glmval(B1,x(:),'logit');
fit2 = glmval(B2,x(:),'probit');
fit3 = glmval(B3,x(:),'comploglog');
```

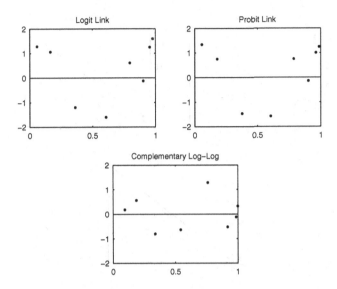

FIGURE 12.9
This shows the residuals plots for the three binomial models from Example 12.6. The fitted response values are shown along the horizontal axis, and the deviance residuals are along the vertical. We do not see indications of outliers or model misspecification.

As an example, we plot the deviance residuals against the fitted values for the logit model using

```
plot(fit1,stats1.residd,'.',[0 1],[0 0]);
```

The corresponding residual plot for all three models is shown in Figure 12.9. Another useful plot for model checking is one that shows the fitted values along the horizontal axis and the observed proportions on the vertical. If the fit is a good one, then the points should fall on the line. We use the following command to create one of these plots for the logit model:

```
plot(fit1,y(:)./n(:),'.',[0 1],[0 1]);
```

The corresponding plots for the three models are shown in Figure 12.10. The plots indicate that the complementary log log produces a better fit.
❏

12.5 Model Selection and Regularization

In Chapter 8, we introduced the standard linear model and discussed how the ordinary least squares approach can be used to estimate the model

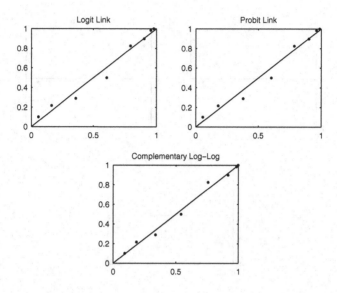

FIGURE 12.10
Each plot in this figure shows the fitted values along the horizontal axis and the observed proportions on the vertical axis. If the fit is a good one, then the points should fall on the straight line shown in the plots. The complementary log log link seems to produce a better model than the other link functions.

parameters by minimizing the residual squared error. The linear model for a response Y and multiple predictors is given by

$$Y = \beta_0 + \beta_1 X_1 + \ldots + \beta_d X_d .$$

We now look at how we might deal with issues, such as over-fitting, correlated terms (or colinearity), and what predictors should be included. We first describe approaches for selecting predictors, including best subset selection and stepwise regression. This is followed by several regularization or shrinkage methods—ridge regression, lasso, and elastic net.

Best Subset Selection

The best subset selection method seeks to find a good model by searching over models that have some specified number of predictors. For instance, we fit all models that contain just one of the predictors, then models with two of the predictors, and so on. The coefficients in each of the models are estimated using least squares, and the final model is chosen using cross-validation or some other approach. The process for models with no interaction terms is listed next [James et al., 2013].

PROCESS – BEST SUBSET SELECTION

1. Start with a model containing just the constant term: $\hat{y} = \hat{\beta}_0$. This results in an estimate based on the sample mean only. We will designate this model as M_0.

2. Set a looping variable $k = 1$. Then estimate all possible models that contain k predictors. We use least squares to build the models.

3. Select the best of these models using a criterion like R^2. This best model for iteration k will be denoted as M_k.

4. Repeat steps 2 and 3 for $k = 1, ..., d$. We now have $d + 1$ models: $M_0, M_1, ...M_d$.

5. Use cross-validation or some other criterion (e.g., adjusted R^2) to choose the best of these models.

In step 2, there are d choose k models. This can be a problem when d is very large. Furthermore, it should be noted that the models in the process above include just the predictors without interactions (product of two predictors) or higher-order terms. If these are also used, then the computational burden becomes even worse.

Stepwise Regression

Besides the computational issues, best subset selection can lead to overfitting. In other words, the model we choose could work well with the training data, but perform poorly with new observations. Stepwise regression is also a subset selection approach, and it attempts to address these issues. However, it is important to note that all non-penalized subset selection methods suffer from overfitting because the criteria used in the selection process tend to favor adding variables or terms to the model.

There are two versions of stepwise regression—forward and backward selection. *Forward stepwise regression* starts with the constant model (no predictors) as we had before. We loop through the predictors, adding a single one at each iteration. The model is fit using least squares. If there is significant evidence that the coefficient for the additional predictor is not zero, then we keep it in the model. We follow the same procedure with interaction terms, which are the product of two predictors. So, we go through all pair-wise interaction terms, adding one at a time to the model and then evaluating it.

Backward stepwise regression does the opposite. In this case, we start with the model containing all single predictors and interaction terms. We then remove individual terms in the current model based on their significance.

With forward selection, we start with the smallest model, so it tends to result in a smaller, less complex model. With backward selection, we start with a large model, and it is likely we will get a larger model than what we get with forward subset selection. Also, both of these approaches use greedy optimization and depend on the starting models. Thus, it is probably a good idea to try both methods and compare the results, as we do in the following example.

Example 12.7

The MATLAB Statistics Toolbox has a function for stepwise regression. It will do either forward regression starting with the constant model or backward regression from the full model. We use a data set downloaded from the UCI Machine Learning Repository, and it is in a file called **airfoil**. The data are from aerodynamic and acoustic tests of airfoil blade sections. The response variable is scaled sound pressure (decibels). The predictors are frequency (Hertz), angle (degrees), chord length (meters), free-stream velocity (meters per second), and suction side displacement thickness (meters). We first load the data file and create a data table.

```
load airfoil
% Create a data table.
air_tab = table(Freq,Angle,Chord,Veloc,...
Suction,Sound);
```

Next, we construct a forward stepwise regression model using the input argument **constant**.

```
% Apply stepwise regression with Sound
% as the response variable.
% This does forward stepwise regression.
model_fwd = stepwiselm(air_tab,'constant',...
'ResponseVar','Sound');
```

This shows the (partial) output from the function, so we can see the process of adding single predictors and pairwise interaction terms to the model.

```
1. Adding Freq, ..., pValue = 5.361777e-56
2. Adding Suction, ..., pValue = 1.51172e-75
3. Adding Chord, ..., pValue = 2.286595e-63
4. Adding Freq:Suction, ..., pValue = 4.450379e-51
5. Adding Freq:Chord, ..., pValue = 2.949524e-42
6. Adding Veloc, ..., pValue = 3.23181e-33
7. Adding Angle, ..., pValue = 1.17462e-13
8. Adding Veloc:Suction, ..., pValue = 0.0018742
9. Adding Chord:Veloc, ..., pValue = 0.0016615
```

Here are the estimated coefficients in the model, and we see that it has all five predictor linear terms and several interaction terms.

```
Estimated Coefficients:
```

	Estimate
(Intercept)	131.11
Freq	-0.00020705
Angle	-0.26303
Chord	-26.572
Veloc	0.039374
Suction	-147.28
Freq:Chord	-0.005075
Freq:Suction	-0.066984
Chord:Veloc	0.24493
Veloc:Suction	2.0927

We saved our model in the output variable **model_fwd**, and we can view the data object by typing the variable name in the workspace and hitting the enter key, just as we can do with any data object. MATLAB displays the final model formula, the estimated coefficients, errors, statistics, p-values, and R^2. This model had $R^2 = 0.632$, and an adjusted R^2 of 0.63. The adjusted R^2 accounts for the number of terms in the model. Now, let's try backwards stepwise regression, where we start with the full model.

```
% This is the backward stepwise regression starting
% with a model that has single predictors and
% all pair-wise interaction terms.
model_back = stepwiselm(air_tab,...
'interactions','ResponseVar','Sound');
```

This is displayed in the command window. We see that only one of the model terms is removed.

> **1. Removing Angle:Veloc, ..., pValue = 0.73542**

This model is much larger, and MATLAB tells us that it has 15 terms. We can view this model object by entering the name at the command line. MATLAB tells us that it has an adjusted R^2 of 0.634. It is only marginally better than the forward stepwise model using this statistic, so we might choose to go with the parsimonious model.

❑

Ridge Regression

The approaches for model-building we just described select subsets of predictors and interaction terms. We now discuss several techniques that impose a penalty for adding terms, which addresses overfitting. There are other methods that fit a model using all of the predictors. These regularize the model coefficients in a way that reduces the variance and improves the prediction performance. Regularization also has the effect of shrinking the estimated coefficients toward zero.

The first method we discuss is *ridge regression*. In ridge regression, we add a term to our objective function used to estimate the coefficients. We seek the estimates $\hat{\beta}_0$ and $\hat{\beta}_j$, $j = 1, ..., d$ that minimize the following

$$\sum_{i=1}^{n}\left(y_i - \beta_0 - \sum_{j=1}^{d}\beta_j x_{ij}\right)^2 + \lambda\sum_{j=1}^{d}\beta_j^2. \tag{12.33}$$

The non-negative parameter λ controls the amount of skrinkage and can be chosen using cross-validation. The second term imposes a penalty that depends on the skrinkage parameter λ. If it is equal to zero, then we have the usual least squares estimates for the model. As λ grows larger, the result approaches the null model with the constant term only. In other words, it shrinks the coefficients for the nonconstant terms to zero.

Ridge regression is particularly useful when terms in the model are correlated or equivalently, the columns of the design matrix $\mathbf{X}$ exhibit linear dependence. Note that highly correlated terms will lead to inflated variance. We discussed previously that the least-squares estimate for the parameters is given by

$$\hat{\beta}_{OLS} = (\mathbf{X}^T\mathbf{X})^{-1}\mathbf{X}^T y, \tag{12.34}$$

where $\mathbf{X}$ is the design matrix. The ridge regression estimate is

$$\hat{\beta}_{ridge} = (\mathbf{X}^T\mathbf{X} + \lambda\mathbf{I})^{-1}\mathbf{X}^T y.$$

If columns of $\mathbf{X}$ are linearly dependent (or close to), then the inverse in the least squares estimate (Equation 12.34) becomes singular. The matrix inverse in the ridge regression estimates is augmented by an amount controlled by the shrinkage parameter, which improves the condition of the matrix. This introduces some bias, but it gives us more efficient lower variance estimators.

Example 12.8

We are going to use a small data set to illustrate the use of ridge regression. The data set includes measurements of twenty-five brands of cigarettes, and it is often used in introductory statistics classes to help students understand multiple regression in the presence of collinearity [McIntyre, 1994]. The data were originally from the Federal Trade Commission who annually measured the weight, tar, nicotine, and carbon monoxide of domestic cigarettes. We are interested in modeling the relationship between three predictor variables—nicotine, tar, and weight and the response variable carbon monoxide. First, we load the data and then explore it using some summary statistics and scatterplots.

```
load cigarette
% Put predictors into a matrix.
X = [Nicotine, Tar, Weight];

% Look at the correlation coefficient matrix
% for the predictors.
corrcoef(X)
```

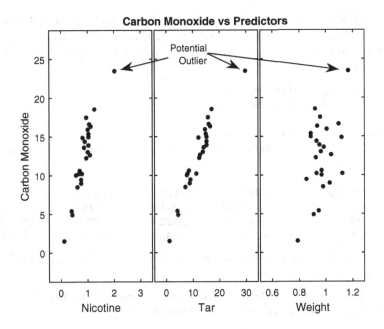

FIGURE 12.11
Here we have scatterplots showing the pairwise relationship between carbon monoxide and each predictor variable. We see a positive correlation between the individual predictors and the response variable. However, carbon monoxide seems to be less linearly correlated with the weight. We also note that McIntyre [1994] discusses the presence of a possible outlier shown in the upper right corner of each plot.

The correlation coefficient matrix for the predictor variables indicates that the variables **Nicotine** and **Tar** are correlated with a coefficient of 0.9766, as we see here.

```
1.0000    0.9766    0.5002
0.9766    1.0000    0.4908
0.5002    0.4908    1.0000
```

Pairwise scatterplots for the carbon monoxide (**CO**) and the three predictors are shown in Figure 12.11. We can see in the plot that carbon monoxide appears to be linearly related to each predictor. The correlation coefficients between the response (**CO**) are shown next.

0.9259   0.9575   0.4640

These indicate a strong positive correlation for **CO** with each of **Nicotine** and **Tar** and less of one with **Weight**, which has a correlation coefficient of 0.4640. Now, we go through the modeling process starting with a basic linear model estimated using least squares regression.

```
% Fit a linear model.
lm = fitlm(X,CO,'linear');
```

The model is shown here, and we see that only the second variable (**Tar**) is significant. Furthermore, the estimated coefficient for **Nicotine** is negative! This does not match the type of relationship we have in Figure 12.11. Here is the estimated model.

```
Linear regression model:
    y ~ 1 + x1 + x2 + x3

Estimated Coefficients:
                 Estimate      SE        tStat        pValue
                 _____    _____    _____    _____

(Intercept)       3.2022     3.4618      0.92502      0.36546
    x1           -2.6317     3.9006     -0.67469      0.50723
    x2            0.96257    0.24224      3.9736      0.00069207
    x3           -0.13048    3.8853     -0.033583     0.97353
```

Next, we try forward stepwise regression. It is no surprise that we get only **Tar** as a predictor.

```
% Stepwise regression - start with constant model.
lm_step1 = stepwiselm(X,CO,'constant');


% Here is the model.
Linear regression model:
    y ~ 1 + x2

Estimated Coefficients:
                 Estimate      SE        tStat        pValue
                 _____    _____    _____    _____

(Intercept)       2.7433     0.67521     4.0629     0.00048117
    x2            0.80098    0.05032     15.918     6.5522e-14
```

Ridge regression is particularly helpful when we have correlated predictors, as we have with these data. We are going to perform many ridge regressions for different values of λ.

```
% Set up some values for lambda.
```

```
% This is called k in the MATLAB documentation.
k = 0:1:30;
lm_ridge1 = ridge(CO,X,k);
```

In Figure 12.12, we plot the standardized coefficients as the ridge parameter λ gets large, so we can look for a value where the coefficients stabilize. It is also interesting to look closely at the smallest values of λ. We see that the coefficient starts out negative for **Nicotine** and then turns positive. You can use the following code to create the plot:

```
h = plot(k,lm_ridge1,'LineWidth',2);
set(h(1),'linestyle','--')
set(h(2),'linestyle','-.')
grid on
xlabel('Ridge Parameter')
ylabel('Standardized Coefficient')
title('{\bf Ridge Trace}')
legend('Nicotine','Tar','Weight')
```

If we extend the ridge traces out to $\lambda = 100$, then we will see the estimated coefficients for **Tar** and **Nicotine** merge to the same value. This is likely due to their high positive linear correlation. The standardized coefficients are useful for finding a value for the ridge parameter, but these cannot readily be used for prediction. The following MATLAB code can be used to find the estimated untransformed coefficients for our chosen λ.

```
% To get the coefficients for prediction purposes
% (i.e., nonstandardized and with the constant term),
% then set the 'scale' flag to 0.
% Let's use a lambda of 15.
lm_ridge2 = ridge(CO,X,15,0);

% These are the coefficients.
 lm_ridge2 =

    1.0762
    4.3257
    0.3119
    3.9686
```

❑

The estimated ridge regression coefficients are not scale invariant [James et al., 2013]. In other words, multiplying a predictor by a constant can change the estimate. Thus, standardizing the predictors to have zero mean and a standard deviation of one is a recommended first step. This is actually the default in the call to **ridge** in MATLAB. The coefficients are estimated after standardizing in this manner, as we saw in the example above.

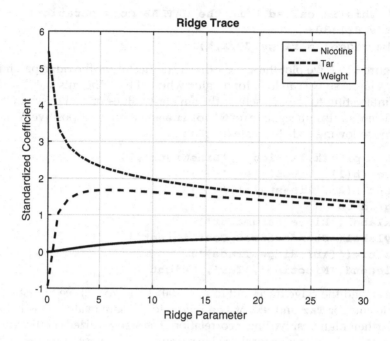

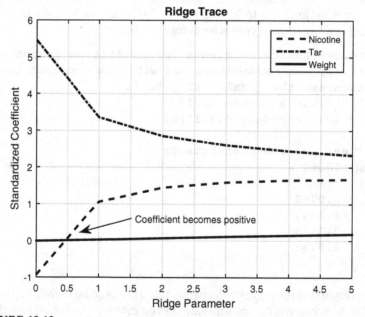

FIGURE 12.12
These are plots of the standardized coefficients versus the ridge parameter λ. We look for a value of λ where the estimated coefficients stabilize. We can see from the plot that ridge regression tends to shrink the coefficients. The values of the coefficients on the y-axis correspond to the least squares solution.

Lasso—Least Absolute Shrinkage and Selection Operator

Ridge regression tends to shrink the estimates of the coefficients to zero, and some can get quite small. However, they will not produce estimated coefficients exactly equal to zero unless λ is infinite. Thus, we keep all of the coefficients or terms in the model, unlike subset selection or stepwise regression. This can be problematic when we have a large number of dimensions d because it degrades the interpretability of the model.

The *lasso* method was developed by Tibshirani [1996] to address some of the problems with ridge regression and subset selection. The estimates using lasso are found using an optimization function, as we have with ordinary least squares and ridge regression. Like ridge regression, lasso includes a penalty term, but the penalty or constraint is based on the absolute value of the non-constant parameters in the model instead of the squared terms. Thus, lasso uses the L_1 penalty, while ridge regression uses the L_2 penalty.

The coefficients in lasso regression are found by minimizing the following

$$\sum_{i=1}^{n}\left(y_i - \beta_0 - \sum_{j=1}^{d}\beta_j x_{ij}\right)^2 + \lambda\sum_{j=1}^{d}|\beta_j|, \qquad (12.35)$$

where λ is a nonnegative parameter that controls the shrinkage. Both ridge regression and the lasso will shrink the coefficients toward zero. However, lasso forces some of the coefficients to equal zero as the shrinkage parameter gets large. Thus, it selects terms like stepwise regression does. It can produce a more parsimonious model, but without the tendency to overfit. Tibshirani's [1996] simulation studies showed that lasso also has some of the good properties of ridge regression like stability.

Example 12.9

We continue with the **cigarette** data to show how to do lasso regression in MATLAB. The following call to **lasso** produces a model and uses ten-fold cross validation, which will be used to find a value of λ.

```
[lm_lasso, fitinfoL] = lasso(X,CO,'CV',10,...
    'PredictorNames',{'Nicotine','Tar','Weight'});
```

We can construct a trace plot. This is somewhat similar to the ridge trace plots in the previous example, in that it helps us understand what is happening with the estimated coefficients as the value of the parameter λ changes. The following MATLAB code creates the plot shown in Figure 12.13.

```
h = lassoPlot(lm_lasso,fitinfoL,...
    'PlotType','Lambda','XScale','log');
set(h.Children(3),'linestyle','--')
set(h.Children(2),'linestyle','-.')
legend(h.Children([1:3]),'Weight','Tar','Nicotine')
```

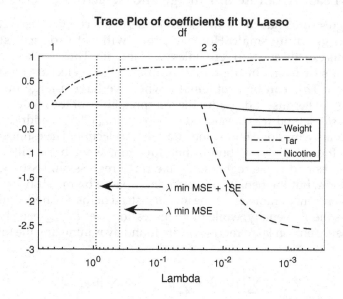

FIGURE 12.13

This is a trace plot produced by the **lassoPlot** *function in the Statistics Toolbox. It is very important to note that the value for* λ *underlines{decreases} from left to right. The estimated coefficients values on the extreme right correspond to the least squares estimates.*

The lasso plot in Figure 12.13 provides a lot of useful information, but it is very important to note that the scale on the horizontal axis flows in a nonstandard direction— in decreasing value for λ. On the left, we have maximum shrinkage with only one nonzero coefficient. On the right side, we have zero shrinkage, and the coefficients correspond to the estimates from least squares. The number of nonzero coefficients (**df**) is shown above the figure box. Because we asked for lasso with cross validation, we also get an optimal value for λ based on the mean squared error (MSE). The vertical line on the right (small λ) is the value that yields the smallest MSE. It is recommended that one choose a λ that is within one standard error, which is the vertical line on the left. We can look at information about the lasso fit by examining the output **fitinfoL**.

```
fitinfoL =


        Intercept: [1x100 double]
           Lambda: [1x100 double]
            Alpha: 1
               DF: [1x100 double]
              MSE: [1x100 double]
   PredictorNames: {}
```

```
         SE: [1x100 double]
LambdaMinMSE: 0.3286
   Lambda1SE: 0.9144
IndexMinMSE: 72
    Index1SE: 83
```

The parameter $\lambda = 0.3286$ corresponds to the smallest MSE, and $\lambda = 0.9144$ is the proposed value that is one standard error from the minimum. This is the value we will use for our model. We can get the estimated coefficients as follows.

```
% Find the index to the minimum
% MSE + 1 SE lambda.
indL = fitinfoL.Index1SE;
lm_lasso(:,indL)
```

We should get something similar to the following in the command window. Cross-validation is based on randomly selected subsets, so results might be different.

```
% This displays in the command window.

 ans =
          0
     0.5620
          0
```

We see from above (and Figure 12.13) that lasso chose **Tar** and shrank the other coefficients to zero. We can get the intercept from **fitinfoL**, as shown here.

```
% Get the intercept term.
fitinfoL.Intercept(indL)

 ans =
     4.9517
```

Our lasso estimated model is therefore given by

$$\hat{y}_{CO} = 4.9517 + 0.562 x_{Tar}.$$

❑

Elastic Net

Zou and Hastie [2005] note some disadvantages with the lasso approach. First, when we have data $d > n$, where the number of observations n is less than the number of dimensions d, then lasso will select at most n variables. If there are groups of predictor variables with high pairwise correlations, then lasso will likely select one in the group. It does not discriminate between them to choose the "best" one. Tibshirani [1996] showed that the prediction accuracy of ridge regression is better than lasso when some predictors are highly correlated.

Zou and Hastie [2005] set out to develop a regularization approach that keeps the best of lasso, but addresses the problems listed above. They call this approach the *elastic net*. Like ridge regression and lasso, elastic net includes a penalty term in the objective function. It actually includes both an L_2 penalty (as in ridge regression) and one based on the absolute value or L_1 (as in lasso).

Using the elastic net approach to estimate the regression coefficients, one solves the following optimization problem for the parameters β_0 and β_j:

$$\sum_{i=1}^{n} \left(y_i - \beta_0 - \sum_{j=1}^{d} \beta_j x_{ij} \right)^2 + \lambda P_\alpha(\beta), \tag{12.36}$$

with

$$P_\alpha(\beta) = \sum_{j=1}^{d} \frac{1-\alpha}{2} \beta_j^2 + \alpha |\beta_j|. \tag{12.37}$$

In elastic net, there is another tuning parameter α, where $0 < \alpha < 1$. The parameter λ is nonnegative, as before.

We see in Equations 12.36 and 12.37, that elastic net approaches lasso when α gets close to the upper bound of 1. Similarly, the estimated coefficients will approach those obtained using ridge regression as it gets close to 0. Zou and Hastie [2005] show that elastic net is likely to perform better than lasso when some predictors are correlated. It also has a grouping effect because it tends to include (or not) groups of correlated variables rather than selecting only one in a group. Finally, it works well in the small n, large d case ($d > n$).

Example 12.10

We are now going to estimate the relationship between carbon monoxide and our predictors in the **cigarette** data using the elastic net approach. We can use the same **lasso** function from the previous example, where we set a value of the **Alpha** input parameter to a value strictly between zero and one. If we use the default value of one in the function, then we get a lasso fit.

```
% Now, let's try elastic net. To do this, we set the
% Alpha parameter to a value between 0 and 1.
% Let's use a value of 0.5.
[lm_elastic, fitinfoE] = lasso(X,CO,'CV',10,...
```

```
    'Alpha',0.5);
```

We can construct a lasso plot as we did before.

```
% The following code constructs a lasso Plot
% for our elastic net fit.
h = lassoPlot(lm_elastic,fitinfoE,...
    'PlotType','Lambda','XScale','log');
set(h.Children(3),'linestyle','--')
set(h.Children(2),'linestyle','-.')
legend(h.Children([1:3]),'Weight','Tar','Nicotine')
```

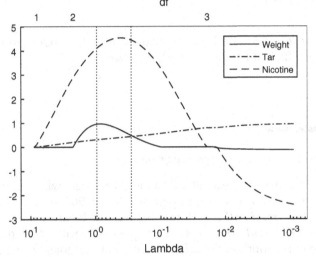

FIGURE 12.14

This is the lasso plot based on our elastic net. Like we had in Figure 12.13, the value for λ decreases moving left to right, and the values of the coefficients on the right side of the plot correspond to ordinary least squares or no regularization.

This is shown in Figure 12.14. The best value for λ (minimum MSE plus one standard error) yields the following model

```
% Find the estimated coefficients for the
% minimum MSE plus one SE.
indE = fitinfoE.Index1SE;
lm_elastic(:,indE)
```

```
ans =

    4.1060
    0.3096
    0.9602
```

The intercept term is extracted separately as shown next.

```
% The intercept term is in the
% field fitinfoE.Intercept.
fitinfoE.Intercept(indE)

 ans =

    4.2158
```

At a value of $\alpha = 0.5$, elastic net kept all terms in the model. However, it did shrink the coefficient for the second term (**Tar**).
❏

12.6 Partial Least Squares Regression

There is another technique available to the user to deal with colinearity in the predictor variables. It was developed by Wold [1966] and is called *partial least squares regression* (PLS). It has also come to be known in the literature as *projection to latent structures*. PLS regression derives a set of inputs that are linear combinations of the original predictor variables. These account for the variance in the inputs (as in principal component analysis) and are also highly correlated with the response variables.

PLS regression combines aspects of other model-building approaches, such as principal component regression and multiple regression. It can also accommodate applications where we have more than one response variable. In part because of this aspect of PLS, it is used quite often in chemometrics, econometrics, bioinformatics, and the social sciences [Rosipal and Kramer, 2006; Geladi and Kowalski, 1986].

Besides handling the issue of colinearity, PLS regression can be employed in some situations where the usual least squares regression fails. For instance, it can be used when the number of predictors d is large, as compared to n. It is also sometimes used for variable selection. However, as with most regression applications, the purpose of PLS is to build a model that allows us to predict a set of response variables given future measurements of our predictor variables.

In our discussion of PLS regression given here, we follow the treatment given in Abdi [2010]. We will describe it using a single response variable to be consistent with previous topics in this book, but keep in mind that it can be used when we have multiple response variables. We first describe a related approach called *principal components* (PC) *regression* and follow it by partial least squares regression.

Principal Component Regression

In PC regression, the first step is to transform the predictor variables to principal components. We then regress the observed predicted values represented in the new principal component space against the original observed responses. This is the approach we will illustrate in Example 12.11.

Alternatively, we could decompose the input matrix X into a product of three matrices using singular value decomposition (SVD), which is a well-known method in linear algebra [Golub and van Loan, 2012; Jackson, 1991]. The SVD of the $n \times d$ matrix X is

$$X = RDV^T.$$

The matrix R is $n \times d$. It contains the left singular vectors. The matrix V is $d \times d$, and it includes the right singular vectors as columns. D is a diagonal $d \times d$ matrix with the singular values along the diagonal. The singular values are the square root of the variance explained by the corresponding singular vector. We order the singular vectors and associated singular values, as we do in PCA. We also impose the following constraints on R and V:

$$R^T R = I_d \quad \text{and} \quad V^T V = I_d,$$

where I_d is the $d \times d$ identity matrix.

The left-singular matrix R is then used in ordinary multiple regression to predict the response variable y [Abdi, 2010]. Because the left singular vectors are orthogonal, we do not have multicolinearity. One could also keep a smaller set of singular vectors that correspond to the largest singular values, similar to what we do with principal component analysis.

It is important to note that there are different conventions for defining the matrices used in the SVD. For example, MATLAB uses the following notation for the SVD:

$$X = USV^T,$$

where U is $n \times n$, V is $d \times d$, and S is an $n \times d$ diagonal matrix.

Example 12.11

The method we just described for principal component regression using the SVD can also be accomplished using PCA [Jackson, 1991]. As described in Chapter 6, we use PCA to find the principal component scores (the observed inputs transformed to the principal component space) and possibly employ a smaller set of components to reduce the number of dimensions. However, we will use all of the principal components in what follows. We continue using the **cigarette** data from the previous section.

```
% First load the cigarette data.
load cigarette
% Form the predictor matrix and response.
X = [Nicotine,Tar,Weight];
y = CO;
```

We could standardize the input variables, but we will not do so in this case. Next, we use the **pca** function to project the observed values of the predictors to the principal component space. These are the **PCAScores** in the function call next.

```
% Get the principal component scores.
% This is the data represented in the PC space.
% They should now be uncorrelated.
[PCComps,PCScores,PCVar] = pca(X);
```

We can calculate the correlation coefficient matrix, which indicates that the inputs are now uncorrelated.

```
% The input data are now uncorrelated.
corrcoef(PCScores)

ans =

    1.0000    0.0000    0.0000
    0.0000    1.0000   -0.0000
    0.0000   -0.0000    1.0000
```

We use the **regress** function to estimate the coefficients using least squares using a model with an intercept term.

```
% We keep all of the components in this example.
% No dimensionality reduction occurs.
coefPCR = regress(y-mean(y), PCScores);

coefPCR =

    0.7994
   -1.9487
   -1.8528
```

The coefficients can be transformed for use with observed measurements of the predictors in the original space. This makes interpretation easier. Note that the default option in the function **pca** is to center the data before PCA, so we need to account for this in our transformation back to the original space.

```
% Transform back to the original space.
beta = PCComps*coefPCR;
% We have to add the intercept.
beta = [mean(y) - mean(X)*beta; beta];
```

Here are the coefficients in our model.

```
beta =

    3.2022
   -2.6317
    0.9626
   -0.1305
```

Looking back at the estimated coefficients we obtained in Example 12.8, we see that the values we get from PC regression are essentially the same.
❑

If we use all of the principal components in building the model, then we will get the same results as ordinary linear least squares regression. So, why use PC regression in this case? It could provide better accuracy if the matrix $X^T X$ is close to singular due to multicolinearity or other problems. It is also possible to use PC regression when the number of predictors is greater than the number of observations (the small n, large d problem). Furthermore, the principal components sometimes have interpretive utility.

One could use fewer principal components to reduce the dimensionality and then estimate the coefficients using least squares. Typically, one would remove the principal components corresponding to the smaller eigenvalues. However, there can be problems doing this in regression. Jackson [1991] has a good discussion on this issue. For example, he cites a case in the literature where the components for the smaller eigenvalues were better predictors of the responses [Jolliffe, 2001]. This can happen because PC regression finds a transformation that explains the variation in the predictors X without regard to their relationship to the response. Partial least squares regression attempts to address this problem.

Partial Least Squares Regression

The results of PLS regression are not invariant to the scale of the input data. So, it is a good idea to transform the input matrix X, such that the predictors

x_j have a mean of zero and variance of one. This means that we subtract the mean from each variable and divide by its standard deviation.

As we saw above, the PCA decomposition of our input matrix X does not account for the response variable—only the variance in X. So, it does not necessarily produce a decomposition that yields good predictions of y. PLS regression seeks to address this by finding components for X that have a high correlation with y, which then improves prediction.

PLS regression finds a simultaneous decomposition of X and y that explains the maximum covariance between the component vectors of both X and y [Jackson, 1991]. We seek a linear decomposition of X and y of the form

$$X = TP + E$$
$$y = UQ + F^*$$

such that the covariance between T and U is maximized. Recall that X has n rows and d columns. Here, we have just one response variable, represented as vector y with n rows. We note again that we could have a response matrix Y with dimensionality $n \times q$, and PLS regression can still be used.

T and U are called *score* matrices, and the columns are often known as *latent* or *score* vectors. The matrices T and U have n rows and k columns, with $k \leq d$. The matrix E contains the residuals of X at the k-th stage, and it is equal to zero (a matrix with all zeros) when $k = d$; i.e., we have an exact representation of X. P and Q contain the *loadings* (or weights) for X and y, respectively. P has dimensions $k \times d$, and Q has dimensions $k \times 1$. The matrix F^* contains the residuals for y at the k-th stage.

There are two popular algorithms for finding the components used in T and U. They are called NIPALS [Wold, 1975] and SIMPLS [de Jong, 1993], and both of them use an iterative approach. A detailed description of them would not add any insights, so we do not include them here. However, they both find the score vectors for X and y—one component at a time.

We also note that PLS regression is sometimes represented as an algorithm, as opposed to a mathematical model, as we use in least squares regression. Thus, it is not an easy matter to make predictions for new observations using the estimated relationship from PLS regression [Jackson, 1991]. Fortunately, software exists for PLS regression, both for estimation and prediction. We show how to use the MATLAB Statistics Toolbox function for PLS regression in the following example.

Example 12.12
We use the **plsregress** function to perform partial least squares regression on the **cigarette** data. We load the data first.

```
% Make sure the data are loaded.
load cigarette
% Form the predictor matrix and response.
```

```
X = [Nicotine,Tar,Weight];
y = CO;
```

The following function call produces the PLS regression for **X** and **y**. The outputs **XL** and **yL** are the loadings for the predictors and response. We can specify the number of components to use ($k \leq d$). The default is to provide all d components, in which case, we will get the same results as least squares.

```
% The following performs PLS regression and
% returns the coefficients in the variable beta.
[XL,yL,XS,yS,beta] = ...
    plsregress(X,y);
```

Looking at **beta**, we see that the estimated coefficients are equal to what we get from ordinary linear regression.

```
beta =

    3.2022
   -2.6317
    0.9626
   -0.1305
```

The **plsregress** function can return some other useful variables. One is the mean-squared errors (MSE) for both the predictors and the response(s). It is a matrix with two rows and $k + 1$ columns, where k is the maximum number of components specified as an input. The first row contains the MSE for the predictor variables, and the second row corresponds to the MSE for the response. The first column is the MSE for the model with zero components. Another useful output we can request is the percentage of variance explained as the number of components used in the model changes. This output is also a two row matrix, but it has k columns. Again, k is either d (the default) or is given by the user as an input to **plsregress**. As with the MSE, the first row corresponds to the percentage of variance explained in **X** and the second row to the response. Here is what we get with the **cigarette** data.

```
% Request two additional output variables.
[XL,yL,XS,yS,beta,pvar,mse] = ...
    plsregress(X,y);
```

The percentage of the variance explained is

```
pvar =

    0.9996    0.0002    0.0002
    0.9167    0.0019    0.0000
```

We see that a significant percentage of the variance in both the predictors and the response is explained by the first component. Let's look at the MSE.

```
mse =

    30.9451      0.0112      0.0053      0.0000
    21.5660      1.7962      1.7562      1.7557
```

The first column corresponds to zero components, so it is not surprising that the MSE is large in this case. There is a slight improvement going from one to two components, as seen in the second and third columns. So, we might use $k = 2$ for our final model.
❑

The **plsregress** function uses the SIMPLS algorithm. It is important to note that it first centers the predictors **X** and the response **y**, but it does not scale the columns. One could do that manually or use the **zscore** function in the Statistics Toolbox, if necessary.

12.7 MATLAB® Code

MATLAB and some associated toolboxes have functions appropriate for the methods discussed in this chapter. The basic MATLAB package also has some tools for estimating functions using spline (e.g., **spline**, **interp1**, etc.) interpolation (something we did not present). We already discussed the **Basic Fitting** tool that is available when one plots x and y values using the **plot** function. This is invoked from the figure window **Tools** menu. One can find various fits, interpolants, statistics, residuals, and plots via this graphical user interface.

The Curve Fitting Toolbox from The MathWorks, Inc. has some functions for spline-based regression, as well as smoothing splines (Chapter 13).

The MATLAB Statistics Toolbox has several functions for regression, including **regstats** and **regress**. They have more output options than the **polyfit** function. For example, **regress** returns the parameter estimates and residuals, along with corresponding confidence intervals. The **polytool** is an interactive demo available in the MATLAB Statistics Toolbox. Among other things, it allows the user to explore the effects of changing the degree of the fit.

We already discussed the functions for generalized linear models that are available in the MATLAB Statistics Toolbox. To summarize, these are **mnrfit** and **mnrval** for multinomial (and logistic) regression and **glmfit** and **glmval** for generalized linear models. Keep in mind that, as usual, the functionality we demonstrate in the text is just an example. So, please consult the MATLAB **help** files for more information on what can be accomplished with these functions.

For another useful source on what can be done with generalized linear models in MATLAB, execute the command **playshow glmdemo** at the command line. This provides a series of plots, explanations, and MATLAB code that can help you learn about the models and ways to assess the goodness of fit.

For the most part, we do not provide code for this chapter because the Statistics Toolbox has the necessary functions. However, we found a function for constructing spline regression fits from MATLAB Central:

http://www.mathworks.com/matlabcentral/

It is called **spfit**, and we include it (with permission) in the Computational Statistics Toolbox. Relevant MATLAB functions are given in Table 12.6.

Another toolbox available for free download is the GLMLAB graphical user interface and functions for MATLAB, written by Peter Dunn. This software does not require the MATLAB Statistics Toolbox, and it has many useful features, including the ability to define your own distribution and link function. It has nice residual diagnostic plots and summary statistics. To download the tools, go to MATLAB Central (link is above) and search for "glmlab." Finally, Dunn and Smyth [1996] present a type of residuals for regression models with independent responses called *randomized quantile residuals*. These produce residuals that are exactly normally distributed. They are available as an option in the GLMLAB Toolbox.

TABLE 12.6

List of Functions for Parametric Models

Purpose	MATLAB Function
MATLAB	
Polynomial curve fit & evaluation	**polyfit, polyval**
1D interpolation	**interp1**
Cubic spline interpolation	**spline**
MATLAB Statistics Toolbox	
Multiple linear regression	**regress**
Interactive polynomial fitting	**polytool**
Polynomial confidence interval	**polyconf**
Inverse prediction for regression	**invpred**
Multinomial logistic regression	**mnrfit, mnrval**
Linear models	**fitlm**
Stepwise regression	**stepwiselm**
Ridge regression	**ridge**
Lasso or elastic net	**lasso**
Partial least squares regression	**plsregress**
Generalized linear models	**glmfit, glmval**
Computational Statistics Toolbox	
Spine regression	**spfit**

12.8 Further Reading

Every introductory statistics textbook has a chapter on regression. However, many do not go into details on the theory and methods for checking the estimated fits. Some good resources for these details are *Applied Regression Analysis* by Draper and Smith [1981] and *Regression Diagnostics* by Belsley, Kuh, and Welsch [1980]. The text called *Regression Analysis by Example* by Chatterjee, Hadi, and Price [2000] is suitable for anyone having an understanding of elementary statistics. It emphasizes data analysis and has discussions of diagnostic plots, time series regression, and logistic regression. Cook and Weisberg [1994] present a graphical approach to regression, and Cook [1998] contains many graphical methods for analyzing the models.

Agresti [1996, 2002] wrote several excellent books on categorical data analysis. Logistic and Poisson regression are two statistical methods often used for this type of analysis. The Agresti texts are easy to understand and contain a wealth of information on understanding and assessing the model fits. For a readable book on logistic regression in the social sciences, see Pampel [2000]. Kleinbaum [1994] published a self-learning text on logistic regression for those in the health sciences. Finally, Menard [2001] discusses applications of logistic regression and provides details on analyzing and interpreting the model fit.

The seminal book on splines is *A Practical Guide to Splines* by Carl de Boor [2001]. The functions in the MATLAB Curve Fitting Toolbox are based on this book. For an easy to read short book on spline regression, see Marsh and Cormier [2002]. Schimek [2000] edited a book containing several chapters on this topic, as well as spline smoothing.

As for generalized linear models, the most influential text is *Generalized Linear Models* by McCullagh and Nelder [1989]. We also highly recommend Gill [2000] and Dobson [2002] for accessible introductions to generalized linear models.

Exercises

12.1. Generate data according to $y = 4x^3 + 6x^2 - 1 + \varepsilon$, where ε represents some noise. Instead of adding noise with constant variance, add noise that is variable and depends on the value of the predictor. So, increasing values of the predictor show increasing variance. Do a polynomial fit and plot the residuals versus the fitted values. Do they show that the constant variance assumption is violated? Use the MATLAB **Basic Fitting** tool to explore your options for fitting a model to these data.

12.2. Generate data as in Problem 12.1, but use noise with constant variance. Fit a first-degree model to it and plot the residuals versus the observed predictor values X_i (residual dependence plot). Do they show that the model is not adequate? Repeat for $d = 2, 3$.

12.3. Repeat Example 12.1. Construct box plots and histograms of the residuals. Do they indicate normality?

12.4. Use the **filip** data and fit a sequence of polynomials of degree $d = 2, 4, 6, 10$. For each fit, construct a residual dependence plot. What do these show about the adequacy of the models?

12.5. Use the MATLAB Statistics Toolbox graphical user interface **polytool** with the **longley** data. Use the tool to find an adequate model.

12.6. Using the slope of the tangent curve estimated in Example 12.4, find the incremental rate of change in the estimated probability for WAIS scores of 6 and 18. Compare these with what we had for a score of 12.

12.7. Fit the WAIS data using probit and the complementary log log link. Compare the results with the logistic regression using the techniques of this chapter to assess model fit.

12.8. Derive the mean and the variance (Equations 12.20 and 12.21) for the Poisson and the binomial using the appropriate $b(\theta)$.

12.9. Repeat Example 12.5 using the weight of the female crab as the predictor variable.

12.10. Convert the number of satellites in the horseshoe crab data to a binomial response variable, keeping the carapace width as the predictor. Set $Y = 1$ if a female crab has one or more satellites, and $Y = 0$ if she has no satellites. Apply the methods of the chapter to fit various models and assess the results.

12.11. Using the data and models from Example 12.6, use a normal probability plot on all available residuals. Assess the results, keeping in mind that the residuals do not have to be normally distributed.

12.12. Use the spline regression function **spfit** to fit linear, quadratic, and cubic splines to the **filip** data. Use knot points that make sense from the scatterplot. Plot the fits in a scatterplot of the data and assess the resulting models.

12.13. The following MATLAB code shows how to fit a linear model to the **cars** data. This is multivariate data with two predictors (weight and horsepower) and a response variable (miles per gallon). We include a constant term and an interaction term in the model (see the MATLAB documentation on **regress**).

```
load cars
% Construct a linear model with a constant term
% and an interaction term.
X = [ones(size(x1)) x1 x2 x1.*x2];
% The next two commands use functions in the
% Statistics Toolbox.
b1 = regress(y,X)
b2 = glmfit(X,y,'normal','constant','off')
% You can also use the backslash operator.
b3 = X\y
```

Note that we used two ways of finding the least-squares estimate of the parameters (**regress** and the backslash operator). We also found estimates of the parameters using **glmfit**. Compare the estimates.

12.14. The **cpunish** data contains $n = 17$ observations [Gill, 2001]. The response variable is the number of times that capital punishment is implemented in a state for the year 1997. The explanatory variables are median per capita income (dollars), the percent of the population living in poverty, the percent of African-American citizens in the population, the log(rate) of violent crimes, a variable indicating whether a state is in the South, and the proportion of the population with a college degree. Use the **glmfit** function to fit a model using the Poisson link function. Note that **glmfit** automatically puts a column of ones for the constant term. Get the deviance (**dev**) and the statistics (**stats**). What are the estimated coefficients and their standard errors? Find the 95% Wald confidence intervals for each estimated coefficient.

12.15. Apply the penalized approaches—ridge regression, lasso, and elastic net— to the **airfoil** data. Compare your results with subset selection (Example 12.7).

12.16. Apply subset selection and penalized approaches to the **environmental** data set. Choose a best model.

Chapter 13

Nonparametric Models

13.1 Introduction

The discussion in the previous chapter illustrates the situation where the analyst assumes a *parametric* form for a model and then estimates the required parameters. We now turn our attention in this chapter to the *nonparametric* case, where we make fewer assumptions about the relationship between the predictor variables and the response. Instead, we let the data provide us with information about how they are related.

A commonly used tool for nonparametric regression is known as the *scatterplot smoother*, and we discuss several of these in the following sections. A scatterplot smoother is a way of conveying (usually visually) the trend of a response as a function of one or more predictor variables. These methods are called smoothers because they provide an estimate of the trend that is a smoothed (less variable) version of the response values.

We first describe several single predictor smoothing methods in Sections 13.2 through 13.4. Some basic smoothing methods such as bin smoothers, the running-mean and running-line smoothers, and the locally-weighted version known as loess are discussed in Section 13.2. We then cover the kernel methods in Section 13.3, followed by smoothing splines in Section 13.4. Techniques for estimating the variability of the smooths and for choosing the value of the smoothing parameter are presented in Section 13.5. We then look at the multivariate case in Sections 13.6 and 13.7, where we discuss two methods—regression trees and additive models. The additive model framework shows how one can use the univariate smoothing methods from the beginning of the chapter as building blocks for multivariate models. We end the chapter with a section on multivariate adaptive regression splines.

As in parametric regression, we assume that we have n measurements of our predictor variable X, given by x_i, along with values of the response Y denoted by y_i. Further, we will assume that for the methods presented in this chapter, the x_i are *ordered* such that $x_1 < \ldots < x_n$. (Note that we are not using the notation for the order statistics $x_{(i)}$ for ease of reading.) This also means that, for the moment, we assume that we have no tied values in the predictor variable. If ties are present, then weighted techniques can be used.

The model we use in nonparametric regression is similar to the parametric case and is given by

$$Y = f(X) + \varepsilon, \tag{13.1}$$

where we assume that $E[\varepsilon] = 0$, the variance of ε is constant, and the errors are independent. Further, we know Equation 13.1 implies that

$$E[Y|X = x] = f(x). \tag{13.2}$$

As we learned from the previous chapter, our goal in regression is to find the estimate of this function, which will be denoted as $\hat{f}(x)$. In contrast to parametric regression, we do not assume a rigid parametric form for $f(X)$. Instead, we assume that the relationship between the predictor and response variables is given by a smooth function. In most cases, we will be looking for a value of the nonparametric regression at an arbitrary target value x_0, where this target value is not necessarily equal to one of our observed predictor values x_i.

13.2 Some Smoothing Methods

We now describe several smoothing techniques, including bin smoothing, the running-mean and running-line smoothers, and the locally-weighted polynomial smoothing technique, commonly known as loess.

The nonparametric regression methods covered in this section are called *scatterplot smooths* because they help to visually convey the relationship between X and Y by graphically summarizing the middle of the data using a smooth function of the points. Besides helping to visualize the relationship, it also provides an estimate or prediction for given values of x. In this section, we cover the single predictor case, where we have an observed sample of a predictor variable x and a response variable, y. These are denoted as $\{x_i, y_i\}$ for $i = 1, ..., n$.

Before we go on to describe the various types of smoothers, we first define what we mean by a *neighborhood* about some target value x_0 in our domain. If our target value is equal to one of the observed predictor values, $x_0 = x_i$, then we can define a *symmetric nearest neighborhood* as one that includes the observation x_i, the m closest points less than x_i, and the m closest points greater than x_i. Thus, the neighborhood consists of $2m + 1$ values. The indices of these points are denoted by $N^S(x_0)$, where the superscript S to remind us that this is a symmetric neighborhood. When we get close to the maximum and minimum values of the predictors, then we will not have m points on both sides of the target value. In this case, we take as many points as we have available to us, realizing that the neighborhoods will no longer be symmetric.

Hastie and Tibshirani [1990] note that defining a symmetric nearest neighborhood at target values other than the observed x_i can be problematic. They recommend finding the fit at the predictor values adjacent to the target value and linearly interpolating.

An alternative to the symmetric neighborhood is to ignore symmetry and include the k nearest points to a target value x_0. This is called the *nearest neighborhood* and is denoted by $N(x_0)$.

We can also represent the size of the neighborhood by the proportion of points in the neighborhood. This is called the *span*. For a symmetric nearest neighborhood, this is given by

$$w = \frac{2m + 1}{n}.$$

The size of the neighborhood determines the smoothness of the resulting relationship between the predictor and the response. If we have very large neighborhoods, then the curves are smoother. In this case, the variance is decreased, but our estimates could be biased. Small neighborhoods tend to produce curves that are more jagged or wiggly, resulting in less bias, but higher variance.

Bin Smoothing

The *bin smoother* [Hastie and Tibshirani, 1990] we are about to describe does not use neighborhoods based on the number of points or the notion of symmetry around some target value. Instead, we divide the x_i into disjoint and exhaustive regions represented by cut points $-\infty = c_0 < \ldots < c_K = \infty$. We can define the indices of the data points belonging to the regions R_k as

$$R_k = \{i; \ (c_k \leq x_i < c_{k+1})\},$$

for $k = 0, \ldots, K - 1$. Then we have the value of the estimated smooth for x_0 in R_k given by the average of the responses in that region:

$$\hat{f}(x_0) = \text{average}_{i \text{ in } R_k}(y_i). \tag{13.3}$$

We illustrate this in Figure 13.1 and Example 13.1. As we see, the resulting curve is piecewise constant, so it is not very smooth. While it does give us a general idea of the trend, the discontinuities at each cut point do not provide a visually appealing picture of the relationship between the variables.

Example 13.1

We will use a data set from Simonoff [1996] to illustrate the bin smoother and several other nonparametric regression fits. It contains the grape yields (in number of lugs) for the harvests from 1989 through 1991. The predictor variable is the row number, and the response is the total lug counts for the three years. First, we load the data and set the response and predictor variables.

```
load vineyard
n = length(row);
y = totlugcount;
x = row;
```

Next, we divide the domain into 10 regions.

```
K = 10;
% The following finds the number of observations
% in each bin to make 10 regions.
nk = ones(1,K)*floor(n/K);
rm = rem(n,K);
if rm ~= 0
    nk(1:rm) = nk(1:rm) + 1;
end
% The next command finds the index of the end
% points' bins.
nk = cumsum(nk);
```

Finally, we sort the data, bin it, and find the smooth as the mean of the values that fall in the bin.

```
% Sort the data.
[xsort,inds] = sort(x);
ys = y(inds);
% Find the c_k and the value of the smooth in each.
ck = xsort(1);
ck(K+1) = xsort(end);
smth = zeros(1,K+1);
smth(1) = mean(ys(1:nk(1)));
for i = 1:(K-1)
    ck(i+1) = mean(xsort(nk(i):nk(i)+1));
    smth(i+1) = mean(ys(nk(i)+1:nk(i+1)));
end
smth(end) = smth(end-1);
```

The following MATLAB® code produces the plot shown in Figure 13.1.

```
plot(x,y,'o');
xlabel('Row'),ylabel('Total Number of Lugs')
hold on, stairs(ck,smth), hold off
```

We see from the plot that the curve is piecewise constant, so it is not smooth. However, it does give some idea of the overall trends.
❑

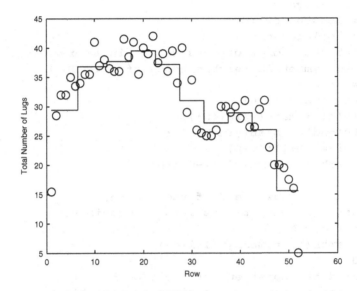

FIGURE 13.1
This shows the nonparametric regression fit obtained via the bin smoother for the **vineyard** *data. Note that the curve is not smooth.*

Running Mean

The *running mean smooth* expands on the idea of the bin smoother by taking overlapping regions or bins. This helps make the estimate smoother. Note that the following definition of a running mean smoother is for target values that are equal to one of our observed x_i:

$$\hat{f}(x_i) = \underset{j \text{ in } N^S(x_i)}{\text{average}}(y_j). \tag{13.4}$$

This smooth is also known as the *moving average* and is used quite often in time series analysis where the observed predictor values are evenly spaced. As we will see in the next example, it is not much better than the bin smoother. Another problem with the running mean is that the smooth can be biased at the boundaries.

Example 13.2

We use the same **vineyard** data set to illustrate the running mean in MATLAB. The following code will fit the running mean to a data set. The resulting scatterplot with the smooth is shown in Figure 13.2.

```
load vineyard
n = length(row);
y = totlugcount;
% Set the number of data points in each window.
% Use symmetric neighborhoods of size 2N+1.
N = 3;
smth = zeros(1,n);
smth(1) = mean(y(1:1+N));
smth(end) = mean(y(n-N:n));
for i = (N+1):(n-N)
    smth(i) = mean(y(i-N:i+N));
end
% Find the lower end of the smooth,
% using as many to the left as possible.
for i = 2:N
    smth(i) = mean(y(1:i+N));
end
% Find the upper end of the smooth,
% using as many to the right as possible.
for i = (n-N+1):(n-1)
    smth(i) = mean(y(i-N:end));
end
```

We see from the plot in Figure 13.2 that the fitted curve is smoother than the bin smooth and provides a better idea of the relationship between our predictor and response variable.
❏

Running Line

The *running line smoother* is a generalization of the running mean, where the estimate of f is found by fitting a local line using the observations in the neighborhood. The definition is given by the following

$$\hat{f}(x_i) = \hat{\beta}_0^i + \hat{\beta}_1^i x_i, \tag{13.5}$$

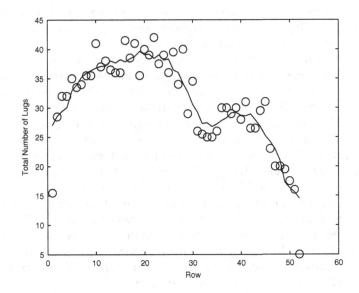

FIGURE 13.2
This shows a nonparametric fit for the **vineyard** *data using the running-mean smoother.*

where $\hat{\beta}_0^i$ and $\hat{\beta}_1^i$ are estimated via ordinary least squares using the data in the neighborhood of x_i.

Hastie and Tibshirani [1990] show that the fitted local line for x_i in the interior of the smooth is largely dominated by the mean and is not greatly affected by the slope of the fitted line. On the other hand, the slope of the fitted line at the end points helps to convey the trend and reduce the bias at those locations.

The MATLAB code in Example 13.2 can be easily adapted to construct the running line smoother, so we do not provide it here. Instead, it is left as an exercise to the reader.

Local Polynomial Regression – Loess

The methods we have discussed so far apply equal weight to all points in a given neighborhood, and points outside are given zero weights. This tends to produce curves that are not very smooth because of the abrupt changes in the weights at the end points of the neighborhoods. The next method we discuss addresses this problem by giving the highest weight to observations closest to our target value and decreasing the weights smoothly as we move further away.

Loess is a method that employs locally weighted regression to smooth a scatterplot and also provides a nonparametric model of the relationship between two variables. It was originally described in Cleveland [1979], and further extensions can be found in Cleveland and McGill [1984] and Cleveland [1993]. Loess is also called *local polynomial regression* [Fox, 2000a; Fox, 2000b] or a *locally-weighted running-line smoother* (when the degree of the local fit is one) [Hastie and Tibshirani, 1990].

The curve obtained from a loess model is governed by two parameters: α and λ. The parameter α is a *smoothing parameter*. We restrict our attention to values of α between zero and one, where high values for α yield smoother curves. Cleveland [1993] addresses the case where α is greater than one. The second parameter λ determines the degree of the local regression. Usually, a first or second degree polynomial is used, so $\lambda = 1$ or $\lambda = 2$. Information on how to set these parameters will be provided in a later section, and other techniques will be explored in the exercises.

The general idea behind loess is the following. To get a value of the curve $\hat{f}$ at a given target point x_0, we first determine a local neighborhood of x_0 based on α. All points in this neighborhood are weighted according to their distance from x_0, with points closer to x_0 receiving larger weight. The estimate $\hat{f}$ at x_0 is obtained by fitting a linear or quadratic polynomial using the weighted points in the neighborhood. This is repeated for a uniform grid of points x in the domain to get the desired curve.

We describe next the steps for obtaining a loess curve [Hastie and Tibshirani, 1990]. The steps of the loess procedure are illustrated in Figures 13.3 through 13.6.

PROCEDURE – LOESS CURVE CONSTRUCTION

1. Let x_i denote a set of n values for a predictor variable and let y_i represent the corresponding response.

2. Choose a value for α such that $0 < \alpha < 1$. Let $k = \lfloor \alpha n \rfloor$, so k is the greatest integer less than or equal to αn.

3. For each x_0, find the k points x_i that are closest to x_0. These x_i comprise a neighborhood of x_0, and this set is denoted by $N(x_0)$.

4. Compute the distance of the x_i in $N_0 = N(x_0)$ that is furthest away from x_0 using

$$\Delta(x_0) = \max_{x_i \in N_0} |x_0 - x_i| .$$

5. Assign a weight to each point (x_i, y_i), x_i in $N(x_0)$, using the tri-cube weight function

$$w_i(x_0) = W\left(\frac{|x_0 - x_i|}{\Delta(x_0)}\right) ,$$

with

$$W(u) = \begin{cases} (1-u^3)^3; & 0 \le u < 1 \\ 0; & \text{otherwise.} \end{cases}$$

6. Obtain the value $\hat{f}$ of the curve at the point x_0 using a weighted least squares fit of the points x_i in the neighborhood $N(x_0)$. (See Equations 13.6 through 13.9.)

7. Repeat steps 3 through 6 for all x_0 of interest.

In step 6, one can fit either a straight line to the weighted points (x_i, y_i), x_i in $N(x_0)$, or a quadratic polynomial can be used. If a line is used as the local model, then $\lambda = 1$. In this case, we minimize the following to find the values of β_0 and β_1:

$$\sum_{i=1}^{k} w_i(x_0)(y_i - \beta_0 - \beta_1 x_i)^2, \tag{13.6}$$

for (x_i, y_i), x_i in $N(x_0)$. Letting $\hat{\beta}_0$ and $\hat{\beta}_1$ be the values that minimize Equation 13.6, the loess fit at x_0 is given by

$$\hat{f}(x_0) = \hat{\beta}_0 + \hat{\beta}_1 x_0. \tag{13.7}$$

When $\lambda = 2$, then we fit a quadratic polynomial using weighted least squares based only on those points in $N(x_0)$. In this case, we find the values for the β_i that minimize

$$\sum_{i=1}^{k} w_i(x_0)(y_i - \beta_0 - \beta_1 x_i - \beta_2 x_i^2)^2. \tag{13.8}$$

Similar to the linear case, if $\hat{\beta}_0$, $\hat{\beta}_1$, and $\hat{\beta}_2$ minimize Equation 13.8, then the loess fit at x_0 is given by

$$\hat{f}(x_0) = \hat{\beta}_0 + \hat{\beta}_1 x_0 + \hat{\beta}_2 x_0^2. \tag{13.9}$$

For more information on weighted least squares see Draper and Smith, [1981].

Example 13.3

In this example, we use a data set that was analyzed in Cleveland and McGill [1984]. These data represent two variables comprising daily measurements of

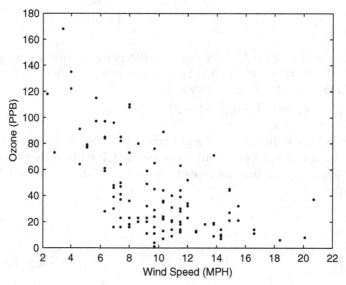

FIGURE 13.3
This shows a scatterplot of ozone and wind speed. It is difficult to tell from this plot what type of relationship exists between these two variables. Instead of using a parametric model, we will try the nonparametric approach.

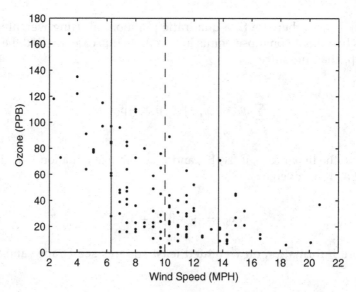

FIGURE 13.4
This shows the neighborhood (solid line) of the point $x_0 = 10$ (dashed line).

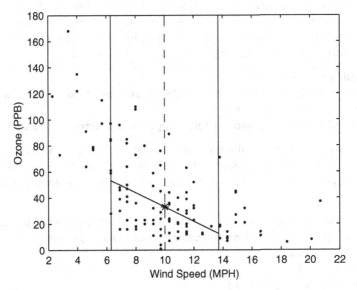

FIGURE 13.5
This shows the local fit at $x_0 = 10$ *using weighted least squares. Here* $\lambda = 1$ *and* $\alpha = (2/3)$.

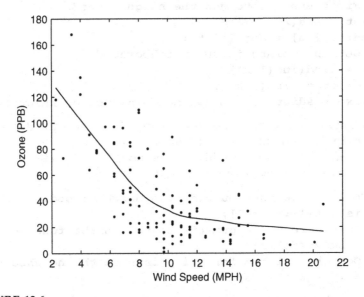

FIGURE 13.6
This shows the scatterplot of ozone and wind speed along with the accompanying loess smooth.

ozone and wind speed in New York City. These quantities were measured on 111 days between May and September 1973. We are interested in understanding the relationship between ozone (the response variable) and wind speed (the predictor variable). The next lines of MATLAB code load the data set and display the scatterplot shown in Figure 13.3.

```
load environ
% Do a scatterplot of the data to see the relationship.
plot(wind,ozone,'k.')
xlabel('Wind Speed (MPH)'),ylabel('Ozone (PPB)')
```

It is difficult to determine the parametric relationship between the variables from the scatterplot, so the loess approach is used. We illustrate how to use the loess procedure to find the estimate of the ozone for a given wind speed of 10 MPH.

```
n = length(wind); % Find the number of data points.
x0 = 10; % Find the estimate at this point.
alpha = 2/3;
lambda = 1;
k = floor(alpha*n);
```

Now that we have the parameters for loess, the next step is to find the neighborhood at $x_0 = 10$.

```
% First step is to get the neighborhood.
dist = abs(x0 - wind);
[sdist,ind] = sort(dist);
% Get the points in the neighborhood.
Nx = wind(ind(1:k));
Ny = ozone(ind(1:k));
delxo = sdist(k);   % Maximum distance of neighborhood
```

The neighborhood of x_0 is shown in Figure 13.4, where the dashed line indicates the point of interest x_0 and the solid line indicates the limit of the local region. All points within this neighborhood receive weights based on their distance from $x_0 = 10$ as shown next.

```
% Delete the ones outside the neighborhood.
sdist((k+1):n) = [];
% These are the arguments to the weight function.
u = sdist/delxo;
% Get the weights for all points in the neighborhood.
w = (1 - u.^3).^3;
```

Using only those points in the neighborhood, we use weighted least squares to get the estimate at x_0.

```
% Now using only those points in the neighborhood,
% do a weighted least squares fit of degree 1.
```

```
% We will follow the procedure in 'polyfit'.
x = Nx(:); y = Ny(:); w = w(:);
W = diag(w);% get weight matrix
A = vander(x);% get right matrix for X
A(:,1:length(x)-lambda-1) = [];
V = A'*W*A;
Y = A'*W*y;
[Q,R] = qr(V,0);
p = R\(Q'*Y);
p = p';% to fit MATLAB convention
% This is the polynomial model for the local fit.
% To get the value at that point, use polyval.
yhat0 = polyval(p,x0);
```

In Figure 13.5, we show the local fit in the neighborhood of x_0. We include a function called **csloess** that will determine the smooth for all points in a given vector. We illustrate its use next.

```
% Now call the loess procedure and plot the result.
% Get a domain over which to evaluate the curve.
x0 = linspace(min(wind),max(wind),50);
yhat = csloess(wind,ozone,x0,alpha,lambda);
% Plot the results.
plot(wind,ozone,'k.',x0,yhat,'k')
xlabel('Wind Speed (MPH)'),ylabel('Ozone (PPB)')
```

The resulting scatterplot with loess smooth is shown in Figure 13.6. The final curve is obtained by linearly interpolating between the estimates from loess.
❏

As we will see in the exercises, fitting curves is an iterative process. Different values for the parameters α and λ should be used to obtain various loess curves. Then the scatterplot with superimposed loess curves and residuals plots can be examined to determine whether or not the model adequately describes the relationship. We will provide more details on this approach and others in a later section.

Robust Loess

Loess is not robust because it relies on the method of least squares. A method is called *robust* if it performs well when the associated underlying assumptions (e.g., normality) are not satisfied [Kotz and Johnson, 1986]. There are many ways in which assumptions can be violated. A common one is the presence of *outliers* or extreme values in the response data. These are points in the sample that deviate from the pattern of the other observations. Least squares regression is vulnerable to outliers, and it takes only one extreme value to unduly influence the result. This is easily seen in Figure

13.7, where there is an outlier in the upper left corner. The dashed line is obtained using least squares with the outlier present, and the solid line is obtained with the outlier removed. It is obvious that the outlier affects the slope of the line and the y intercept. So, the predictions that one gets from the two models (with and without the outlier) would be different.

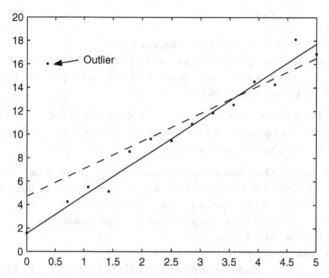

FIGURE 13.7
This is an example of what can happen with the least squares method when an outlier is present. The dashed line is the fit with the outlier present, and the solid line is the fit with the outlier removed. The slope of the line and the y-intercept are different when the outlier is included.

Cleveland [1993, 1979] and Cleveland and McGill [1984] present a method for smoothing a scatterplot using a robust version of loess. This technique uses the bisquare method [Hoaglin, Mosteller, and Tukey, 1983; Mosteller and Tukey, 1977; Huber, 1973; Andrews, 1974] to add robustness to the weighted least squares step in loess. The idea behind the bisquare is to reweight points based on their residuals. If the residual for a given point in the neighborhood is large (i.e., it has a large deviation from the model), then the weight for that point should be decreased, since large residuals tend to indicate outlying observations. On the other hand, if the point has a small residual, then it should be weighted more heavily.

Before showing how the bisquare method can be incorporated into loess, we first describe the general bisquare least squares procedure. First a linear regression is used to fit the data, and the residuals $\hat{\varepsilon}_i$ are calculated from

$$\hat{\varepsilon}_i = Y_i - \hat{f}_i. \tag{13.10}$$

The residuals are used to determine the weights from the bisquare function given by

$$B(u) = \begin{cases} (1-u^2)^2; & |u| < 1 \\ 0; & \text{otherwise.} \end{cases} \tag{13.11}$$

The **robust weights** are obtained from

$$r_i = B\left(\frac{\hat{\varepsilon}_i}{6\hat{q}_{0.5}}\right), \tag{13.12}$$

where $\hat{q}_{0.5}$ is the median of $|\hat{\varepsilon}_i|$. A weighted least squares regression is performed using r_i as the weights.

To add bisquare to loess, we first fit the loess smooth, using the same procedure as before. We then calculate the residuals using Equation 13.10 and determine the robust weights from Equation 13.12. The loess procedure is repeated using weighted least squares, but the weights are now $r_i w_i(x_0)$. Note that the points used in the fit are the ones in the neighborhood of x_0. This is an iterative process and is repeated until the loess curve converges or stops changing. Cleveland and McGill [1984] suggest that two or three iterations are sufficient to get a reasonable model.

PROCEDURE – ROBUST LOESS

1. Fit the data using the loess procedure with weights w_i.
2. Calculate the residuals, $\hat{\varepsilon}_i = y_i - \hat{f}_i$ for each observation.
3. Determine the median of the absolute value of the residuals, $\hat{q}_{0.5}$.
4. Find the robustness weight (Equation 13.12).
5. Repeat the loess procedure using weights of $r_i w_i$.
6. Repeat steps 2 through 5 until the loess curve converges.

In essence, the robust loess iteratively adjusts the weights based on the residuals. We illustrate the robust loess procedure in the next example.

Example 13.4

We turn to a data set from Simonoff [1996] to demonstrate robust local polynomial regression. This data set contains measurements of the annual spawners and recruits for sockeye salmon in the Skeena River for the years 1940 through 1967. We will construct nonparametric regression curves using the robust and nonrobust versions of loess and then display them on a scatterplot. We do not show all of the steps for these methods, but instead illustrate how one can use the two loess functions that are included in the Computational Statistics Toolbox. Both of these functions take the following input arguments: the observed values of the predictor variable, the observed values of the response variable, the target values x_0 where one wants to evaluate the fit, the smoothing parameter α, and the degree of the polynomial λ.

```
load salmon
% Get the predictor values and the response.
x = salmon(:,1);
y = salmon(:,2);
% Set up the target values.
x0 = linspace(min(x),max(x),20);
% Get the regular loess fit.
% This function is included in the
% Computational Statistics Toolbox.
alpha = 0.5;
lambda = 2;
yloess = csloess(x,y,x0,alpha,lambda);
```

A function called **csloessr** that implements the robust version of loess is included with the text. We now use this function to get the loess curve.

```
% Get the robust loess fit.
yloessr = csloessr(x,y,x0,alpha,lambda);
% Now construct the scatterplot with smooth.
plot(x,y,'o',x0,yloess,x0,yloessr,':')
```

The resulting smooth is shown in Figure 13.8, where we see that the two loess curves are different. It is recommended that one try both versions of loess in exploring the relationship between the predictor and the response.
❑

13.3 Kernel Methods

This section follows the treatment of kernel smoothing methods given in Wand and Jones [1995]. We first discussed kernel methods in Chapter 9, where we applied them to the problem of estimating a probability density function in a nonparametric setting. We now present a class of smoothing methods based on kernel estimators that are similar in spirit to loess, in that they fit the data in a local manner. These are called *local polynomial kernel estimators*. We first define these estimators in general and then present two special cases: the *Nadaraya-Watson estimator* and the *local linear kernel estimator*.

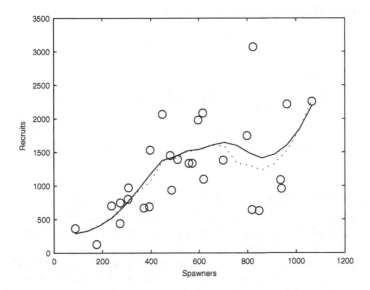

FIGURE 13.8
*This shows a scatterplot of the **salmon** data. The curve represented by the solid line is the fit obtained using the regular version of loess. The curve shown with the dotted line is the smooth found using the robust version of loess. We see that there is a difference in the fit when we use the robust loess [Simonoff, 1996].*

With local polynomial kernel estimators, we obtain an estimate $\hat{f}_0$ at a point x_0 by fitting a d-th degree polynomial using weighted least squares. As with loess, we want to weight the points based on their distance to x_0. Those points that are closer should have greater weight, while points further away have less weight. To accomplish this, we use weights that are given by the height of a kernel function that is centered at x_0.

As with probability density estimation, the kernel has a bandwidth or smoothing parameter represented by h. This controls the degree of influence points will have on the local fit. If h is small, then the curve will be wiggly (i.e., high variability) because the estimate will depend heavily on points closest to x_0. In this case, the model is trying to fit to local values (our "neighborhood" is small), and we have overfitting. Larger values for h means that points further away will have a similar influence on the fit as points that are close to x_0 (the "neighborhood" is large). With a large enough h, we would be fitting the line to the whole data set.

We now give the expression for the local polynomial kernel estimator. Let d represent the degree of the polynomial that we fit at a point x. We obtain the estimate $\hat{y} = \hat{f}(x)$ by fitting the polynomial

$$\beta_0 + \beta_1(X_i - x) + \dots + \beta_d(X_i - x)^d \tag{13.13}$$

using the points (X_i, Y_i) and utilizing the weighted least squares procedure. The weights are given by the kernel function

$$K_h(X_i - x) = \frac{1}{h} K\left(\frac{X_i - x}{h}\right). \tag{13.14}$$

The value of the estimate at a point x is $\hat{\beta}_0$, where the $\hat{\beta}_i$ minimize

$$\sum_{i=1}^{n} K_h(X_i - x)(Y_i - \beta_0 - \beta_1(X_i - x) - \ldots - \beta_d(X_i - x)^d)^2. \tag{13.15}$$

Because the points that are used to estimate the model are all centered at x (see Equation 13.13), the estimate at x is obtained by setting the argument in the model equal to zero. Thus, the only parameter left is the constant term $\hat{\beta}_0$.

The attentive reader will note that the argument of the K_h is backwards from what we had in probability density estimation using kernels. There, the kernels were centered at the random variables X_i. We follow the notation of Wand and Jones [1995] that shows explicitly that we are centering the kernels at the points x where we want to obtain the estimated value of the function.

We can write this weighted least squares procedure using matrix notation. According to standard weighted least squares theory [Draper and Smith, 1981], the solution can be written as

$$\hat{\beta} = (\mathbf{X}_x^T \mathbf{W}_x \mathbf{X}_x)^{-1} \mathbf{X}_x^T \mathbf{W}_x \mathbf{Y}, \tag{13.16}$$

where $\mathbf{Y}$ is the $n \times 1$ vector of responses,

$$\mathbf{X}_x = \begin{bmatrix} 1 & X_1 - x & \ldots & (X_1 - x)^d \\ \vdots & \vdots & \ldots & \vdots \\ 1 & X_n - x & \ldots & (X_n - x)^d \end{bmatrix}, \tag{13.17}$$

and $\mathbf{W}_x$ is an $n \times n$ matrix with the weights along the diagonal. These weights are given by

$$w_{ii}(x) = K_h(X_i - x). \tag{13.18}$$

Some of these weights might be zero depending on the kernel that is used. The estimator $\hat{f}(x)$ is the intercept coefficient $\hat{\beta}_0$ of the local fit, so we can obtain the value from

$$\hat{f}(x) = \mathbf{e}_1^T(\mathbf{X}_x^T\mathbf{W}_x\mathbf{X}_x)^{-1}\mathbf{X}_x^T\mathbf{W}_x\mathbf{Y} \tag{13.19}$$

where $\mathbf{e}_1^T$ is a vector of dimension $(d+1) \times 1$ with a one in the first place and zeroes everywhere else.

Nadaraya–Watson Estimator

Some explicit expressions exist when $d = 0$ and $d = 1$. When d is zero, we fit a constant function locally at a given point x. This estimator was developed separately by Nadaraya [1964] and Watson [1964]. The Nadaraya–Watson estimator is given next.

NADARAYA–WATSON KERNEL ESTIMATOR:

$$\hat{f}_{NW}(x) = \frac{\displaystyle\sum_{i=1}^{n} K_h(X_i - x)Y_i}{\displaystyle\sum_{i=1}^{n} K_h(X_i - x)}. \tag{13.20}$$

Note that this is for the case of a *random design*. When the design points are fixed, then the X_i is replaced by x_i, but otherwise the expression is the same [Wand and Jones, 1995].

There is an alternative estimator that can be used in the *fixed design* case. This is called the Priestley-Chao kernel estimator [Simonoff, 1996].

PRIESTLEY–CHAO KERNEL ESTIMATOR:

$$\hat{f}_{PC}(x) = \frac{1}{h}\sum_{i=1}^{n} (x_i - x_{i-1})K\left(\frac{x - x_i}{h}\right)y_i, \tag{13.21}$$

where the x_i, $i = 1, ..., n$, represent a fixed set of *ordered* nonrandom numbers. The Nadarya–Watson estimator is illustrated in Example 13.5, while the Priestley–Chao estimator is saved for the exercises.

Example 13.5

We show how to implement the Nadarya–Watson estimator in MATLAB. As in the previous example, we generate data that follows a sine wave with added noise.

```
% Generate some noisy data.
x = linspace(0, 4 * pi,100);
y = sin(x) + 0.75*randn(size(x));
```

The next step is to create a MATLAB **inline** function so we can evaluate the weights. Note that we are using the normal kernel.

```
% Create an inline function to evaluate the weights.
mystrg='(2*pi*h^2)^(-1/2)*exp(-0.5*((x - mu)/h).^2)';
wfun = inline(mystrg);
```

We now get the estimates at each value of x.

```
% Set up the space to store the estimated values.
% We will get the estimate at all values of x.
yhatnw = zeros(size(x));
n = length(x);
% Set the window width.
h = 1;
% find smooth at each value in x
for i = 1:n
 w = wfun(h,x(i),x);
 yhatnw(i) = sum(w.*y)/sum(w);
end
```

The smooth from the Nadarya-Watson estimator is shown in Figure 13.9.
❑

Local Linear Kernel Estimator

When we fit a straight line at a point x, then we are using a local linear estimator. This corresponds to the case where $d = 1$, so our estimate is obtained as the solutions $\hat{\beta}_0$ and $\hat{\beta}_1$ that minimize the following,

$$\sum_{i=1}^{n} K_h(X_i - x)\{Y_i - \beta_0 - \beta_1(X_i - x)\}^2.$$

We give an explicit formula for the estimator next.

LOCAL LINEAR KERNEL ESTIMATOR:

$$\hat{f}_{LL}(x) = \frac{1}{n}\sum_{i=1}^{n} \frac{\{\hat{s}_2(x) - \hat{s}_1(x)(X_i - x)\}K_h(X_i - x)Y_i}{\hat{s}_2(x)\hat{s}_0(x) - \hat{s}_1(x)^2}, \qquad (13.22)$$

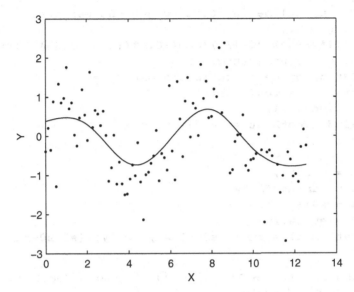

FIGURE 13.9
This figure shows the smooth obtained from the Nadarya–Watson estimator with h = 1.

where

$$\hat{s}_r(x) = \frac{1}{n}\sum_{i=1}^{n}(X_i - x)^r K_h(X_i - x).$$

The fixed design case is obtained by replacing the random variable X_i with the fixed point x_i.

When using the kernel smoothing methods, problems can arise near the boundary or extreme edges of the sample. This happens because the kernel window at the boundaries has missing data. In other words, we have weights from the kernel, but no data to associate with them. Wand and Jones [1995] show that the local linear estimator behaves well in most cases, even at the boundaries. If the Nadaraya-Watson estimator is used, then modified kernels are needed [Scott, 2015; Wand and Jones, 1995].

Example 13.6

The local linear estimator is applied to the same generated sine wave data. The entire procedure is implemented here.

```
% Generate some data.
x = linspace(0, 4 * pi,100);
y = sin(x) + 0.75*randn(size(x));
```

```
h = 1;
deg = 1;
% Set up inline function to get the weights.
mystrg = ...
    '(2*pi*h^2)^(-1/2)*exp(-0.5*((x - mu)/h).^2)';
wfun = inline(mystrg);
% Set up space to store the estimates.
yhatlin = zeros(size(x));
n = length(x);
% Find smooth at each value in x.
for i = 1:n
  w = wfun(h,x(i),x);
  xc = x-x(i);
  s2 = sum(xc.^2.*w)/n;
  s1 = sum(xc.*w)/n;
  s0 = sum(w)/n;
  yhatlin(i) = sum(((s2-s1*xc).*w.*y)/(s2*s0-s1^2))/n;
end
```

The resulting smooth is shown in Figure 13.10. Note that the curve seems to behave well at the boundary.
❑

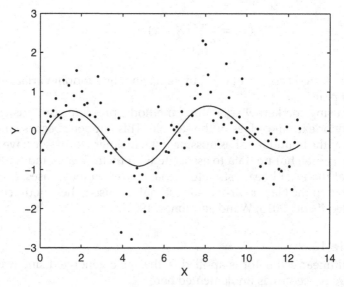

FIGURE 13.10
This figure shows the smooth obtained from the local linear kernel estimator.

13.4 Smoothing Splines

We discussed regression splines in Chapter 12, and we saw that they are a flexible way to model the relationship between X and Y. In this section, we present a method of nonparametric regression or smoothing called *smoothing splines*. This method is discussed in several books, and we provide pointers to some of these at the end of the chapter. We use the treatment found in Green and Silverman [1994] for the description of smoothing splines given next.

Smoothing splines relax the linear assumptions of classical regression by taking a roughness penalty approach. This means that we find our estimate $\hat{f}$ by optimizing an objective function that includes a penalty for the roughness of $\hat{f}$. Given any twice-differentiable function f defined on the interval $[a, b]$, we define the penalized sum of squares by

$$S(f) = \sum_{i=1}^{n} \{Y_i - f(x_i)\}^2 + \alpha \int_a^b \{f''(t)\}^2 dt, \qquad (13.23)$$

where $\alpha > 0$ is the smoothing parameter. The estimate $\hat{f}$ is defined to be the minimizer of $S(f)$ over the class of all twice-differentiable functions f. The only constraint we have on the interval $[a, b]$ is that it must contain all of the observed x_i [Green and Silverman, 1994].

The term on the left in Equation 13.23 is the familiar residual sum of squares. Recall that we minimize this quantity in least-squares regression, and it determines the goodness-of-fit to the data. The term on right is a measure of roughness in the fit, and it is our roughness penalty term. The smoothing parameter α governs the tradeoff between smoothness and goodness-of-fit. For large values of α, we will have little variability or a very smooth curve. When α approaches infinity, the curve $\hat{f}$ approaches the fit we get from linear regression. As α gets small, we have a curve that is allowed to be more variable, so we have an estimate that tries to fit the data better. As α approaches zero, we get closer to interpolating the data. As with other smoothing methods, we see the same bias-variability tradeoff issues we had before.

Natural Cubic Splines

Splines and their role in regression were presented in the previous chapter. Recall that splines are piecewise polynomials that are joined at knot points t_i with various conditions that have to be satisfied at the knots. We now return to this type of model since it plays an important role in smoothing splines.

A function f defined on some interval $[a, b]$ is a **cubic spline** with knots given by $a < t_1 < \ldots < t_n < b$ if both of the following are true:

1. The function f is a cubic polynomial on each of the intervals (a, t_1), (t_1, t_2), ..., (t_n, b).
2. The polynomials connect at the knots such that f and its first two derivatives are continuous at each t_i.

The second condition implies that the entire function f is continuous over the interval $[a, b]$ [Green and Silverman, 1994].

A **natural cubic spline** is a cubic spline with additional conditions at the boundary that must be satisfied. A cubic spline over the interval $[a, b]$ is a *natural* cubic spline if its second and third derivatives are zero at the end points a and b. This means that the function f is linear on the two boundary intervals $[a, t_1]$ and $[t_n, b]$.

In what follows, we will assume that $n \geq 3$. Green and Silverman [1994] show that a natural cubic spline can be represented by the value of f and the second derivative at each of the knots t_i. If f is a natural cubic spline with knots $t_1 < \ldots < t_n$, then we define the value at the knot as

$$f_i = f(t_i),$$

and the second derivative at the knot as

$$\gamma_i = f''(t_i),$$

for $i = 1, \ldots, n$. We know from the definition of a natural cubic spline that the second derivatives of the end points are zero, so we have $\gamma_1 = \gamma_n = 0$. We gather the values of the function and the second derivative into vectors $\mathbf{f}$ and γ, where $\mathbf{f} = (f_1, \ldots, f_n)^T$ and $\gamma = (\gamma_2, \ldots, \gamma_{n-1})^T$. It is important to note that the vector γ is defined and indexed in a nonstandard way.

We now go on to define two band matrices $\mathbf{Q}$ and $\mathbf{R}$ that will be used in the algorithm for smoothing splines. A **band matrix** is one whose nonzero entries are contained in a band along the main diagonal. The matrix entries outside this band are zero. First, we let $h_i = t_{i+1} - t_i$, for $i = 1, \ldots, n-1$. Then $\mathbf{Q}$ is the matrix with nonzero entries q_{ij}, given by

$$q_{j-1,j} = \frac{1}{h_{j-1}} \qquad q_{jj} = -\frac{1}{h_{j-1}} - \frac{1}{h_j} \qquad q_{j+1,j} = \frac{1}{h_j}, \qquad (13.24)$$

for $j = 2, \ldots, n-1$. Since this is a band matrix, the elements q_{ij} are zero for index values where $|i - j| \geq 2$. From this definition, we see that the columns of the matrix $\mathbf{Q}$ are indexed in a nonstandard way with the top left element given by q_{12} and that the size of $\mathbf{Q}$ is $n \times (n-2)$.

The matrix **R** is an $(n-2) \times (n-2)$ symmetric matrix with nonzero elements given by

$$r_{ii} = \frac{h_{i-1} + h_i}{3}; \qquad i = 2, ..., n-1, \qquad (13.25)$$

and

$$r_{i,i+1} = r_{i+1,i} = \frac{h_i}{6}; \qquad i = 2, ..., n-2. \qquad (13.26)$$

The elements r_{ij} are zero for $|i-j| \geq 2$.

It turns out that not all vectors **f** and **γ** (as defined above) will represent natural cubic splines. Green and Silverman [1994] show that the vectors **f** and **γ** specify a natural cubic spline f if and only if

$$\mathbf{Q}^T\mathbf{g} = \mathbf{R}\boldsymbol{\gamma}. \qquad (13.27)$$

We will use these matrices and relationships to help us find the estimated curve $\hat{f}$ that minimizes the penalized sum of squares in Equation 13.23.

Reinsch Method for Finding Smoothing Splines

Reinsch [1967] proved a remarkable result that the function $\hat{f}$ that uniquely minimizes the penalized sum of squares is a natural cubic spline with knots given by the observed values x_i [Hastie and Tibshirani, 1990; Green and Silverman, 1994]. So, we will use the observations x_i in place of our knot notation in the following discussion. Recall that our assumption is that the x_i are *ordered*.

The Reinsch algorithm finds the smoothing splines by forming a system of linear equations for the γ_i at the knot locations. We can then obtain the values of the smoothing spline in terms of the γ_i and the observations y_i. See Green and Silverman [1994] for details on the derivation of the algorithm.

REINSCH PROCEDURE FOR SMOOTHING SPLINES

1. Given the observations $\{x_i, y_i\}$, where $a < x_1 < ... < x_n < b$ and $n \geq 3$, we collect them into vectors

$$\mathbf{x} = (x_1, ..., x_n)^T \text{ and } \mathbf{y}_i = (y_1, ..., y_n)^T.$$

2. Find the matrices **Q** and **R**, using Equations 13.24 through 13.26.

3. Form the vector $\mathbf{Q}^T\mathbf{y}$ and the matrix $\mathbf{R} + \alpha\mathbf{Q}^T\mathbf{Q}$.

4. Solve the following equation for γ:

$$\mathbf{R} + \alpha\mathbf{Q}^T\mathbf{Q}\gamma = \mathbf{Q}^T\mathbf{y}. \qquad (13.28)$$

5. Find $\hat{\mathbf{f}}$ from

$$\hat{\mathbf{f}} = \mathbf{y} - \alpha\mathbf{Q}\gamma. \qquad (13.29)$$

Note that we are using the "hat" notation in step 5 to indicate that this is our estimated function obtained by optimizing the penalized sum of squares.

Example 13.7

We will use the **vineyard** data set to illustrate the Reinsch method given above. First we load the data and set the smoothing parameter to $\alpha = 5$.

```
load vineyard
% Sort the x values, so they can be knots.
[x,ind] = sort(row);
y = totlugcount(ind);
x = x(:); y = y(:);
n = length(x);
alpha = 5;
```

The next step is to find the $\mathbf{Q}$ and $\mathbf{R}$ matrices using Equations 13.24 through 13.26. Note that we keep the matrices at a full size of n by n to keep the indices correct. We then remove the columns or rows that are not needed.

```
qDs = -hinv(1:n-2) - hinv(2:n-1);
I = [1:n-2, 2:n-1, 3:n];
J = [2:n-1,2:n-1,2:n-1];
S = [hinv(1:n-2), qDs, hinv(2:n-1)];
% Create a sparse matrix.
Q = sparse(I,J,S,n,n);
% Delete the first and last columns.
Q(:,n) = []; Q(:,1) = [];
% Now find the R matrix.
I = 2:n-2;
J = I + 1;
tmp = sparse(I,J,h(I),n,n);
t = (h(1:n-2) + h(2:n-1))/3;
R = tmp'+tmp+sparse(2:n-1,2:n-1,t,n,n);
% Get rid of the rows/cols that are not needed.
R(n,:) = []; R(1,:) = [];
R(:,n) = []; R(:,1) = [];
```

The final step is to find the smoothing spline using steps 3 through 5 of the Reinsch method.

```
% Get the smoothing spline.
S1 = Q'*y;
S2 = R + alpha*Q'*Q;
% Solve for gamma;
gam = S2\S1;
% Find f^hat.
fhat = y - alpha*Q*gam;
```

The smoothing spline is shown in Figure 13.11. We see that this is a much smoother curve than the bin smoother and the running-mean fits we had previously.
❑

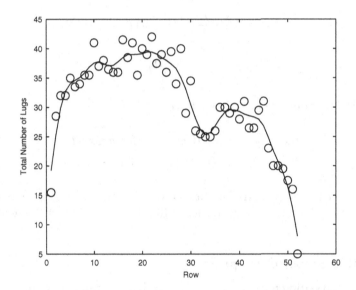

FIGURE 13.11
This shows the cubic smoothing spline for the **vineyard** *data. The smoothing parameter is α = 5.*

Values for a Cubic Smoothing Spline

We now have our smoothing spline at the knot locations. However, we are also interested in getting values of the spline at arbitrary target values so we can make predictions or plot the curve. Green and Silverman [1994] show how one can plot a full cubic spline using the vectors **f** and γ and knots

$t_1 < \ldots < t_n$. Note that while we use the more general notation here, this can be used to plot the function $\hat{f}$ in Equation 13.29.

First, we give the expression for finding the value of the cubic spline for any value of x in intervals with endpoints given by the knots:

$$f(x) = \frac{(x - t_i)f_{i+1} + (t_{i+1} - x)f_i}{h_i}$$

$$- \frac{1}{6}(x - t_i)(t_{i+1} - x)\left\{ \left(1 + \frac{x - t_i}{h_i}\right)\gamma_{i+1} + \left(1 + \frac{t_{i+1} - x}{h_i}\right)\gamma_{ii} \right\} \tag{13.30}$$

for $t_i \leq x \leq t_{i+1}$, $i = 1, \ldots, n - 1$. So, we see that Equation 13.30 is valid over the intervals between each of the knots.

To find the value of the cubic spline outside of the interval $[t_1, t_n]$, we first need the first derivative at the outside knots:

$$f'(t_1) = \frac{f_2 - f_1}{t_2 - t_1} - \frac{1}{6}(t_2 - t_1)\gamma_2$$

$$f'(t_n) = \frac{f_n - f_{n-1}}{t_n - t_{n-1}} + \frac{1}{6}(t_n - t_{n-1})\gamma_{n-1} . \tag{13.31}$$

Since f is linear outside $[t_1, t_n]$, we have

$$f(x) = f_1 - (t_1 - x)f'(t_1); \qquad \text{for } a \leq x \leq t_1$$

$$f(x) = f_n + (x - t_n)f'(t_n); \qquad \text{for } t_n \leq x \leq b. \tag{13.32}$$

One can write a MATLAB function to obtain the value of the smoothing spline for any target value x_0; we leave this as an exercise for the reader.

Weighted Smoothing Spline

Recall that in setting the stage for nonparametric regression, we made certain assumptions about the distribution of the errors and the predictors. In particular, we assume the errors ε_i are independent with zero mean and constant variance. As for the observed predictor values, we made the assumption that ties were not present. We now show how to find nonparametric regression fits when these assumptions are violated. We first focus on the case of smoothing splines, but the concept of weighted smoothing is easily extended to other nonparametric regression techniques [Hastie and Tibshirani, 1990; Green and Silverman, 1994].

The penalized sum of squares in Equation 13.23 includes a term that corresponds to the ordinary sum of squares. We can add strictly positive weights to that term, which yields the weighted penalized sum of squares. This is given by

$$S_w(f) = \sum_{i=1}^{n} w_i \{Y_i - f(x_i)\}^2 + \alpha \int_a^b \{f''(t)\}^2 dt, \qquad (13.33)$$

where $w_i > 0$ for $i = 1, ..., n$.

When we have unequal variances with $\text{var}[\varepsilon_i] = \sigma_i^2$, then one can use weights that are given by $w_i = 1/\sigma_i^2$. If the variances are not known, then they can be estimated using an iterative weighted-smoothing method [Hastie and Tibshirani, 1990].

The situation where we have ties is somewhat more complicated. Say we have m_i predictor values x_i that are equal. Along with each of the x_i, we also have corresponding observed values of the response:

$$\{x_i, y_i^{(1)}\}, \{x_i, y_i^{(2)}\}, ..., \{x_i, y_i^{(m_i)}\}.$$

The superscript refers to the j-th independent observed response at x_i. We then take the mean of these responses as follows:

$$\bar{y}_i = \frac{1}{m_i} \sum_{j=1}^{m_i} y_i^{(j)}.$$

Note that $\bar{y}_i = y_i$ when there are no ties. We can now define the weighted penalized sum of squares for tied predictors as

$$S_w(f) = \sum_i m_i \{\bar{Y}_i - f(x_i)\}^2 + \alpha \int_a^b \{f''(t)\}^2 dt.$$

The algorithm used for the weighted smoothing spline is given next [Green and Silverman, 1994].

REINSCH PROCEDURE FOR WEIGHTED SMOOTHING SPLINES

1. Construct a diagonal matrix using strictly positive weights w_i.
2. Find the matrices $\mathbf{Q}$ and $\mathbf{R}$, using Equations 13.24 through 13.26.
3. Form the vector $\mathbf{Q}^T\mathbf{y}$ and the matrix $\mathbf{R} + \alpha \mathbf{Q}^T\mathbf{W}^{-1}\mathbf{Q}$.
4. Solve the following equation for γ:

$$\mathbf{R} + \alpha \mathbf{Q}^T \mathbf{W}^{-1} \mathbf{Q} \gamma = \mathbf{Q}^T \mathbf{y}.$$ (13.34)

5. Find $\hat{\mathbf{f}}$ from

$$\hat{\mathbf{f}} = \mathbf{y} - \alpha \mathbf{W}^{-1} \mathbf{Q} \gamma.$$ (13.35)

So, we can see from this procedure that the smoothing spline is easily adapted for the weighted case.

With local nonparametric regression techniques, such as the running-mean and running-line smoother, we could define the weighted smooth using weighted means and lines. The local polynomial smoother (loess) uses weights in the neighborhood of the target value. If we want to incorporate other weights, then we can multiply the local weights by the observation weights. Hastie and Tibshirani [1990] note that in these cases, the neighborhood sizes should be defined in terms of the total weight rather than the number of observations.

13.5 Nonparametric Regression – Other Details

In this section, we turn our attention to some details on the nonparametric regression methods presented in this chapter. First is a discussion on how one can choose the neighborhood, the smoothing parameter, or any quantity that drives the smoothness of our estimate. Next, we cover the notion of degrees of freedom in smoothing and how we can estimate the variance of the residuals. Finally, we will present several techniques for conveying the variability in the fit by using confidence bands and constructing variability plots.

Choosing the Smoothing Parameter

All of the nonparametric regression methods we have described use some parameter that determines the smoothness of the estimate $\hat{f}$. For example, this might be the number of regions used in the bin smoother method, the size of the neighborhood in the running line smoother, the value for h in kernel smoothing methods, or the smoothing parameter α in loess and smoothing splines. In what follows, we will use the term *smoothing parameter* and the notation s to represent all of these quantities that drive the smoothness in our estimated curve.

Recall that the smoothing parameter s can take on a range of values that depends on the method that is used. When values of s at one end of the range are used, the curve will more closely follow the data, yielding a rougher or wiggly curve. As the value of the parameter changes towards the other end of the range of smoothing parameter values, the curve becomes smoother. This means that we might get estimates that show different types of structures (e.g., modes, trends, etc.) as we change the smoothing parameter.

Thus, the choice of the smoothing parameter is an important one, and there are two main approaches one can use to make this decision. One approach is more exploratory in nature, while the other can be considered somewhat automatic.

In the exploratory approach, one would construct estimates $\hat{f}$ for different values of the smoothing parameter and plot the curve on the scatterplot. This yields a sequence of curves and plots that are then examined by the analyst. It is possible that these estimates $\hat{f}$ with different degrees of smoothing will yield different structures or features (e.g., a bump). It is also possible that if one sees the *same* structure in many of the curves (i.e., the structure has persistence), then it is likely that the structure is *not* an artifact of the smoothing parameter that is used.

However, some data analysts advocate an approach that is more automatic and data-driven. This second approach uses cross-validation based on the prediction error. Recall that in cross-validation, part of the data set is removed before we find the nonparametric regression (in this application). This subset of the data is used for testing the accuracy of our predictions using $\hat{f}$. Usually, we leave only one data point out at a time, so we will denote the estimated function at x_i with smoothing parameter s and leaving out the i-th observation as $\hat{f}_s^{(-i)}(x_i)$. Then, the **cross-validation sum of squares** (sometimes called the **cross-validation score function**) as a function of the smoothing parameter can be written as

$$CV(s) = \frac{1}{n} \sum_{i=1}^{n} \left\{ y_i - \hat{f}_s^{(-i)}(x_i) \right\}^2. \tag{13.36}$$

To estimate the smoothing parameter, we would then minimize the cross-validation function given in Equation 13.36. The procedure is to first select a set of suitable values for s. Then, we find $CV(s)$ for each value of s and select the value of s that minimizes $CV(s)$.

Keep in mind that this procedure produces an *estimate* of the smoothing parameter s that optimizes the prediction error. Thus, it is subject to sampling variability, just like any estimate. Fox [2000b] points out that using the $CV(s)$ to choose a value for s can produce estimates that are too small when the sample size n is small. Also, it is important to keep in mind that the value for s from cross-validation might be just a starting point for an exploratory procedure as we described earlier. Regardless of what method is used (exploratory or automatic), the decision on the amount of smoothing is ultimately a subjective one.

We can always use the cross-validation procedure and the expression in Equation 13.36, but a faster way to find the value of $CV(s)$ is available [Hastie and Tibshirani, 1990; Green and Silverman, 1994]. This alternative expression for the cross-validation function is valid when the smoothers are linear. A *linear smoother* is one that can be written in terms of a smoother matrix $\mathbf{S}$, as follows

$$\hat{\mathbf{f}} = \mathbf{S}(s)\mathbf{y},$$

where we have the fit $\hat{\mathbf{f}}$ at the observations x_i. $\mathbf{S}$ is called the *smoother matrix* with n rows and n columns, where the notation indicates its dependence on the smoothing parameter s.

The following smoothers are linear: running-mean, running-line, cubic smoothing spline, kernel methods, and loess. The robust version of loess is nonlinear. Another example of a nonlinear smoother would be the running-median, where the median is used in each window instead of the mean. See Hastie and Tibshirani [1990] and Loader [1999] for information on how to construct the elements of the smoother matrix for the nonrobust version of loess. In the case of smoothing splines, the smoother matrix is [Green and Silverman, 1994]:

$$\mathbf{S}(\alpha) = (\mathbf{I} + \alpha \mathbf{Q}\mathbf{R}^{-1}\mathbf{Q}^T)^{-1},$$

where we have $s = \alpha$.

The cross-validation score for smoothing parameter s can be written as

$$CV(s) = \frac{1}{n}\sum_{i=1}^{n}\left\{\frac{y_i - \hat{f}(x_i)}{1 - S_{ii}(s)}\right\}^2, \tag{13.37}$$

where $S_{ii}(s)$ is the diagonal elements of the smoother matrix, and $\hat{f}(x_i)$ is the value of the nonparametric regression with smoothing parameter s at x_i using the *full data set*. This gives us a tremendous computational savings because we do not have to find n regression fits for each value of the smoothing parameter s. See Green and Silverman [1994] for a proof of Equation 13.37. We will illustrate the cross-validation function for smoothing splines in the next example.

Example 13.8

We will show how to calculate the cross-validation function for the case of cubic smoothing splines. Note that the code given here uses a function called **cssplinesmth** that is included in the Computational Statistics Toolbox. This function returns the smoother matrix for smoothing splines, so we can use the diagonal values in the cross-validation function. We will use the **vineyard** data in this example.

```
load vineyard
x = row;
y = totlugcount;
% Set some possible values for alpha.
alpha = 0.5:0.25:10;
CV = zeros(1,length(alpha));
for i = 1:length(CV);
    [x0,fhat,S] = cssplinesmth(x,y,alpha(i));
    num = y - fhat;
    den = 1 - diag(S);
    CV(i) = mean((num./den).^2);
end
% Find the minimum value of CV(alpha).
[m,ind] = min(CV);
% Find the corresponding alpha.
cvalpha = alpha(ind);
```

The value of the smoothing parameter α that corresponds to the minimum of the cross-validation function is $\alpha = 2.25$. A plot of the cross-validation function for this example is shown in Figure 13.12.
❑

There is another form of cross-validation called the *generalized cross-validation* function [Craven and Wahba, 1979]. One of the motivations for using generalized cross-validation was to save on computations because it does not require the individual diagonal elements of the smoother matrix. The basic change is in the denominator of Equation 13.37, where we replace the $S_{ii}(s)$ with their average value. This can be calculated by finding the trace of the smoother matrix, so we have

$$GCV(s) = \frac{1}{n} \frac{\sum_{i=1}^{n} \{y_i - \hat{f}(x_i)\}^2}{\{1 - n^{-1} tr\mathbf{S}(s)\}^2}, \tag{13.38}$$

where $tr\mathbf{S}(s)$ denotes the trace of the smoother matrix.

The generalized cross-validation is used to find the smoothing parameter s in the same way as before. Thus, we choose values of the smoothing parameter, evaluate the generalized cross-validation score, and choose the value of the smoothing parameter that minimizes Equation 13.38.

We can obtain the general cross-validation score function for the weighted nonparametric regression case by making a logical change to Equation 13.36, and we have

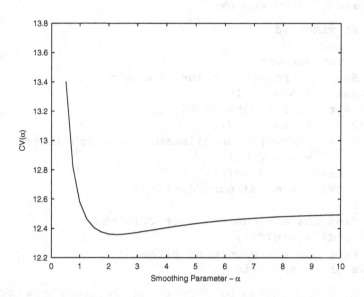

FIGURE 13.12
This shows the cross-validation function for the **vineyard** *data. We see that the minimum of the cross-validation function corresponds to* $\alpha \approx 2$.

$$CV_w(s) = \sum_{i=1}^{n} w_i \{ y_i - \hat{f}_{s,w}^{(-i)}(x_i) \}^2 , \tag{13.39}$$

where $\hat{f}_{s,w}^{(-i)}(x_i)$ is the value of the fit that is found with the i-th point removed.

In the case of smoothing splines with tied predictor values, we can write the cross-validation function in a simpler form. Let N be the total number of data points, with n representing the number of unique values of x_i. We then define the following sum of squares at predictor value x_i with m_i tied values:

$$a_i^2 = \sum_{j=1}^{m_i} \{ y_{ij} - \bar{y}_i \}^2 .$$

Green and Silverman [1994] show that the cross-validation score function for smoothing splines with tied predictors can be written as

$$CV_T(\alpha) = \frac{1}{N} \sum_{i=1}^{n} \frac{m_i \{ \bar{y}_i - \hat{f}(x_i) \}^2 + a_i^2}{\{ 1 - m_i^{-1} S_{ii}^w(\alpha) \}} , \tag{13.40}$$

where $S_{ii}^w(\alpha)$ is the i-th diagonal element of the weighted smoother matrix. This is given by

$$S^w(\alpha) = (\mathbf{W} + \alpha\mathbf{Q}\mathbf{R}^{-1}\mathbf{Q}^T)^{-1}\mathbf{W}.$$

Estimation of the Residual Variance

It is sometimes useful to have an estimate of the residual variance, as we will see in the next section. Several methods for obtaining these estimates have been proposed in the literature [Green and Silverman, 1994], and we outline a few of them here.

Rice [1984] proposed a method based on the first differences of the y_i, which is given by

$$\hat\sigma_R^2 = \frac{1}{2(n-1)}\sum_{i=1}^{n-1}(y_{i+1} - y_i)^2. \tag{13.41}$$

This estimator is based on the idea that the first difference for a smooth curve has a squared mean that is small relative to its variance $2\sigma^2$. If the gradient of the curve is not large, then this is a reasonable approximation.

Another, possibly more familiar approach, is to use the residual sum of squares about our nonparametric regression fit. For a fit that uses smoothing parameter s, we have the following estimator

$$\hat\sigma_s^2 = \frac{\sum\{y_i - \hat{f}_s(x_i)\}^2}{tr\{\mathbf{I} - \mathbf{S}(s)\}}. \tag{13.42}$$

In parametric regression, a similar estimator for the residual variance is to divide the residual sum of squares by the degrees of freedom for the noise. Thus, the value in the denominator in Equation 13.42 is known as the *equivalent degrees of freedom* for noise. See Buja, Hastie, and Tibshirani [1989], Hastie and Tibshirani [1990], and Green and Silverman [1994] for an extensive discussion of equivalent degrees of freedom.

Variability of Smooths

We now turn our attention to ways in which we can better understand the variability in our nonparametric regression or smooths. The methods we present include upper and lower smooths and a method based on the bootstrap. See Fox [2000b] and Hastie and Tibshirani [1990] for information on constructing a pointwise confidence band using the smoother matrix.

The smoothing methods we have discussed so far provide a model of the middle of the distribution of Y given X. This can be extended to give us upper and lower smooths [Cleveland and McGill, 1984], where the distance between the upper and lower smooths indicates the spread. The procedure for obtaining the upper and lower smooths follows.

PROCEDURE – UPPER AND LOWER SMOOTHS

1. Compute the fitted values $\hat{f}_i$ using nonparametric regression.
2. Calculate the residuals $\hat{\varepsilon}_i = y_i - \hat{y}_i$.
3. Find the positive residuals $\hat{\varepsilon}_i^+$ and the corresponding x_i and $\hat{y}_i$ values. Denote these pairs as $(x_i^+, \hat{y}_i^+)$.
4. Find the negative residuals $\hat{\varepsilon}_i^-$ and the corresponding x_i and $\hat{y}_i$ values. Denote these pairs as $(x_i^-, \hat{y}_i^-)$.
5. Smooth the $(x_i^+, \hat{\varepsilon}_i^+)$ and add the fitted values from that smooth to $\hat{y}_i^+$. This is the upper smoothing.
6. Smooth the $(x_i^-, \hat{\varepsilon}_i^-)$ and add the fitted values from this smooth to $\hat{y}_i^-$. This is the lower smoothing.

Example 13.9

In this example, we generate some data to show how to get the upper and lower loess smooths. These data are obtained by adding noise to a sine wave. We then use the function called **csloessenv** that comes with the Computational Statistics Toolbox. The inputs to this function are the same as the other loess functions.

```
% Generate some x and y values.
x = linspace(0, 4 * pi,100);
y = sin(x) + 0.75*randn(size(x));
% Use loess to get the upper and lower smooths.
[yhat,ylo,xlo,yup,xup] = csloessenv(x,y,x,0.5,1,0);
% Plot the smoots and the data.
plot(x,y,'k.',x,yhat,'k',xlo,ylo,'k',xup,yup,'k')
```

The resulting middle, upper, and lower smooths are shown in Figure 13.13, and we see that the smooths do somewhat follow a sine wave. It is also interesting to note that the upper and lower smooths indicate the symmetry of the noise and the constancy of the spread.

❑

We learned that the bootstrap method can be used for statistical inference when analytical methods are not available or distributional assumptions are not possible. Thus, in this case, we might want to use the bootstrap

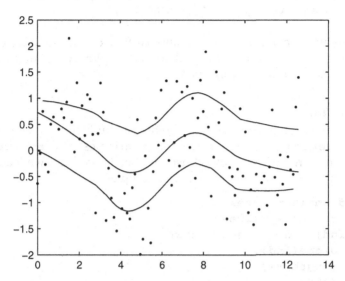

FIGURE 13.13
The data for this example are generated by adding noise to a sine wave. The middle curve
• *is the usual loess smooth, while the other curves are obtained using the upper and lower loess smooths.*

methodology to help us estimate confidence bands for our smooths [Fox, 2000b; Simonoff, 1996; Hastie and Tibshirani, 1990].

Recall that the bootstrap involves repeated sampling *with replacement* from the data set, where each one of these is called a bootstrap sample. We will fit the nonparametric regression for each bootstrap sample and use these to construct the confidence bands. The procedure is detailed next.

PROCEDURE – BOOTSTRAP CONFIDENCE BANDS

1. Set the number of bootstrap samples B to some large number.
2. Select the b-th bootstrap sample by sampling with replacement from the data set. Denote the $\{x, y\}$ pairs for the b-th bootstrap sample as $\{x_1^*, y_1^*\}, \{x_2^*, y_2^*\}, ..., \{x_n^*, y_n^*\}$. Note that the asterisks indicate that the subscripts do not necessarily correspond to the associated $\{x_i, y_i\}$ in the original sample.
3. Fit the nonparametric regression curve using the b-th bootstrap sample. The fit is obtained for n target values. These could be given by the original observations or they could be arbitrary values. Denote these fitted values by $\hat{f}_{bi}^*$.
4. We repeat steps 2 and 3, B times.
5. For each of the i target values, order the $\hat{f}_{bi}^*$ and denote these as $\hat{f}_{(b)i}^*$.

6. The end points of a 95% confidence envelope for the i-th target value are given by $\hat{f}^*_{(0.025 \times B)i}$ and $\hat{f}^*_{(0.975 \times B)i}$.

The procedure above constructs a pointwise 95% confidence envelope, and the use of the bootstrap means that we are implicitly treating the x_i as random rather than fixed design points.

Example 13.10

We show how to obtain the bootstrap pointwise confidence envelopes for the **environmental** data, which has more variables than **environ**. The nonparametric regression method used will be loess. First, we load up the data and specify the number of bootstrap samples.

```
load environmental
% Sort the x values.
[x,ind] = sort(Temperature);
y = Ozone(ind);
n = length(x);
B = 2000;
```

The following code implements the bootstrap procedure outlined above.

```
% Use arbitrary target values for the fit.
x0 = linspace(min(x),max(x),40);
yhat = csloess(x,y,x0,2/3,1);
% Create a matrix to keep the bootstrap
% regression fits. Each row will be a fit
% evaluated at x_i.
Bmat = zeros(B,length(x0));
for i = 1:B
    % Get the bootstrap sample.
    indB = unidrnd(n,1,n); % Statistics Toolbox
    yhatB = csloess(x(indB),y(indB),x0,2/3,1);
    Bmat(i,:) = yhatB;
end
% Sort the values in each column.
Bmats = sort(Bmat);
low = Bmats(0.025*B,:);
up = Bmats(0.975*B,:);
```

A scatterplot of the data with the nonparametric regression fit and confidence envelopes is shown in Figure 13.14.
❑

We can also construct pointwise standard-error bands using the smoother matrix and the residual variance. These are given by ±2 times the square root of the diagonal elements of $SS^T\sigma^2$. In most cases, we will not know the

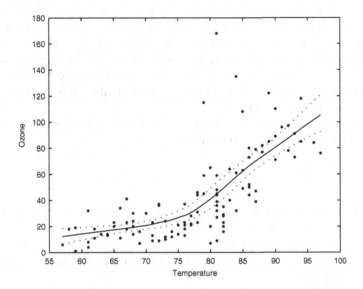

FIGURE 13.14

Here we show the scatterplot smooth using loess with $\alpha = 2/3$ *(solid line) for the* **environmental** *data. The dashed lines represent the pointwise 95% confidence bands obtained using the bootstrap procedure.*

residual variance, but we can estimate it using one of the methods described earlier.

13.6 Regression Trees

The tree-based approach to nonparametric regression is useful when one is trying to understand the structure or interaction among the predictor variables. As we stated earlier, one of the main uses of modeling the relationship between variables is to be able to make predictions given future measurements of the predictor variables. Regression trees accomplish this purpose and also provide insight into the structural relationships and the possible importance of the variables. Much of the information about classification trees, covered in Chapter 10, applies in the regression case. So, the reader is encouraged to read that chapter first, where the procedure is covered in more detail.

In this section, we move to the multivariate situation where we have a response variable Y along with a set of predictors $X = (X_1, ..., X_d)$. Using a procedure similar to classification trees, we will examine all predictor variables for a best split, such that the two groups are homogeneous with respect to the response variable Y. The procedure examines all possible splits and chooses the split that yields the smallest within-group variance in the two groups. The result is a binary tree, where the predicted responses are given by the average value of the response in the corresponding terminal node. To predict the value of a response given an observed set of predictors $\mathbf{x} = (x_1, \ldots, x_d)$, we drop $\mathbf{x}$ down the tree, and assign to $\hat{y}$ the value of the terminal node that it falls into. Thus, we are estimating the function using a piecewise constant surface.

Before we go into the details of how to construct regression trees, we provide the notation that will be used.

NOTATION – REGRESSION TREES

- $D(\mathbf{x})$ represents the prediction rule that takes on real values. Here D will be our regression tree, T.
- $\mathbf{L}$ is the learning sample of size n. Each case in the learning sample comprises a set of measured predictors and the associated response.
- $\mathbf{L}_v, v = 1, \ldots, V$ is the v-th partition of the learning sample $\mathbf{L}$ in cross-validation. This set of cases is used to calculate the prediction error in $D^{(v)}(\mathbf{x})$.
- $\mathbf{L}^{(v)} = \mathbf{L} - \mathbf{L}_v$ is the set of cases used to grow a sequence of subtrees.
- $(\mathbf{x}_i, y_i)$ denotes one case.
- $R^*(D)$ is the true mean squared error of predictor $D(\mathbf{x})$.
- $\hat{R}^{TS}(D)$ is the estimate of the mean squared error of D using the independent test sample method.
- $\hat{R}^{CV}(D)$ denotes the estimate of the mean squared error of D using cross-validation.
- T is the regression tree.
- T_{max} is an overly large tree that is grown.
- $T_{max}^{(v)}$ is an overly large tree grown using the set $L^{(v)}$.
- T_k is one of the nested subtrees from the pruning procedure.
- t is a node in the tree T.
- t_L and t_R are the left and right child nodes.
- $\widehat{T}$ is the set of terminal nodes in tree T.
- $|\widehat{T}|$ is the number of terminal nodes in tree T.
- $n(t)$ represents the number of cases that are in node t.
- $\bar{y}(t)$ is the average response of the cases that fall into node t.

- $R(t)$ represents the weighted within-node sum-of-squares at node t.
- $R(T)$ is the average within-node sum-of-squares for the tree T.
- $\Delta R(s, t)$ denotes the change in the within-node sum-of-squares at node t using split s.

To construct a regression tree, we proceed as we did with classification trees. We seek to partition the space for the predictor values using a sequence of binary splits so that the resulting nodes are better in some sense than the parent node. Once we grow the tree, we use the minimum error complexity pruning procedure to obtain a sequence of nested trees with decreasing complexity. Once we have the sequence of subtrees, independent test samples or cross-validation can be used to select the best tree.

Growing a Regression Tree

We need a criterion that measures node impurity in order to grow a regression tree. We measure this impurity using the squared difference between the predicted response from the tree and the observed response. First, note that the predicted response when a case falls into node t is given by the average of the responses that are contained in that node,

$$\bar{y}(t) = \frac{1}{n(t)} \sum_{x_i \in t} y_i. \tag{13.43}$$

The squared error in node t is given by

$$R(t) = \frac{1}{n} \sum_{x_i \in t} (y_i - \bar{y}(t))^2. \tag{13.44}$$

Note that Equation 13.44 is the average error with respect to the entire learning sample. If we add up all of the squared errors in all of the terminal nodes, then we obtain the mean squared error for the tree. This is also referred to as the total within-node sum-of-squares, and is given by

$$R(T) = \sum_{t \in \tilde{T}} R(t) = \frac{1}{n} \sum_{t \in \tilde{T}} \sum_{x_i \in t} (y_i - \bar{y}(t))^2. \tag{13.45}$$

The regression tree is obtained by iteratively splitting nodes so that the decrease in $R(T)$ is maximized. Thus, for a split s and node t, we calculate the change in the mean squared error as

$$\Delta R(s, t) = R(t) - R(t_L) - R(t_R),   \qquad (13.46)$$

and we look for the split s that yields the largest $\Delta R(s, t)$.

We could grow the tree until each node is pure in the sense that all responses in a node are the same, but that is an unrealistic condition. Breiman et al. [1984] recommend growing the tree until the number of cases in a terminal node is five.

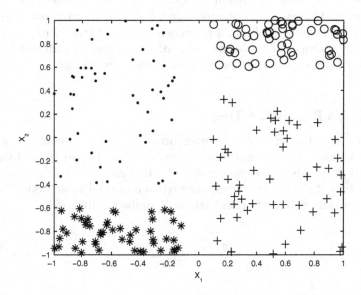

FIGURE 13.15
This shows the bivariate data used in Example 13.11. The observations in the upper right corner have response $y = 2$ ("o"); the points in the upper left corner have response $y = 3$ ("."); the points in the lower left corner have response $y = 10$ (""); and the observations in the lower right corner have response $y = -10$ ("+"). The response variables have no noise, and the regions are well-separated, so we should be able to perfectly partition the space with a small tree.*

Example 13.11

We show how to grow a regression tree using a simple example with generated data. We use bivariate data such that the response in each region is constant (with no added noise). We are using this simple toy example to illustrate the concept of a regression tree. In the next example, we will add noise to make the problem a *little* more realistic.

```
% Generate bivariate data.
X(1:50,1) = unifrnd(0.1,1,50,1);
X(1:50,2) = unifrnd(0.6,1,50,1);
y(1:50) = 2;
```

```
X(51:100,1) = unifrnd(-1,-.1,50,1);
X(51:100,2) = unifrnd(-0.4,1,50,1);
y(51:100) = 3;
X(101:150,1) = unifrnd(-1,-.1,50,1);
X(101:150,2) = unifrnd(-1,-0.6,50,1);
y(101:150) = 10;
X(151:200,1) = unifrnd(0.1,1,50,1);
X(151:200,2) = unifrnd(-1,0.4,50,1);
y(151:200) = -10;
```

These data are shown in Figure 13.15. The next step is to use the function **treefit** to get a tree. This function is available in the Statistics Toolbox.

```
% Now grow the tree.
tree = treefit(X,y(:))
```

This is the tree information we get back from the above command:

```
tree =
Decision tree for regression
1  if x1<-0.0107576 then node 2 else node 3
2  if x2<-0.498968 then node 4 else node 5
3  if x2<0.464465 then node 6 else node 7
4  fit = 10
5  fit = 3
6  fit = -10
7  fit = 2
```

We can use the following command to display the tree:

```
treedisp(tree); % Displays the tree
```

The tree is shown in Figure 13.16, and the partition view is given in Figure 13.17. Notice that the response at each node is exactly right because there is no noise. We see that the first split is at x_1, where values of x_1 less than zero (approximately) go to the left branch. Each resulting node from this split is partitioned based on x_2. The response of each terminal node is given in Figure 13.16, and we see that the tree does yield the correct response.
❑

Pruning a Regression Tree

Once we grow a large tree, we can prune it back using the same procedure that was presented for classification trees. We define an error-complexity measure as follows:

$$R_\alpha(T) = R(t) + \alpha|\widehat{T}| \qquad (13.47)$$

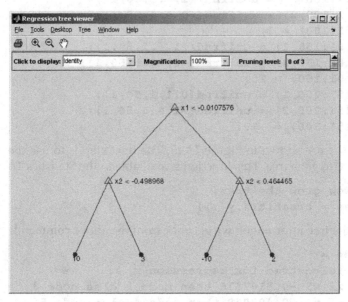

FIGURE 13.16
This is the regression tree for Example 13.11.

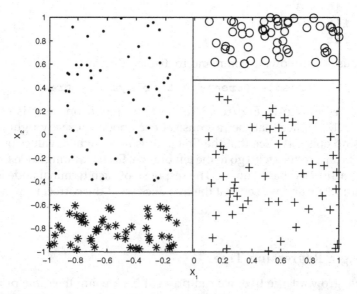

FIGURE 13.17
This shows the partition view of the regression tree from Example 13.11. Since we have bivariate data, it is easier to see how the space is partitioned with this display. The method first splits the region based on variable x_1. The left side of the space is then partitioned at $x_2 = -0.49$, and the right side of the space is partitioned at $x_2 = 0.46$.

From this we obtain a sequence of nested trees

$$T_{max} > T_1 > \ldots > T_K = \{t_1\},$$

where $\{t_1\}$ denotes the root of the tree. Along with the sequence of pruned trees, we have a corresponding sequence of values for α, such that

$$0 = \alpha_1 < \alpha_2 < \ldots < \alpha_k < \alpha_{k+1} < \ldots < \alpha_K.$$

Recall that for $\alpha_k \leq \alpha < \alpha_{k+1}$, the tree T_k is the smallest subtree that minimizes $R_\alpha(T)$.

Selecting a Tree

Once we have the sequence of pruned subtrees, we wish to choose the best tree such that the complexity of the tree and the estimation error $R(T)$ are both minimized. We could obtain minimum estimation error by making the tree very large, but this increases the complexity. Thus, we must make a trade-off between these two criteria.

To select the right sized tree, we must have honest estimates of the true error $R^*(T)$. This means that we should use cases that were not used to create the tree to estimate the error. As before, there are two possible ways to accomplish this. One is through the use of independent test samples and the other is cross-validation. We briefly discuss both methods, and the reader is referred to Chapter 10 for more details on the procedures. The independent test sample method is illustrated in Example 13.12.

To obtain an estimate of the error $R^*(T)$ using the independent test sample method, we randomly divide the learning sample $\mathbf{L}$ into two sets $\mathbf{L}_1$ and $\mathbf{L}_2$. The set $\mathbf{L}_1$ is used to grow the large tree and to obtain the sequence of pruned subtrees. We use the set of cases in $\mathbf{L}_2$ to evaluate the performance of each subtree, by presenting the cases to the trees and calculating the error between the actual response and the predicted response. If we let $D_k(\mathbf{x})$ represent the predictor corresponding to tree T_k, then the estimated error is

$$\hat{R}^{TS}(T_k) = \frac{1}{n_2} \sum_{(\mathbf{x}_i, y_i) \in \mathbf{L}_2} (y_i - D_k(\mathbf{x}_i))^2, \tag{13.48}$$

where the number of cases in $\mathbf{L}_2$ is n_2.

We first calculate the error given in Equation 13.48 for all subtrees and then find the tree that corresponds to the smallest estimated error. The error is an estimate, so it has some variation associated with it. If we pick the tree with the smallest error, then it is likely that the complexity will be larger than it should be. Therefore, we desire to pick a subtree that has the fewest number of nodes, but is still in keeping with the prediction accuracy of the tree with the smallest error [Breiman, et al. 1984].

First we find the tree that has the smallest error and call the tree T_0. We denote its error by $\hat{R}_{min}^{TS}(T_0)$. Then we find the standard error for this estimate, which is given by [Breiman, et al., 1984, p. 226]

$$\hat{SE}(\hat{R}_{min}^{TS}(T_0)) = \frac{1}{\sqrt{n_2}}\left[\frac{1}{n_2}\sum_{i=1}^{n_2}(y_i - D(\mathbf{x}_i))^4 - (\hat{R}_{min}^{TS}(T_0))^2\right]^{\frac{1}{2}}. \qquad (13.49)$$

We then select the smallest tree T_k^*, such that

$$\hat{R}^{TS}(T_k^*) \le \hat{R}_{min}^{TS}(T_0) + \hat{SE}(\hat{R}_{min}^{TS}(T_0)). \qquad (13.50)$$

Equation 13.50 says that we should pick the tree with minimal complexity that has accuracy equivalent to the tree with the minimum error.

If we are using cross-validation to estimate the prediction error for each tree in the sequence, then we divide the learning sample $\mathbf{L}$ into sets $\mathbf{L}_1, \ldots, \mathbf{L}_V$. It is best to make sure that the V learning samples are all the same size or nearly so. Another important point mentioned in Breiman, et al. [1984] is that the samples should be kept balanced with respect to the response variable Y. They suggest that the cases be put into levels based on the value of their response variable and that stratified random sampling be used to get a balanced sample from each stratum.

We let the v-th learning sample be represented by $\mathbf{L}^{(v)} = \mathbf{L} - \mathbf{L}_v$, so that we reserve the set $\mathbf{L}_v$ for estimating the prediction error. We use each learning sample to grow a large tree and to get the corresponding sequence of pruned subtrees. Thus, we have a sequence of trees $T^{(v)}(\alpha)$ that represent the minimum error-complexity trees for given values of α.

At the same time, we use the entire learning sample $\mathbf{L}$ to grow the large tree and to get the sequence of subtrees T_k and the corresponding sequence of α_k. We would like to use cross-validation to choose the best subtree from this sequence. To that end, we define

$$\alpha'_k = \sqrt{\alpha_k \alpha_{k+1}}, \qquad (13.51)$$

and use $D_k^{(v)}(\mathbf{x})$ to denote the predictor corresponding to the tree $T^{(v)}(\alpha'_k)$. The cross-validation estimate for the prediction error is given by

$$\hat{R}^{CV}(T_k(\alpha'_k)) = \frac{1}{n}\sum_{v=1}^{V}\sum_{(\mathbf{x}_i, y_i) \in L_v}(y_i - D_k^{(v)}(\mathbf{x}_i))^2. \qquad (13.52)$$

We use each case from the test sample $\mathbf{L}_v$ with $D_k^{(v)}(\mathbf{x})$ to get a predicted response, and we then calculate the squared difference between the predicted response and the true response. We do this for every test sample and all n cases. From Equation 13.52, we take the average value of these errors to estimate the prediction error for a tree.

We use the same rule as before to choose the best subtree. We first find the tree that has the smallest estimated prediction error. We then choose the tree with the smallest complexity such that its error is within one standard error of the tree with minimum error.

We obtain an estimate of the standard error of the cross-validation estimate of the prediction error using

$$\hat{SE}(\hat{R}^{CV}(T_k)) = \sqrt{\frac{s^2}{n}}, \tag{13.53}$$

where

$$s^2 = \frac{1}{n} \sum_{(\mathbf{x}_i, y_i)} [(y_i - D_k^{(v)}(\mathbf{x}_i))^2 - \hat{R}^{CV}(T_k)]^2. \tag{13.54}$$

Once we have the estimated errors from cross-validation, we find the subtree that has the smallest error and denote it by T_0. Finally, we select the smallest tree T_k^*, such that

$$\hat{R}^{CV}(T_k^*) \le \hat{R}_{min}^{CV}(T_0) + \hat{SE}(\hat{R}_{min}^{CV}(T_0)) \tag{13.55}$$

Since the procedure is somewhat complicated for cross-validation, we list the procedure next.

PROCEDURE – CROSS-VALIDATION METHOD

1. Given a learning sample $\mathbf{L}$, obtain a sequence of trees T_k with associated parameters α_k.

2. Determine the parameter $\alpha'_k = \sqrt{\alpha_k \alpha_{k+1}}$ for each subtree T_k.

3. Partition the learning sample $\mathbf{L}$ into V partitions, $\mathbf{L}_v$. These will be used to estimate the prediction error for trees grown using the remaining cases.

4. Build the sequence of subtrees $T_k^{(v)}$ using the observations in all $\mathbf{L}^{(v)} = \mathbf{L} - \mathbf{L}_v$.

5. Now find the prediction error for the subtrees obtained from the entire learning sample $\mathbf{L}$. For tree T_k and α'_k, find all equivalent trees $T_k^{(v)}$, $v = 1, ..., V$ by choosing trees $T_k^{(v)}$ such that

$$\alpha'_k \in [\alpha_k^{(v)}, \alpha_{k+1}^{(v)}).$$

6. Take all cases in $\mathbf{L}_v, v = 1, ..., V$ and present them to the trees found in step 5. Calculate the error as the squared difference between the predicted response and the true response.

7. Determine the estimated error for the tree $\hat{R}^{CV}(T_k)$ by taking the average of the errors from step 6.

8. Repeat steps 5 through 7 for all subtrees T_k to find the prediction error for each one.

9. Find the tree T_0 that has the minimum error,

$$\hat{R}_{min}^{CV}(T_0) = \min_k \left\{ \hat{R}^{CV}(T_k) \right\}.$$

10. Determine the standard error for tree T_0 using Equation 13.53.

11. For the final model, select the tree that has the fewest number of nodes and whose estimated error is within one standard error of $\hat{R}_{min}^{CV}(T_0)$.

The MATLAB Statistics Toolbox has a function called **treetest** that computes the cost of a tree using one of three methods: (1) the resubstitution method in which the testing is done with the same training sample, (2) the use of an independent test set, and (3) cross-validation. The cost returned by this function for a regression tree is the average squared error over the observations in the node. It will also return the standard error of each value, the number of terminal nodes for each subtree, and the best level of pruning. If the best level of pruning is zero, then this means that the full tree is best (i.e., no pruning). The best level is the one within one standard error of the sub-tree with minimum error, as we described earlier.

Example 13.12

We return to the same data that was used in the previous example, where we now add random noise to the responses. We generate the data as follows.

```
% We add noise to the response variables, as follows:
y(1:50) = 2+sqrt(2)*randn(1,50);
y(51:100) = 3+sqrt(2)*randn(1,50);
y(101:150) = 10+sqrt(2)*randn(1,50);
y(151:200) = -10+sqrt(2)*randn(1,50);
```

The next step is to grow the tree. The T_{max} that we get from this tree should be larger than the one in Example 13.11.

```
% Fit the regression tree.
% Note that we do not have to use the 'method'
% argument because the responses are
% continuous indicating it is regression.
tree = treefit(X,y(:));
```

The tree we get has a total of 93 nodes, with 47 terminal nodes. We can use the MATLAB function called **treetest** to get the cost of the tree as described earlier. We will get the cross-validation error in this example and ask the reader to explore the other options as part of the exercises.

```
% Now find the error using cross-validation.
[cost,secost,nt,best] = ...
     treetest(tree,'cross',X,y(:));
```

The function uses 10-fold cross-validation. This is the default value and can be changed by the user. Note that the variable **best** is an index indicating the best level tree. It starts at zero, indicating no pruning should be done. If one wants to access the cost, number of terminal nodes, and the standard error for the best tree, then the index needs to be increased by one. The following code shows how to plot the cost as a function of the number of terminal nodes in the tree.

```
plot(nt,cost,nt(best+1),cost(best+1),'o')
xlabel('Number of Terminal Nodes')
ylabel('Cost')
```

This plot is shown in the top of Figure 13.18. The circle indicates the best tree, which has four terminal nodes. We can prune the tree using

```
treebest = treeprune(tree,'level',best)
```

This tree is shown in the bottom part of Figure 13.18, and we see that it is a reasonable fit given the previous example.
❏

13.7 Additive Models

We saw that regression trees are a way of obtaining nonparametric regression fits in the multivariate case. Another approach would be to extend the univariate local polynomial regression (loess) to the multivariate case, but we do not cover it here. Cleveland [1993], Fox [2000a], and Martinez et al. [2010] provide details on this method. Another example of nonparametric multiple regression is called projection pursuit regression [Friedman and Stuetzle, 1981; Fox, 2000a]. We will first briefly describe these methods and the problems associated with them, which will be followed by a brief review

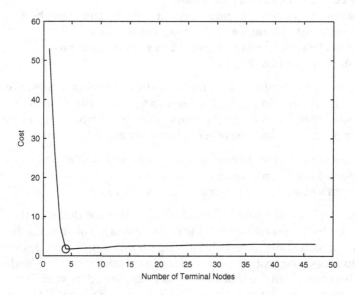

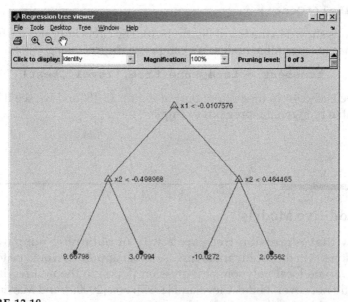

FIGURE 13.18

The top part of this figure shows a plot of the estimated error using cross-validation. Note that there is a sharp minimum for $|\hat{T}_k| = 4$. The bottom part of the figure shows the pruned tree in which the best level was selected using cross-validation. We see that the result is reasonable when we compare this with the tree in Example 13.11.

of the linear multiple regression model. The section concludes with a description of the additive model for nonparametric regression in multiple dimensions.

One of the issues with extending the univariate smoothers like loess to higher dimensions is the difficulty in defining the local neighborhood. There is also a problem with the curse of dimensionality [Bellman, 1961], which states that as the number of predictor variables increases, the number of points in the local neighborhood of a target value decreases. In other words, the neighborhoods have to become very large to include the same number of points. Another problem has to do with interpreting the results. It becomes more difficult to understand and to graphically convey the resulting model as the dimensionality increases.

Projection pursuit regression is one way of dealing with regression in the multivariate case. It fits a model where each of the additive components is a function of linear combinations of the x's. One advantage of this approach is that the model is more general than the additive regression model because it can capture certain interactions among the predictors. On the other hand, the projection pursuit model can be more difficult to interpret, since arbitrary linear combinations might not represent meaningful quantities. Also, finding a projection pursuit regression fit is not as straightforward as additive regression.

Before we explain additive models, we first review the linear case for multiple regression and discuss some of the limitations involved with estimating and interpreting the fit. We start off with n values of our response variable collected in a vector $\mathbf{y} = (y_i, ..., y_n)^T$. Each y_i has a corresponding d-dimensional vector that contains the d observed predictor values. This is denoted by $\mathbf{x}_i = (x_{i1}, ..., x_{id})^T$, and these might be fixed or realizations of a random variable.

Recall that the standard linear multiple regression model is given by

$$Y = \alpha + \beta_1 X_1 + ... + \beta_d X_d + \varepsilon, \tag{13.56}$$

where the errors ε are independent of the X_j, the expected value of ε is zero, and the variance of ε is σ^2. As with simple univariate regression, the goal is to estimate the coefficients β_i given the data, so we can use the fitted model to describe the relationship between variables and to make predictions.

The model for multiple regression given in Equation 13.56 makes an assumption that the dependence of the expected value of the response Y, $E[Y]$, on each of the predictors is linear. A consequence of this model is the ease of interpretation that results because it is additive in the predictors. Once the estimated regression is obtained, we can look at the predictor effects separately. We seek the same type of interpretability with the additive model.

The nonparametric **additive model** retains this property of interpretability. Following the treatment in Hastie and Tibshirani [1990], the additive model is given by

$$Y = \alpha + \sum_{j=1}^{d} f_j(X_j) + \varepsilon. \tag{13.57}$$

where the errors ε have the same properties as before, and we assume that $E[f_j(X_j)] = 0$. Note that we have separate univariate functions for each predictor $f_j(X_j)$. Usually, the $f_j(X_j)$ are univariate and smooth, but they do not have to be.

The additive model and the procedure to fit them can accommodate component functions that are quite general, such as multivariate functions, linear or polynomial regression, splines, logistic regression, etc. See Hastie and Tibshirani [1990] for a thorough discussion of the possibilities with the additive model.

We fit the additive model using a procedure called the **backfitting algorithm**. This is a general algorithm that allows us to fit the additive model using any regression fitting method to find the component functions $f_j(X_j)$. It is an iterative procedure, and convergence is guaranteed in many common cases [Hastie and Tibshirani, 1990].

If the additive model in Equation 13.57 is valid, then for the j-th individual component function, we should be able to remove the effects of the other predictor variables from the response to get the **partial residual**. Then we can fit the partial residual against the x_{ij} to estimate f_j. We continue to do this in an iterative fashion until the functions do not change. This, in essence, is the backfitting algorithm.

To address the issue of the constant α in the model, we will find the nonparametric regression fit to get the component function and then subtract off the average of the smooth. This will ensure that the component functions have zero mean at each stage of the procedure [Hastie and Tibshirani, 1990].

PROCEDURE – BACKFITTING ALGORITHM FOR ADDITIVE MODELS

1. We are given n observations consisting of a response variable y_i and the corresponding d predictors $\mathbf{x}_i = (x_{i1}, ..., x_{id})^T$, and we want to fit the model

$$y_i = \alpha + f_1(x_{i1}) + ... + f_d(x_{id}) + \varepsilon_i$$

for $i = 1, ..., n$ and $j = 1, ..., d$.

2. Initialize the algorithm by finding starting estimates $\hat{f}_j$. A logical starting fit can be found using linear regression on:

$$y_i - \bar{y} = \beta_1 x_{i1} + \dots + \beta_d x_{id} + e.$$

So, we now have initial estimates given by

$$\hat{f}_j^{(0)}(x_{ij}) = \hat{\beta}_j x_{ij}; \qquad j = 1, \dots, d.$$

3. At each iteration, form the estimated partial residual for the *j*-th predictor

$$e_{ij} = y_i - \bar{y} - \sum_{j \neq k} f_k.$$

Apply any desired nonparametric regression method, such as loess or smoothing splines, on the partial residual e_{ij} against the x_{ij} to get a revised estimate $\hat{f}_j$.

4. Subtract off the average of the fit found in step 3.
5. Continue cycling through the functions $\hat{f}_j, j = 1, \dots, d, 1, \dots, d, \dots$, until they converge.

The resulting additive model will be an approximation to the true regression surface. However, it is hopefully a useful one as it allows us to more easily interpret the results by examining 2D scatterplots of partial residuals with their corresponding curves. Also, note that these relationships for the partial residual versus a single predictor are conditional, given all the other predictors in the model. The backfitting procedure to get additive models is shown in the next example.

Example 13.13

We will use an example from Simonoff [1996] to demonstrate the backfitting algorithm for additive models. This data set has gasoline consumption (measured in tens of millions of dollars) as the response variable. The predictors are the price index for gasoline (dollars), the price index for used cars (dollars), and the per capita disposable income (dollars). The data span the years 1960 through 1986, and the dollars are measured in 1967 dollars. As usual, we first load the data and specify the variables.

```
load gascons
% Put all of the observations into one data matrix.
% Center the data first.
x1 = dispinc;
x2 = pricegas;
x3 = pricecars;
X = [x1(:),x2(:),x3(:)];
```

```
[n,d] = size(X);
yc = gascons - mean(gascons);
% Initialize some other arrays.
fcng = zeros(n,d);
r = zeros(n,d);
```

We show a scatterplot of the gasoline consumption and price index for gasoline in Figure 13.19. Superimposed on the scatterplot is a loess fit, and this shows a nonlinear relationship where demand initially increases as the cost of gasoline increases. This does not fit our common understanding that the demand should decrease as price increases. Simonoff [1996] shows that additive models provides some further insight when we account for the other predictor variables. The following MATLAB code implements the backfitting algorithm for additive models and applies it to the gasoline consumption data.

```
% Get the initial estimate using least-squares.
bhat = X\yc;
% Now use these as starting points for the individual
% component functions in additive models.
b = repmat(bhat',n,1);
fold = b.*X;
% Now start the iteration.
delta = 1;
I = 0;
col = [1 0 0;0 1 0; 0 0 1];
while delta >= 0.001
    I = I + 1;
    disp(['Iteration number ', int2str(I)])
    % Loop through the component functions and
    % smooth the partial residuals.
    for i = 1:d
        J = 1:d;
        % Form the i-th partial residual.
        J(i) = [];
        r(:,i) = yc(:) - sum(fold(:,J),2);
        % Smooth r_i against x_i.
        fnew = csloess(X(:,i),r(:,i),X(:,i),2/3,1);
        % Now subtract the mean of the smooth.
        fnew = fnew - mean(fnew);
        fcng(:,i) = abs(fnew(:) - fold(:,i));
        fold(:,i) = fnew(:);
    end
    % Now check for convergence.
    delta = max(max(fcng));
    disp(['Max error is ', num2str(delta)]);
end
```

The results are shown in Figures 13.20 through 13.22. Each plot shows the contributions of the various predictors to the additive model estimate of gasoline consumption. We see that the conditional per capita disposable income given the other variables has an approximately linear relationship with consumption increasing as the disposable income increases. Also, we see the familiar relationship between price and demand in Figure 13.21, where we now have an inverse relationship between the price index of gasoline and consumption, given per capita disposable income and the price index of used cars.

❑

Hastie and Tibshirani [1990] discuss an extension to generalized linear models (and additive models) called *generalized additive models*. This approach extends generalized linear models in the same way as additive models extended linear regression, by replacing the linear form in generalized linear models with the additive form. The procedure to estimate generalized additive models is similar to the linear model case. The basic change that has to be made in the procedure is the use of the desired weighted additive fit instead of the weighted linear regression. We will not discuss the generalized additive models further, and we refer the reader to Hastie and Tibshirani [1990] for details.

13.8 Multivariate Adaptive Regression Splines

Multivariate adaptive regression splines was developed by Friedman [1991a] as a nonparametric approach to regression and model-building. As such, it makes no assumptions about the relationship between the predictors and the response variable. *Multivariate adaptive regression splines* is a flexible method that can accommodate various levels of interactions, as well as non-linearities. It is particularly useful in situations where we have very high-dimensional data. As we see, it has some similarities to other nonparametric methods, such as additive models and regression trees.

We follow the notation and description of multivariate adaptive regression splines given in Hastie, Tibshirani, and Friedman [2009]. The model is given by

$$f(X) = \beta_0 + \sum_{m=1}^{M} \beta_m h_m(X),$$
(13.58)

where h_m is a basis function, which can be a hinge function (defined next) or a product of two or more functions (interactions).

A *hinge function* is defined as

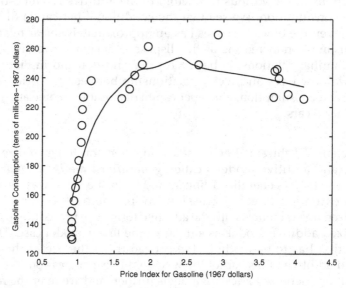

FIGURE 13.19
This is a scatterplot of total gasoline consumption in tens of millions of 1967 dollars versus the price index for gasoline. This encompasses the years 1960 through 1986. The relationship seems to be nonlinear with the consumption actually increasing as the price increases. The fit shown here is obtained using loess.

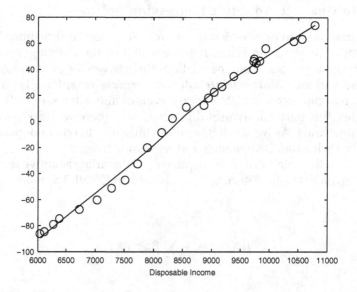

FIGURE 13.20
Here we have the contributions of the disposable income to the additive model estimate of gas consumption, given the remaining variables.

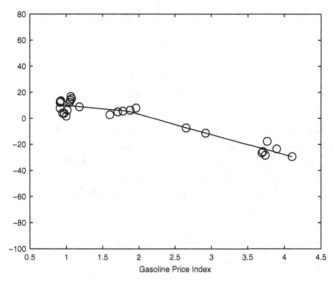

FIGURE 13.21
This figure shows the contribution of the price index for gasoline to the additive model, given the per capita disposable income and price index of used cars. We now see the familiar inverse relationship between price and demand.

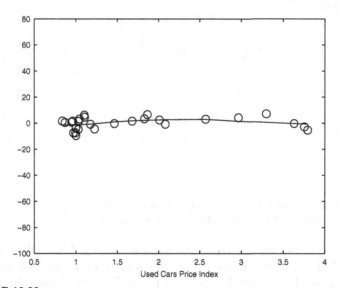

FIGURE 13.22
This is the contribution of the price index for used cars to the additive model, given per capita disposable income and the price index for gasoline. We see that the relationship is somewhat nonlinear.

$$(X - t)_+ = \begin{cases} X - t & \text{if } X > t \\ 0 & \text{otherwise} \end{cases}$$

and   (13.59)

$$(t - X)_+ = \begin{cases} t - X & \text{if } X < t \\ 0 & \text{otherwise} \end{cases}$$

The "+" notation means that we take only the positive part, and the constant t denotes the knot. Hinge functions are also called two-sided truncated functions, and we can see from the definition that they are mirrors of each other. An example of hinge functions with a knot at $t = 0$ is shown in Figure 13.23. The both of them together are called a ***reflected pair***.

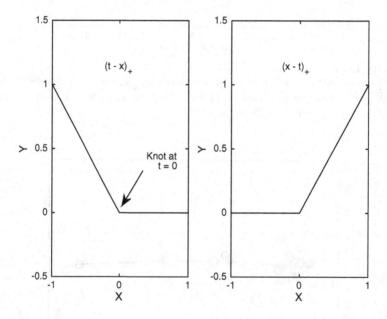

FIGURE 13.23
This shows a reflected pair of hinge functions $(t - X)_+$ and $(X - t)_+$ for knot $t = 0$. Note that the function is zero over part of its domain because we take only the positive part.

The multivariate adaptive regression splines approach uses these reflected pairs of hinge functions (Equation 13.59) to form a set of ***basis functions*** from which we build our model (Equation 13.58). The set of basis functions is given by

$$C = \{(X_j - t)_+, (t - X_j)_+\},$$ (13.60)

where $j = 1, ..., d$. In Equation 13.60, the possible knot points t are taken from the unique observed values for the predictors, so $t \in \{x_{1j}, x_{2j}, ..., x_{nj}\}$. The basis functions $h_m(X)$ in Equation 13.58 are elements of C. They can also be a product of two or more elements of C.

The model estimation process is similar to classification and regression trees in that we construct a rather large model and then remove terms using some optimality criterion. We start the process by including the constant term in the model. At this point, we have

$$f(X) = \beta_0,$$

and $h_0(X) = 1$ is the basis function.

Next, we search for new basis functions to add to the model by looking at products of the functions already in our model with reflected pairs from C. The reflected pair of hinge functions that yields the largest reduction in training error is added to the model, and all β_m and β_0 are estimated using the ordinary least squares method. The forward model-building process for multivariate adaptive regression splines is outlined next.

PROCEDURE – MULTIVARIATE ADAPTIVE REGRESSION SPLINES

1. Start with the basis function $h_0(X) = 1$ and set $M = 0$.

2. Form the product of the basis functions currently in the model with each of the reflected pairs in C. This will produce terms with the following form

$$\beta_{M+1} h_k(X)(X_j - t)_+ + \beta_{M+2} h_k(X)(t - X_j)_+ ,$$

where the knots t are taken over all unique observed values of x_{ij}, and $h_k(X)$ is a basis function in the current model.

3. Estimate the $M + 3$ coefficients (includes the constant term) in the proposed models using least squares.

4. Add the product of terms that results in the largest decrease in the training error.

5. Set $M = M + 2$ and repeat steps 2 through 4 until some maximum number of terms is reached.

This yields a potentially large model, which likely overfits the data. As with classification and regression trees, we do some pruning of the model. This is accomplished by removing the term yielding the smallest increase in the residual squared error. In this way, we get an estimated model for a given number of terms λ. Hastie, Tibshirani, and Friedman [2009] use generalized cross-validation to choose the number of terms λ, as given here

$$GCV(\lambda) = \frac{\sum_{i=1}^{n}(y_i - \hat{f}_\lambda(x_i))^2}{(1 - M(\lambda)/n)^2},$$ (13.61)

where $\hat{f}_\lambda(x)$ is the estimated model with λ terms, and $M(\lambda)$ is the effective number of terms in the model. This is calculated using

$$M(\lambda) = r + cK,$$ (13.62)

with K denoting the number of knots resulting from the model-building process, and r is the number of linearly independent basis functions in the model $\hat{f}_\lambda(x)$ (includes the intercept term). Hastie, Tibshirani, and Friedman [2009] note that the number of effective terms should account for the selection of knots. They propose a value of two or three for the constant c [Friedman, 1991a]. The model with the smallest GCV is chosen.

It might help to look at the forward model-building process in action, where we consider a model with no interactions (i.e., the additive model). In step 1, we have just the constant term $h_0(X) = 1$ in the model. In step 2, we multiply this (i.e., functions in the current model) by reflected pairs of hinge functions from C. This will result in terms like

$$\beta_1(X_j - t)_+ + \beta_2(t - X_j)_+,$$

for knots $t \in \{x_{ij}\}$. We will represent the winning product from this search as

$$\beta_1(X_J - x_{IJ})_+ + \beta_2(x_{IJ} - X_J)_+,$$

where the knot producing the best decrease in the training error is from the I-th observed value of predictor variable J (i.e., $t = x_{IJ}$).

We are now back at step 2 for the second iteration, and this time our current basis functions in the model are

$$h_0(X) = 1$$
$$h_1(X) = (X_J - x_{IJ})_+$$
$$h_2(X) = (x_{IJ} - X_J)_+ .$$

Note that we leave off the coefficients above because they are not part of the basis function and are estimated in step 3 as we consider additional terms for the model.

We multiply each of the model terms above by the candidate functions in C. The resulting products will be of the following forms:

Products with $h_0(X)$: $\beta_3(X_j - t)_+ + \beta_4(t - X_j)_+$

Products with $h_1(X)$: $(X_J - x_{IJ})_+[\beta_3(X_j - t)_+ + \beta_4(t - X_{j+})_+]$

Products with $h_2(X)$: $(x_{IJ} - X_J)_+[\beta_3(X_j - t)_+ + \beta_4(t - X_j)_+]_.$

There are some additional aspects of the forward-building process we should consider. As we see from the procedure above, one can specify the maximum number of terms to add, which stops the forward process. One can also set the highest number of interactions (products) allowed. Setting a limit of one (no interactions) yields an additive model. Finally, there is a further constraint applied to the products. Higher-order powers of an input are not allowed, which reduces the computational burden. Based on the example above, this means that the variable X_J cannot be used in the candidate pair of hinge functions.

The multivariate adaptive regression splines approach to model-building has some attractive properties. The method can handle data sets with many observations (large n) and predictors (large d), as well as data sets with both continuous and categorical data [Friedman, 1991b]. There is a faster process for building these models described in Friedman [1993], which can help with massive data sets.

The final models from multivariate adaptive regression splines tend to be interpretable, especially for interaction limits of one (additive) and two (pairwise). The models can also be parsimonious, as compared to trees and random forests. Thus, they save on storage space for the model and can be fast when used for making predictions.

There is a constant c used in the calculation of the $M(\lambda)$ in the GCV. Some theory [Friedman, 1991a; Jekabsons, 2015] suggests values from two to four should be used. If $c = 0$, then GCV is not penalizing for the number of knots—only the number of terms. When c is large, then the penalty for knots increases, and fewer knots are used, resulting in a simpler model. Hastie, Tibshirani, and Friedman [2009] recommend a value of $c = 1$ when the additive model is specified (uses maximum interactions of one).

Example 13.14

Jekabsons [2015] published a toolbox for multivariate adaptive regression splines called ARESLab. It provides several useful functions for building and analyzing the models. We illustrate just a few of them in this example, but all are summarized in Table 13.3. We use the **peaks** function to generate noise-free data, as shown here. Since this is for illustration purposes, we did not add noise to the response **Y**.

```
% Use the peaks function to generate
% data with no noise.
[X,Y,Z] = peaks(100);
% Create a matrix of predictors.
X = [X(:), Y(:)];
```

```
% Set the response.
Y = Z(:);
```

There is an optional function called **aresparams**, which is used to set one or more of sixteen parameters for constructing the model. The user does not have to provide values for all of the parameters because default values are provided. To build our model, we are going to request a maximum of 300 basis functions for the forward stage, no penalty for knots ($c = 0$), linear basis functions, and pairwise interactions. We only have to list the first six input arguments because the default value of the rest of the inputs are sufficient.

```
maxfunc = 300;
c = 0;
maxprod = 2;
params = aresparams(maxfunc,c,false,...
    [],[],maxprod);
```

Note that we use the empty matrix **[ ]** where we want to use the defaults. We set the maximum number of basis functions for the forward process to a large number. Jekabsons [2015] implemented two other mechanisms, which might halt the process before the maximum number of basis functions is reached. This happens if the number of coefficients in the current model is larger than the number of observations or the improvement in the training error is less than some threshold (can be set in **aresparams**). We set the value of c used in the calculation of the effective number of parameters (Equation 13.62) to zero. This means that we are not penalizing the number of knots, only the number of terms. The implementation of multivariate adaptive regression splines in ARESLab provides the option of using piecewise-cubic functions instead of linear basis functions in the model-building process [Friedman, 1991]. The default is to first use piecewise-linear functions for the forward and backward phases, and then to use cubic functions for the final model. We want linear basis functions for the final model, so we use **false** for the third argument. Finally, we allow a maximum interaction of two, as given by the last argument specified in **aresparams**. We now construct the model based on these parameters using the **aresbuild** function in ARESLab.

```
% Build the model.
[ARES, time] = aresbuild(X,Y,params);
```

As we see from the following output displayed in the command window that the process took a long time, mostly because we set a large maximum value for the number of terms in the forward process. It is interesting to note that the lack of improvement in the training error stopped the forward process, not the number of functions allowed in the model.

```
% This appears in the command window
Building ARES model...
Forward phase  ................................
```

```
Termination condition is met:
R2 improvement is below threshold.
Backward phase ...............................

Number of basis functions in the final model: 95
Total effective number of parameters: 95.0
Highest degree of interactions in the final model: 2
Execution time: 1648.63 seconds
```

We get some other useful information, such as the number of basis functions (95) and the highest degree of interactions (2) in the final model. We can plot the estimated function using **aresplot**.

```
% View the estimated relationship.
aresplot(ARES)
xlabel('X_1')
ylabel('X_2')
zlabel('Y')
```

The resulting plot is shown in Figure 13.24, where we see the multivariate adaptive regression splines approach does a reasonable job estimating the relationship. Of course, the data are noise-free, so we would expect a good result. We can see the effect of using piecewise-linear basis functions, especially in the peaks. We could set the arguments to the default, which is to use cubic splines after the forward and backward passes. This would give a smoother relationship in the final model.

❑

13.9 MATLAB® Code

The basic MATLAB software does not have functions for nonparametric regression. There is a Curve Fitting Toolbox from The MathWorks, Inc. that will construct smoothing splines. It also has functions to fit parametric and nonparametric models to the data. This toolbox also has capabilities for preprocessing and postprocessing the data, to obtain statistics to help analyze the models, and graphical user interfaces to make it easier to explore and understand the relationships between the variables.

The Statistics Toolbox continues to improve by adding tools for building models. As we mentioned earlier in relation to classification trees, the toolbox includes functions that create and use regression trees. These functions are **treefit, treedisp, treeval, treeprune,** and **treetest**. The tree functions we used in this chapter will work with recent versions of the MATLAB Statistics Toolbox, but some may be removed in the future.

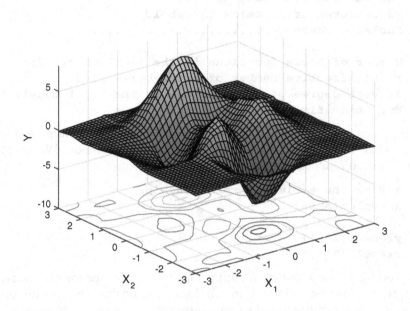

FIGURE 13.24
Here is our estimated surface of the noise-free **peaks** *function using the multivariate adaptive regression spline approach implemented in ARESLab [Jekabsons, 2015]. Linear basis functions are used in all phases of the model-building process, and we can see the effects in the peaks. In spite of this, the method finds the right overall structure.*

The toolbox includes a function **classregtree** that creates a tree *object* for use in predicting response values from given predictor values. There are some additional functions for operating on these tree objects. These are **view**, **eval**, **test**, and **prune**. We list relevant functions in Table 13.1. Since some of these functions might be removed in later versions of MATLAB, they recommend using the **fitrtree** function instead.

The latest version of the Statistics Toolbox includes two additional nonparametric techniques not described in this book. The first is **fitrgp** that fits Gaussian process models. The second function is **fitrsvm**. This fits regression models using support vector machines.

The smoothing techniques described in *Visualizing Data* [Cleveland, 1993] have been implemented in MATLAB and are available at

http://www.datatool.com/resources.html

for free download. A MATLAB function from this tool that might be of particular interest to readers is one that will produce a 2D nonparametric regression fit using local polynomial regression. We provide several functions in the Computational Statistics Toolbox for local polynomial smoothing, loess, regression trees, and others. These are listed in Table 13.2.

TABLE 13.1

Functions in the MATLAB Statistics Toolbox

Purpose	MATLAB Function
These functions are used for fitting, testing, pruning, evaluating, and displaying trees.	`fitrtree` `treefit` `treetest` `treeprune` `treeval` `treedisp`
This function creates a tree class.	`classregtree`
These functions are used for testing, pruning, evaluating, and displaying trees objects.	`test` `prune` `eval` `view`

TABLE 13.2

List of Functions from Chapter 13 Included in the Computational Statistics Toolbox

Purpose	MATLAB Function
These functions are used for smoothing.	`csloess` `csloessr` `csbinsmth` `csrmeansmth`
This function does local polynomial smoothing.	`cslocpoly`
These functions provide confidence bands for various smooths.	`csloessenv` `csbootband`
This function performs nonparametric regression using kernels.	`csloclin`
This function will construct the cubic smoothing spline.	`cssplinesmth`
This function will return the additive model using smoothing splines.	`csadmodel`

The ARESLab Adaptive Regression Splines Toolbox for MATLAB was downloaded from

http://www.cs.rtu.lv/jekabsons/regression.html

The toolbox has a very detailed reference manual. It is included in the toolbox, but can also be downloaded separately at the above link. It has some good examples showing how to use the functions in ARESLab. Most of the information in the manual explaining the functions are passed to the command window when one uses the help on the ARESLab functions. We list the toolbox functions in Table 13.3.

TABLE 13.3

List of Functions in the ARESLab Toolbox [Jekabsons, 2015]

Purpose	MATLAB Function
Sets or gets the parameters for an ARES model	**aresparams**
Constructs the model—forward and backward passes	**aresbuild**
Gets an estimate of the optimal value for the parameter c used in the GCV	**arescvc**
Manual deletion of a basis function from the model	**aresdel**
Make predictions using the model	**arespredict**
Test the performance of the model	**arestest**
	arescv
Display the model	**aresplot**
	areseq
Perform ANOVA	**aresanova**
	aresanovareduce

13.10 Further Reading

For more information on loess, Cleveland's book *Visualizing Data* [1993] is an excellent resource. It contains many examples and is easy to read and understand. In this book, Cleveland describes many other ways to visualize data, including extensions of loess to multivariate predictors. The paper by Cleveland and McGill [1984] discusses other smoothing methods such as polar smoothing, sum-difference smooths, and scale-ratio smoothing.

For a more theoretical treatment of smoothing methods, the reader is referred to Simonoff [1996], Wand and Jones [1995], Bowman and Azzalini [1997], Green and Silverman [1994], Scott [2015], and Hastie and Tibshirani [1990]. The text by Loader [1999] describes other methods for local regression and likelihood that are not covered in our book.

One smoothing technique that we did not discuss in this book is exponential smoothing. This method is typically used in applications where the predictor represents time (or something analogous), and measurements are taken over equally spaced intervals. This type of smoothing is typically covered in introductory texts; one possible resource for the interested reader is Wadsworth [1990].

For a discussion of boundary problems with kernel estimators, see Wand and Jones [1995] and Scott [2015]. Both of these references also compare the performance of various kernel estimators for nonparametric regression. When we discussed approaches for probability density estimation in Chapter 9, we presented some results from Scott [2015] regarding the integrated squared error that can be expected with various kernel estimators. Since the local kernel estimators are based on density estimation techniques, expressions for the squared error can be derived. See Scott [2015], Wand and Jones [1995], and Simonoff [1996] for details.

The *SiZer method* of Chaudhuri and Marron [1999] might be of interest to readers. SiZer stands for statistically SIgnificant ZERo crossings of derivatives, and it incorporates several important concepts from *scale space* ideas that arise in computer vision [Lindeberg, 1994]. In the smoothing context, *scale* corresponds to the amount of smoothing. SiZer attempts to answer the question in smoothing (or kernel density estimation): "What observed features are due to significant structure, and which ones cannot be distinguished from background sampling variability?"

SiZer looks at a many smooths at different levels of smoothing. Features or structures are determined to be statistically significant by looking at the derivatives of the smooths (with respect to x). SiZer has been implemented in MATLAB by Chaudhuri and Marron, and it can be used with kernel density estimation or nonparametric regression. The SiZer code includes two plots to help in the analysis. One is a family of smooths superimposed over the scatterplot, and the other is called a SiZer map.

The SiZer map is a color map where each pixel represents a point in scale space. The index along the horizontal axis of the SiZer map corresponds to the x axis, and the vertical represents the degree of smoothing. A confidence interval for the derivative is determined (see Chaudhuri and Marron, [1999] for details), and this is an indication of the slope of the curve. If both end points of this interval are above zero, then the smooth is significantly increasing. In this case, the pixel is shown in a dark shade or in blue. If the interval is below zero, the smooth is significantly decreasing, and the pixel is shown in red or in a light shade. If the interval contains zero, then the pixel is displayed in gray or purple. The code and documentation are available at

http://www.unc.edu/~marron/marron.html#redirect

For information on smoothing and additive models, we highly recommend *Generalized Additive Models* by Hastie and Tibshirani [1990]. This book includes some useful information on other approaches for univariate and multivariate nonparametric regression.

Other excellent sources on nonparametric regression with splines is Green and Silverman [1994]. This text is easy to understand, provides a lot of the underlying theory, and addresses the computational issues. Fox [2000a] is a useful book for readers who want an overview of smoothing methods, additive models, regression trees, and projection pursuit regression. Finally, while not the main topic of them, the books by Simonoff [1996] and Scott [2015] have discussions on additive models and the backfitting algorithm.

Exercises

13.1.   Fit a loess curve to the **environ** data using $\lambda = 1, 2$ and various values for α. Compare the curves. What values of the parameters seem to be the best? In making your comparison, look at residual plots and smoothed scatterplots. One thing to look for is excessive structure (wiggliness) in the loess curve that is not supported by the data.

13.2.   Write a MATLAB function that implements the Priestley-Chao estimator.

13.3.   Repeat Example 13.5 for various values of the smoothing parameter h. What happens to your curve as h goes from very small values to very large ones?

13.4.   The **human** data set [Hand, et al., 1994; Mazess, et al., 1984] contains measurements of percent fat and age for 18 normal adults (males and females). Use loess or one of the other smoothing methods to determine how percent fat is related to age.

13.5.   The data set called **anaerob** has two variables: oxygen uptake and the expired ventilation [Hand, et al., 1994; Bennett, 1988]. Use loess to describe the relationship between these variables.

13.6.   The **brownlee** data contains observations from 21 days of a plant operation for the oxidation of ammonia [Hand, et al., 1994; Brownlee, 1965]. The predictor variables are X_1 is the air flow, X_2 is the cooling water inlet temperature (degrees C), and X_3 is the percent acid concentration. The response variable Y is the stack loss (the percentage of the ingoing ammonia that escapes). Use a regression tree to determine the relationship between these variables. Pick the best tree using cross-validation.

13.7.   The **abrasion** data set has 30 observations, where the two predictor variables are hardness and tensile strength. The response variable is abrasion loss [Hand, et al., 1994; Davies and Goldsmith, 1972]. Construct a regression tree using cross-validation to pick a best tree.

13.8. The data in **helmets** contains measurements of head acceleration (in g) and times after impact (milliseconds) from a simulated motorcycle accident [Hand, et al., 1994; Silverman, 1985]. Do a loess smooth on these data. Include the upper and lower envelopes. Is it necessary to use the robust version?

13.9. Use regression trees on the **boston** data set. Choose the best tree using an independent test sample (taken from the original set) and cross-validation.

13.10. Discuss the case of $n = 1$ and $n = 2$ in minimizing the penalized sum of squares in Equation 13.23.

13.11. Explain why the fit from the smoothing spline approaches the fit from linear regression as the smoothing parameter goes to infinity.

13.12. Write a MATLAB function that implements a running-median smoother. This is similar to the running-mean, but the median is calculated using the data in each window rather than the mean. Use it to smooth the **vineyard** data and compare the results with other methods from this chapter.

13.13. Write MATLAB code that will find and plot the value of the cross-validation function for the running mean and the locally-weighted running-line smoothing parameters.

13.14. Write a function that will construct the running line smoother, using the MATLAB code in Example 13.2 (or write your own version). Smooth the scatterplot for the **vineyard** data and compare with the smooths obtained using the other methods.

13.15. Smooth the **vineyard** data using loess and the kernel estimator. Compare with the smooths obtained using the other methods.

13.16. Repeat the smooth of the **vineyard** data using the running mean smoother (Example 13.2) with $N = 5, 10, 15, 20$. Discuss what happens with the smooth.

13.17. Write a MATLAB function that will return a value for the smoothing spline given the target value x_0, the knots t_i, γ_i, and $f(t_i)$.

13.18. Write MATLAB code that will find the pointwise confidence bands for smoothing splines using ± 2 times the square root of the diagonal elements of $\mathbf{SS}^T \sigma^2$ and the estimate of the variance in Equation 13.42. Apply this to the environmental data of Example 13.10 and compare to the bands obtained using the bootstrap.

13.19. Use regression trees to model the relationship between the predictors and the response in the **gascons** data set. Use cross-validation and resubstitution error estimation to choose the best tree. Compare and discuss the results.

13.20. Find the error for the tree of Example 13.12 using resubstitution and an independent test sample. Compare the results with cross-validation.

13.21. Explore some of the functions in the ARESLab toolbox. First, generate some noisy **peaks** data by adding normal random variables to the response **Y**. Use **mvnrnd** to generate noise with mean zero and an appropriate covariance. Vary some of the parameters for building the models, e.g., value for c, using cubic functions, trying different values for the maximum number of basis functions in the forward pass, etc. Assess the resulting models using the testing functions **arestest** and **arescv**.

13.22. Construct a multivariate adaptive regression spline model of the **environmental** data. The response variable is **Ozone**, and the predictor variables are **Temperature** and **WindSpeed**. Try various parameter values to build several models. Use **aresplot**, **arestest**, and **arescv** to choose the final model.

Chapter 14

Markov Chain Monte Carlo Methods

14.1 Introduction

In many applications of statistical modeling, the data analyst would like to use a more complex model for a data set, but is forced to resort to an over-simplified model in order to use available techniques. Markov chain Monte Carlo (MCMC) methods are simulation-based and enable the statistician or engineer to examine data using realistic statistical models.

We start off with the following example taken from Raftery and Akman [1986] and Roberts [2000] that looks at the possibility that a change-point has occurred in a Poisson process. Raftery and Akman [1986] show that there is evidence for a change-point by determining Bayes factors for the change-point model versus other competing models. These data are a time series that indicate the number of coal mining disasters per year from 1851 to 1962. A plot of the data is shown in Figure 14.8, and it does appear that there has been a reduction in the rate of disasters during that time period. Some questions we might want to answer using the data are

- What is the most likely year in which the change occurred?
- Did the rate of disasters increase or decrease after the change-point?

Example 14.8, presented later on, answers these questions using Bayesian data analysis and Gibbs sampling.

The main application of the MCMC methods that we present in this chapter is to generate a sample from a distribution. This sample can then be used to estimate various characteristics of the distribution such as moments, quantiles, modes, the density, or other statistics of interest.

In Section 14.2, we provide some background information to help the reader understand the concepts underlying MCMC. Because much of the recent developments and applications of MCMC arise in the area of Bayesian inference, we provide a brief introduction to this topic. This is followed by a discussion of Monte Carlo integration, since one of the applications of MCMC methods is to obtain estimates of integrals. In Section 14.3, we present several Metropolis-Hastings algorithms, including the random-walk Metropolis sampler and the independence sampler. A widely used special case of the general Metropolis-Hastings method called the Gibbs sampler is covered in Section 14.4. An important consideration with MCMC is whether or not the chain has converged to the desired distribution. So, some convergence diagnostic techniques are discussed in Section 14.5. Sections 14.6 and 14.7 contain references to MATLAB® code and references for the theoretical underpinnings of MCMC methods.

14.2 Background

Bayesian Inference

Bayesians represent uncertainty about unknown parameter values by probability distributions and proceed as if parameters were random quantities [Gilks, et al., 1996a]. If we let D represent the data that are observed and θ represent the model parameters, then to perform any inference, we must know the joint probability distribution $P(D, \theta)$ over all random quantities. Note that we allow θ to be multi-dimensional. From Chapter 2, we know that the joint distribution can be written as

$$P(D, \theta) = P(\theta)P(D|\theta),$$

where $P(\theta)$ is called the *prior*, and $P(D|\theta)$ is called the *likelihood*. Once we observe the data D, we can use Bayes' theorem to get the *posterior distribution* as follows

$$P(\theta|D) = \frac{P(\theta)P(D|\theta)}{\int P(\theta)P(D|\theta)d\theta}. \tag{14.1}$$

Equation 14.1 is the distribution of θ conditional on the observed data D. Since the denominator of Equation 14.1 is not a function of θ (since we are integrating over θ), we can write the posterior as being proportional to the prior times the likelihood,

$$P(\theta|D) \propto P(\theta)P(D|\theta) = P(\theta)L(\theta;D).$$

We can see from Equation 14.1 that the posterior is a conditional distribution for the model parameters, given the observed data. Understanding and using the posterior distribution is at the heart of Bayesian inference, where one is interested in making inferences using various features of the posterior distribution (e.g., moments, quantiles, etc.). These quantities can be written as posterior expectations of functions of the model parameters as follows

$$E[f(\theta)|D] = \frac{\int f(\theta)P(\theta)P(D|\theta)d\theta}{\int P(\theta)P(D|\theta)d\theta}. \tag{14.2}$$

Note that the denominator in Equations 14.1 and 14.2 is a constant of proportionality to make the posterior integrate to one. If the posterior is non-standard, then this can be very difficult, if not impossible, to obtain. This is especially true when the problem is high dimensional because there are a lot of parameters to integrate over. Analytically performing the integration in these expressions has been a source of difficulty in applications of Bayesian inference, and often simpler models would have to be used to make the analysis feasible. Monte Carlo integration using MCMC is one answer to this problem.

Because the same problem also arises in frequentist applications, we will change the notation to make it more general. We let X represent a vector of d random variables, with distribution denoted by $\pi(x)$. To a frequentist, X would contain data, and $\pi(x)$ is called a likelihood. For a Bayesian, X would be comprised of model parameters, and $\pi(x)$ would be called a posterior distribution. For both, the goal is to obtain the expectation

$$E[f(X)] = \frac{\int f(x)\pi(x)dx}{\int \pi(x)dx}. \tag{14.3}$$

As we will see, with MCMC methods we only have to know the distribution of X up to the constant of normalization. This means that the denominator in Equation 14.3 can be unknown. It should be noted that in what follows we assume that X can take on values in a d-dimensional Euclidean space. The methods can be applied to discrete random variables with appropriate changes.

Monte Carlo Integration

As stated before, most methods in statistical inference that use simulation can be reduced to the problem of finding integrals. This is a fundamental part of the MCMC methodology, so we provide a short explanation of classical

Monte Carlo integration. References that provide more detailed information on this subject are given in the last section of the chapter.

Monte Carlo integration estimates the integral $E[f(X)]$ of Equation 14.3 by obtaining samples X_t, $t = 1, ..., n$ from the distribution $\pi(x)$ and calculating

$$E[f(X)] \approx \frac{1}{n} \sum_{t=1}^{n} f(X_t). \tag{14.4}$$

The notation t is used here because there is an ordering or sequence to the random variables in MCMC methods. We know that when the X_t are independent, then the approximation can be made as accurate as needed by increasing n. We will see in the following sections that with MCMC methods, the samples are not independent in most cases. That does not limit their use in finding integrals using approximations such as Equation 14.4. However, care must be taken when determining the variance of the estimate in Equation 14.4 because of dependence [Gentle, 1998; Robert and Casella, 1999]. We illustrate the method of Monte Carlo integration in the next example.

Example 14.1

For a distribution that is exponential with $\lambda = 1$, we find $E[\sqrt{X}]$ using Equation 14.4. We generate random variables from the required distribution, take the square root of each one and then find the average of these values. This is implemented next in MATLAB.

```
% Generate 1000 exponential random
% variables with lambda = 1.
% This is a Statistics Toolbox function.
x = exprnd(1,1,1000);
% Take square root of each one.
xroot = sqrt(x);
% Take the mean - Equation 14.4
exroothat = mean(xroot);
```

From this, we get an estimate of 0.889. We can use MATLAB to find the value using numerical integration, as shown here.

```
% Now get it using numerical integration.
strg = 'sqrt(x).*exp(-x)';
myfun = inline(strg);
% quadl is a base MATLAB function.
exroottru = quadl(myfun,0,50);
```

From this, we get a value of 0.886, which is close to the Monte Carlo method.
❑

The samples X_t do not have to be independent as long as they are generated using a process that obtains samples from the "entire" domain of $\pi(\mathbf{x})$ and in the correct proportions [Gilks, et al., 1996a]. This can be done by constructing a Markov chain that has $\pi(\mathbf{x})$ as its stationary distribution. We now give a brief description of Markov chains.

Markov Chains

A Markov chain is a sequence of random variables such that the next value or state of the sequence depends only on the previous one. Thus, we are generating a sequence of random variables, $X_0, X_1, \ldots$ such that the next state X_{t+1} with $t \geq 0$ is distributed according to $P(X_{t+1}|X_t)$, which is called the *transition kernel*. A realization of this sequence is also called a Markov chain. We assume that the transition kernel does not depend on t, making the chain time-homogeneous.

One issue that must be addressed is how sensitive the chain is to the starting state X_0. Given certain conditions [Robert and Casella, 1999], the chain will forget its initial state and will converge to a stationary distribution, which is denoted by ψ. As the sequence grows larger, the sample points X_t become dependent samples from ψ. The reader interested in knowing the conditions under which this happens and for associated proofs of convergence to the stationary distribution is urged to read the references given in Section 14.7.

Say the chain has been run for m iterations, and we can assume that the sample points X_t, $t = m + 1, \ldots, n$ are distributed according to the stationary distribution ψ. We can discard the first m iterations and use the remaining $n - m$ samples along with Equation 14.4 to get an estimate of the expectation as follows

$$E[f(X)] \approx \frac{1}{n-m} \sum_{t=m+1}^{n} f(X_t). \tag{14.5}$$

The number of samples m that are discarded is called the *burn-in*. The size of the burn-in period is the subject of current research in MCMC methods. Diagnostic methods to help determine m and n are described in Section 14.5. Geyer [1992] suggests that the burn-in can be between 1% and 2% of n, where n is large enough to obtain adequate precision in the estimate given by Equation 14.5.

So now we must answer the question: how large should n be to get the required precision in the estimate? As stated previously, estimating the variance of the estimate given by Equation 14.5 is difficult because the samples are not independent. One way to determine n via simulation is to run several Markov chains in parallel, each with a different starting value. The estimates from Equation 14.5 are compared, and if the variation between them is too great, then the length of the chains should be increased [Gilks, et al., 1996b]. Other methods are given in Roberts [1996], Raftery and Lewis [1996], and in the general references mentioned in Section 14.7.

Analyzing the Output

We now discuss how the output from the Markov chains can be used in statistical analysis. An analyst might be interested in calculating means, standard deviations, correlations, and marginal distributions for components of X. If we let $X_{t,j}$ represent the j-th component of X_t at the t-th step in the chain, then using Equation 14.5, we can obtain the marginal means and variances from

$$\overline{X}_{,j} = \frac{1}{n-m} \sum_{t=m+1}^{n} X_{t,j},$$

and

$$S^2_{,j} = \frac{1}{n-m-1} \sum_{t=m+1}^{n} (X_{t,j} - \overline{X}_{,j})^2.$$

These estimates are simply the componentwise sample mean and sample variance of the sample points X_t, $t = m + 1, ..., n$. Sample correlations are obtained similarly. Estimates of the marginal distributions can be obtained using the techniques of Chapter 8.

One last problem we must deal with to make Markov chains useful is the stationary distribution ψ. We need the ability to construct chains such that the stationary distribution of the chain is the one we are interested in: $\pi(x)$. In the MCMC literature, $\pi(x)$ is often referred to as the *target distribution*. It turns out that this is not difficult and is the subject of the next two sections.

14.3 Metropolis-Hastings Algorithms

The Metropolis-Hastings method is a generalization of the Metropolis technique of Metropolis, et al. [1953], which had been used for many years in the physics community. The paper by Hastings [1970] further generalized the technique in the context of statistics. The Metropolis sampler, the independence sampler and the random-walk are all special cases of the

Metropolis-Hastings method. Thus, we cover the general method first, followed by the special cases.

These methods share several properties, but one of the more useful properties is that they can be used in applications where $\pi(\mathbf{x})$ is known up to the constant of proportionality. Another property that makes them useful in a lot of applications is that the analyst does not have to know the conditional distributions, which is the case with the Gibbs sampler. While it can be shown that the Gibbs sampler is a special case of the Metropolis-Hastings algorithm [Robert and Casella, 1999], we include it in the next section because of this difference.

Metropolis-Hastings Sampler

The Metropolis-Hastings sampler obtains the state of the chain at $t + 1$ by sampling a *candidate point Y* from a *proposal distribution* $q(.|X_t)$. Note that this depends only on the previous state X_t and can have any form, subject to regularity conditions [Roberts, 1996]. An example for $q(.|X_t)$ is the multivariate normal with mean X_t and fixed covariance matrix. One thing to keep in mind when selecting $q(.|X_t)$ is that the proposal distribution should be easy to sample from.

The required regularity conditions for $q(.|X_t)$ are irreducibility and aperiodicity [Chib and Greenberg, 1995]. *Irreducibility* means that there is a positive probability that the Markov chain can reach any nonempty set from all starting points. *Aperiodicity* ensures that the chain will not oscillate between different sets of states. These conditions are usually satisfied if the proposal distribution has a positive density on the same support as the target distribution. They can also be satisfied when the target distribution has a restricted support. For example, one could use a uniform distribution around the current point in the chain.

The candidate point is accepted as the next state of the chain with probability given by

$$\alpha(X_t, Y) = \min\left\{1, \frac{\pi(Y)q(X_t|Y)}{\pi(X_t)q(Y|X_t)}\right\}. \tag{14.6}$$

If the point Y is not accepted, then the chain does not move and $X_{t+1} = X_t$. The steps of the algorithm are outlined next. It is important to note that the distribution of interest $\pi(\mathbf{x})$ appears as a ratio, so the constant of proportionality cancels out. This is one of the appealing characteristics of the Metropolis-Hastings sampler, making it appropriate for a wide variety of applications.

PROCEDURE – METROPOLIS-HASTINGS SAMPLER

1. Initialize the chain to X_0 and set $t = 0$.
2. Generate a candidate point Y from $q(.|X_t)$.
3. Generate U from a uniform $(0, 1)$ distribution.
4. If $U \le \alpha(X_t, Y)$ (Equation 14.6) then set $X_{t+1} = Y$, else set $X_{t+1} = X_t$.
5. Set $t = t + 1$ and repeat steps 2 through 5.

The Metropolis-Hastings procedure is implemented in Example 14.2, where we use it to generate random variables from a standard Cauchy distribution. As we will see, this implementation is one of the special cases of the Metropolis-Hastings sampler described later.

Example 14.2

We show how the Metropolis-Hastings sampler can be used to generate random variables from a standard Cauchy distribution given by

$$f(x) = \frac{1}{\pi(1 + x^2)}; \qquad -\infty < x < \infty.$$

From this, we see that

$$f(x) \propto \frac{1}{1 + x^2}.$$

We will use the normal as our proposal distribution, with a mean given by the previous value in the chain and a standard deviation given by σ. We start by setting up **inline** MATLAB functions to evaluate the densities for Equation 14.6.

```
% Set up an inline function to evaluate the Cauchy.
% Note that in both of the functions,
% the constants are canceled.
strg = '1./(1+x.^2)';
cauchy = inline(strg,'x');
% set up an inline function to evaluate the Normal pdf
strg = '1/sig*exp(-0.5*((x-mu)/sig).^2)';
norm = inline(strg,'x','mu','sig');
```

We now generate $n = 10000$ samples in the chain.

```
% Generate 10000 samples in the chain.
% Set up the constants.
n = 10000;
```

```
sig = 2;
x = zeros(1,n);
x(1) = randn(1);% generate the starting point
for i = 2:n
  % generate a candidate from the proposal distribution
  % which is the normal in this case. This will be a
  % normal with mean given by the previous value in the
  % chain and standard deviation of 'sig'
  y = x(i-1) + sig*randn(1);
  % generate a uniform for comparison
  u = rand(1);
  alpha = min([1, cauchy(y)*norm(x(i-1),y,sig)/...
           (cauchy(x(i-1))*norm(y,x(i-1),sig))]);
  if u <= alpha
    x(i) = y;
  else
    x(i) = x(i-1);
  end
end
```

We can plot a density histogram along with the curve corresponding to the true probability density function. We discard the first 500 points for the burn-in period. The plot is shown in Figure 14.1.
❏

Metropolis Sampler

The Metropolis sampler refers to the original method of Metropolis, et al. [1953], where only symmetric distributions are considered for the proposal distribution. Thus, we have that

$$q(Y|X) = q(X|Y).$$

for all X and Y. As before, a common example of a distribution like this is the normal distribution with mean X and fixed covariance. Because the proposal distribution is symmetric, those terms cancel out in the acceptance probability yielding

$$\alpha(X_t, Y) = \min\left\{1, \frac{\pi(Y)}{\pi(X_t)}\right\}. \tag{14.7}$$

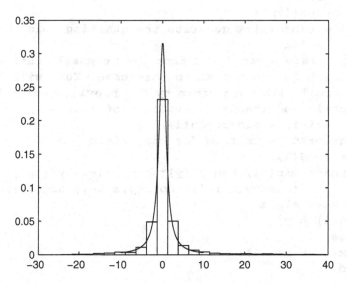

FIGURE 14.1
We generated 10,000 variates from the Cauchy distribution using the Metropolis-Hastings sampler. This shows a density histogram of the random variables after discarding the first 500 points. The curve corresponding to the true probability density function is superimposed over the histogram. We see that the random variables do follow the standard Cauchy distribution.

PROCEDURE – METROPOLIS SAMPLER

1. Initialize the chain to X_0 and set $t = 0$.
2. Generate a candidate point Y from $q(.|X_t)$.
3. Generate U from a uniform $(0, 1)$ distribution.
4. If $U \leq \alpha(X_t, Y)$ (Equation 14.7) then set $X_{t+1} = Y$, else set $X_{t+1} = X_t$.
5. Set $t = t + 1$ and repeat steps 2 through 5.

When the proposal distribution is such that $q(Y|X) = q(|X - Y|)$, then it is called the **random-walk Metropolis**. This amounts to generating a candidate point $Y = X_t + Z$, where Z is an increment random variable from the distribution q.

We can gain some insight into how this algorithm works by looking at the conditions for accepting a candidate point as the next sample in the chain. In the symmetric case, the probability of moving is $\pi(Y)/\pi(X_t)$. If $\pi(Y) \geq \pi(X_t)$, then the chain moves to Y because $\alpha(X_t, Y)$ will be equal to 1. This means that a move that climbs up the curve given by the target distribution is always accepted. A move that is worse (i.e., one that goes downhill) is accepted with probability given by $\pi(Y)/\pi(X_t)$. These concepts are illustrated in Figure 14.2. This is the basic algorithm proposed by Metropolis, et al. [1953], and it is the foundation for other optimization algorithms such as simulated annealing [Kirkpatrick, Gelatt, and Vechi, 1983; Aarts and Korst, 1989].

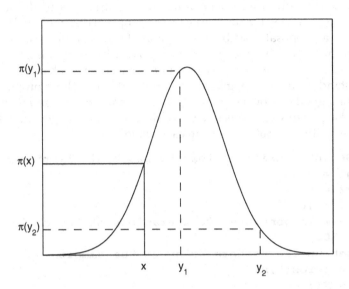

FIGURE 14.2
This shows what happens when a candidate point is selected and the proposal distribution is symmetric [Chib and Greenberg, 1995]. In this case, the probability of moving to another point is based on the ratio $\pi(y)/\pi(x)$. If $\pi(y) \geq \pi(x)$, then the chain moves to the candidate point y. If $\pi(y) < \pi(x)$, then the chain moves to y with probability $\pi(y)/\pi(x)$. So we see that a move from x to y_1 would be automatically accepted, but a move to y_2 would be accepted with probability $\pi(y_2)/\pi(x)$.

When implementing any of the Metropolis-Hastings algorithms, it is important to understand how the scale of the proposal distribution affects the efficiency of the algorithm. This is especially apparent with the random-walk version and is illustrated in the next example. If a proposal distribution takes small steps, then the acceptance probability given by Equation 14.7 will be high, yielding a higher rate at which we accept candidate points. The problem here is that the chain will mix slowly, meaning that the chain will take longer to get to the stationary distribution. On the other hand, if the proposal distribution generates large steps, then the chain could move to the tails, resulting in low acceptance probabilities. Again, the chain fails to mix quickly.

Example 14.3

In this example, we show how to implement the random-walk version of the Metropolis-Hastings sampler [Gilks, et al., 1996a] and use it to generate variates from the standard normal distribution (the target distribution). Of course, we do not have to resort to MCMC methods to generate random variables from this target distribution, but it serves to illustrate the importance of picking the right scale for the proposal distribution. We use the normal as a proposal distribution to generate the candidates for the next value in the chain. The mean of the proposal distribution is given by the current value in the chain x_t. We generate three chains with different values for the standard deviation, given by: $\sigma = 0.5, 0.1, 10$. These provide chains that exhibit good mixing, poor mixing due to small step size and poor mixing due to a large step size, respectively. We show next how to generate the three sequences with $n = 500$ variates in each chain.

```
% Get the variances for the proposal distributions.
sig1 = 0.5;
sig2 = 0.1;
sig3 = 10;
% We will generate 500 iterations of the chain.
n = 500;
% Set up the vectors to store the samples.
X1 = zeros(1,n);
X2 = X1;
X3 = X1;
% Get the starting values for the chains.
X1(1) = -10;
X2(1) = 0;
X3(1) = 0;
```

Now that we have everything initialized, we can obtain the chains.

```
% Run the first chain.
for i = 2:n
  % Generate variate from proposal distribution.
  y = randn(1)*sig1 + X1(i-1);
  % Generate variate from uniform.
  u = rand(1);
  % Calculate alpha.
  alpha = normpdf(y,0,1)/normpdf(X1(i-1),0,1);
  if u <= alpha
    % Then set the chain to the y.
    X1(i) = y;
  else
    X1(i) = X1(i-1);
  end
end
```

```
% Run second chain.
for i = 2:n
  % Generate variate from proposal distribution.
  y = randn(1)*sig2 + X2(i-1);
  % Generate variate from uniform.
  u = rand(1);
  % Calculate alpha.
  alpha = normpdf(y,0,1)/normpdf(X2(i-1),0,1);
  if u <= alpha
    % Then set the chain to the y.
    X2(i) = y;
  else
    X2(i) = X2(i-1);
  end
end
% Run the third chain.
for i = 2:n
  % Generate variate from proposal distribution.
  y = randn(1)*sig3 + X3(i-1);
  % Generate variate from uniform.
  u = rand(1);
  % Calculate alpha.
  alpha = normpdf(y,0,1)/normpdf(X3(i-1),0,1);
  if u <= alpha
    % Then set the chain to the y.
    X3(i) = y;
  else
    X3(i) = X3(i-1);
  end
end
```

Plots of these sequences are illustrated in Figure 14.3, where we also show horizontal lines at ±2. These lines are provided as a way to determine if most values in the chain are mixing well (taking on many different values) within two standard deviations of zero, since we are generating standard normal variates. Note that the first one converges quite rapidly and exhibits good mixing in spite of an extreme starting point. The second one with $\sigma = 0.1$ (small steps) is mixing very slowly and does not seem to have converged to the target distribution in these 500 steps of the chain. The third sequence, where large steps are taken, also seems to be mixing slowly, and it is easy to see that the chain sometimes does not move. This is due to the large steps taken by the proposal distribution.

❑

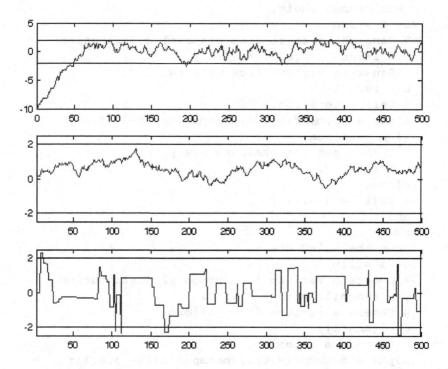

FIGURE 14.3

These are the three sequences from Example 14.3. The target distribution is the standard normal. For all three sequences, the proposal distribution is normal with the mean given by the previous element in the sequence. The standard deviations of the proposal distribution are $\sigma = 0.5, 0.1, 10$. Note that the first sequence approaches the target distribution after the first 50 – 100 iterations. The other two sequences are slow to converge to the target distribution because of slow mixing due to the poor choice of σ.

Independence Sampler

The independence sampler was proposed by Tierney [1994]. This method uses a proposal distribution that does not depend on X; i.e., it is generated independently of the previous value in the chain. The proposal distribution is of the form $q(Y|X) = q(Y)$, so Equation 14.6 becomes

$$\alpha(X_t, Y) = \min\left\{1, \frac{\pi(Y)q(X_t)}{\pi(X_t)q(Y)}\right\}. \tag{14.8}$$

This is sometimes written in the literature as

$$\alpha(X_t, Y) = \min\left\{1, \frac{w(Y)}{w(X_t)}\right\}$$

where $w(X) = \pi(X)/q(X)$.

Caution should be used when implementing the independence sampler. In general, this method will not work well unless the proposal distribution q is very similar to the target distribution π. Gilks, et al. [1996a] show that it is best if q is heavier-tailed than π. Note also that the resulting sample is still not independent, even though we generate the candidate points independently of the previous value in the chain. This is because the acceptance probability for the next value X_{t+1} depends on the previous one. For more information on the independence sampler and the recommended usage, see Roberts [1996] or Robert and Casella [1999].

Autoregressive Generating Density

Another choice for a candidate generating density is proposed by Tierney [1994] and described by Chib and Greenberg [1995]. This is represented by an autoregressive process of order 1 and is obtained by generating candidates as follows

$$Y = a + B(X_t - a) + Z, \tag{14.9}$$

where a is a vector and B is a matrix, both of which are conformable in terms of size with X_t. The vector Z has a density given by q. If $B = -I$, then the chains are produced by reflecting about the point a, yielding negative correlation between successive values in the sequence. The autoregressive generating density is described in the next example.

Example 14.4

We show how to use the Metropolis-Hastings sampler with the autoregressive generating density to generate random variables from a target distribution given by a bivariate normal with the following parameters:

$$\mu = \begin{bmatrix} 1 \\ 2 \end{bmatrix} \qquad \Sigma = \begin{bmatrix} 1 & 0.9 \\ 0.9 & 1 \end{bmatrix}.$$

Variates from this distribution can be easily generated using the techniques of Chapter 4, but it serves to illustrate the concepts. In the exercises, the reader is asked to generate a set of random variables using those techniques and compare them to the results obtained in this example. We generate a sequence of $n = 6000$ points and use a burn-in of 4000.

```
% Set up some constants and arrays to store things.
n = 6000;
xar = zeros(n,2); % to store samples
mu = [1;2];   % Parameters - target distribution.
covm = [1 0.9;  0.9 1];
```

We now set up a MATLAB **inline** function to evaluate the required probabilities.

```
% Set up the function to evaluate alpha
% for this problem. Note that the constant
% has been canceled.
strg = 'exp(-0.5*(x-mu)''*inv(covm)*(x-mu))';
norm = inline(strg,'x','mu','covm');
```

The following MATLAB code sets up a random starting point and obtains the elements of the chain.

```
xar(1,:) = randn(1,2); % Generate starting point.
for i = 2:n
  % Get the next variate in the chain.
  y = mu - (xar(i-1,:)'-mu) + (-1+2*rand(2,1));
  u = rand(1);
  % Uses inline function 'norm' from above.
  alpha = min([1,norm(y,mu,covm)/...
          norm(xar(i-1,:)',mu,covm)]);
  if u <= alpha
    xar(i,:) = y';
  else
    xar(i,:) = xar(i-1,:);
  end
end
```

A scatterplot of the last 2000 variates is given in Figure 14.4, and it shows that they do follow the target distribution. To check this further, we can get the sample covariance matrix and the sample mean using these points. The result is

$$\bar{\mathbf{x}} = \begin{bmatrix} 1.04 \\ 2.03 \end{bmatrix} \qquad \mathbf{S} = \begin{bmatrix} 1 & 0.899 \\ 0.899 & 1 \end{bmatrix},$$

from which we see that the sample does reflect the target distribution.
❑

Example 14.5
This example shows how the Metropolis-Hastings method can be used with an example in Bayesian inference [Roberts, 2000]. This is a genetic linkage

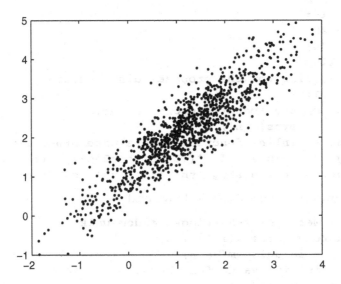

FIGURE 14.4
This is a scatterplot of the last 2000 elements of a chain generated using the autoregressive generating density of Example 14.4 (random walk version).

example, looking at the genetic linkage of 197 animals. The animals are divided into four categories with frequencies given by

$$Z = (z_1, z_2, z_3, z_4) = (125, 18, 20, 34),$$

with corresponding cell probabilities of

$$\left(\frac{1}{2} + \frac{\theta}{4}, \frac{1}{4}(1-\theta), \frac{1}{4}(1-\theta), \frac{\theta}{4} \right).$$

From this, we get a posterior distribution of θ, given the data Z, of

$$P(\theta|Z) = \pi(\theta) \propto (2+\theta)^{z_1}(1-\theta)^{z_2+z_3}\theta^{z_4}.$$

We would like to use this to observe the behavior of the parameter θ (i.e., what are likely values for θ) given the data. Note that any constants in the denominator in $\pi(\theta)$ have been eliminated because they cancel in the Metropolis-Hastings sampler. We use the random-walk version where the step is generated by the uniform distribution over the interval $(-a, a)$. Note that we set up a MATLAB **inline** function to get the probability of accepting the candidate point.

```
% Set up the preliminaries.
z1 = 125;
z2 = 18;
z3 = 20;
z4 = 34;
n = 1100;
% Step size for the proposal distribution.
a = 0.1;
% Set up the space to store values.
theta = zeros(1,n);
% Get an inline function to evaluate probability.
strg = '((2+th).^z1).*((1-th).^(z2+z3)).*(th.^z4)';
ptheta = inline(strg,'th','z1','z2','z3','z4');
```

We can now generate the chain as shown next.

```
% Use Metropolis-Hastings random-walk
% where y = theta(i-1) + z
% and z is uniform(-a,a).
% Get initial value for theta.
theta(1) = rand(1);
for i = 2:n
 % Generate from proposal distribution.
 y = theta(i-1) - a + 2*a*rand(1);
 % Generate from uniform.
 u = rand(1);
 alpha = min([ ptheta(y,z1,z2,z3,z4)/...
     ptheta(theta(i-1),z1,z2,z3,z4),1]);
 if u <= alpha
   theta(i) = y;
 else
   theta(i) = theta(i-1);
 end
end
```

We set the burn-in period to 100, so only the last 1000 elements are used to produce the density histogram estimate of the posterior density of θ given in Figure 14.5.

❑

14.4 The Gibbs Sampler

Although the Gibbs sampler can be shown to be a special case of the Metropolis-Hastings algorithm [Gilks, et al., 1996b; Robert and Casella,

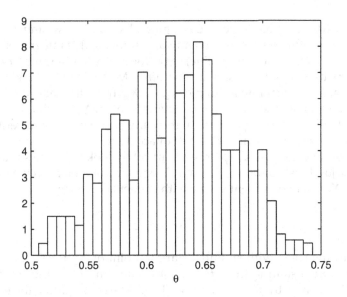

FIGURE 14.5
This shows the density histogram estimate of the posterior density of θ *given the observed data.*

1999], we include it in its own section because it is different in some fundamental ways. The two main differences between the Gibbs sampler and Metropolis-Hastings are

1) We always accept a candidate point.
2) We must know the full conditional distributions.

In general, the fact that we must know the full conditional distributions makes the algorithm less applicable.

The Gibbs sampler was originally developed by Geman and Geman [1984], where it was applied to image processing and the analysis of Gibbs distributions on a lattice. It was brought into mainstream statistics through the articles of Gelfand and Smith [1990] and Gelfand, et al. [1990].

In describing the Gibbs sampler, we follow the treatment in Casella and George [1992]. Let's assume that we have a joint density that is given by $f(x, y_1, \ldots, y_d)$, and we would like to understand more about the marginal density. For example, we might want to know the shape, the mean, the variance or some other characteristic of interest.

The marginal density is given by

$$f(x) = \int \dots \int f(x, y_1, \dots, y_d) dy_1 \dots dy_d. \qquad (14.10)$$

Equation 14.10 says that to get the marginal distribution, we must integrate over all of the other variables. In many applications, this integration is very difficult (and sometimes impossible) to perform. The Gibbs sampler is a way to get $f(x)$ by simulation. As with the other MCMC methods, we use the Gibbs sampler to generate a sample $X_1, \dots, X_m$ from $f(x)$ and then use the sample to estimate the desired characteristic of $f(x)$. Casella and George [1992] note that if m is large enough, then any population characteristic can be calculated with the required degree of accuracy.

To illustrate the Gibbs sampler, we start off by looking at the simpler case where the joint distribution is $f(x_1, x_2)$. Using the notation from the previous sections, X_t is a two element vector with elements given by

$$X_t = (X_{t,1}, X_{t,2})$$

We start the chain with an initial starting point of $X_0 = (X_{0,1}, X_{0,2})$. We then generate a sample from $f(x_1, x_2)$ by sampling from the conditional distributions given by $f(x_1|x_2)$ and $f(x_2|x_1)$. At each iteration, the elements of the random vector are obtained one at a time by alternately generating values from the conditional distributions. We illustrate this in the procedure given next.

PROCEDURE – GIBBS SAMPLER (BIVARIATE CASE)

1. Generate a starting point $X_0 = (X_{0,1}, X_{0,2})$. Set $t = 0$.
2. Generate a point $X_{t,1}$ from

$$f(X_{t,1}|X_{t,2} = x_{t,2}).$$

3. Generate a point $X_{t,2}$ from

$$f(X_{t,2}|X_{t+1,1} = x_{t+1,1}).$$

4. Set $t = t + 1$ and repeat steps 2 through 4.

Note that the conditional distributions are conditioned on the current or most recent values of the other components of X_t. Example 14.6 shows how this is done in a simple case taken from Casella and George [1992].

Example 14.6

To illustrate the Gibbs sampler, we consider the following joint distribution

$$f(x, y) \propto \binom{n}{x} y^{x + \alpha - 1} (1 - y)^{n - x + \beta - 1} ,$$

where $x = 0, 1, ..., n$ and $0 \le y \le 1$. Let's say our goal is to estimate some characteristic of the marginal distribution $f(x)$ of X. By ignoring the overall dependence on n, α, and β, we find that the conditional distribution $f(x|y)$ is binomial with parameters n and y, and the conditional distribution $f(y|x)$ is a beta distribution with parameters $x + \alpha$ and $n - x + \beta$ [Casella and George, 1992]. The MATLAB commands given next use the Gibbs sampler to generate variates from the joint distribution.

```
% Set up preliminaries.
% Here we use k for the chain length because n
% is used for the number of trials in a binomial.
k = 1000;      % generate a chain of size 1000
m = 500;       % burn-in will be 500
a = 2;    % chosen
b = 4;
x = zeros(1,k);
y = zeros(1,k);
n = 16;
```

We are now ready to generate the elements in the chain. We start off by generating a starting point.

```
% Pick a starting point.
x(1) = binornd(n,0.5,1,1);
y(1) = betarnd(x(1) + a, n - x(1) + b,1,1);
for i = 2:k
  x(i) = binornd(n,y(i-1),1,1);
  y(i) = betarnd(x(i)+a, n-x(i)+b, 1, 1);
end
```

Note that we do not have to worry about whether or not we will accept the next value in the chain. With Gibbs sampling every candidate is accepted. We can estimate the marginal using the following

$$\hat{f}(x) = \frac{1}{k - m} \sum_{i = m + 1}^{k} f(x|y_i) .$$

This says that we evaluate the probability conditional on the values of y_i that were generated after the burn-in period. This is implemented in MATLAB as follows:

```
% Get the marginal by evaluating the conditional.
% Use MATLAB's Statistics Toolbox.
```

```
% Find the P(X=x|Y's)
fhat = zeros(1,17);
for i = 1:17
  fhat(i) = mean(binopdf(i-1,n,y(500:k)));
end
```

The true marginal probability mass function is [Casella and George, 1992]

$$f(x) = \binom{n}{x} \frac{\Gamma(\alpha + \beta)}{\Gamma(\alpha)\Gamma(\beta)} \frac{\Gamma(x + \alpha)\Gamma(n - x + \beta)}{\Gamma(\alpha + \beta + n)},$$

for $x = 0, 1, ..., n$. We plot the estimated probability mass function along with the true marginal in Figure 14.6. This shows that the estimate is very close to the true function.
❑

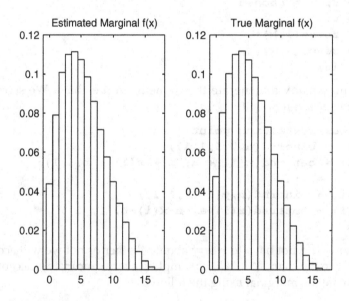

FIGURE 14.6
On the left, we have the estimated probability mass function for the marginal distribution $f(x)$. *The mass function on the right is from the true probability mass function. We see that there is close agreement between the two.*

Casella and George [1992] and Gelfand and Smith [1990] recommend that K different sequences be generated, each one with length n. Then the last element of each sequence is used to obtain a sample of size K that is approximately independent for large enough K. We do note that there is some disagreement in the literature regarding the utility of running one really long chain to get better convergence to the target distribution or many shorter chains to get independent samples [Gilks, et al., 1996b]. Most researchers in this field observe that one long run would often be used for exploratory analysis and a few moderate size runs is preferred for inferences.

The procedure given next for the general Gibbs sampler is for one chain only. It is easier to understand the basic concepts by looking at one chain, and it is simple to expand the algorithm to multiple chains.

PROCEDURE – GIBBS SAMPLER

1. Generate a starting point $X_0 = (X_{0,1}, ..., X_{0,d})$. Set $t = 0$.
2. Generate a point $X_{t,1}$ from

$$f(X_{(t+1),1}|X_{t,2} = x_{t,2}, ..., X_{t,d} = x_{t,d}).$$

Generate a point $X_{(t+1),2}$ from

$$f(X_{(t+1),2}|X_{t+1,1} = x_{t+1,1}, X_{t,3} = x_{t,3}, ..., X_{t,d} = x_{t,d}).$$

. . .

Generate a point $X_{(t+1),d}$ from

$$f(X_{(t+1),d}|X_{t+1,1} = x_{t+1,1}, ..., X_{t+1,d-1} = x_{t+1,d-1}).$$

3. Set $t = t + 1$ and repeat steps 2 through 3.

Example 14.7

We show another example of Gibbs sampling as applied to bivariate normal data. Say we have the same model as we had in Example 14.4, where we wanted to generate samples from a bivariate normal with the following parameters

$$\mu = \begin{bmatrix} \mu_1 \\ \mu_2 \end{bmatrix} = \begin{bmatrix} 1 \\ 2 \end{bmatrix} \qquad \Sigma = \begin{bmatrix} 1 & \rho \\ \rho & 1 \end{bmatrix} = \begin{bmatrix} 1 & 0.9 \\ 0.9 & 1 \end{bmatrix}.$$

From Gelman, et al. [1995] we know that $f(x_1|x_2)$ is univariate normal with mean $\mu_1 + \rho(x_2 - \mu_2)$ and standard deviation $1 - \rho^2$. Similarly, $f(x_2|x_1)$ is univariate normal with mean $\mu_2 + \rho(x_1 - \mu_1)$ and standard deviation $1 - \rho^2$. With this information, we can implement the Gibbs sampler to generate the random variables.

```
% Set up constants and arrays.
n = 6000;
xgibbs = zeros(n,2);
rho = 0.9;
y = [1;2];% This is the mean.
sig = sqrt(1-rho^2);
% Initial point.
xgibbs(1,:) = [10 10];
% Start the chain.
for i = 2:n
    mu = y(1) + rho*(xgibbs(i-1,2)-y(2));
    xgibbs(i,1) = mu + sig*randn(1);
    mu = y(2) + rho*(xgibbs(i,1) - y(1));
    xgibbs(i,2) = mu + sig*randn(1);
end
```

Notice that the next element in the chain is generated based on the current values for x_1 and x_2. A scatterplot of the last 2000 variates generated with this method is shown in Figure 14.7.
❑

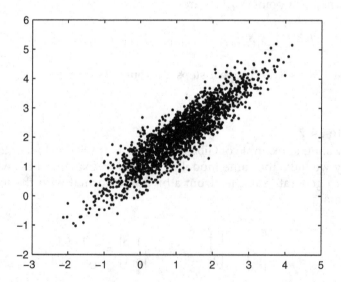

FIGURE 14.7
This is a scatterplot of the bivariate normal variates generated using Gibbs sampling. Note that the results are similar to Figure 14.4.

We return now to our example described at the beginning of the chapter, where we are investigating the hypothesis that there has been a reduction in coal mining disasters over the years 1851 to 1962. To understand this further, we follow the model given in Roberts [2000]. This model assumes that the number of disasters per year follows a Poisson distribution with a mean rate of θ until the k-th year. After the k-th year, the number of disasters is distributed according to the Poisson distribution with a mean rate of λ. This is represented as

$$Y_i \sim \text{Poisson}(\theta) \qquad i = 1, ..., k$$
$$Y_i \sim \text{Poisson}(\lambda) \qquad i = k+1, ..., n,$$

where the notation "~" means *"is distributed as."*

A Bayesian model is given by the following:

$$\theta \sim \text{Gamma}(a_1, b_1)$$
$$\lambda \sim \text{Gamma}(a_2, b_2)$$
$$b_1 \sim \text{Gamma}(c_1, d_1)$$
$$b_2 \sim \text{Gamma}(c_2, d_2)$$

and the k is discrete uniform over $\{1, ..., 112\}$ (since there are 142 years). Note that θ, λ, and k are all independent of each other.

This model leads to the following conditional distributions:

$$\theta | Y, \lambda, b_1, b_2, k \sim \text{Gamma}\left(a_1 + \sum_{i=1}^{k} Y_i, k + b_1\right)$$

$$\lambda | Y, \theta, b_1, b_2, k \sim \text{Gamma}\left(a_2 + \sum_{i=k+1}^{n} Y_i, n - k + b_2\right)$$

$$b_1 | Y, \theta, \lambda, b_2, k \sim \text{Gamma}(a_1 + c_1, \theta + d_1)$$
$$b_2 | Y, \theta, \lambda, b_1, k \sim \text{Gamma}(a_2 + c_2, \lambda + d_2)$$
$$f(k | Y, \theta, \lambda, b_1, b_2) = \frac{L(Y;k, \theta, \lambda)}{\sum_{j=1}^{n} L(Y;j, \theta, \lambda)}$$

The likelihood is given by

$$L(Y;k, \theta, \lambda) = \exp\{k(\lambda - \theta)\}(\theta/\lambda)^{\sum_{i=1}^{k} Y_i}.$$

We use Gibbs sampling to simulate the required distributions and examine the results to explore the change-point model. For example, we could look at the posterior densities of θ, λ, and k to help us answer the questions posed at the beginning of the chapter.

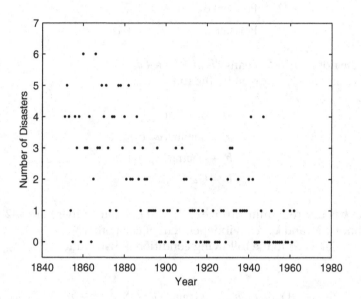

FIGURE 14.8
This shows a scatterplot of the **coal** *data. It does appear that there was a reduction in the rate of disasters per year, after a certain year. Estimating that year is the focus of this example.*

Example 14.8

A plot of the time series for the **coal** data is shown in Figure 14.8, where we see graphical evidence supporting the hypothesis that a change-point does occur [Raftery and Akman, 1986] and that there has been a reduction in the rate of coal mine disasters over this time period.

We set up the preliminary data needed to implement Gibbs sampling as follows:

```
% Set up preliminaries.
load coal
% y contains number of disasters.
% year contains the year.
n = length(y);
m = 1100;     % number in chain
% The values for the parameters are the same
% as in Roberts[2000].
```

```
a1 = 0.5;
a2 = 0.5;
c1 = 0;
c2 = 0;
d1 = 1;
d2 = 1;
theta = zeros(1,m);
lambda = zeros(1,m);
k = zeros(1,n);
% Holds probabilities for k.
like = zeros(1,n);
```

We are now ready to implement the Gibbs sampling. We will run the chain for 1100 iterations and use a burn-in period of 100.

```
% Get starting points.
k(1) = unidrnd(n,1,1);
% Note that k will indicate an index to the year
% that corresponds to a hypothesized change-point.
theta(1) = 1;
lambda(1) = 1;
b1 = 1;
b2 = 1;
% Start the Gibbs Sampler.
for i = 2:m
    kk = k(i-1);
    % Get parameters for generating theta.
    t = a1 + sum(y(1:kk));
    lam = kk + b1;
    % Generate the variate for theta.
    theta(i) = gamrnd(t,1/lam,1,1);
    % Get parameters for generating lambda.
    t = a2 + sum(y) - sum(y(1:kk));
    lam = n-kk+b2;
    % Generate the variate for lambda.
    lambda(i) = gamrnd(t,1/lam,1,1);
    % Generate the parameters b1 and b2.
    b1 = gamrnd(a1+c1,1/(theta(i)+d1),1,1);
    b2 = gamrnd(a2+c2,1/(lambda(i)+d2),1,1);
    % Now get the probabilities for k.
    for j = 1:n
        like(j) = exp((lambda(i)-theta(i))*j)*...
            (theta(i)/lambda(i))^sum(y(1:j));
    end
    like = like/sum(like);
    k(i) = cssample(1:n,like,1); % Sample for k.
end
```

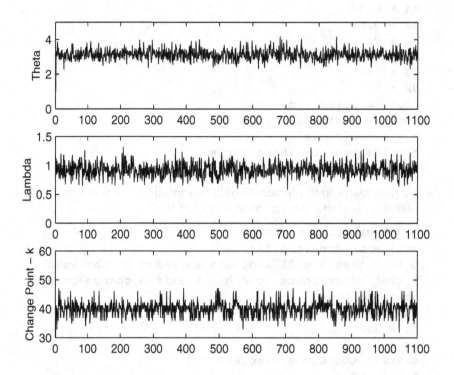

FIGURE 14.9
This shows the sequences that were generated using the Gibbs sampler.

The sequences for θ, λ, and k are shown in Figure 14.9, where we can see that a burn-in period of 100 is reasonable. In Figure 14.10, we plot the frequencies for the estimated posterior distribution using the generated k variates. We see evidence of a posterior mode at $k = 41$, which corresponds to the year 1891. So, we suspect that the change-point most likely occurred around 1891. We can also look at density histograms for the posterior densities for θ and λ. These are given in Figure 14.11. They indicate that the mean rate of disasters did decrease after the change-point.
❏

14.5 Convergence Monitoring

The problem of deciding when to stop the chain is an important one and is the topic of current research in MCMC methods. After all, the main purpose of using MCMC is to get a sample from the target distribution and explore its

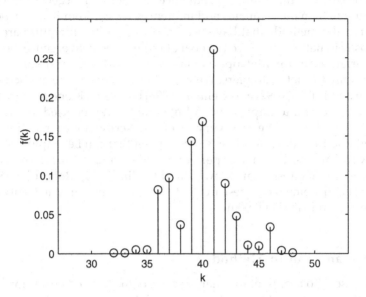

FIGURE 14.10
This is the frequency histogram for the random variables k generated by the Gibbs sampler of Example 14.8. Note the mode at k = 41 corresponding to the year 1891.

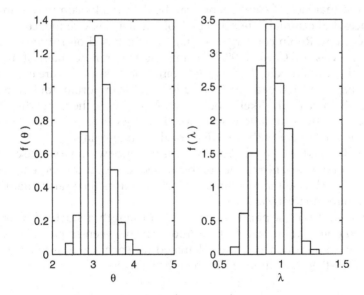

FIGURE 14.11
This figure shows density histograms for the posterior distributions for θ and λ, and there seems to be evidence showing that there was a reduction in the mean rate of disasters per year.

characteristics. If the resulting sequence has not converged to the target distribution, then the estimates and inferences we get from it are suspect.

Most of the methods that have been proposed in the literature are really diagnostic in nature and have the goal of monitoring convergence. Some are appropriate only for Metropolis-Hastings algorithms and some can be applied only to Gibbs samplers. We will discuss in detail one method due to Gelman and Rubin [1992] and Gelman [1996] because it is one of the simplest to understand and to implement. Additionally, it can be used in any of the MCMC algorithms. We also very briefly describe another widely used method due to Raftery and Lewis [1992, 1996] that can be employed within the MCMC method. Other papers that review and compare the various convergence diagnostics are Cowles and Carlin [1996], Robert [1995], and Brooks [1998]. Some recent research in this area can be found in Canty [1999] and Brooks and Giudici [2000].

Gelman and Rubin Method

We will use v to represent the characteristic of the target distribution (mean, moments, quantiles, etc.) in which we are interested. One obvious way to monitor convergence to the target distribution is to run multiple sequences of the chain and plot v versus the iteration number. If they do not converge to approximately the same value, then there is a problem. Gelman [1996] points out that lack of convergence can be detected by comparing multiple sequences, but cannot be detected by looking at a single sequence.

The Gelman-Rubin convergence diagnostic is based on running multiple chains. Cowles and Carlin [1996] recommend ten or more chains if the target distribution is unimodal. The starting points for these chains are chosen to be widely dispersed in the target distribution. This is important for two reasons. First, it will increase the likelihood that most regions of the target distribution are visited in the simulation. Additionally, any convergence problems are more likely to appear with over-dispersed starting points.

The method is based on the idea that the variance within a single chain will be less than the variance in the combined sequences, if convergence has not taken place. The Gelman-Rubin approach monitors the scalar quantities of interest in the analysis (i.e., v).

We start off with k parallel sequences of length n starting from over-dispersed points in the target distribution. The between-sequence variance B and the within-sequence W are calculated for each scalar summary v. We denote the j-th scalar summary in the i-th chain by

$$v_{ij}; \qquad i = 1, \ldots, k, \quad j = 1, \ldots, n.$$

Thus, the subscript j represents the position in the chain or sequence, and i denotes which sequence it was calculated from.

The between-sequence variance is given as

$$B = \frac{n}{k-1} \sum_{i=1}^{k} (\bar{v}_{i.} - \bar{v}_{..})^2, \tag{14.11}$$

where

$$\bar{v}_{i.} = \frac{1}{n} \sum_{j=1}^{n} v_{ij}, \tag{14.12}$$

and

$$\bar{v}_{..} = \frac{1}{k} \sum_{i=1}^{k} \bar{v}_{i.} . \tag{14.13}$$

Equation 14.12 is the mean of the n values of the scalar summary in the i-th sequence, and Equation 14.13 is the average across sequences.

The within-sequence variance is determined by

$$W = \frac{1}{k} \sum_{i=1}^{k} s_i^2, \tag{14.14}$$

with

$$s_i^2 = \frac{1}{n-1} \sum_{j=1}^{n} (v_{ij} - \bar{v}_{i.})^2. \tag{14.15}$$

Note that Equation 14.15 is the sample variance of the scalar summary for the i-th sequence, and Equation 14.14 is the average variance for the k sequences.

Finally, W and B are combined to get an overall estimate of the variance of v in the target distribution:

$$\hat{\text{var}}(v) = \frac{n-1}{n} W + \frac{1}{n} B. \tag{14.16}$$

Equation 14.16 is a conservative estimate of the variance of v, if the starting points are over-dispersed [Gelman, 1996]. In other words, it tends to over estimate the variance.

Alternatively, the within-sequence variance given by W is an underestimate of the variance of v. This should make sense considering the fact that finite sequences have not had a chance to travel all of the target distribution resulting in less variability for v. As n gets large, both $\hat{\text{var}}(v)$ and W approach the true variance of v, one from above and one from next.

The Gelman-Rubin approach diagnoses convergence by calculating

$$\sqrt{\hat{R}} = \sqrt{\frac{\hat{\text{var}}(v)}{W}}. \tag{14.17}$$

This is the ratio between the upper bound on the standard deviation of v and the lower bound. It estimates the factor by which $\hat{\text{var}}(v)$ might be reduced by further iterations. The factor given by Equation 14.17 is called the ***estimated potential scale reduction***. If the potential scale reduction is high, then the analyst is advised to run the chains for more iterations. Gelman [1996] recommends that the sequences be run until $\hat{R}$ for all scalar summaries are less than 1.1 or 1.2.

Example 14.9

We return to Example 14.3 to illustrate the Gelman-Rubin method for monitoring convergence. Recall that our target distribution is the univariate standard normal. This time our proposal distribution is univariate normal with $\mu = X_t$ and $\sigma = 5$. Our scalar summary v is the mean of the elements of the chain. We implement the Gelman-Rubin method using four chains.

```
% Set up preliminaries.
sig = 5;
% We will generate 500 iterations of the chain.
n = 5000;
numchain = 4;
% Set up the vectors to store the samples.
% This is 4 chains, 5000 samples.
X = zeros(numchain,n);
% This is 4 sequences (rows) of summaries.
nu = zeros(numchain,n);
% Track the rhat for each iteration:
rhat = zeros(1,n);
% Get the starting values for the chain.
% Use over-dispersed starting points.
X(1,1) = -10;
X(2,1) = 10;
X(3,1) = -5;
X(4,1) = 5;
```

The following implements the chains. Note that each column of our matrices **X** and **nu** is one iteration of the chains, and each row contains one of the chains. The **X** matrix keeps the chains, and the matrix **nu** is the sequence of scalar summaries for each chain.

```
% Run the chain.
for j = 2:n
    for i = 1:numchain
        % Generate variate from proposal distribution.
        y = randn(1)*sig + X(i,j-1);
        % Generate variate from uniform.
        u = rand(1);
        % Calculate alpha.
        alpha = normpdf(y,0,1)/normpdf(X(i,j-1),0,1);
        if u <= alpha
            % Then set the chain to the y.
            X(i,j) = y;
        else
            X(i,j) = X(i,j-1);
        end
    end
    % Get the scalar summary - means of each row.
    nu(:,j') = mean(X(:,1:j)')';
    rhat(j) = csgelrub(nu(:,1:j));
end
```

The function **csgelrub** will return the estimated $\hat{R}$ for a given set of sequences of scalar summaries. We plot the four sequences for the summary statistics of the chains in Figure 14.12. From these plots, we see that it might be reasonable to assume that the sequences have converged, since they are getting close to the same value in each plot. In Figure 14.13, we show a plot of $\hat{R}$ for each iteration of the sequence. This seems to confirm that the chains are getting close to convergence. Our final value of $\hat{R}$ at the last iteration of the chain is 1.05.
❑

One of the advantages of the Gelman-Rubin method is that the sequential output of the chains does not have to be examined by the analyst. This can be difficult, especially when there are a lot of summary quantities that must be monitored. The Gelman-Rubin method is based on means and variances, so it is especially useful for statistics that approximately follow the normal distribution. Gelman, et al. [1995] recommend that in other cases, extreme quantiles of the between and within sequences should be monitored.

Raftery and Lewis Method

We briefly describe this method for two reasons. First, it is widely used in applications. Secondly, it is available in MATLAB code through the Econometrics Toolbox (see Section 14.6 for more information) and in Fortran from StatLib. So, the researcher who needs another method besides the one of Gelman and Rubin is encouraged to download these and try them. The

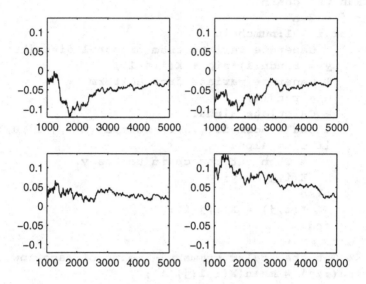

FIGURE 14.12
Here are the sequences of summary statistics in Example 14.9. We are tracking the mean of sequences of variables generated by the Metropolis-Hastings sampler. The target distribution is a univariate standard normal. It appears that the sequences are close to converging, since they are all approaching the same value.

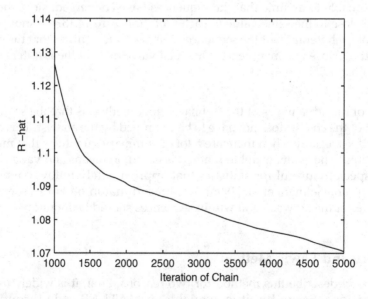

FIGURE 14.13
This sequence of values for $\hat{R}$ indicates that it is very close to one, showing near convergence.

article by Raftery and Lewis [1996] is another excellent resource for information on the theoretical basis for the method and for advice on how to use it in practice.

This technique is used to detect convergence of the chain to the target distribution and also provides a way to bound the variance of the estimates obtained from the samples. To use this method, the analyst first runs one chain of the Gibbs sampler for N_{min}. This is the minimum number of iterations needed for the required precision, given that the samples are independent. Using this chain and other quantities as inputs (the quantile to be estimated, the desired accuracy, the probability of getting that accuracy, and a convergence tolerance), the Raftery-Lewis method yields several useful values. Among them are the total number of iterations needed to get the desired level of accuracy and the number of points in the chain that should be discarded (i.e., the burn-in).

14.6 MATLAB® Code

The Statistics Toolbox for MATLAB has a function called **mhsample** that implements the Markov chain Metropolis-Hastings sampler. Also, the examples given in this text can be adapted to fit most applications by simply changing the proposal and target distributions.

There is an Econometrics Toolbox that contains M-files for the Gibbs sampler and the Raftery-Lewis convergence diagnostic. The software can be freely downloaded at

http://www.spatial-econometrics.com

Extensive documentation for the procedures in the Econometrics Toolbox is also available at the website. The Raftery-Lewis method for S-plus and Fortran can be downloaded at:

- S-plus: **http://lib.stat.cmu.edu/S/gibbsit**
- Fortran: **http://lib.stat.cmu.edu/general/gibbsit**

There are several user-contributed M-files for MCMC available for download at The MathWorks website:

http://www.mathworks.com/matlabcentral/

Just search on "MCMC" or "Markov chains" to find the relevant tools.

For those who do not use MATLAB, another resource for software that will do Gibbs sampling and Bayesian analysis is the WinBUGS (Bayesian inference Using Gibbs Sampling) software. The software and manuals can be downloaded at `http://www.mrc-bsu.cam.ac.uk/software/bugs/`

There is a MATLAB interface to WinBUGS to make it easier for statisticians and data analysts to use the methods. The MATBUGS interface can be downloaded here

`https://code.google.com/p/matbugs/`

In the Computational Statistics Toolbox, we provide an M-file function called **csgelrub** that implements the Gelman-Rubin diagnostic. It returns R for given sequences of scalar summaries. We also include a function that implements a demo of the Metropolis-Hastings sampler where the target distribution is standard bivariate normal. This runs four chains, and the points are plotted as they are generated so the user can see what happens as the chain grows. The M-file functions pertaining to MCMC that we provide are summarized in Table 14.1.

TABLE 14.1

List of Functions from Chapter 14 Included in the Computational Statistics Toolbox

Purpose	MATLAB Function
Gelman-Rubin convergence diagnostic given sequences of scalar summaries	**csgelrub**
Graphical demonstration of what happens in the Metropolis-Hastings sampler	**csmcmcdemo**

14.7 Further Reading

For an excellent introduction to Markov chain Monte Carlo methods, we recommend the book *Markov Chain Monte Carlo in Practice* [Gilks, et al., 1996b]. This contains a series of articles written by leading researchers in the area and describes most aspects of MCMC from the theoretical to the practical. For a complete theoretical treatment of MCMC methods and many examples, the reader is referred to Robert and Casella [1999]. This book also contains a description of many of the hybrid MCMC methods that have been developed. The text by Tanner [1996] gives an introduction to computational algorithms for Bayesian and likelihood inference.

Most recent books on random number generation discuss the Metropolis-Hastings sampler and the Gibbs sampler. Gentle [1998] has a good discussion of MCMC methods and includes some examples in MATLAB. Ross [1997]

has a chapter on MCMC and also discusses the connection between Metropolis-Hastings and simulated annealing. Ross [2000] also covers the topic of MCMC.

The monograph by Lindley [1995] gives an introduction and review of Bayesian statistics. For an overview of general Markov chain theory, see Tierney [1996], Meyn and Tweedie [1993], or Norris [1997]. For the reader who needs more information on general Bayesian methods, we highly recommend *An Introduction to Bayesian Statistics* by Bolstad [2007]. This presents the usual introductory statistics topics from a Bayesian perspective.

If the reader would like more information on Bayesian data analysis, then the book *Bayesian Data Analysis* [Gelman, et al., 1995] is a good place to start. This text also contains some information and examples about the MCMC methods discussed in this chapter. Most of these books also include information on Monte Carlo integration methods, including importance sampling and variance reduction.

Marin and Robert [2007] have an excellent text on computational Bayesian statistics that uses R as the computing platform. This book includes information on many of the techniques from this chapter, including MCMC. It takes an operational approach rather than a theoretical one, so readers should find it helpful.

Besides simulated annealing, a connection between MCMC methods and the finite mixtures EM algorithm has been discussed in the literature. For more information on this, see Robert and Casella [1999]. There is also another method that, while not strictly an MCMC method, seems to be grouped with them. This is called Sampling Importance Resampling [Rubin, 1987, 1988]. A good introduction to this can be found in Ross [1997], Gentle [1998], and Albert [1993].

Exercises

14.1. The von Mises distribution is given by

$$f(x) = \frac{1}{2\pi I_0(b)} e^{b\cos(x)} \qquad -\pi \le x \le \pi,$$

where I_0 is the modified Bessel function of the first kind and order zero. Letting $b = 3$ and a starting point of 1, use the Metropolis random-walk algorithm to generate 1000 random iterations of the chain. Use the uniform distribution over the interval $(-1, 1)$ to generate steps in the walk. Plot the output from the chain versus the iteration number. Does it look like you need to discard the initial values in the chain for this example? Plot a histogram of the sample [Gentle, 1998].

14.2. Use the Metropolis-Hastings algorithm to generate samples from the beta distribution. Try using the uniform distribution as a candidate distribution. Note that you can simplify by canceling constants.

14.3. Use the Metropolis-Hastings algorithm to generate samples from the gamma distribution. What is a possible candidate distribution? Simplify the ratio by canceling constants.

14.4. Repeat Example 14.3 to generate a sample of standard normal random variables using different starting values and burn-in periods.

14.5. Let's say that $X_{,1}$ and $X_{,2}$ have conditional distributions that are exponential over the interval $(0, B)$, where B is a known positive constant. Thus,

$$f(x_{,1}|x_{,2}) \propto x_{,2} e^{-x_{,2} x_{,1}} \qquad 0 < x_{,1} < B < \infty$$

$$f(x_{,2}|x_{,1}) \propto x_{,1} e^{-x_{,1} x_{,2}} \qquad 0 < x_{,2} < B < \infty$$

Use Gibbs sampling to generate samples from the marginal distribution $f(x_{,1})$. Choose your own starting values and burn-in period. Estimate the marginal distribution. What is the estimated mean, variance, and skewness coefficient for $f(x_{,1})$? Plot a histogram of the samples obtained after the burn-in period and the sequential output. Start multiple chains from over-dispersed starting points and use the Gelman-Rubin convergence diagnostics for the mean, variance and skewness coefficient [Casella and George, 1992].

14.6. Explore the use of the Metroplis-Hastings algorithm in higher dimensions. Generate 1000 samples for a trivariate normal distribution centered at the origin and covariance equal to the identity matrix. Thus, each coordinate direction should be a univariate standard normal distribution. Use a trivariate normal distribution with covariance matrix $\Sigma = 9 \cdot I$, (i.e., 9's are along the diagonal and 0's everywhere else) and mean given by the current value of the chain x_t. Use $x_{0,i} = 10$, $i = 1, ..., 3$ as the starting point of the chain. Plot the sequential output for each coordinate. Construct a histogram for the first coordinate direction. Does it look like a standard normal? What value did you use for the burn-in period [Gentle, 1998]?

14.7. A joint density is given by

$$f(x_{,1}, x_{,2}, x_{,3}) = C\exp\{-(x_{,1} + x_{,2} + x_{,3} + x_{,1}x_{,2} + x_{,1}x_{,3} + x_{,2}x_{,3})\},$$

where $x_{,i} > 0$. Use one of the techniques from this chapter to simulate samples from this distribution and use them to estimate $E[X_{,1}X_{,2}X_{,3}]$. Start multiple chains and track the estimate to monitor the convergence [Ross, 1997].

14.8. Use Gibbs sampling to generate samples that have the following density

$$f(x_{,1}, x_{,2}, x_{,3}) = kx_{,1}^4 x_{,2}^3 x_{,3}^2 (1 - x_{,1} - x_{,2} - x_{,3})$$

where $x_{,i} > 0$ and $x_{,1} + x_{,2} + x_{,3} < 1$. Let $B(a, b)$ represent a beta distribution with parameters a and b. We can write the conditional distributions as

$$\begin{aligned}
X_{,1} | X_{,2}, X_{,3} &\sim (1 - X_{,2} - X_{,3})Q & Q &\sim B(5, 2) \\
X_{,2} | X_{,1}, X_{,3} &\sim (1 - X_{,1} - X_{,3})R & R &\sim B(4, 2) \\
X_{,3} | X_{,1}, X_{,2} &\sim (1 - X_{,1} - X_{,2})S & S &\sim B(3, 2)
\end{aligned}$$

where the notation $Q \sim B(a, b)$ means Q is from a beta distribution. Plot the sequential output for each $x_{,i}$ [Arnold, 1993].

14.9. Let's say that we have random samples $Z_1, \ldots, Z_n$ that are independent and identically distributed from the normal distribution with mean θ and variance 1. In the notation of Equation 14.1, these constitute the set of observations D. We also have a prior distribution on θ such that

$$P(\theta) \propto \frac{1}{1 + \theta^2},$$

We can write the posterior as follows

$$P(\theta | D) \propto P(\theta)L(\theta; D) = \frac{1}{1 + \theta^2} \times \exp\left\{\frac{-n(\theta - \bar{z})^2}{2}\right\}.$$

Let the true mean be $\theta = 0.06$ and generate a random sample of size $n = 20$ from the normal distribution to obtain the z_i. Use Metropolis-Hastings to generate random samples from the posterior distribution and use them to estimate the mean and the variance of the posterior distribution. Start multiple chains and use the

Gelman-Rubin diagnostic method to determine when to stop the chains.

14.10. Generate a set of $n = 2000$ random variables for the bivariate distribution given in Example 14.4 using the technique from Chapter 4. Create a scatterplot of these data and compare to the set generated in Example 14.4.

14.11. For the bivariate distribution of Example 14.4, use a random-walk generating density ($Y = X_t + Z$) where the increment random variable Z is distributed as bivariate uniform. Generate a sequence of 6000 elements and construct a scatterplot of the last 2000 values. Compare to the results of Example 14.4.

14.12. For the bivariate distribution of Example 14.4, use a random-walk generating density ($Y = X_t + Z$) where the increment random variables Z are bivariate normal with mean zero and covariance

$$\Sigma = \begin{bmatrix} 0.6 & 0 \\ 0 & 0.4 \end{bmatrix}.$$

Generate a sequence of 6000 elements and construct a scatterplot of the last 2000 values. Compare to the results of Example 14.4.

14.13. Use the Metropolis-Hastings sampler to generate random samples from the lognormal distribution

$$f(x) = \frac{1}{x\sqrt{2\pi}} \exp\left\{ -\frac{(\ln x)^2}{2} \right\}$$

$$f(x) \propto \frac{1}{x} \exp\left\{ -\frac{(\ln x)^2}{2} \right\}.$$

Use the independence sampler and the gamma as a proposal distribution, being careful about the tails. Plot the sample using the density histogram and superimpose the true probability density function to ensure that your random variables are from the desired distribution.

Appendix A

MATLAB® Basics

The purpose of this appendix is to provide some introductory information to help you get started using MATLAB. It has been adapted from *Statistics in MATLAB: A Primer* [Martinez and Cho, 2014]. We will describe:

- The desktop environment
- How to get help from several sources
- Ways to get your data into and out of MATLAB
- The different data types in MATLAB
- How to work with arrays
- Functions and commands
- Simple plotting functions

A.1 Desktop Environment

This section will provide information on the desktop environment. Figure A.1 shows the desktop layout for MATLAB, where we chose to view just the Command Window. We made the window smaller, which causes some of the sections—CODE, ENVIRONMENT, and RESOURCES, in this case—to be collapsed. Simply click on the arrow buttons for the sections to see what tools are available or make the window bigger by resizing with the mouse.

The default desktop layout includes the following panels or windows: Command Window, Current Folder, Workspace, and Command History. You can personalize your desktop layout by choosing those panels or sub-windows that you need for your work, dragging the panels to different places within the desktop, or re-sizing them.

The Command Window is the main interface to communicate with MATLAB. The window shows the MATLAB prompt, which is shown as a double arrow: >>. This is where you enter commands, functions, and other code. MATLAB will also display information in this spot in response to certain commands.

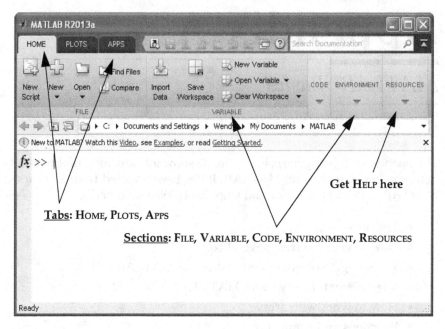

FIGURE A.1

This is a screenshot of the main MATLAB window or desktop, where we have chosen to view only the Command Window in our layout. The desktop layout can be changed by clicking on the ENVIRONMENT section and the LAYOUT option. MATLAB now uses the familiar ribbon interface common to Microsoft® applications.

MATLAB now uses the typical ribbon interface found in recent Microsoft Windows® applications. The interface for MATLAB is comprised of tabs and sections, as highlighted in Figure A.1. The ribbon interface will be different depending on the selected tab (HOME, PLOTS, or APPS).

We will provide more detail on the Figure window in a later section, but we mention it here briefly. Plots are displayed in a separate Figure window with its own user interface, which is comprised of the more familiar menu options and toolbar buttons. The main menu items include FILE, EDIT, TOOLS, HELP, and more. See Figure A.6 for an example.

The Workspace sub-window provides a listing of the variables that are in the current workspace, along with information about the type of variable, the size, and summary information. You can double-click on the VARIABLE icon to open it in a spreadsheet-like interface. Using this additional interface, you can change elements, delete and insert rows, select and plot columns, and much more.

The Script Editor is a useful tool. This can be opened by clicking the NEW SCRIPT button on the ribbon interface. This editor has its own ribbon interface with helpful options for programming in MATLAB. The editor also has some nice features to make it easier to write error-free MATLAB code, such as tracking parentheses, suggesting the use of a semi-colon, and warnings about erroneous expressions or syntax.

A.2 Getting Help and Other Documentation

This appendix is meant to be a brief introduction to MATLAB—just enough to get you started. So, the reader is encouraged to make use of the many sources of assistance available via the online documentation, command line tools, help files, and the user community. A starting point could be to select the HELP button found on the HOME ribbon and the RESOURCES section; see Figure A.2.

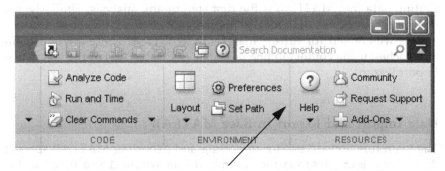

FIGURE A.2
This shows the options available from the HELP button on the RESOURCES section of the HOME tab. Click the question mark to open a documentation window or click the arrow button to access more options.

There are several options to get help at the command line. They are the easiest and quickest way to get some help, especially once you become familiar with MATLAB.

You can find a shortened version of the documentation for a function by typing

help *functionname*

at the command line. This will provide information on the syntax for the function, definitions of input/output arguments, and examples. The command

doc *functionname*

will open the documentation for ***functionname*** in a separate window.

You can access MATLAB documentation in several ways. One is via the HELP button (see Figure A.2) on the HOME ribbon. This opens a window that has links to the documentation for your installed toolboxes. There is also a link to .pdf documentation on all toolboxes (scroll down to the bottom of the list on the website). Clicking on that link will take you to the Documentation Center at the MathWorks, Inc. website.

There is a vast MATLAB user community. This is a great resource for finding answers, obtaining user-written code, and following news groups. The main portal for this community is MATLAB Central, and it can be found at this link:

http://www.mathworks.com/matlabcentral/

A.3 Data Import and Export

Getting data into MATLAB is the first step in any analysis. Similarly, we might need to export data for use in other software (e.g., R) or to save the objects we created for future analysis in MATLAB. There are two main approaches: using command line functions (see Table A.1) or the interactive Import Wizard.

Data Import and Export in Base MATLAB

The main functions to use for importing and exporting MATLAB specific data files (**.mat** files) via the command line are the **load** function for importing and the **save** function for exporting. These functions can also be used with an ASCII text file.

MATLAB has several options for reading in ASCII text files. If the file has numerical entries separated by white space, then the **load** (for exporting use **save -ascii**) command can be employed. If the values are separated by a different character, such as commas or tabs, then **dlmread** can be used. The character used to separate the values is inferred from the file or it can be specified as an argument. A related function called **dlmwrite** is used for exporting data in an ASCII file with different delimiters.

Most software packages for data analysis provide an option to import and export files where the data entries are separated by commas (**.csv** files). MATLAB has functions **csvread** and **csvwrite** to handle this type of file. Note that the file must contain only numeric data to use these functions.

You can import and export spreadsheet files with the **.xls** or **.xlsx** extensions using **xlsread** and **xlswrite**. The **xlsread** function will also read in OpenDocument™ spreadsheets that have the **.ods** file extension. For more information on OASIS OpenDocument formats, see

http://www.opendocumentformat.org/.

Sometimes a file can contain numerical and text data. There are two command-line options for reading in these types of files. One is the **importdata** function. This function can handle headers for ASCII files and spreadsheets, but the rest of the data should be in tabular form and has to be numeric. If you have an ASCII file with columns of non-numeric data (characters or formatted dates/times), then you can use the **textscan** function or the Import Wizard described next.

TABLE A.1

Common Data I/O Functions in Base MATLAB®

load, save	Read and write **.mat** files
	Read and write text files using the **-ascii** flag
dlmread, dlmwrite	Handles text files with specified delimiter
csvread, csvwrite	Use for comma delimited or **.csv** files
xlsread, xlswrite	Read and write spreadsheet files
importdata, textscan	Use for files that have a mixture of text and numbers

The Import Wizard can also be used to import data. It is started by clicking on the IMPORT DATA button located on the MATLAB desktop HOME ribbon's VARIABLE section; see Figure A.3. You can also start it by typing **uiimport** at the command line. The wizard guides you through the process of importing data from many different recognized file types. It allows you to view the contents of a file, to select the variables and the observations to import, to specify a value for unimportable cells, and more.

FIGURE A.3
Select the IMPORT DATA button to start the Import Wizard.

Data Import and Export with the Statistics Toolbox

There are some special functions for importing and exporting data in the Statistics Toolbox; see Table A.2 for a summary of the functions. These are particularly useful for statisticians and data analysts, and they include functions for importing and exporting tabular data, as well as data stored in the SAS XPORT format.

The functions **tblread** and **tblwrite** will import and export data in a tabular or matrix-like format. The data file must have the variable names on the first row, and the case names (or record identification variable) are in the first column. The data entries would start in the cell corresponding to the second row and second column (position $(2, 2)$ in a matrix).

The basic syntax to *interactively* select a file to load is:

$$[\texttt{data, vnames, cnames}] = \texttt{tblread},$$

where **vnames** contains the variable names (first row of the file) and **cnames** has the names of the observations (first column of the file). The variable **data** is a numeric matrix, where rows correspond to the observations, and the columns are the variables or characteristics.

Calling **tblread** with no input argument opens a window interface for selecting files. You can also specify a file name by using an input argument, as follows:

$$[\texttt{data, vnames, cnames}] = \texttt{tblread}(\textit{filename}).$$

The following function call

$$\texttt{tblwrite(data,vnames,cnames,}\textit{filename}\texttt{,delim)}$$

will export the data to a file with the delimiter specified by **delim**. Use **help tblwrite** at the command line for delimiter options.

Statisticians and data analysts often have to read in SAS files. The function **xptread** can be used to import files in the SAS XPORT transport format. The file can be selected interactively by leaving out the file name as the input argument or you can use the following syntax to read in a specific file:

$$\texttt{data} = \texttt{xptread}(\textit{filename}).$$

The documentation for the Statistics Toolbox describes the functions we identified above and some additional ones for reading and exporting case names (**caseread, casewrite**). MATLAB documentation and help pages can be accessed from the desktop environment by clicking on the HELP button (on the HOME ribbon), selecting STATISTICS TOOLBOX ... EXPLORATORY DATA ANALYSIS ... DATA IMPORT AND EXPORT.

TABLE A.2

Data I/O Functions in the Statistics Toolbox

`tblread, tblwrite`	Data in tabular format with variable names on the first row and case names in the first column
`xptread`	SAS XPORT (transport) format files
`caseread, casewrite`	Import and export text files with one case name per line
`export`	Write a dataset array to a tab-delimited file
`tdfread`	Import a tab-delimited file that has text and numeric data

A.4 Data in MATLAB®

We now describe the basic data types in MATLAB and the Statistics Toolbox. We also discuss how to merge data sets and to create special arrays that might prove useful in data analysis. We conclude with a short introduction to object-oriented programming constructs and how they are used in MATLAB.

Data Objects in Base MATLAB®

One can consider two main aspects of a data object in MATLAB—the object class and what it contains. We can think of the object *class* as the type of container that holds the data. The most common ones are arrays, cell arrays, and structures. The *content* of objects in MATLAB can be numeric (e.g., double precision floating point or integers) or characters (e.g., text or strings). We now describe the common types or classes of objects in the base MATLAB software.

The word *array* is a general term encompassing scalars, vectors, matrices, and multi-dimensional arrays. All of these objects have a *dimension* associated with them. You can think of the dimension as representing the number of indexes you need to specify to access elements in the array. We will represent this dimension with the letter k.

- **Scalar:** A scalar is just a single number (or character), and it has dimension $k = 0$. You do not need to specify an address because it is just a single element.

- **Vector**: A vector is usually a column of values, but MATLAB also has row vectors. A vector has dimension $k = 1$ because you have to provide one value to address an element in the column (or row) vector.

- **Matrix**: A matrix is an object that has rows and columns—like a table. To extract or access an element in a matrix, you have to specify what row it is in and also the column. Thus, the dimension of a matrix is $k = 2$.

- **Multi-dimensional array**: A multi-dimensional array has dimension $k > 2$. For example, we could think of a three-dimensional array being organized in pages (the third dimension), where each page contains a matrix. So, to access an element in such an array, we need to provide the row, column, and page number.

As a data analyst, you will usually import data using **load** or some other method we described previously. However, you will likely also need to type in or construct arrays for testing code, entering parameters, or getting the arrays into the right form for calling functions. We now cover some of the ways to build small arrays.

Commas or spaces concatenate elements (or other arrays) as columns. In other words, it puts them together as a row. Thus, the following MATLAB code will produce a row vector:

$$x = [1, 4, 5]$$

Or, we can concatenate two column vectors **a** and **b** to create one matrix, as shown here:

$$Y = [a \; b]$$

The semi-colon will stack elements as rows. So, we would obtain a column vector from this command:

$$z = [1; 4; 5]$$

As another example, we could put three row vectors together to get a matrix, as shown next:

$$Y = [a; b; c]$$

It is important to note that the building blocks of your arrays have to be conformal in terms of the number of elements in the sub-arrays when using the comma or semi-colon to merge data sets. Otherwise, you will get an error.

We might need to generate a regular sequence of values when working in MATLAB. We can accomplish this by using the colon. We get a sequence of values from one to ten with the following syntax:

$$x = 1:10$$

Other step sizes can be used, too. For instance, this will give us a sequence from one to ten in steps of 0.5:

$$x = 1:0.5:10$$

and this will yield a decreasing sequence:

$$x = 10:-1:1$$

We can create an array of all zeros or all ones. Arrays of this type are often used in data analysis. Here is how we can create a 3×3 matrix of zeros:

$$Z = zeros(3,3)$$

This next function call will produce a multidimensional array with $k = 3$:

$$O = ones(2,4,3)$$

The array **O** has three pages (third dimension), and each page has a matrix with two rows and four columns.

There is a special type of array in MATLAB called the *empty array*. An empty array is one that contains no elements. Thus, it does not have any dimensions associated with it, as we mentioned above with other arrays. The empty array is designated by closed square brackets, as shown here: **[ ]**. It can be used to delete elements of an array at the command line, and it is sometimes returned in response to function calls and logical expressions.

Here is an example of the first case, where we show how to delete an element from a vector **x**.

```
% Create a vector x.
x = [2, 4, 6];
% Delete the second element.
x(2) = [];
% Display the vector x.
disp(x)
      2       6
```

A *cell array* is a useful data object, especially when working with strings. The elements of a cell array are called *cells*. Each cell provides a flexible container for our data because it can hold data of any type—even other cell arrays. Furthermore, each element of the cell array can have a different size.

The cell array has an overall structure that is similar to basic numeric or character data arrays covered previously, and as such, they have to be conformal in their overall arrangement. For example, the cells are arranged in rows, columns, pages, etc., as with arrays. If we have a cell array with two rows, then each of its rows has to have the same number of cells. However, the content of the cells can be different in terms of data type and size.

We can create an empty cell array using the function **cell**, as shown here, where we set up a $2 \times 4 \times 3$ array of cells. Each of the cells in the following cell array object is empty:

```
            cell_array = cell(2,4,3)
```

You can also construct a cell array and fill it with data, as shown in the code given next. Note that curly braces are used to denote a cell array.

```
% Create a cell array, where one cell contains
% numbers and another cell element is a string.
cell_array2 = {[1,2], 'This is a string'};
```

Like cell arrays, *structures* allow one to combine dissimilar data into a single variable. The basic syntax to create a structure is

```
S = struct('field1',data1,'field2',data2,...).
```

Note that the structure can have one or more fields, along with the associated data values. Let's use this to create a small structure.

```
% Create a structure called author with four fields.
author = struct(...
    'name',{{'Wendy','Angel'}},...
    'area',{{'Clustering','Visualization'}},...
    'deg',{{'PhD','PhD'}});
```

A dot notation is used to extract or access a field. Suppose we want to get all of the names in our **author** structure, then we can use

```
        all_names = author.name
```

to get all of the entries in the **name** field. We will discuss how to access individual data elements and records in the next section.

A *table* is a type of data object in the base MATLAB software. The table object is like cell arrays and structures. We can combine data of different data types in one object. However, it has a table-like format that is familiar to statisticians and data analysts. The rows of a table object would contain the observations or cases, and the columns correspond to the characteristics or features.

We can use the **table** function to create a table object employing variables that are in the workspace. The basic syntax is

```
% Create a table using two objects.
mytab = table(var1,var2)
```

You can import a file as a table object using the **readtable** function. This function works with delimited text files (**.txt**, **.dat**, or **.csv**). It will also read in an Excel spreadsheet file with **.xls** or **.xlsx** extensions.

Accessing Data Elements

In this section, we demonstrate how you can identify elements of arrays, cell objects, and structures. This is useful in data analysis because we often need to analyze subsets of our data or to create new data sets by combining others. Table A.3 provides some examples of how to access elements of arrays. These can be numeric, string, or cell arrays. In the case of cell arrays, the notation is used to access the cell elements, but not the *contents* of the cells. Curly braces **{ }** are used to get to the data that are inside the cells. For example, **A{1,1}** would give us the contents of the cell, which would have a numeric or character type. Whereas, **A(1,1)** is the cell, and it has a type (or class) of cell.

These two notations can be combined to access part of the contents of a cell. To get the first two elements of the vector contents of cell **A(1,1)**, we can use

$$A\{1,1\}(1:2)$$

The curly braces in **A{1,1}** tells MATLAB to go inside the cell in position **(1,1)**, and the **(1:2)** points to elements 1 and 2 inside the cell.

TABLE A.3

Examples of Accessing Elements of Arrays

Notation	Usage
a(i)	Access the *i*th element (cell) of a row or column vector array (cell array)
a(3:5)	Access elements 3 through 5 of a vector or cell array
A(:,i)	Access the *i*th column of a matrix or cell array. In this case, the colon in the row dimension tells MATLAB to access all rows.
A(i,:)	Access the *i*th row of a matrix or cell array. The colon tells MATLAB to gather all of the columns.
A(2:4,1:2)	Access the elements in the second, third, and fourth rows and the first two columns
A(1,3,4)	Access the element in the first row, third column on the fourth entry of dimension 3 (sometimes called the page).

Recall that we can access entire fields in a structure using the dot notation. We can extract partial content from the fields by using the techniques we described for numeric and cell arrays, as illustrated next.

```
% Display Wendy's degree.
author.deg(1)

ans = 'PhD'
```

The **ans** object above is actually a one-cell array. To get the contents of the cell (as a string object), we use the curly braces, as shown here.

```
author.deg{1}
```

```
ans = PhD
```

The **ans** result is now a **char** array. The dot notation is used to access the field. We then specify the elements using the notation for arrays (Table A.3).

The techniques for manipulating subsets of data in table objects are similar to structures and arrays, but they have some additional options because of the column or variable names. This shows how to create a sub-table with the first three records and all variables.

```
% Get a sub-table by extracting the first 3 rows.
newtab = mytab(1:3,:)
```

We are able to extract a column of the table using the dot notation that we had with structures.

```
% Get second variable in the table we
% created in the previous section.
vt = mytab.var2;
```

A *dataset array* is a special data object that is included in the Statistics Toolbox, and it can be used to store variables of different data types. As an example, you can combine numeric, logical, character, and categorical data in one array. Each row of the dataset array corresponds to an observation, and each column corresponds to a variable. Therefore, each column has to have elements that are of the same data type. However, the individual columns can be different. For instance, one column can be numeric, and another can be text.

A dataset array can be created using variables that exist in the workspace or as a result of importing data. The function **dataset** is used in either case, as shown here. The basic syntax for creating a dataset array from various file types is shown next.

```
% Create from a tab-delimited text file.
ds = dataset('File','filename.txt')
```

```
% Create from a .csv file.
ds = dataset('File','filename.csv','Delimiter',',')
```

```
% Create from an Excel file.
ds = dataset('XLSFile','filename.xlsx')
```

There are several options for creating a dataset array from variables in the workspace. They are listed here.

```
% Create by combining three different variables.
ds = dataset(var1, var2, var3);

% Create by converting a numeric matrix called data.
ds = mat2dataset(data);
```

A dataset array has its own set of defined operations, and you cannot operate on this type of array in the same manner as a numeric array. We will discuss this idea in more detail shortly, when we cover object-oriented programming.

The dataset array might be removed in future versions of MATLAB, but it was still available in version 2015a. Because of this, it is recommended that you use the **table** object in base MATLAB instead of a dataset array.

Object-Oriented Programming

Certain aspects of MATLAB are *object-oriented*, which is a programming approach based on three main ideas:

1. **Classes and objects**: A *class* is a description or definition of a programming construct. An *object* is a specific instance of a class.

2. **Properties**: These are aspects of the object that can be manipulated or extracted.

3. **Methods**: These are behaviors, operations, or functions that are defined for the class.

The benefit of object-oriented programming is that the computer code for the class and the methods are defined once, and the same method can be applied to different instances of the class without worrying about the details.

Every data object has a class associated with it. There is a function called **class** that will return the class of an object. This can be very helpful when trying to understand how to access elements and to perform other data analytic tasks. You will encounter some special object classes throughout this book. There are several instances of unique classes that are defined in the Statistics Toolbox. Some examples of these include probability distributions, models, and trees.

A.5 Workspace and Syntax

In this section, we cover some additional topics you might find helpful when using MATLAB. These include command line functions for managing your workspace and files, punctuation, arithmetic operators, and functions.

File and Workspace Management

You can enter MATLAB expressions interactively at the command line or save them in an M-file. This special MATLAB file is used for saving scripts or writing functions. We described the Script Editor in an earlier section, which is a very handy tool for writing and saving MATLAB code.

As stated previously, we will not be discussing how to write your own programs or functions, but you might find it helpful to write *script* M-files. These script files are just text files with the **.m** extension, and they contain any expressions or commands you want to execute. Thus, it is important to know some commands for file management. There are lots of options for directory and file management on the desktop. In a previous section, we briefly mentioned some interactive tools to interface with MATLAB. Table A.4 provides some commands to list, view, and delete files.

TABLE A.4

File Management Commands

Command	Usage
dir, ls	Shows the files in the present directory
delete *filename*	Deletes *filename*
pwd	Shows the present directory
cd *dir*	Changes the directory. There is also a pop-up menu and button on the desktop that allows the user to change directory, as we show here.
edit *filename*	Brings up *filename* in the editor
type *filename*	Displays the contents of the file in the command window
which *filename*	Displays the path to *filename*. This can help determine whether a file is part of base MATLAB.
what	Lists the **.m** files and **.mat** files that are in the current directory

Variables created in a session (and not deleted) live in the MATLAB *workspace*. You can recall the variable at any time by typing in the variable name with no punctuation at the end. Note that variable names in MATLAB are case sensitive, so **Temp**, **temp**, and **TEMP** are different variables.

As with file management, there are several tools on the desktop to help you manage your workspace. For example, there is a **VARIABLE** section on the

TABLE A.5

Commands for Workspace Management

Command	Usage
who	Lists all variables in the workspace.
whos	Lists all variables in the workspace along with the size in bytes, array dimensions, and object type.
clear	Removes all variables from the workspace.
clear x y	Removes variables **x** and **y** from the workspace.

desktop ribbon interface that allows you to create a variable, open current variables in a spreadsheet-like interface, save the workspace, and clear it. Some commands to use for workspace management are given in Table A.5.

Syntax in MATLAB®

Punctuation and syntax are important in any programming language. If you get this wrong, then you will either get errors or your results will not be what you expect. Some of the common punctuation characters used in MATLAB are described in Table A.6.

MATLAB has the usual arithmetic operators for addition, subtraction, multiplication, division, and exponentiation. These are designated by +, −, *, /, ^ , respectively. It is important to remember that MATLAB interprets these operators in a linear algebra sense and uses the corresponding operation definition for arrays, also.

For example, if we multiply two matrices **A** and **B**, then they must be dimensionally correct. In other words, the number of columns of **A** must equal the number of rows of **B**. Similarly, adding and subtracting arrays requires the same number of elements and configuration for the arrays. Thus, you can add or subtract *row* vectors with the same number of elements, but you will get an error of you try to add or subtract a row and column vector, even if they have the same number of elements. See any linear algebra book for more information on these concepts and how the operations are defined with arrays [Strang, 1993].

Addition and subtraction operations are defined element-wise on arrays, as we have in linear algebra. In some cases, we might find it useful to perform other element-by-element operations on arrays. For instance, we might want to square each element of an array or multiply two arrays element-wise. To do this, we change the notation for the multiplication, division, and exponentiation operators by adding a period before the operator. As an example, we could square each element of **A**, as follows:

TABLE A.6

List of MATLAB® Punctuation

Punctuation	Usage
%	A percent sign denotes a comment line. Information after the **%** is ignored.
,	When used to separate commands on a single line, a comma tells MATLAB to display the results of the preceding command. When used to combine elements or arrays, a comma or a blank space groups elements along a row. A comma also has other uses, including separating function arguments and array subscripts.
;	When used after a line of input or between commands on a single line, a semicolon tells MATLAB not to display the results of the preceding command. When used to combine elements or arrays, a semicolon stacks them in a column.
...	Three periods denote the continuation of a statement onto the next line.
:	The colon specifies a range of numbers. For example, **1:10** means the numbers 1 through 10. A colon in an array dimension accesses all elements in that dimension.

A.^2

A summary of these element-by-element operators are given in Table A.7.

MATLAB follows the usual order of operations we are familiar with from mathematics and computer programming. The precedence can be changed by using parentheses.

TABLE A.7

List of Element-by-Element Operators

Operator	Usage
.*	Multiply element by element
./	Divide element by element
.^	Raise each element to a power

Functions in MATLAB®

MATLAB is a powerful computing environment, but one can also view it as a programming language. Most computer programming languages have mini-programs; these are called *functions* in MATLAB.

In most cases, there are two different ways to call or invoke functions in MATLAB: *function syntax* or *command syntax*, as we describe next. What type of syntax to use is up to the user and depends on what you need to accomplish. For example, you would typically use the function syntax option when the output from the function is needed for other tasks.

Function syntax

The function syntax approach works with input arguments and/or output variables. The basic syntax includes the function name and is followed by arguments enclosed in parentheses. Here is an illustration of the syntax:

```
functionname(arg1, ..., argk)
```

The statement above does not return any output from the function to the current workspace. The output of the function can be assigned to one or more output variables enclosed in square brackets:

```
[out1,...,outm] = functionname(arg1,...,argk)
```

You do not need the brackets, if you have only one output variable. The number of inputs and outputs you use depends on the definition of the function and what you want to accomplish. Always look at the **help** pages for a function to get information on the definitions of the arguments and the possible outputs. There are many more options for calling functions than what we describe in this book.

Command Syntax

The main difference between command and function syntax is how you designate the input arguments. With command syntax, you specify the function name followed by arguments separated by spaces. There are no parentheses with command syntax. The basic form of a command syntax is shown here:

```
functionname arg1 ... arg2
```

The other main difference with command syntax pertains to the outputs from the function. You cannot obtain any output values with commands; you must use function syntax for that purpose.

A.6 Basic Plot Functions

In this section, we discuss the main functions for plotting in two and three dimensions. We also describe some useful auxiliary functions to add content to the graph. Type **help graph2d** or **help graph3d** for a list of plotting functions in the base MATLAB software. Table A.8 contains a list of common plotting functions in base MATLAB. We will examine several of these functions starting off with **plot** for creating 2D graphics.

TABLE A.8

List of Plotting Functions in Base MATLAB®

area	Plot curve and fill in the area
bar, bar3	2-D and 3-D bar plots
contour, contour3	Display isolines of a surface
errorbar	Plot error bars with curve
hist	Histogram
image	Plot an image
pie, pie3	Pie charts
plot, plot3	2-D and 3-D lines
plotmatrix	Matrix of scatterplots
scatter, scatter3	2-D and 3-D scatterplots
stem, stem3	Stem plot for discrete data

Plotting 2D Data

The main function for creating a 2D plot is called **plot**. When the function **plot** is called, it opens a new Figure window. It scales the axes to fit the limits of the data, and it plots the points, as specified by the arguments to the function. The default is to plot the points and connect them with straight lines. If a Figure window is already available, then it produces the plot in the current Figure window, replacing what is there.

The main syntax for **plot** is

$$\texttt{plot(x,y,'color\_linestyle\_marker')}$$

where **x** and **y** are vectors of the same size. The **x** values correspond to the horizontal axis, and the **y** values are represented on the vertical axis.

Several pairs of vectors (for the horizontal and vertical axes) can be provided to **plot**. MATLAB will plot the values given by the pairs of vectors on the same set of axes in the Figure window. If just one vector is provided as an argument, then the function plots the values in the vector against the index $1 \ldots n$, where n is the length of the vector. For example, the following command plots two curves

$$\text{plot}(x1,y1,x2,y2)$$

The first curve plots **y1** against **x1**, and the second shows **y2** versus the values in **x2**.

Many arguments can be used with the **plot** function giving the user a lot of control over the appearance of the graph. Most of them require the use of MATLAB's Handle Graphics® system, which is beyond the scope of this introduction. The interested reader is referred to Marchand and Holland [2002] for more details on Handle Graphics.

Type **help plot** at the command line to see what can be done with **plot**. We present some of the basic options here. The default line style in MATLAB is a solid line, but there are other options as listed in Table A.9. The first column in the table contains the notation for the line style that is used in the **plot** function. For example, one would use the notation

$$\text{plot}(x,y,':')$$

to create a dotted line with the default marker style and color. Note that the specification of the line style is given within single quotes (denoting a string) and is placed immediately after the vectors containing the observations to be graphed with the line style.

We can also specify different colors and marker (or point) styles within the single quotes. The predefined colors are given in Table A.10. There are thirteen marker styles, and they are listed in Table A.11. They include circles, asterisks, stars, x-marks, diamonds, squares, triangles, and more.

The following is a brief list of common plotting tasks:

- Solid green line with no markers or plotting with just points:

$$\text{plot}(x,y,'g'), \ \text{plot}(x,y,'.')$$

- Dashed blue line with points shown as an asterisk:

$$\text{plot}(x,y,'b--*')$$

- Two lines with different colors and line styles (see Figure A.4):

$$\text{plot}(x,y,'r:',x2,y2,'k.-o')$$

TABLE A.9

Line Styles for Plots

Notation	Line Type
–	solid line
:	dotted line
–.	dash-dot line
––	dashed line

TABLE A.10

Line Colors for Plots

Notation	Color
b	blue
g	green
r	red
c	cyan
m	magenta
y	yellow
k	black
w	white

TABLE A.11

Marker Styles for Plots

Notation	Marker Style
.	point
o	circle
x	x-mark
+	plus
*	star
s	square
d	diamond
v	down triangle
^	up triangle
<	left triangle
>	right triangle
p	pentagram
h	hexagram

It is always good practice to include labels for all axes and a title for the plot. You can add these to your plot using the functions **xlabel**, **ylabel**, and **title**. The basic input argument for these functions is a text string:

xlabel('*text*'),ylabel('*text*'),title('*text*')

In the bulleted list above, we showed how to plot two sets of **x** and **y** pairs, and this can be extended to plot any number of lines on one plot. There is another mechanism to plot multiple lines that can be useful in loops and other situations. The command to use is **hold on**, which tells MATLAB to apply subsequent graphing commands to the current plot. To unfreeze the current plot, use the command **hold off**.

We sometimes want to plot different lines (or any other type of graph) in a Figure window, but we want each one on their own set of axes. We can do this through the use of the **subplot** function, which creates a matrix of plots in a Figure window. The basic syntax is

$$\texttt{subplot(m, n, p)}$$

This produces **m** rows and **n** columns of plots in one Figure window. The third argument denotes what plot is active. Any plotting commands after this function call are applied to the **p**-th plot. The axes are numbered from top to bottom and left to right.

Plotting 3D Data

We can plot three variables in MATLAB using the **plot3** function, which works similarly to **plot**. In this case, we have to specify three arguments that correspond to the three axes. The basic syntax is

$$\texttt{plot3(x,y,z)}$$

where **x**, **y**, and **z** are vectors with the same number of elements. The function **plot3** will graph a line in 3D through the points with coordinates given by the elements of the vectors. There is a **zlabel** function to add an axes label to the third dimension.

Sometimes, one of our variables is a response or dependent variable, and the other two are predictors or independent variables. Notationally, this situation is given by the relationship

$$z = f(x, y).$$

The z values define a surface given by points above the x–y plane.

We can plot this type of relationship in MATLAB using the **mesh** or **surf** functions. Straight lines are used to connect adjacent points on the surface. The **mesh** function shows the surface as a wireframe with colored lines, where the color is proportional to the height of the surface. The **surf** function fills in the surface facets with color.

The **surf** and **mesh** functions require three matrix arguments, as shown here

$$\texttt{surf(X,Y,Z), mesh(X,Y,Z)}$$

The **X** and **Y** matrices contain repeated rows and columns corresponding to the domain of the function. If you do not have these already, then you can generate the matrices using a function called **meshgrid**. This function takes two vectors **x** and **y** that specify the domains, and it creates the matrices for constructing the surface. The **x** vector is copied as rows of the **X** matrix, and the vector **y** is copied as columns of the **Y** matrix.

As stated previously, the color is mapped to the height of the surface using the default color map. The definition of the color map can be changed using the function **colormap**. See the **help** on **graph3d** for a list of built-in color maps. You can also add a bar to your graph that illustrates the scale associated with the colors by using the command **colorbar**.

MATLAB conveys 3D surfaces on a 2D screen, but this is just one view of the surface. Sometimes interesting structures are hidden from view. We can change the view interactively via a toolbar button in the Figure window, as shown in Figure A.4. We can also use the **view** function on the command line, as shown here

<code>view(azimuth, elevation)</code>

The first argument **azimuth** defines the horizontal rotation. The second argument **elevation** corresponds to the vertical rotation. Both of these are given in degrees. For example, we are looking directly overhead (2D view) if the azimuth is equal to zero, and the elevation is given by 90 degrees.

Sometimes, it is easier to rotate the surface interactively to find a good view. We can use the ROTATE 3D button in the Figure window, as shown in Figure A.4. When the button is pushed, the cursor changes to a curved arrow. At this point, you can click in the plot area and rotate it while holding the left mouse button. The current azimuth and elevation is given in the lower left corner, as you rotate the axes.

FIGURE A.4
Click on the ROTATE 3D *button to rotate 3D plots. The current azimuth and elevation is indicated in the lower left corner as the axes rotate.*

Scatterplots

The scatterplot is one of the main tools a statistician should use before doing any analysis of the data or modeling. A *scatterplot* is a plot of one variable against another, where (x, y) pairs are plotted as points. The points are not connected by lines. This type of plot can be used to explore the distribution of multivariate data, to assess the relationship between two variables, or to look for groups and other structure. These were also discussed in Chapter 5, but we provide additional information here for completeness.

We can easily use the functions **plot** and **plot3** to create 2D and 3D scatterplots. We just specify the desired marker or symbol, as shown here:

<code>plot(x,y,'o'), plot3(x,y,z,'*')</code>

The call to **plot** shows the markers as open circles, and the call to **plot3** uses the asterisk as the plotting symbol.

The **plot** and **plot3** functions are best for the case where you are using only one or two marker styles and/or symbol colors. MATLAB has two special functions for scatterplots that provide more control over the symbols used in the plot. These are called **scatter** and **scatter3**. The syntax is

scatter(x,y,s,c), scatter3(x,y,z,s,c)

The first two (or three for **scatter3**) arguments represent the coordinates for the points. The optional arguments **s** (marker size) and **c** (marker color) allow one to control the appearance of the symbols. These inputs can be a single value, in which case, they are applied to all markers. Alternatively, one could assign a color and/or size to each point.

The default symbol is an open circle. An additional input argument specifying the marker (see **help** on **plot** for a list) can be used to get a different plotting symbol. If we assign a different size to each of the open circles, possibly corresponding to some other variable, then this is known as a *bubble plot*.

Scatterplot Matrix

We often have data with many variables or dimensions. In this case, we can use the *scatterplot matrix* to look at all 2D scatterplots, which gives us an idea of the pair-wise relationships or distributions in the data. A scatterplot matrix is a matrix of plots, where each one is a 2D scatterplot.

MATLAB provides a function called **plotmatrix** that takes care of the plotting commands for us, and we do not have to worry about multiple uses of **subplot**. The syntax for **plotmatrix** is:

plotmatrix(X), plotmatrix(X,Y)

The first of these plots the columns of **X** as scatterplots, with histograms of the columns on the diagonal. This is the version used most often by statisticians. The second plots the columns of **Y** against the columns of **X**.

GUIs for Graphics

MATLAB has several graphical user interfaces (GUIs) to make plotting data easier. We describe the main tools in this section, including simple edits using menu options in the Figure window, the plotting tools interface, and the PLOTS tab on the desktop ribbon.

We often want to add simple graphics objects to our plot, such as labels, titles, arrows, rectangles, circles, and more. We can do most of these tasks via the command line, but it can be frustrating trying to get them to appear just the way we want them to. We can use the INSERT menu on the Figure window, as shown in Figure A.5, to help with these tasks. Just select the desired option, and interactively add the object to the plot.

Perhaps the most comprehensive GUIs for working with graphics are the plotting tools. This is an interactive set of tools that work similarly to the

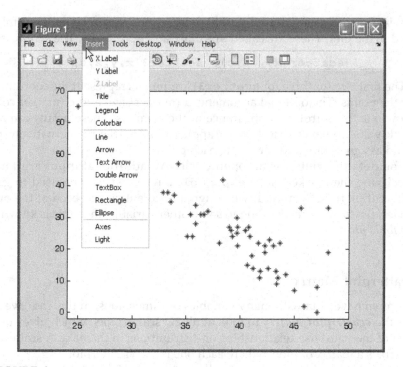

FIGURE A.5
Select the **INSERT** *menu on the Figure window to add objects to your plot.*

main MATLAB desktop environment. In other words, they can be part of an expanded Figure window, or they can be undocked by clicking the downward arrow in the upper right corner of the tool. Look at **help** on **plottools** for more information and examples of using these tools.

The plotting tool GUIs consist of three panels or editors, as listed here:

- **Property Editor**: This provides access to some of the properties of the graphics objects in a figure. This includes the Figure window, the axes, line objects, and text. The editor can be started by using the command **propertyeditor**. It can also be opened via the TOOLS > EDIT PLOT menu item in a Figure window. Double-click on a highlighted graphics object to open the editor.

- **Figure Palette**: This tool allows the user to add and position axes, plot variables from the workspace, and annotate the graph. The command **figurepalette** will open the editor.

- **Plot Browser**: The browser is used to add data to the plot and to control the visibility of existing objects, such as the axes. This can be opened using the command **plotbrowser**.

The plotting tools can be opened in different ways, and we specified some of them in the list given above. One option is to use the command

plottools. This will add panels to an existing Figure window, or it will open a new Figure window and the tools, if one is not open already. The most recently used plotting tools are opened for viewing.

One could also click on the highlighted toolbar button shown in Figure A.6. The button on the right shows the plotting tools, and the one on the left closes them. Finally, the tools can be accessed using the VIEW menu on the Figure window. Clicking on the desired tool will toggle it on or off.

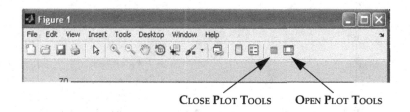

CLOSE PLOT TOOLS   OPEN PLOT TOOLS

FIGURE A.6
This shows the toolbar buttons for a Figure window. The buttons on the right will open and close the interactive plotting tools.

An example of a Figure window with the plotting tools open is given in Figure A.7. Only the Property Editor is opened because that was the last tool we used. Select other editors using the VIEW menu.

Another GUI option to create plots is available via the PLOTS tab on the main MATLAB ribbon interface. The Workspace browser has to be open because this is used to select variables for plotting. Click on the desired variables in the browser, while holding the CTRL key. The variables will appear in the left section of the PLOTS tab.

Next, select the type of plot by clicking the corresponding picture. There is a downward arrow button that provides access to a complete gallery of plots. It will show only those plots that are appropriate for the types of variables selected for plotting. See Figure A.8 for an example.

A.7 Summary and Further Reading

MATLAB has more graphing functions as part of the base software. The main ones were described in this chapter, and we provide an expanded list in Table A.8. Use the **help** *functionname* at the command line for information on how to use them and to learn about related functions. Table A.12 has a list of auxiliary functions for enhancing your plots.

We already recommended the Marchand and Holland book [2002] for more information on Handle Graphics. This text also has some useful tips and ideas for creating plots and building GUIs. For more background on how

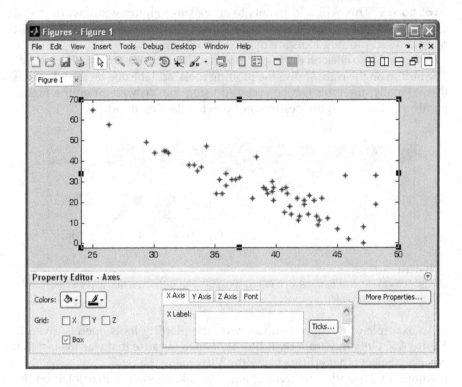

FIGURE A.7
Here is an example of a Figure window with the potting tools open. Only the Property Editor is viewed because this is the last one that was used. Select other editors using the **VIEW** *menu.*

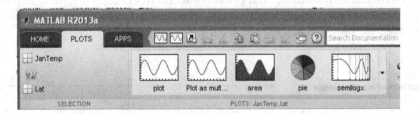

FIGURE A.8
This is a portion of the **PLOTS** *ribbon in the main MATLAB desktop environment. Click the* **PLOTS** *tab to access it. Variables for plotting are selected in the Workspace browser, and they appear in the left section of the ribbon. Click on one of the icons in the right section of the* **PLOTS** *ribbon to create the plot. More options are available using the downward arrow button in the plots section.*

TABLE A.12

List of Auxiliary Plotting Functions in Base MATLAB®

`axis`	Change axes scales and appearance
`box`	Draw a box around the axes
`grid`	Add grid lines at the tick marks
`gtext`	Add text interactively
`hidden`	Remove hidden lines in **mesh** plots
`hold`	Hold the current axes
`legend`	Insert a legend
`plotedit`	Tools for annotation and editing
`rotate`	Rotate using given angles
`subplot`	Include multiple axes in figure window
`text`	Insert text at locations
`title`	Put a title on the plot
`xlabel, ylabel, zlabel`	Label the axes
`view`	Specify the view for a 3-D plot
`zoom`	Zoom in and out of the plot

to visualize data in MATLAB, please see the book *Exploratory Data Analysis with MATLAB* by Martinez et al. [2010]. Finally, you should always consult the MATLAB documentation and help files for examples. For example, there is a section called Graphics in the MATLAB documentation center. Recall that you get to the documentation by clicking the HELP button in the RESOURCES section of the HOME ribbon and selecting the MATLAB link.

We now provide some references to books that describe scientific and statistical visualization, in general. One of the earliest ones in this area is called the *Semiology of Graphics: Diagrams, Networks, Maps* [Bertin, 1983]. This book discusses rules and properties of graphics. For examples of graphical mistakes, we recommend the book by Wainer [1997]. Wainer also published a book called *Graphic Discovery: A Trout in the Milk and Other Visual Adventures* [2004] detailing some of the history of graphical displays in a very thought provoking and entertaining way. The book *Visualizing Data* [Cleveland, 1993] includes descriptions of visualization tools, the relationship of visualization to classical statistical methods, and some of the cognitive aspects of data visualization and perception. Another excellent resource on graphics for data analysis is Chambers et al. [1983]. Finally, we highly recommend Naomi Robbins' [2005] book called *Creating More Effective Graphs*. This text provides a wonderful introduction on ways to convey data correctly and effectively.

There is a *Graphics* section in the online documentation for base MATLAB. Recall that you can access this documentation via the HELP button on the RESOURCES tab. The Statistics Toolbox documentation has a chapter called *Exploratory Data Analysis*, under which is a section on *Statistical Visualization*. This has details about univariate and multivariate plots.

Appendix B

Projection Pursuit Indexes

In this appendix, we list several indexes for projection pursuit [Posse, 1995b]. Since structure is considered to be departures from normality, these indexes are developed to detect non-normality in the projected data. There are some criteria that we can use to assess the usefulness of projection indexes. These include affine invariance [Huber, 1985], speed of computation, and sensitivity to departure from normality in the core of the distribution rather than the tails. The last criterion ensures that we are pursuing structure and not just outliers.

B.1 Friedman-Tukey Index

This projection pursuit index [Friedman and Tukey, 1974] is based on interpoint distances and is calculated using the following:

$$PI_{FT}(\alpha, \beta) = \sum_{i=1}^{n} \sum_{j=1}^{n} (R^2 - r_{ij}^2)^3 I(R^2 - r_{ij}^2) ,$$

where $R = 2.29n^{-1/5}$, $r_{ij}^2 = (z_i^\alpha - z_j^\alpha)^2 + (z_i^\beta - z_j^\beta)^2$, and $I()$ is the indicator function for positive values,

$$I(x) = \begin{cases} 1; & x > 0 \\ 0; & x \le 0. \end{cases}$$

This index has been revised from the original to be affine invariant [Swayne, Cook, and Buja, 1991] and has computational order $O(n^2)$.

B.2 Entropy Index

This projection pursuit index [Jones and Sibson, 1987] is based on the entropy and is given by

$$PI_E(\alpha, \beta) = \frac{1}{n}\sum_{i=1}^{n} \log\left[\frac{1}{nh_\alpha h_\beta}\sum_{j=1}^{n}\phi_2\left(\frac{(z_i^\alpha - z_j^\alpha)}{h_\alpha}, \frac{(z_i^\beta - z_j^\beta)}{h_\beta}\right)\right] + \log(2\pi e),$$

where ϕ_2 is the bivariate standard normal density. The bandwidths $h_\gamma, \gamma = \alpha, \beta$ are obtained from

$$h_\gamma = 1.06n^{-1/5}\left(\sum_{i=1}^{n}\left\{z_i^\gamma - \sum_{j=1}^{n}z_j^\gamma/n\right\}^2/(n-1)\right)^{\frac{1}{2}}.$$

This index is also $O(n^2)$.

B.3 Moment Index

This index was developed in Jones and Sibson [1987] and uses bivariate third and fourth moments. This is very fast to compute, so it is useful for large data sets. However, a problem with this index is that it tends to locate structure in the tails of the distribution. It is given by

$$PI_M(\alpha, \beta) = \frac{1}{12}\left\{\kappa_{30}^2 + 3\kappa_{21}^2 + 3\kappa_{12}^2 + \kappa_{03}^2 + \frac{1}{4}(\kappa_{40}^2 + 4\kappa_{31}^2 + 6\kappa_{22}^2 + 4\kappa_{13}^2 + \kappa_{04}^2)\right\},$$

where

$$\kappa_{30} = \frac{n}{(n-1)(n-2)}\sum_{i=1}^{n}(z_i^\alpha)^3 \qquad \kappa_{03} = \frac{n}{(n-1)(n-2)}\sum_{i=1}^{n}(z_i^\beta)^3$$

$$\kappa_{31} = \frac{n(n+1)}{(n-1)(n-2)(n-3)} \sum_{i=1}^{n} (z_i^\alpha)^3 z_i^\beta$$

$$\kappa_{13} = \frac{n(n+1)}{(n-1)(n-2)(n-3)} \sum_{i=1}^{n} (z_i^\beta)^3 z_i^\alpha$$

$$\kappa_{04} = \frac{n(n+1)}{(n-1)(n-2)(n-3)} \left\{ \sum_{i=1}^{n} (z_i^\beta)^4 - \frac{3(n-1)^3}{n(n+1)} \right\}$$

$$\kappa_{40} = \frac{n(n+1)}{(n-1)(n-2)(n-3)} \left\{ \sum_{i=1}^{n} (z_i^\alpha)^4 - \frac{3(n-1)^3}{n(n+1)} \right\}$$

$$\kappa_{22} = \frac{n(n+1)}{(n-1)(n-2)(n-3)} \left\{ \sum_{i=1}^{n} (z_i^\alpha)^2 (z_i^\beta)^2 - \frac{(n-1)^3}{n(n+1)} \right\}$$

$$\kappa_{21} = \frac{n}{(n-1)(n-2)} \sum_{i=1}^{n} (z_i^\alpha)^2 z_i^\beta \qquad \kappa_{12} = \frac{n}{(n-1)(n-2)} \sum_{i=1}^{n} (z_i^\beta)^2 z_i^\alpha.$$

B.4 L^2 Distances

Several indexes estimate the L^2 distance between the density of the projected data and a bivariate standard normal density. The L^2 projection indexes use orthonormal polynomial expansions to estimate the marginal densities of the projected data. One of these proposed by Friedman [1987] uses Legendre polynomials with J terms. Note that MATLAB® has a function for obtaining these polynomials called **legendre**.

$$PI_{Leg}(\alpha, \beta) = \frac{1}{4} \left\{ \sum_{j=1}^{J} (2j+1) \left(\frac{1}{n} \sum_{i=1}^{n} P_j y_i^\alpha \right)^2 \right.$$

$$+ \sum_{k=1}^{J} (2k+1) \left(\frac{1}{n} \sum_{i=1}^{n} P_k y_i^\beta \right)^2$$

$$+ \left. \sum_{j=1}^{J} \sum_{k=1}^{J-j} (2j+1)(2k+1) \left(\frac{1}{n} \sum_{i=1}^{n} P_j(y_i^\alpha) P_k(y_i^\beta) \right)^2 \right\}$$

where $P_a(\ )$ is the Legendre polynomial of order a. This index is not affine invariant, so Morton [1989] proposed the following revised index. This is based on a conversion to polar coordinates as follows:

$$\rho = (z^\alpha)^2 + (z^\beta)^2 \qquad \theta = \text{atan}\left(\frac{z^\beta}{z^\alpha}\right).$$

We then have the following index, where Fourier series and Laguerre polynomials are used:

$$PI_{LF}(\alpha, \beta) = \frac{1}{\pi} \sum_{l=0}^{L} \sum_{k=1}^{K} \left[\left(\frac{1}{n} \sum_{i=1}^{n} L_l(\rho_i) \exp(-\rho_i/2) \cos(k\theta_i) \right)^2 \right.$$

$$\left. + \left(\frac{1}{n} \sum_{i=1}^{n} L_l(\rho_i) \exp(-\rho_i/2) \sin(k\theta_i) \right)^2 \right] + \frac{1}{2\pi} \sum_{l=0}^{L} \left(\frac{1}{n} \sum_{i=1}^{n} L_l(\rho_i) \exp(-\rho_i/2) \right)^2$$

$$- \frac{1}{2\pi n} \sum_{i=1}^{n} \exp(-\rho_i/2) + \frac{1}{8\pi} ,$$

where L_a represents the Laguerre polynomial of order a. Two more indexes based on the L^2 distance using expansions in Hermite polynomials are given in Posse [1995b].

Appendix C

Data Sets

C.1 Introduction

In this appendix, we list the data sets that are used in the book and are included with the Computational Statistics Toolbox. For the most part, these data are available for download in either text format (`.txt`) or MATLAB® binary format (`.mat`).

C.2 Descriptions

abrasion

The **abrasion** data set has 30 observations, where the two predictor variables are hardness and tensile strength (**x**). The response variable is abrasion loss (**y**) [Hand, et al., 1994; Davies and Goldsmith, 1972]. The first column of **x** contains the hardness and the second column contains the tensile strength.

airfoil

These data set were obtained from the UCI Machine Learning Repository. The data correspond to the aerodynamic and acoustic tests of airfoil blade sections. The response variable is scaled sound pressure (decibels). The predictors are frequency (Hertz), angle (degrees), chord length (meters), free-stream velocity (meters per second), and suction side displacement thickness (meters).

anaerob

A subject performs an exercise, gradually increasing the level of effort. The data set called **anaerob** has two variables based on this experiment: oxygen uptake and the expired ventilation [Hand, et al., 1994; Bennett, 1988]. The oxygen uptake is contained in the variable **x** and the expired ventilation is in **y**.

anscombe
These data were taken from Hand, et al. [1994]. They were originally from Anscombe [1973], where he created these data sets to illustrate the importance of graphical exploratory data analysis. This file contains four sets of **x** and **y** measurements.

bank
This file contains two matrices, one corresponding to features taken from 100 forged Swiss bank notes (**forge**) and the other comprising features from 100 genuine Swiss bank notes (**genuine**) [Flury and Riedwyl, 1988]. There are six features: length of the bill, left width of the bill, right width of the bill, width of the bottom margin, width of the top margin, and length of the image diagonal.

bankdata
This data set was downloaded from the UCI Machine Learning Repository. The **.mat** file contains two data objects—**BankDat** and **Class**. The data comprise features that were extracted from images of genuine and forged bank notes. Wavelet transforms were used to obtain the features.

beetle
These data came from Dobson [2002]. They represent the number of dead beetles after they have been exposed to different concentrations of gaseous carbon disulphide for five hours [Bliss, 1935].

biology
The **biology** data set contains the number of research papers (**numpaps**) for 1534 biologists [Tripathi and Gupta, 1988; Hand, et al., 1994]. The frequencies are given in the variable **freqs**.

bodmin
These data represent the locations of granite tors on Bodmin Moor [Pinder and Witherick, 1977; Upton and Fingleton, 1985; Bailey and Gatrell, 1995]. The file contains vectors **x** and **y** that correspond to the coordinates of the tors. The two-column matrix **bodpoly** contains the vertices to the region.

boston
The **boston** data set contains data for 506 census tracts in the Boston area, taken from the 1970 Census [Harrison and Rubinfeld, 1978]. The predictor variables are (1) per capita crime rate, (2) proportion of residential land zoned for lots over 25,000 sq.ft., (3) proportion of nonretail business acres, (4) Charles River dummy variable (1 if tract bounds river; 0 otherwise), (5) nitric oxides concentration (parts per 10 million), (6) average number of rooms per

dwelling, (7) proportion of owner-occupied units built prior to 1940, (8) weighted distances to five Boston employment centers, (9) index of accessibility to radial highways, (10) full-value property-tax rate per $10,000, (11) pupil-teacher ratio, (12) proportion of African-Americans, and (13) lower status of the population. These are contained in the variable **x**. The response variable **y** represents the median value of owner-occupied homes in $1000's. These data were downloaded from

> `http://www.stat.washington.edu/raftery/Courses/`

brownlee
The **brownlee** data contains observations from 21 days of a plant operation for the oxidation of ammonia [Hand, et al., 1994; Brownlee, 1965]. The predictor variables are the air flow, the cooling water inlet temperature (degrees C), and the percent acid concentration. The response variable Y is the stack loss (the percentage of the ingoing ammonia that escapes). The matrix **x** contains the observed predictor values and the vector **y** has the corresponding response variables.

cardiff
This data set has the locations of homes of juvenile offenders in Cardiff, Wales in 1971 [Herbert, 1980]. The file contains vectors **x** and **y** that correspond to the coordinates of the homes. The two-column matrix **cardpoly** contains the vertices to the region.

cars
These data were extracted from a data set that is in the MATLAB Statistics Toolbox. The **cars** data includes two measurements (predictor variables) on small model cars: weight (**x2**) and horsepower (**x3**). The response variable is the miles per gallon (**y**).

cement
Manufacturers of cement are interested in the tensile strength of their product. The strength depends on many factors, one of which is the length of time the cement is dried. An experiment is conducted where different batches of cement are tested for tensile strength after different drying times. Engineers would like to determine the relationship between drying time and tensile strength of the cement [Hand, et al., 1994; Hald, 1952]. The **cement** data file contains a vector **x** representing the drying times and a vector **y** that contains the tensile strength.

cereal
These data were obtained from ratings of eight brands of cereal [Chakrapani and Ehrenberg, 1981; Venables and Ripley, 1994]. The **cereal** file contains a matrix where each row corresponds to an observation and each column

represents one of the variables or the percent agreement to statements about the cereal. It also contains a cell array of strings (**labs**) for the type of cereal.

cigarette

The data set includes measurements of twenty-five brands of cigarettes, and it is often used in introductory statistics classes to help students understand multiple regression in the presence of collinearity [McIntyre, 1994]. The data were originally from the Federal Trade Commission who annually measured the weight, tar, nicotine, and carbon monoxide of domestic cigarettes. We are interested in modeling the relationship between three predictor variables—nicotine, tar, and weight and the response variable carbon monoxide.

coal

The **coal** data set contains the number of coal mining disasters (**y**) over 112 years (**year**) [Raftery and Akman, 1986].

counting

In the **counting** data set, we have the number of scintillations in 72 second intervals arising from the radioactive decay of polonium [Rutherford and Geiger, 1910; Hand, et al., 1994]. There are a total of 10097 scintillations and 2608 intervals. Two vectors, **count** and **freqs**, are included in this file.

cpunish

The **cpunish** data contains $n = 17$ observations [Gill, 2001]. The response variable (**y**) is the number of times that capital punishment is implemented in a state for the year 1997. The explanatory variables (**X**) are median per capita income (dollars), the percent of the population living in poverty, the percent of African-American citizens in the population, the log(rate) of violent crimes per 1,000,000 residents in 1996, a variable indicating whether a state is in the South, and the proportion of the population with a college degree. The number of people with a college degree contains those with an AA degree and above.

crab

The original **crab** data set was analyzed by Agresti [1996] and came from a study by Brockmann [1996]. We downloaded the data set from

    http://lib.stat.cmu.edu/datasets/agresti

and extracted two of the explanatory variables: carapace width (cm) and weight (kg). These are contained in the matrix **X**. The response variable is the number of male crabs (satellites) surrounding the nesting female/male couple. This is saved in the **satellites** vector.

elderly

The **elderly** data set contains the height measurements (in centimeters) of 351 elderly females [Hand, et al., 1994]. The variable that is loaded is called **heights**.

environ/environmental

This data set was analyzed in Cleveland and McGill [1984]. They represent two variables comprising daily measurements of **ozone** and **wind** speed in New York City. These quantities were measured on 111 days between May and September 1973. One might be interested in understanding the relationship between ozone (the response variable) and wind speed (the predictor variable). The **environmental** data set contains more variables and data.

filip

These data are used as a standard to test the results of least squares calculations. The file contains two vectors **x** and **y**.

flea

The **flea** data set [Hand, et al., 1994; Lubischew, 1962] contains measurements on three species of flea beetle: *Chaetocnema concinna* (**conc**), *Chaetocnema heikertingeri* (**heik**), and *Chaetocnema heptapotamica* (**hept**). The features for classification are the maximal width of aedeagus in the forepart (microns) and the front angle of the aedeagus (units are 7.5 degrees).

forearm

These data [Hand, et al., 1994; Pearson and Lee, 1903] consist of 140 measurements of the length (in inches) of the forearm of adult males. The vector **x** contains the measurements.

gascons

This data Simonoff [1996] set has gasoline consumption (measured in tens of millions of dollars) as the response variable. The predictors are the price index for gasoline (dollars), the price index for used cars (dollars), and the per capita disposable income (dollars). The data span the years 1960 through 1986, and the dollars are measured in 1967 dollars.

geyser

These data represent the waiting times (in minutes) between eruptions of the Old Faithful geyser at Yellowstone National Park [Hand, et al, 1994; Scott, 1992]. This contains one vector called **geyser**.

glassdata

These data belong to a Glass Identification Database and were downloaded from the UCI Machine Learning Repository [Newman, et al., 1998]. This data set was created by B. German of Berkshire, England in 1987. There are 214 observations, with 9 variables: refractive index, sodium, magnesium, aluminum, silicon, potassium, calcium, barium, and iron. The data set also includes a class label: building windows float processed, building windows not float processed, vehicle windows float processed, building windows not float processed, containers, tableware, and head lamps. One could use these data to develop a classifier that could be used in forensic science applications.

helmets
The data in **helmets** contain measurements of head acceleration (in g) (**accel**) and times after impact (milliseconds) (**time**) from a simulated motorcycle accident [Hand, et al., 1994; Silverman, 1985].

household
The **household** [Hand, et al., 1994; Aitchison, 1986] data set contains the expenditures for housing, food, other goods, and services (four expenditures) for households comprised of single people. The observations are for single **women** and single **men**.

human
The **human** data set [Hand, et al., 1994; Mazess, et al., 1984] contains measurements of percent fat and age for 18 normal adults (**males** and **females**).

icaexample
This data set is used in Chapter 6 to illustrate independent component analysis. It was created by loading the **chirp** and **gong** signals that are included in the basic MATLAB software package. Each of these is a column the matrix **G**, and the signal mixtures are columns in the matrix **x**. The spherized version of the mixture is found in **z**. The mixing coefficients are in **mixmat**. Consult the example file for details on how these were created.

insect
In this data set, we have three variables measured on ten insects from each of three species [Hand, et al.,1994]. The variables correspond to the width of the first joint of the first tarsus, the width of the first joint of the second tarsus and the maximal width of the aedeagus. All widths are measured in microns. When **insect** is loaded, you get one 30×3 matrix called **insect**. Each group of ten rows belongs to one of the insect species.

insulate
The **insulate** data set [Hand, et al., 1994] contains observations corresponding to the average outside temperature in degrees Celsius (first column) and the amount of weekly gas consumption measured in 1000 cubic feet (second column). One data set is before insulation (**befinsul**) and the other corresponds to measurements taken after insulation (**aftinsul**).

iris

The **iris** data were collected by Anderson [1935] and were analyzed by Fisher [1936] (and many statisticians since then!). The data consist of 150 observations containing four measurements based on the petals and sepals of three species of iris. The three species are *Iris setosa, Iris virginica,* and *Iris versicolor*. When the **iris** data are loaded, you get three 50 × 4 matrices, one corresponding to each species.

law/lawpop

The **lawpop** data set [Efron and Tibshirani, 1993] contains the average scores on the LSAT (**lsat**) and the corresponding average undergraduate grade point average (**gpa**) for the 1973 freshman class at 82 law schools. Note that these data constitute the entire population. The data contained in **law** comprise a random sample of 15 of these classes, where the **lsat** score is in the first column and the **gpa** is in the second column.

longley

The data in **longley** were used by Longley [1967] to verify the computer calculations from a least squares fit to data. The data set (**X**) contains measurements of 6 predictor variables and a column of ones representing the constant term. The observed responses are contained in **Y**.

measure

The **measure** [Hand, et al., 1994] data contain 20 measurements of chest, waist, and hip data. Half of the measured individuals are women and half are men.

moths

The **moths** data represent the number of moths caught in a trap over 24 consecutive nights [Hand, et al., 1994].

nfl

The **nfl** data [Csorgo and Welsh, 1989; Hand, et al., 1994] contain bivariate measurements of the game time to the first points scored by kicking the ball between the end posts (X_1), and the game time to the first points scored by moving the ball into the end zone (X_2). The times are in minutes and seconds.

okblack and okwhite

These data represent locations where thefts occurred in Oklahoma City in the late 1970's [Bailey and Gatrell, 1995]. The file **okwhite** contains the data for Caucasian offenders, and the file **okblack** contains the data for African-American offenders. The boundary for the region is not included with these data.

peanuts
The **peanuts** data set [Hand, et al., 1994; Draper and Smith, 1981] contains measurements of the average level of alfatoxin (**X**) of a batch of peanuts and the corresponding percentage of noncontaminated peanuts in the batch (**Y**).

posse
The **posse** file contains several data sets generated for simulation studies in Posse [1995b]. These data sets are called **croix** (a cross), **struct2** (an L-shape), **boite** (a donut), **groupe** (four clusters), **curve** (two curved groups), and **spiral** (a spiral). Each data set has 400 observations in 8-D. These data can be used in PPEDA.

quakes
The **quakes** data [Hand, et al., 1994] contain the time in days between successive earthquakes.

remiss
The **remiss** data set contains the remission times for 42 leukemia patients. Some of the patients were treated with the drug called 6-mercaptopurine (**mp**), and the rest were part of the control group (**control**) [Hand, et al., 1994; Gehan, 1965].

salmon
This is a data set from Simonoff [1996]. It contains measurements of the annual spawners and recruits for sockeye salmon in the Skeena River for the years 1940 through 1967.

snowfall
The Buffalo **snowfall** data [Scott, 1992] represent the annual snowfall in inches in Buffalo, New York over the years 1910-1972. This file contains one vector called **snowfall**.

spatial
These data came from Efron and Tibshirani [1993]. Here we have a set of measurements of 26 neurologically impaired children who took a test of spatial perception called test A.

steam
In the **steam** data set, we have a sample representing the average atmospheric temperature (**x**) and the corresponding amount of steam (**y**) used per month [Draper and Smith, 1981]. We get two vectors **x** and **y** when these data are loaded.

thrombos
The **thrombos** data set contains measurements of urinary-thromboglobulin excretion in 12 **normal** and 12 **diabetic** patients [van Oost, et al.; 1983; Hand, et al., 1994].

tibetan
This file contains the heights of 32 Tibetan skulls [Hand, et al. 1994; Morant, 1923] measured in millimeters. These data comprise two groups of skulls collected in Tibet. One group of 17 skulls comes from graves in Sikkim and nearby areas of Tibet and the other 15 skulls come from a battlefield in Lhasa. The original data contain five measurements for the 32 skulls. When you load this file, you get a 32×5 matrix called **tibetan**.

uganda
This data set contains the locations of crater centers of 120 volcanoes in west Uganda [Tinkler, 1971, Bailey and Gatrell, 1995]. The file has vectors **x** and **y** that correspond to the coordinates of the craters. The two-column matrix **ugpoly** contains the vertices to the region.

vineyard
This is a data set from Simonoff [1996]. It contains the grape yields (in number of lugs) for the harvests from 1989 through 1991. The predictor variable is the row number, and the response is the total lug counts for the three years.

wais
This data set was taken from Dobson [2002] and is used to demonstrate logistic regression. These data comprise a sample of older people who were tested using a subset of the Wechsler Adult Intelligent Scale (WAIS). The testers recorded whether or not symptoms of senility were present (success, $y = 1$) or not (failure, $y = 0$). The explanatory or predictor variable is the WAIS score.

whisky
In 1961, 16 states owned the retail liquor stores (**state**). In 26 others, the stores were owned by private citizens (**private**). The data contained in **whisky** reflect the price (in dollars) of a fifth of Seagram 7 Crown Whisky from these 42 states. Note that this represents the population, not a sample [Hand, et al., 1994].

Appendix D

Notation

In this appendix, we briefly describe the general notation used in the text. While we try to be consistent and to use unique symbols, we will sometimes fail at doing this. So, we hope that this appendix will help with any confusion that might happen as a result.

In general, Greek symbols will refer to parameters, and English letters will be used for statistics and other quantities. Matrices will be referred to using uppercase bold letters (**A**), with corresponding lowercase unbold letters and subscripts denoting elements of the matrices (a_{ij}). Vectors are usually shown by lowercase bold letters (**x**).

The exception to this *bold* rule is when we use Greek letters to denote arrays. They are sometimes shown in regular font, but it should be clear from the context that they represent an array.

The hat ("^") notation refers to an estimate of the corresponding parameter or quantity. So, $\hat{f}$ is an estimate of a function, $\hat{y}$ is an estimate of y, $\hat{\mu}$ is an estimate of the mean, and so on.

For the most part, we try to use uppercase letters to denote random variables and to use the corresponding lowercase letter to represent realizations of the random variable.

Finally, MATLAB® code, function names, and websites are given in Courier bold font. The end of examples are indicated by a small box: ❑.

D.1 Observed Data

d	Dimensionality of the data, $d \geq 1$
n	Sample size
X	Matrix of size $n \times d$ with each row corresponding to an observation
X_c	The data matrix X centered at the origin
x	Usually a column vector of size $n \times 1$

$\mathbf{X}_c$ Data matrix where observations are centered about the sample mean

$\mathbf{Z}$ The data matrix $\mathbf{X}$ that has been spherized

D.2 Greek Letters

μ Mean of a discrete or continuous random variable

σ^2 Variance of a discrete or continuous random variable

μ'_r The r-th moment

μ_r The r-th central moment

σ_{ij} The ij-th element of the covariance matrix

Σ Covariance matrix for the population

D.3 Functions and Distributions

$P(X)$ Probability of an event

$f(x)$ Probability density or mass function, where x could be discrete or continuous

$F(x)$ Cumulative distribution function

$E[*]$ Expected value

$N(\mu, \sigma^2)$ Normal distribution (univariate) mean μ and variance σ^2

$\Phi(x)$ Cumulative distribution function for a normal random variable

$\phi(x)$ Probability density function for a normal random variable

$L(\theta|\mathbf{x})$ Likelihood function

$L_R(\mathbf{x})$ Likelihood ratio

$z^{(\alpha/2)}$ The $\alpha/2$-th quantile of the standard normal distribution

D.4 Matrix Notation

$|*|$ The matrix determinant

$\mathbf{A}^T$ Transpose of matrix $\mathbf{A}$

D.5 Statistics

$\overline{X}$ Sample mean

μ Alternative notation for the sample mean

M'_r Sample moment

M_r Sample central moment

Σ General notation for an estimate of the covariance matrix

$\mathbf{S}$ An unbiased estimate of the covariance matrix. If $\mathbf{X}_c$ is the data centered at the origin, then $\mathbf{S}$ is

$$\mathbf{S} = \frac{1}{n-1}\mathbf{X}_c^T\mathbf{X}_c$$

s^2 Sample standard deviation

$X_{(i)}$ The i-th order statistic

References

Aarts, E. and J. Korst. 1989. *Simulated Annealing and Boltzmann Machines*, New York: John Wiley & Sons.

Abdi, Herve. 2010. "Partial least squares regression and projection on latent structure regression (PLS Regression)," *WIREs Computational Statistics*, 2: 97–106.

Agresti, Alan. 1996. *An Introduction to Categorical Data Analysis*, New York: John Wiley & Sons.

Agresti, Alan. 2002. *Categorical Data Analysis, 2nd Edition*. New York: John Wiley & Sons.

Aitchison, J. 1986. *The Statistical Analysis of Compositional Data*, London: Chapman and Hall.

Albert, James H. 1993. "Teaching Bayesian statistics using sampling methods and MINITAB," *The American Statistician*. 47:182-191.

Anderberg, Michael R. 1973. *Cluster Analysis for Applications*, New York: Academic Press.

Anderson, E. 1935. "The irises of the Gaspe Peninsula," *Bulletin of the American Iris Society*, 59:2-5.

Andrews, D. F. 1972. "Plots of high-dimensional data," *Biometrics*, 28:125-136.

Andrews, D. F. 1974. "A robust method of multiple linear regression," *Technometrics*, 16:523-531.

Andrews, D. F. and A. M. Herzberg. 1985. *Data: A Collection of Problems from Many Fields for the Student and Research Worker*, New York: Springer-Verlag.

Anscombe, F. J. 1973. "Graphs in statistical analysis," *The American Statistician*, 27:17-21.

Arlinghaus, S. L. (ed.). 1996. *Practical Handbook of Spatial Statistics*, Boca Raton, FL: CRC Press.

Arnold, Steven F. 1993. "Gibbs sampling," in *Handbook of Statistics, Vol 9, Computational Statistics*, C. R. Rao, ed., The Netherlands: Elsevier Science Publishers, pp. 599-625.

Ash, Robert. 1972. *Real Analysis and Probability*, New York: Academic Press.

Asimov, Daniel. 1985. "The grand tour: a tool for viewing multidimensional data," *SIAM Journal of Scientific and Statistical Computing*, 6:128-143.

Bailey, T. A. and R. Dubes. 1982. "Cluster validity profiles," *Pattern Recognition*, 15:61-83.

Bailey, T. C. and A. C. Gatrell. 1995. *Interactive Spatial Data Analysis*, London: Longman Scientific & Technical.

Bain, L. J. and M. Engelhardt. 1992. *Introduction to Probability and Mathematical Statistics, Second Edition,* Boston: PWS-Kent Publishing Company.

Banfield, A. D. and A. E. Raftery. 1993. "Model-based Gaussian and non-Gaussian clustering," *Biometrics,* **49**:803-821.

Banks, Jerry, John Carson, Barry Nelson, and David Nicol. 2001. *Discrete-Event Simulation, Third Edition,* New York: Prentice Hall.

Bauer, Eric and Ron Kohavi. 1999. "An empirical comparison of voting classification algorithms: Bagging, boosting, and variants," *Machine Learning,* **36**:105-139.

Bellman, R. E. 1961. *Adaptive Control Processes,* Princeton University Press.

Belsley, D. A., E. Kuh, and R. E. Welsch. 1980. *Regression Diagnostics: Identifying Influential Data and Sources of Collinearity,* New York: John Wiley & Sons.

Benjamini, Yoav. 1988. "Opening the box of a boxplot," *The American Statistician,* **42**: 257-262.

Bennett, G. W. 1988. "Determination of anaerobic threshold," *Canadian Journal of Statistics,* **16**:307-310.

Bennett, K. P. and C. Campbell. 2000. "Support vector machines: Hype or Hallelujah?" *ACM SIGKDD Explorations Newsletter,* **2**:1-13.

Bensmail, H., G. Celeux, A. E. Raftery, and C. P. Robert. 1997. "Inference in model-based cluster analysis," *Statistics and Computing,* **7**:1-10.

Besag, J. and P. J. Diggle. 1977. "Simple Monte Carlo tests for spatial patterns," *Applied Statistics,* **26**:327-333.

Bickel, Peter J. and Kjell A. Doksum. 2001. *Mathematical Statistics: Basic Ideas and Selected Topics, Vol 1, Second Edition,* New York: Prentice Hall.

Billingsley, Patrick. 1995. *Probability and Measure, 3rd Edition,* New York: John Wiley & Sons.

Binder, D. A. 1978. "Bayesian cluster analysis," *Biometrika,* **65**:31-38.

Bishop, Christopher M. 2006. *Pattern Recognition and Machine Learning,* New York: Springer-Verlag.

Bliss, C. I. 1935. "The calculation of the dose-mortality curve," *Annals of Applied Biology,* **22**:134-167.

Bock, H. 1996. "Probabilistic models in cluster analysis," *Computational Statistics and Data Analysis,* **23**:5-28.

Bolstad, William M. 2007. *An Introduction to Bayesian Statistics, Second Edition,* New York: John Wiley & Sons.

Bolton, R. J. and W. J. Krzanowski. 1999. "A characterization of principal components for projection pursuit," *The American Statistician,* **53**:108-109.

Boos, D. D. and J. Zhang. 2000. "Monte Carlo evaluation of resampling-based hypothesis tests," *Journal of the American Statistical Association,* **95**:486-492.

Borg, Ingwer and Patrick Groenen. 1997. *Modern Multidimensional Scaling: Theory and Applications,* New York: Springer.

Bowman, A. W. and A. Azzalini. 1997. *Applied Smoothing Techniques for Data Analysis: The Kernel Approach with S-Plus Illustrations,* Oxford: Oxford University Press.

Breiman, Leo. 1992. *Probability.* Philadelphia: Society for Industrial and Applied Mathematics.

Breiman, Leo. 1996a. "Bagging predictors," *Machine Learning,* **24**:123-140.

Breiman, Leo. 1996b. "The heuristics of instability in model selection," *Annals of Statistics*, **24**:2350-2383.

Breiman, Leo. 1996c. "Out-of-bag estimation," Department of Statistics, University of California (`ftp://ftp.stat.berkeley.edu/pub/users/breiman/`).

Breiman, Leo. 1998. "Arcing classifiers," *The Annals of Statistics*, **26**:801-849.

Breiman, Leo. 2001. "Random forests," *Machine Learning*, **45**:5-32.

Breiman, Leo, Jerome H. Friedman, Richard A. Olshen and Charles J. Stone. 1984. *Classification and Regression Trees*, New York: Wadsworth, Inc.

Brockmann, H. J. 1996. "Satellite male groups in horseshoe crabs. *Limulus polyphemus*," *Ethology*, **102**:1-21.

Brooks, S. P. 1998. "Markov chain Monte Carlo and its application," *The American Statistician*, **47**:69-100.

Brooks, S. P. and P. Giudici. 2000. "Markov chain Monte Carlo convergence assessment via two-way analysis of variance," *Journal of Computational and Graphical Statistics*, **9**:266-285.

Brownlee, K. A. 1965. *Statistical Theory and Methodology in Science and Engineering, Second Edition*, London: John Wiley & Sons.

Buja, A., T. J. Hastie, and R. J. Tibshirani. 1989. "Linear smoothers and additive models (with discussion)," *Annals of Statistics*, **17**:453-555.

Burges, Christopher J. C. 1998. "A tutorial on support vector machines for pattern recognition," *Data Mining and Knowledge Discovery*, **2**:121-167.

Cacoullos, T. 1966. "Estimation of a multivariate density," *Annals of the Institute of Statistical Mathematics*, **18**:178-189.

Calinski, R. and J. Harabasz. 1974. "A dendrite method for cluster analysis," *Communications in Statistics*, **3**:1-27.

Canty, A. J. 1999. "Hypothesis tests of convergence in Markov chain Monte Carlo," *Journal of Computational and Graphical Statistics*, **8**:93-108.

Carr, D., R. Littlefield, W. Nicholson, and J. Littlefield. 1987. "Scatterplot matrix techniques for large N," *Journal of the American Statistical Association*, **82**:424-436.

Carter, R. L. and K. Q. Hill. 1979. *The Criminals' Image of the City*, Oxford: Pergamon Press.

Casella, George and Roger L. Berger. 1990. *Statistical Inference*, New York: Duxbury Press.

Casella, George, and E. I. George. 1992. "An introduction to Gibbs sampling," *The American Statistician*, **46**:167-174.

Cattell, R. B. 1966. "The scree test for the number of factors," *Journal of Multivariate Behavioral Research*, **1**:245-276.

Celeux, G. and G. Govaert. 1995. "Gaussian parsimonious clustering models," *Pattern Recognition*, **28**:781-793.

Cencov, N. N. 1962. "Evaluation of an unknown density from observations," *Soviet Mathematics*, **3**:1559-1562.

Chakrapani, T. K. and A. S. C. Ehrenberg. 1981. "An alternative to factor analysis in marketing research — Part 2: Between group analysis," *Professional Marketing Research Society Journal*, **1**:32-38.

Chambers, John. 1999. "Computing with data: Concepts and challenges," *The American Statistician*, **53**:73-84.

Chambers, John and Trevor Hastie. 1992. *Statistical Models in S*, New York: Wadsworth & Brooks/Cole Computer Science Series.

Chatterjee, S., A. S. Hadi, and B. Price. 2000. *Regression Analysis by Example, 3rd Edition*, New York: John Wiley & Sons.

Chatterjee, S., M. S. Handcock, and J. S. Simonoff. 1995. *A Casebook for a First Course in Statistics and Data Analysis*, New York: John Wiley & Sons.

Chaudhuri, P. and J. S. Marron. 1999. "SiZer for exploration of structures in curves," *Journal of the American Statistical Association*, **94**:807-823.

Chernick, M. R. 1999. *Bootstrap Methods: A Practitioner's Guide*, New York: John Wiley & Sons.

Chernoff, Herman. 1973. "The use of faces to represent points in k-dimensional space graphically," *Journal of the American Statistical Association*, **68**:361-368.

Chib, S., and E. Greenberg. 1995. "Understanding the Metropolis-Hastings Algorithm," *The American Statistician*, **49**:327-335.

Cleveland, W. S. 1979. "Robust locally weighted regression and smoothing scatterplots," *Journal of the American Statistical Association*, **74**,829-836.

Cleveland, W. S. 1993. *Visualizing Data*, New York: Hobart Press.

Cleveland, W. S. and Robert McGill. 1984. "The many faces of a scatterplot," *Journal of the American Statistical Association*, **79**:807-822.

Cliff, A. D. and J. K. Ord. 1981. *Spatial Processes: Models and Applications*, London: Pion Limited.

Cook, Dianne, A. Buha, J. Cabrera, and C. Hurley. 1995. "Grand tour and projection pursuit," *Journal of Computational and Graphical Statistics*, **4**:155-172.

Cook, Dianne and Deborah F. Swayne. 2007. *Interactive and Dynamic Graphics for Data Analysis: With R and GGobi (Use R)*, New York: Springer-Verlag.

Cook, R. Dennis. 1998. *Regression Graphics: Ideas for Studying Regressions through Graphics*. New York: John Wiley & Sons.

Cook, R. D. and S. Weisberg. 1994. *An Introduction to Regression Graphics*, New York: John Wiley & Sons.

Cormack, R. M. 1971. "A review of classification," *Journal of the Royal Statistical Society, Series A*, **134**:321-367.

Cowles, M. K. and B. P. Carlin. 1996. "Markov chain Monte Carlo convergence diagnostics: a comparative study," *Journal of the American Statistical Association*, **91**:883–904.

Cox, Trevor F. and Michael A. A. Cox. 2001. *Multidimensional Scaling, 2nd Edition*, Boca Raton, FL: Chapman & Hall/CRC.

Craven, P. and G. Wahba. 1979. "Smoothing noisy data with spline functions," *Numerische Mathematik*, **31**:377-403.

Crawford, Stuart. 1991. "Genetic optimization for exploratory projection pursuit," *Proceedings of the 23rd Symposium on the Interface*, **23**:318-321.

Cressie, Noel A. C. 1993. *Statistics for Spatial Data, Revised Edition*. New York: John Wiley & Sons.

Cristianini, N. and J. Shawe-Taylor. 2000. *An Introduction to Support Vector Machines and Other Kernel-based Learning Methods*. New York: Cambridge University Press.

Csorgo, S. and A. S. Welsh. 1989. "Testing for exponential and Marshall-Olkin distributions," *Journal of Statistical Planning and Inference*, **23**:278-300.

Dasgupta, A. and A. E. Raftery. 1998. "Detecting features in spatial point processes with clutter via model-based clustering," *Journal of the American Statistical Association*, **93**:294-302.

David, Herbert A. 1981. *Order Statistics, 2nd edition*, New York: John Wiley & Sons.

Davies, O. L. and P. L. Goldsmith (eds.). 1972. *Statistical Methods in Research and Production, 4th Edition*, Edinburgh: Oliver and Boyd.

Day, N. E. 1969. "Estimating the components of a mixture of normal distributions," *Biometrika*, **56**:463-474.

Dempster, A. P., N. M. Laird, and D. B. Rubin. 1977. "Maximum likelihood from incomplete data via the EM algorithm (with discussion)," *Journal of the Royal Statistical Society: B*, **39**:1-38.

de Boor, Carl. 2001. *A Practical Guide to Splines, Revised Edition*, New York: Springer-Verlag.

de Jong, Sijmen. 1993. "SIMPLS: An Alternative Approach to Partial Least Squares Regression," *Chemometrics and Intelligent Laboratory Systems*, **18**:251–263.

Deng, L. and D. K. J. Lin. 2000. "Random number generation for the new century," *The American Statistician*, **54**:145-150.

Devroye, Luc. and L. Gyorfi. 1985. *Nonparametric Density Estimation: the L_1 View*, New York: John Wiley & Sons.

Devroye, Luc, Laszlo Gyorfi and Gabor Lugosi. 1996. *A Probabilistic Theory of Pattern Recognition*, New York: Springer-Verlag.

Dietterich, Thomas G. 2000. "An experimental comparison of three methods for constructing ensembles of decision trees: Bagging, boosting, and randomization," *Machine Learning*, **40**:139-157.

Diggle, Peter J. 1981. "Some graphical methods in the analysis of spatial point patterns," in *Interpreting Multivariate Data*, V. Barnett, ed., New York: John Wiley & Sons, pp. 55-73.

Diggle, Peter J. 1983. *Statistical Analysis of Spatial Point Patterns*, New York: Academic Press.

Diggle, P. J. and R. J. Gratton. 1984. "Monte Carlo methods of inference for implicit statistical models," *Journal of the Royal Statistical Society: B*, **46**:193–227.

Dobson, Annette J. 2002. *An Introduction to Generalized Linear Models, Second Edition*, Boca Raton, FL: Chapman & Hall/CRC.

Donoho, D. L. and C. Grimes. 2003. "Hessian eigenmaps: Locally linear embedding techniques for high-dimensional data," *Proceedings of the National Academy of Science*, **100**:5591-5596.

Draper, N. R. and H. Smith. 1981. *Applied Regression Analysis, 2nd Edition*, New York: John Wiley & Sons.

du Toit, S. H. C., A. G. W. Steyn, and R. H. Stumpf. 1986. *Graphical Exploratory Data Analysis*, New York: Springer-Verlag.

Dubes, R. and A. K. Jain. 1980. "Clustering methodologies in exploratory data analysis," *Advances in Computers, Vol. 19*, New York: Academic Press.

Duda, Richard O. and Peter E. Hart. 1973. *Pattern Classification and Scene Analysis*, New York: John Wiley & Sons.

Duda, Richard O., Peter E. Hart, and David G. Stork. 2001. *Pattern Classification, Second Edition*, New York: John Wiley & Sons.

Dunn, P. K. and Gordon K. Smyth. 1996. "Randomized quantile residuals," *Journal of Computational and Graphical Statistics*, **5**:1-10.

Durrett, Richard. 1994. *The Essentials of Probability*, New York: Duxbury Press.

Edwards, A. W. F. and L. L. Cavalli-Sforza. 1965. "A method for cluster analysis," *Biometrics*, **21**:362-375.

Efron, B. 1979. "Computers and the theory of statistics: thinking the unthinkable," *SIAM Review*, **21**:460-479.

Efron, B. 1981. "Nonparametric estimates of standard error: the jackknife, the bootstrap, and other methods," *Biometrika*, **68**:589-599.

Efron, B. 1982. *The Jackknife, the Bootstrap, and Other Resampling Plans*, Philadelphia: Society for Industrial and Applied Mathematics.

Efron, B. 1983. "Estimating the error rate of a prediction rule: improvement on cross-validation," *Journal of the American Statistical Association*, **78**:316-331.

Efron, B. 1985. "Bootstrap confidence intervals for a class of parametric problems," *Biometrika*, **72**:45–58.

Efron, B. 1986. "How biased is the apparent error rate of a prediction rule?" *Journal of the American Statistical Association*, **81**:461-470.

Efron, B. 1987. "Better bootstrap confidence intervals' (with discussion)," *Journal of the American Statistical Association*, **82**:171-200.

Efron, B. 1990. "More efficient bootstrap computations, *Journal of the American Statistical Association*, **85**:79-89.

Efron, B. 1992. "Jackknife-after-bootstrap standard errors and influence functions," *Journal of the Royal Statistical Society: B*, **54**:83-127.

Efron, B. and G. Gong. 1983. "A leisurely look at the bootstrap, the jackknife and cross-validation," *The American Statistician*, **37**:36-48.

Efron, B. and R. J. Tibshirani. 1991. "Statistical data analysis in the computer age," *Science*, **253**:390-395.

Efron, B. and R. J. Tibshirani. 1993. *An Introduction to the Bootstrap*, London: Chapman and Hall.

Egan, J. P. 1975. *Signal Detection Theory and ROC Analysis*, New York: Academic Press.

Embrechts, P. and A. Herzberg. 1991. "Variations of Andrews' plots," *International Statistical Review*, **59**:175-194.

Epanechnikov, V. K. 1969. "Non-parametric estimation of a multivariate probability density," *Theory of Probability and its Applications*, **14**:153-158.

Estivill-Castro, Vladimir. 2002. "Why so many clustering algorithms - A position paper," *SIGKDD Explorations*, **4**:65-75.

Esty, Warren W. and Jeffrey D. Banfield. 2003. "The box-percentile plot," *Journal of Statistical Software*, **8**, **http://www.jstatsoft.org/v08/i17**.

Everitt, Brian S. 1993. *Cluster Analysis, Third Edition,* New York: Edward Arnold Publishing.

Everitt, B. S. and D. J. Hand. 1981. *Finite Mixture Distributions,* London: Chapman and Hall.

Everitt, B. S., S. Landau, and M. Leese. 2001. *Cluster Analysis, Fourth Edition,* New York: Edward Arnold Publishing.

Fienberg, S. 1979. "Graphical methods in statistics," *The American Statistician,* **33:**165-178.

Fisher, R. A. 1936. "The use of multiple measurements in taxonomic problems," *Annals of Eugenics,* **7:**179-188.

Flick, T., L. Jones, R. Priest, and C. Herman. 1990. "Pattern classification using projection pursuit," *Pattern Recognition,* **23:**1367-1376.

Flury, B. and H. Riedwyl. 1988. *Multivariate Statistics: A Practical Approach,* London: Chapman and Hall.

Fodor, Imola K. 2002. "A survey of dimension reduction techniques," Lawrence Livermore National Laboratory Technical Report, UCRL-ID-148494.

Fortner, Brand. 1995. *The Data Handbook: A Guide to Understanding the Organization and Visualization of Technical Data, Second Edition,* New York: Springer-Verlag.

Fortner, Brand and Theodore E. Meyer. 1997. *Number by Colors: A Guide to Using Color to Understand Technical Data,* New York: Springer-Verlag.

Fowlkes, E. B. and C. L. Mallows. 1983. "A method for comparing two hierarchical clusterings," *Journal of the American Statistical Association,* **78:**553-584.

Fox, John. 2000a. *Multiple and Generalized Nonparametric Regression,* Thousand Oaks, CA: Sage Publications, Inc.

Fox, John. 2000b. *Nonparametric Simple Regression,* Thousand Oaks, CA: Sage Publications, Inc.

Fraley, C. 1998. "Algorithms for model-based Gaussian hierarchical clustering," *SIAM Journal on Scientific Computing,* **20:**270-281.

Fraley, C. and A. E. Raftery. 1998. "How many clusters? Which clustering method? Answers via model-based cluster analysis," *The Computer Journal,* **41:**578-588.

Fraley, C. and A. E. Raftery. 2002. "Model-based clustering, discriminant analysis, and density estimation: MCLUST," *Journal of the American Statistical Association,* **97:**611-631.

Fraley, C. and A. E. Raftery. 2003. "Enhanced software for model-based clustering, discriminant analysis, and density estimation: MCLUST," *Journal of Classification,* **20:**263-286.

Freedman, D. and P. Diaconis. 1981. "On the histogram as a density estimator: L_2 theory," *Zeitschrift fur Wahrscheinlichkeitstheorie und verwandte Gebiete,* **57:**453-476.

Frees, Edward W. and Emiliano A. Valdez. 1998. "Understanding relationships using copulas," *North American Actuarial Journal,* **2:**1-25.

Freund, Yoav. 1995. "Boosting a weak learning algorithm by majority," *Information and Computation,* **121:**256-285.

Freund, Yoav and Robert E. Schapire. 1997. "A decision-theoretic generalization of on-line learning and an application to boosting," *Journal of Computer and System Sciences,* **55:**119-139.

Friedman, J. 1987. "Exploratory projection pursuit," *Journal of the American Statistical Association*, **82**:249-266.

Friedman, J. 1991a. "Multivariate adaptive regression splines," *The Annals of Statistics*, **19**:1-141.

Friedman, J. 1991b. "Estimating functions of mixed ordinal and categorical variables using adaptive splines," Department of Statistics Technical Report No. 108, Stanford University.

Friedman, J. 1993. "Fast MARS," Department of Statistics Technical Report No. 110, Stanford University.

Friedman, J. 2006. "Recent advances in predictive (machine) learning," *Journal of Classification*, **23**:175-197.

Friedman, J., T. Hastie, and R. J. Tibshirani. 2000. "Additive logistic regression: A statistical view of boosting," *The Annals of Statistics*, **38**:337-374.

Friedman, J., T. Hastie, and R. J. Tibshirani. 2001. *The Elements of Statistical Learning: Data Mining, Inference, and Prediction*, New York: Springer-Verlag.

Friedman, J. and W. Stuetzle. 1981. "Projection pursuit regression," *Journal of the American Statistical Association*, **76**:817-823.

Friedman, J. and John Tukey. 1974. "A projection pursuit algorithm for exploratory data analysis," *IEEE Transactions on Computers*, **23**:881-889.

Friedman, J., W. Stuetzle, and A. Schroeder. 1984. "Projection pursuit density estimation," *Journal of the American Statistical Association*, **79**:599-608.

Friendly, Michael. 2000. *Visualizing Categorical Data*. Cary, NC: SAS Publishing.

Frigge, M., D. C. Hoaglin, and B. Iglewicz. 1989. "Some implementations of the boxplot," *The American Statistician*, **43**:50-54.

Fukunaga, Keinosuke. 1990. *Introduction to Statistical Pattern Recognition, Second Edition*, New York: Academic Press.

Gehan, E. A. 1965. "A generalized Wilcoxon test for comparing arbitrarily single-censored samples," *Biometrika*, **52**:203-233.

Geladi, Paul and Bruce Kowalski. 1986. "Partial least squares regression: A tutorial," *Analytica Chimica Acta*, **185**:1-17.

Gelfand, A. E. and A. F. M. Smith. 1990. "Sampling-based approaches to calculating marginal densities," *Journal of the American Statistical Association*, **85**:398-409.

Gelfand, A. E., S. E. Hills, A. Racine-Poon, and A. F. M. Smith. 1990. "Illustration of Bayesian inference in normal data models using Gibbs sampling," *Journal of the American Statistical Association*, **85**:972-985.

Gelman, A. 1996. "Inference and monitoring convergence," in *Markov Chain Monte Carlo in Practice*, W. R. Gilks, S. Richardson, and D. T. Spiegelhalter, eds., London: Chapman and Hall, pp. 131-143.

Gelman, A. and D. B. Rubin. 1992. "Inference from iterative simulation using multiple sequences (with discussion)," *Statistical Science*, **7**:457–511.

Gelman, A., J. B. Carlin, H. S. Stern, and D. B. Rubin. 1995. *Bayesian Data Analysis*, London: Chapman and Hall.

Geman, S. and D. Geman. 1984. "Stochastic relaxation, Gibbs distributions, and the Bayesian restoration of images," *IEEE Transactions PAMI*, **6**:721-741.

Genest, Christian and Anne-Catherine Favre. 2007. "Everything you always wanted to know about copula modeling but were afraid to ask," *Journal of Hydrologic Engineering*, **12**:347-368.

Genest, Christian and Louis-Paul Rivest. 1993. "Statistical inference procedures for bivariate Archimedean copulas," *Journal of the American Statistical Association*, **88**:1034-1043.

Gentle, James E. 1998. *Random Number Generation and Monte Carlo Methods*, New York: Springer-Verlag.

Gentle, James E. 2005. *Elements of Computational Statistics*, New York: Springer-Verlag.

Gentle, James E. 2009. *Computational Statistics*, New York: Springer-Verlag.

Gentle, James E., Wolfgang Härdle, and Yuichi Mori (editors). 2004. *Handbook of Computational Statistics*, New York: Springer-Verlag.

Geyer, C. J. 1992. "Practical Markov chain Monte Carlo," *Statistical Science*, **7**:473-511.

Gilks, W. R., S. Richardson, and D. J. Spiegelhalter. 1996a. "Introducing Markov chain Monte Carlo," in *Markov Chain Monte Carlo in Practice*, W. R. Gilks, S. Richardson, and D. T. Spiegelhalter, eds., London: Chapman and Hall, pp. 1-19.

Gilks, W. R., S. Richardson, and D. J. Spiegelhalter (eds.). 1996b. *Markov Chain Monte Carlo in Practice*, London: Chapman and Hall.

Gill, Jeff. 2001. *Generalized Linear Models: A Unified Approach*, Sage University Papers Series on Quantitative Applications in the Social Sciences, 07-134, Thousand Oaks, CA: Sage.

Golub, Gene and Charles van Loan. 2012. *Matrix Computations, 4th Edition*, Baltimore, MD: Johns Hopkins University Press.

Gordon, A. D. 1994. "Identifying genuine clusters in a classification," *Computational Statistics and Data Analysis*, **18**:561-581.

Gordon, A. D. 1999. *Classification*, London: Chapman and Hall/CRC.

Gower, J. C. 1966. "Some distance properties of latent root and vector methods in multivariate analysis," *Biometrika*, **53**:325-338.

Green P. J. and B. W. Silverman. 1994. *Nonparametric Regression and Generalized Linear Models: A Roughness Penalty Approach*, Boca Raton, FL: Chapman and Hall/CRC.

Groenen, P. 1993. *The Majorization Approach to Multidimensional Scaling: Some Problems and Extensions*, Leiden, The Netherlands: DSWO Press.

Haining, Robert. 1993. *Spatial Data Analysis in the Social and Environmental Sciences*, Cambridge: Cambridge University Press.

Hair, Joseph, Rolph Anderson, Ronald Tatham, and William Black. 1995. *Multivariate Data Analysis, Fourth Edition*, New York: Prentice Hall.

Hald, A. 1952. *Statistical Theory with Engineering Applications*, New York: John Wiley & Sons.

Hall, P. 1992. *The Bootstrap and Edgeworth Expansion*, New York: Springer-Verlag.

Hall, P. and M. A. Martin. 1988. "On bootstrap resampling and iteration," *Biometrika*, **75**:661-671.

Hand, D., F. Daly, A. D. Lunn, K. J. McConway, and E. Ostrowski. 1994. *A Handbook of Small Data Sets*, London: Chapman and Hall.

Hanley, J. A. and K. O. Hajian-Tilaki. 1997. "Sampling variability of nonparametric estimates of the areas under receiver operating characteristic curves: An update," *Academic Radiology*, 4:49-58.

Hanley, J. A. and B. J. McNeil. 1983. "A method of comparing the areas under receiver operating characteristic curves derived from the same cases," *Radiology*, **148**:839-843.

Hanselman, D. and B. Littlefield. 2011. *Mastering MATLAB® 8*, New Jersey: Prentice Hall.

Harrison, D. and D. L. Rubinfeld. 1978. "Hedonic prices and the demand for clean air," *Journal of Environmental Economics and Management*, **5**:81-102.

Hartigan, J. 1975. *Clustering Algorithms*, New York: Wiley-Interscience.

Hartigan, J. A. 1985. "Statistical theory in clustering," *Journal of Classification*, **2**:63-76.

Hastie, T. J. and R. J. Tibshirani. 1990. *Generalized Additive Models*, London: Chapman and Hall.

Hastie, T.J., R. J. Tibshirani, and J. Friedman. 2009. *The Elements of Statistical Learning: Data Mining, Inference, and Prediction, 2nd Edition*. New York: Springer-Verlag, http://statweb.stanford.edu/~tibs/ElemStatLearn/

Hastings, W. K. 1970. "Monte Carlo sampling methods using Markov chains and their applications," *Biometrika*, **57**:97-109.

Herbert, D. T. 1980 "The British experience," in *Crime: a Spatial Perspective*, D. E. Georges-Abeyie and K. D. Harries, eds., New York: Columbia University Press.

Hjorth, J. S. U. 1994. *Computer Intensive Statistical Methods: Validation Model Selection and Bootstrap*, London: Chapman and Hall.

Ho, T. K. 1998. "The random subspace method for constructing decision forests," *IEEE Transactions on Pattern Analysis and Machine Intelligence*, **20**:832-844.

Hoaglin, D. C. and D. F. Andrews. 1975. "The reporting of computation-based results in statistics," *The American Statistician*, **29**:122-126.

Hoaglin, D. and John Tukey. 1985. "Checking the shape of discrete distributions," in *Exploring Data Tables, Trends and Shapes*, D. Hoaglin, F. Mosteller, J. W. Tukey, eds., New York: John Wiley & Sons.

Hoaglin, D. C., F. Mosteller, and J. W. Tukey (eds.). 1983. *Understanding Robust and Exploratory Data Analysis*, New York: John Wiley & Sons.

Hogg, Robert. 1974. "Adaptive robust procedures: a partial review and some suggestions for future applications and theory (with discussion)," *The Journal of the American Statistical Association*, **69**:909-927.

Hogg, Robert and Allen Craig. 1978. *Introduction to Mathematical Statistics, 4th Edition*, New York: Macmillan Publishing Co.

Hope, A. C. A. 1968. "A simplified Monte Carlo Significance test procedure," *Journal of the Royal Statistical Society, Series B*, **30**:582-598.

Huber, P. J. 1973. "Robust regression: asymptotics, conjectures, and Monte Carlo," *Annals of Statistics*, **1**:799-821.

Huber, P. J. 1981. *Robust Statistics*, New York: John Wiley & Sons.

Huber, P. J. 1985. "Projection pursuit (with discussion)," *Annals of Statistics*, **13**:435-525.

Hubert, L. J. and P. Arabie. 1985. "Comparing partitions," *Journal of Classification*, **2**:193-218.

Hunter, J. Stuart. 1988. "The digidot plot," *The American Statistician*, **42**:54-54.

Hyvärinen, Aapo. 1999. "Survey on independent component analysis," *Neural Computing Surveys*, **2**:94-128.

Hyvärinen, A., J. Karhunen, and E. Oja. 2001. *Independent Component Analysis*, New York: John Wiley & Sons.

Hyvärinen, A. and E. Oja. 2000. "Independent component analysis: Algorithms and applications," *Neural Networks*, **13**:411-430.

Inselberg, Alfred. 1985. "The plane with parallel coordinates," *The Visual Computer*, **1**:69-91.

Isaaks. E. H. and R. M. Srivastava. 1989. *An Introduction to Applied Geo-statistics*, New York: Oxford University Press.

Izenman, A. J. 1991. 'Recent developments in nonparametric density estimation," *Journal of the American Statistical Association*, **86**:205-224.

Jackson, J. Edward. 1991. *A User's Guide to Principal Components*, New York: John Wiley & Sons.

Jain, Anil K. and Richard C. Dubes. 1988. *Algorithms for Clustering Data*, New York: Prentice Hall.

Jain, Anil K., Robert P. W. Duin, and J. Mao. 2000. "Statistical pattern recognition: A review," *IEEE Transactions on Pattern Analysis and Machine Intelligence*, **22**:4-37.

Jain, Anil K., M. N. Murty, and P. J. Flynn. 1999. "Data clustering: A review," *ACM Computing Surveys*, **31**:264-323.

James, G., D. Witten, T. Hastie, and R. Tibshirani. 2013. *An Introduction to Statistical Learning: with Applicatoins in R*, New York: Springer Texts in Statistics.

Jeffreys, H. 1935. "Some tests of significance, treated by the theory of probability," *Proceedings of the Cambridge Philosophy Society*, **31**:203-222.

Jeffreys, H. 1961. *Theory of Probability, Third Edition*, Oxford, U. K.: Oxford University Press.

Jekabsons, G. 2015. *ARESLab: Adaptive Regression Splines toolbox for MATLAB/Octave*, http://www.cs.rtu.lv/jekabsons/

Joeckel, K. 1991. "Monte Carlo techniques and hypothesis testing," *The Frontiers of Statistical Computation, Simulation, and Modeling, Volume 1 of the Proceedings ICOSCO-I*, pp. 21-41.

Johnson, Mark E. 1987. *Multivariate Statistical Simulation*, New York: John Wiley & Sons.

Jolliffe, I. T. 2002. *Principal Component Analysis*, New York: Springer-Verlag.

Jones, M. C., J. S. Marron, and S. J. Sheather. 1996. *Journal of the American Statistical Association*, **91**:401-406.

Jones, M. C. and R. Sibson. 1987. "What is projection pursuit (with discussion)," *Journal of the Royal Statistical Society, Series A*, **150**:1–36.

Journel, A. G. and C. J. Huijbregts. 1978. *Mining Geostatistics*, London: Academic Press.

Kalos, Malvin H. and Paula A. Whitlock. 1986. *Monte Carlo Methods, Volume 1: Basics*, New York: Wiley Interscience.

Kaplan, D. T. 1999. *Resampling Stats in MATLAB®*, Arlington, VA: Resampling Stats®, Inc.

Kass, R. E. and A. E. Raftery. 1995. "Bayes factors," *Journal of the American Statistical Association*, **90**:773-795.

Kaufman, Leonard and Peter J. Rousseeuw. 1990. *Finding Groups in Data: An Introduction to Cluster Analysis*, New York: John Wiley & Sons.

Keating, Jerome, Robert Mason, and Pranab Sen. 1993. *Pitman's Measure of Closeness - A Comparison of Statistical Estimators*, New York: SIAM Press.

Kirkpatrick, S., C. D. Gelatt Jr., and M. P. Vecchi. 1983. "Optimization by simulated annealing," *Science*, **220**:671-680.

Kleinbaum, David G. 1994. *Logistic Regression: A Self-learning Text*, New York: Springer-Verlag.

Kotz, Samuel, N. Balakrishnan, and N. L. Johnson. 2000. *Continuous Multivariate Distributions, Volume 1, Models and Applications*, New York, Wiley Interscience.

Kotz, Samuel and Norman L. Johnson (eds.). 1986. *Encyclopedia of Statistical Sciences*, New York: John Wiley & Sons.

Krzanowski, W. J. and Y. T. Lai. 1988. "A criterion for determining the number of groups in a data set using sum-of-squares clustering," *Biometrics*, **44**:23-34.

Launer, R. and G. Wilkinson (eds.). 1979. *Robustness in Statistics*, New York: Academic Press.

Lee, Thomas C. M. 2002. "On algorithms for ordinary least squares regression spline fitting: A comparative study," *Journal of Statistical Computation and Simulation*, **72**:647-663.

Lehmann, E. L. 1994. *Testing Statistical Hypotheses*, London: Chapman and Hall.

Lehmann, E. L. and G. Casella. 1998. *Theory of Point Estimation, Second Edition*, New York: Springer-Verlag.

LePage, R. and L. Billard (eds.). 1992. *Exploring the Limits of the Bootstrap*, New York: John Wiley & Sons.

Levy, Paul S. and Stanley Lemeshow. 1999. *Sampling of Populations: Methods and Applications*, New York: John Wiley & Sons.

Li, G. and Z. Chen. 1985. "Projection-pursuit approach to robust dispersion matrices and principal components: primary theory and Monte Carlo," *Journal of the American Statistical Association*, **80**:759-766.

Lindeberg, T. 1994. *Scale Space Theory in Computer Vision*, Boston: Kluwer Academic Publishers.

Lindgren, Bernard W. 1993. *Statistical Theory, Fourth Edition*, London: Chapman and Hall.

Lindley, D. V. 1995. *Bayesian Statistics, A Review*, Philadelphia: Society for Industrial and Applied Mathematics.

Lindsey, J. C., A. M. Herzberg, and D. G. Watts. 1987. "A method for cluster analysis based on projections and quantile-quantile plots," *Biometrics*, **43**:327-341.

Loader, Clive. 1999. *Local Regression and Likelihood*, New York: Springer-Verlag.

Loh, W. Y. 1987. "Calibrating confidence coefficients," *Journal of the American Statistical Association*, **82**:155-162.

Loh, W. Y. 2014. "Fifty years of classification and regression trees," *International Statistical Review*, **82**:329-348.

Longley, J. W. 1967. "An appraisal of least squares programs for the electronic computer from the viewpoint of the user," *Journal of the American Statistical Association*, **62**:819-841.

Lubischew, A. A. 1962. "On the use of discriminant functions in taxonomy," *Biometrics*, **18**:455-477.

Lusted, L. B. 1971. "Signal detectability and medical decision-making," *Science*, **171**:1217-1219.

Manly, B. F. J. 2004. *Multivariate Statistical Methods - A Primer, Third Edition*, London: Chapman & Hall.

Marchand, Patrick and O. Thomas Holland. 2003. *Graphics and GUIs with MATLAB®, Third Edition*, Boca Raton, FL: CRC Press.

Marin, Jean-Michel and Christian P. Robert. 2007. *Bayesian Core: A Practical Approach to Computational Bayesian Statistics*, New York: Springer-Verlag.

Marsh, L. C. and D. R. Cormier. 2002. *Spline Regression Models*, Sage University Papers Series on Quantitative Applications in the Social Sciences, 07-137, Thousand Oaks, CA: Sage Publications, Inc.

Martinez, Wendy L. and MoonJung Cho. 2014. *Statistics in MATLAB®: A Primer*, Boca Raton, FL: CRC Press.

Martinez, Wendy L., Angel R. Martinez, and Jeffrey. L. Solka. 2010. *Exploratory Data Analysis with MATLAB®, Second Edition*, Boca Raton, FL: CRC Press.

Mazess, R. B., W. W. Peppler, and M. Gibbons. 1984. "Total body composition by dualphoton (^{153}Gd) absorptiometry," *American Journal of Clinical Nutrition*, **40**:834-839.

McCullagh, P. and John A. Nelder. 1989. *Generalized Linear Models, 2nd Edition*, Boca Raton, FL: Chapman & Hall/CRC Press.

McGill, Robert, John Tukey, and Wayne Larsen. 1978. "Variations of box plots," *The American Statistician*, **32**:12-16.

McIntyre, Lauren. 1994. "Using cigarette data dor an introduction to multiple regression," *Journal of Statistics Education*, **2**(1), http://www.amstat.org/publications/jse/v2n1/datasets.mcintyre.html

McLachlan, G. J. and K. E. Basford. 1988. *Mixture Models: Inference and Applications to Clustering*, New York: Marcel Dekker.

McLachlan, G. J. and T. Krishnan. 1997. *The EM Algorithm and Extensions*, New York: John Wiley & Sons.

McLachlan, G. J. and D. Peel. 2000. *Finite Mixture Models*, New York: John Wiley & Sons.

McNeil, B. J., E. Keeler, and S. J. Adelstein. 1975. "Primer on certain elements of medical decision making," *New England Journal of Medicine*, **293**:211-215.

Meeker, William and Luis Escobar. 1998. *Statistical Methods for Reliability Data*, New York: John Wiley & Sons.

Meir, Ron and Gunnar Rätsch. 2003. "An introduction to boosting and leveraging," in *Advanced Lectures on Machine Learning, LNCS*, 119-184.

Menard, Scott. 2001. *Applied Logistic Regression Analysis*, Sage University Papers Series on Quantitative Applications in the Social Sciences, 07-106, Thousand Oaks, CA: Sage.

Metropolis, N., A. W. Rosenbluth, M. N. Rosenbluth, A. H. Teller, and E. Teller. 1953. "Equations of state calculations by fast computing machine," *Journal of Chemistry and Physics*, **21**:1087-1091.

Meyn, S. P. and R. L. Tweedie. 1993. *Markov Chains and Stochastic Stability*, New York: Springer-Verlag.

Milligan, G. W. and M. C. Cooper. 1985. "An examination of procedures for determining the number of clusters in a data set," *Psychometrika*, **50**:159-179.

Mingers, John. 1989a. "An empirical comparison of selection measures for decision tree induction," *Machine Learning*, **3**:319-342.

Mingers, John. 1989b. "An empirical comparison of pruning methods for decision tree induction," *Machine Learning*, **4**:227-243.

Minnotte, M. and R. West. 1998. "The data image: a tool for exploring high dimensional data sets," *Proceedings of the ASA Section on Statistical Graphics*.

Mojena, R. 1977. "Hierarchical grouping methods and stopping rules: An evaluation," *Computer Journal*, **20**:359-363.

Montanari, Angela and Laura Lizzani. 2001. "A projection pursuit approach to variable selection," *Computational Statistics and Data Analysis*, **35**:463-473.

Montgomery, Douglas C., George C. Runger, and Norma F. Hubele. 1998. *Engineering Statistics*, New York: John Wiley & Sons.

Mood, Alexander, Franklin Graybill, and Duane Boes. 1974. *Introduction to the Theory of Statistics, Third Edition*, New York: McGraw-Hill Publishing.

Mooney, C. Z. 1997. *Monte Carlo Simulation*, London: Sage Publications.

Mooney, C. Z. and R. D. Duval. 1993. *Bootstrapping: A Nonparametric Approach to Statistical Inference*, London: Sage University Press.

Morant, G. M. 1923. "A first study of the Tibetan skull," *Biometrika*, **14**:193-260.

Morton, S. 1989. "Interpretable projection pursuit," Technical Report 106, Stanford University, Laboratory for Computational Statistics.

Mosteller, F. and J. W. Tukey. 1977. *Data Analysis and Regression: A Second Course in Statistics*, Reading, Massachusetts: Addison-Wesley.

Mosteller, F. and D. L. Wallace. 1964. *Inference and Disputed Authorship: The Federalist Papers*, Reading, MA: Addison-Wesley.

Murdoch, Duncan J. 2000. "Markov chain Monte Carlo," *Chance*, **13**:48-51.

Musa, J. D., A. Iannino, and K. Okumoto. 1987. *Software Reliability: Measurement, Prediction, Application*, New York: McGraw-Hill.

Nadarajah, Saralees and Samuel Kotz. 2005. "Probability integrals of the multivariate *t* distribution," *Canadian Applied Mathematics Quarterly*, **13**:43-87

Nadaraya, E. A. 1964. "On estimating regression," *Theory of Probability and its Applications*, **10**:186-190.

Nason, Guy. 1995. "Three-dimensional projection pursuit," *Applied Statistics*, **44**:411–430.

Nelder, J. A. 1990. "Nearly parallel lines in residual plots," *The American Statistician*, **44**:221-222.

Nelder, J. A. and R. W. M. Wedderburn. 1972. "Generalized linear models," *Journal of the Royal Statistical Society, Series A*, **135**:370-384.

Newman, D. J., S. Hettich, C. L. Blake, and C. J. Merz. 1998. *UCI Repository of Machine Learning Databases* [**http://www.ics.uci.edu/~mlearn/MLRepository.html**], University of California, Department of Information and Computer Science, Irivine, CA.

Norris, J. 1997. *Markov Chains*, Cambridge: Cambridge University Press.

Pampel, Fred C. 2000. *Logistic Regression: A Primer*, Sage University Papers Series on Quantitative Applications in the Social Sciences, 07-132, Thousand Oaks, CA: Sage Publications, Inc.

Parzen, E. 1962. "On estimation of probability density function and mode," *Annals of Mathematical Statistics*, **33**:1065-1076.

Pearson, K. and A. Lee. 1903. "On the laws of inheritance in man. I. Inheritance of physical characters," *Biometrika*, **2**:357-462.

Perisic, Igor and Christian Posse. 2005. "Projection pursuit indices based on the empirical distribution function," *Journal of Computational and Graphical Statistics*, **14**:700-715.

Pierce, Donald A. and Daniel W. Schafer. 1986. "Residuals in generalized linear models," *Journal of the American Statistical Association*, **81**:977-986.

Pinder, D. A. and M. E. Witherick. 1977. "The principles, practice, and pitfalls of nearest neighbor analysis," *Geography*, **57**:277–288.

Polansky, Alan M. 1999. "Upper bounds on the true coverage of bootstrap percentile type confidence intervals," *The American Statistician*, **53**:362-369.

Politis, D. N., J. P. Romano, and M. Wolf. 1999. *Subsampling*, New York: Springer-Verlag.

Port, Sidney C. 1994. *Theoretical Probability for Applications*, New York: John Wiley & Sons.

Posse, Christian. 1995a. "Projection pursuit exploratory data analysis," *Computational Statistics and Data Analysis*, **29**:669–687.

Posse, Christian. 1995b. "Tools for two-dimensional exploratory projection pursuit," *Journal of Computational and Graphical Statistics*, **4**:83–100.

Priebe, C. E. 1993. *Nonparametric maximum likelihood estimation with data-driven smoothing*, Ph.D. Dissertation, Fairfax, VA: George Mason University.

Priebe, C. E. 1994. "Adaptive mixture density estimation," *Journal of the American Statistical Association*, **89**:796-806.

Priebe, C. E., R. A. Lori, D. J. Marchette, J. L. Solka, and G. W. Rogers. 1994. "Nonparametric spatio-temporal change point analysis for early detection in mammography," *Proceedings of the Second International Workshop on Digital Mammography, SIWDM*, pp. 111-120.

Priebe, C. E. and D. J. Marchette. 2000. "Alternating kernel and mixture density estimates," *Computational Statistics and Data Analysis*, **35**:43-65.

Quinlan, J. R. 1987. "Simplifying decision trees," *International Journal of Man-Machine Studies*, **27**:221-234.

Quinlan, J. R. 1993. *C4.5: Programs for Machine Learning*, San Mateo: Morgan Kaufmann.

Quinlan, J. R. 1996. "Bagging, boosting, and C4.5," in *Proceedings of the Thirteenth National Conference on Artificial Intelligence*, AAAI Press and the MIT Press, 725-730.

Quenouille, M. 1949. "Approximate tests of correlation in time series," *Journal of the Royal Statistical Society, Series B*, **11**:18-44.

Quenouille, M. 1956. "Notes on bias estimation," *Biometrika*, **43**:353-360.

Rafterty, A. E. and V. E. Akman. 1986. "Bayesian analysis of a Poisson process with a change-point," *Biometrika*, **85**:85-89.

Raftery, A. E. and S. M. Lewis. 1992. "How many iterations in the Gibbs sampler?", in *Bayesian Statistics 4*, J. M. Bernardo, J. Berger, A. P. Dawid, and A. F. M. Smith, eds., Oxford: Oxford University Press, pp. 763-773.

Raftery, A. E. and S. M. Lewis. 1996. "Implementing MCMC," in *Markov Chain Monte Carlo in Practice*, W. R. Gilks, S. Richardson, and D. J. Spiegelhalter, eds., London: Chapman and Hall, pp. 115-130.

Raftery, A. E., M. A. Tanner, and M. T. Wells (editors). 2002. *Statistics in the 21st Century*, Boca Raton, FL: CRC Press.

Rand, William M. 1971. "Objective criteria for the evaluation of clustering methods," *Journal of the American Statistical Association*, **66**:846-850.

Rao, C. R. 1993. *Computational Statistics*, The Netherlands: Elsevier Science Publishers.

Reinsch, C. 1967. "Smoothing by spline functions," *Numerical Mathematics*, **10**:177-183.

Redner, A. R. and H. F. Walker. 1984. "Mixture densities, maximum likelihood, and the EM algorithm," *SIAM Review*, **26**:195-239.

Rice, J. 1984. "Bandwidth choice for nonparametric regression," *Annals of Statistics* **12**:1215-1230.

Ripley, B. D. 1976. "The second-order analysis of stationary point processes," *Journal of Applied Probability*, **13**:255-266.

Ripley, B. D. 1981. *Spatial Statistics*, New York: John Wiley & Sons.

Ripley, Brian D. 1996. *Pattern Recognition and Neural Networks*, Cambridge: Cambridge University Press.

Robert, C. P. 1995. "Convergence control techniques for Markov chain Monte Carlo algorithms," *Statistical Science*, **10**:231-253.

Robert, C. P. and G. Casella. 1999. *Monte Carlo Statistical Methods*, New York: Springer-Verlag.

Roberts, G. O. 1996. "Markov chain concepts related to sampling algorithms," in *Markov Chain Monte Carlo in Practice*, W. R. Gilks, S. Richardson, and D. J. Spiegelhalter, eds., London: Chapman and Hall, pp. 45-57.

Roberts, G. O. 2000. *Computer Intensive Methods*, Course Notes, Lancaster University, UK, **www.maths.lancs.ac.uk/~robertgo/notes.ps**.

Robbins, Naomi B. 2013. *Creating More Effective Graphs*, New York: Chart House.

Rohatgi, V. K. 1976. *An Introduction to Probability Theory and Mathematical Statistics* by New York: John Wiley & Sons.

Rohatgi, V. K. and A. K. Md. Ehsanes Saleh. 2000. *An Introduction to Probability and Statistics*, New York: John Wiley & Sons.

Rosenblatt, M. 1956. "Remarks on some nonparametric estimates of a density function," *Annals of Mathematical Statistics*, **27**:832-837.

Rosipal, Roman and Nicole Kramer. 2006. "Overview and Recent Advances in Partial Least Squares," in *Subspace, Latent Structure and Feature Selection, Lecture Notes in Computer Science 3940*, Berlin, Germany: Springer-Verlag, pp. 34–51.

Ross, Sheldon. 1994. *A First Course in Probability, Fourth Edition*. New York: Macmillan College Publishing.

Ross, Sheldon. 1997. *Simulation, Second Edition*, New York: Academic Press.

Ross, Sheldon. 2000. *Introduction to Probability Models, Seventh Edition*, San Diego: Academic Press.

Rousseeuw, P. J. and A. M. Leroy. 1987. *Robust Regression and Outlier Detection*, New York: John Wiley & Sons.

Rousseeuw, P. J., I. Ruts, and J. W. Tukey. 1999. "The bagplot: A bivariate boxplot," *The American Statistician*, **53**:382-387.

Roweis, S. T. and L. K. Saul. 2000. "Nonlinear dimensionality reduction by locally linear embedding," *Science*, **290**:2323-2326.

Rubin, Donald B. 1987. "Comment on Tanner and Wong: The calculation of posterior distributions by data augmentation," *Journal of the American Statistical Association*, **82**:543-546.

Rubin, Donald B. 1988. "Using the SIR algorithm to simulate posterior distributions (with discussion)," in *Bayesian Statistics 3*, J. M. Bernardo, M. H. DeGroot, D. V. Lindley, and A. F. M. Smith, eds., Oxford: Oxford University Press, pp. 395-402.

Rubinstein, Reuven Y. 1981. *Simulation and the Monte Carlo Method*, New York: John Wiley & Sons.

Rutherford, E. and M. Geiger. 1910. "The probability variations in the distribution of alpha-particles," *Philosophical Magazine*, Series 6, **20**:698-704.

Safavian, S. R. and D. A. Landgrebe. 1991. "A survey of decision tree classifier methodology," *IEEE Transactions on Systems, Man, and Cybernetics*, **21**:660-674.

Sasieni, Peter and Patrick Royston. 1996. "Dotplots," *Applied Statistics*, **45**:219-234.

Schapire, Robert E. 1990. "The strength of weak learnability," *Machine Learning*, **5**:197-227.

Schapire, R. E., Y. Freund, P. Bartlett, and W. S. Lee. 1998. "Boosting the margin: A new explanation for the effectiveness of voting methods," *The Annals of Statistics*, **26**:1651-1686.

Schimek, Michael G. (ed.). 2000. *Smoothing and Regression: Approaches, Computational, and Application*, New York: John Wiley & Sons.

Scott, David W. 1979. "On optimal and data-based histograms," *Biometrika*, **66**:605-610.

Scott, David W. 1985. "Frequency polygons," *Journal of the American Statistical Association*, **80**:348-354.

Scott, David W. 2015. *Multivariate Density Estimation: Theory, Practice, and Visualization, Second Edition*, New York: John Wiley & Sons.

Scott, A. J. and M. J. Symons. 1971. "Clustering methods based on likelihood ratio criteria," *Biometrics*, **27**:387-397.

Schwarz, G. 1978. "Estimating the dimension of a model," *The Annals of Statistics*, **6**:461-464.

Seber, G. A. F. 1984. *Multivariate Observations*, New York: John Wiley & Sons.

Shao, J. and D. Tu. 1995. *The Jackknife and Bootstrap*, New York: Springer-Verlag.

Silverman, B. W. 1985. "Some aspects of the spline smoothing approach to nonparametric curve fitting," *Journal of the Royal Statistical Society, Series B*, **47**:1-52.

Silverman, B. W. 1986. *Density Estimation for Statistics and Data Analysis*, London: Chapman and Hall.

Simon, J. 1999. *Resampling: The New Statistics*, Arlington, VA: Resampling Stats, Inc.

Simonoff, J. S. 1996. *Smoothing Methods in Statistics*, New York: Springer-Verlag.

Skurichina, Marina and Robert P. W. Duin. 1998. "Bagging for linear classifiers," *Pattern Recognition*, **31**:909-930.

Snedecor, G. W. and G. C. Cochran. 1967. *Statistical Methods, Sixth Edition*, Ames: Iowa State University Press.

Snedecor, G. W. and G. C. Cochran. 1980. *Statistical Methods, Seventh Edition*, Ames: Iowa State University Press.

Solka, J., W. L. Poston, and E. J. Wegman. 1995. "A visualization technique for studying the iterative estimation of mixture densities," *Journal of Computational and Graphical Statistics*, **4**:180-198.

Solka, J. 1995. *Matching Model Information Content to Data Information*, Ph.D. Dissertation, Fairfax, VA: George Mason University.

Spath, Helmuth. 1980. *Cluster Analysis Algorithms for Data Reduction and Classification of Objects*, New York: Halsted Press.

Stone, James V. 2004. *Independent Component Analysis: A Tutorial Introduction*, Cambridge, MA: The MIT Press.

Stork, David G. and Elad Yom-Tov. 2004. *Computer Manual in MATLAB® to Accompany Pattern Classification, 2nd Edition*. New York: Wiley-Interscience.

Strang, Gilbert. 2003. *Introduction to Linear Algebra, Third Edition*, Wellesley, MA: Wellesley-Cambridge Press.

Strang, Gilbert. 2005. *Linear Algebra and its Applications, Fourth Edition*, New York: Brooks Cole.

Sugar, Catherine A. and Gareth M. James. 2003. "Finding the number of clusters in a dataset: An information-theoretic approach," *Journal of the American Statistical Association*, **98**:750-763.

Swayne, D. F., D. Cook, and A. Buja. 1991. "XGobi: Interactive dynamic graphics in the X window system with a link to S," *ASA Proceedings of the Section on Statistical Graphics*. pp. 1-8.

Tanner, Martin A. 1996. *Tools for Statistical Inference: Methods for the Exploration of Posterior Distributions and Likelihood Functions, Third Edition*, New York: Springer-Verlag.

Tapia, R. A. and J. R. Thompson. 1978. *Nonparametric Probability Density Estimation*, Baltimore: Johns Hopkins University Press.

Teichrow, D. 1965. "A history of distribution sampling prior to the era of the computer and its relevance to simulation," *Journal of the American Statistical Association*, **60**:27-49.

Tenenbaum, J. B., V. de Silva, and J. C. Langford. 2000. "A global geometric framework for nonlinear dimensionality reduction," *Science*, **290**:2319-2323.

Terrell, G. R. 1990. "The maximal smoothing principle in density estimation," *Journal of the American Statistical Association*, **85**:470-477.

Thisted, R. A. 1988. *Elements of Statistical Computing*, London: Chapman and Hall.

Tibshirani, R. J. 1988. "Variance stabilization and the bootstrap," *Biometrika*, **75**:433-444.

Tibshirani, R. J. 1996. "Regression shrinkage and selection via the lasso," *Journal of the Royal Statistical Society, Series B*, **58**:267-288.

Tibshirani, R. J., G. Walther, D. Botstein, and P. Brown. 2001. "Cluster validation by prediction strength," Technical Report, Stanford University.

Tibshirani, R. J., Guenther Walther, and Trevor Hastie. 2001. "Estimating the number of clusters in a data set via the gap statistic," *Journal of the Royal Statistical Society, B*, **63**:411-423.

Tierney, L. 1994. "Markov chains for exploring posterior distributions (with discussion)," *Annals of Statistics*, **22**:1701-1762.

Tierney, L. 1996. "Introduction to general state-space Markov chain theory," in *Markov Chain Monte Carlo in Practice*, W. R. Gilks, S. Richardson, and D. J. Spiegelhalter, eds., London: Chapman and Hall, pp. 59-74.

Tinkler, K. J. 1971. "Statistical analysis of tectonic patterns in areal volcanism: the Bunyaruguru volcanic field in west Uganda," *Mathematical Geology*, **3**:335–355.

Titterington, D. M., A. F. M. Smith, and U. E. Makov. 1985. *Statistical Analysis of Finite Mixture Distributions*, New York: John Wiley & Sons.

Torgerson, W. S. 1952. "Multidimensional scaling: 1. Theory and method," *Psychometrika*, **17**:401-419.

Tripathi, R. C. and R. C. Gupta. 1988. "Another generalization of the logarithmic series and the geometric distribution," *Communications in Statistics - Theory and Methods*, **17**:1541-1547.

Tufte, E. 1990. *Envisioning Information*, Cheshire, CT: Graphics Press.

Tufte, E. 1997. *Visual Explanations*, Cheshire, CT: Graphics Press.

Tufte, E. 2001. *The Visual Display of Quantitative Information*, (second edition), Cheshire, CT: Graphics Press.

Tufte, E. 2006. *Beautiful Evidence*, Cheshire, CT: Graphics Press.

Tukey, John W. 1958. "Bias and confidence in not quite large samples," *Annals of Mathematical Statistics*, **29**: pp. 614.

Tukey, John W. 1977. *Exploratory Data Analysis*, New York: Addison-Wesley.

Upton, G. and B. Fingleton. 1985. *Spatial Data Analysis by Example: Volume I: Point Pattern and Quantitative Data*, New York: John Wiley & Sons.

Utts, Jessica. 1996. *Seeing Through Statistics*, New York: Duxbury Press.

van der Heijden, Ferdi, R. P. W. Duin, D. de Ridder, and D. M. J. Tax. 2004. *Classification, parameter estimation, and state estimation - An engineering approach using MATLAB®*, New York: John Wiley & Sons.

van Oost, B. A., B. Veldhayzen, A. P. M. Timmermans, and J. J. Sixma. 1983. "Increased urinary β-thromoglobulin excretion in diabetes assayed with a modified RIA kit-technique," *Thrombosis and Haemostasis*, **9**:18-20.

Vapnik, V. N. 1998. *Statistical Learning Theory*, New York: John Wiley & Sons.

Vapnik, V. N. 1999. "An overview of statistical learning theory," *IEEE Transactions on Neural Networks*, **10**:988-999.

Venables, W. N. and B. D. Ripley. 1994. *Modern Applied Statistics with S-Plus*, New York: Springer-Verlag.

Vezhnevets, A. and V. Vezhnevets. 2005. "Modest AdaBoost - teaching AdaBoost to generalize better," *Graphicon 2005*.

Wadsworth, H. M. (ed.). 1990. *Handbook of Statistical Methods for Engineers and Scientists*, New York: McGraw-Hill.

Wahba, Grace. 1990. *Spline Functions for Observational Data*, CBMS-NSF Regional Conference series, Philadelphia: SIAM.

Wainer, H. 1997. *Visual Revelations: Graphical Tales of Fate and Deception from Napoleon Bonaparte to Ross Perot*, New York: Copernicus/Springer-Verlag.

Wainer, H. 2004. *Graphic Discovery: A Trout in the Milk and Other Visual Adventures*, Princeton, NJ: Princeton University Press.

Walpole, R. E. and R. H. Myers. 1985. *Probability and Statistics for Engineers and Scientists*, New York: Macmillan Publishing Company.

Wand, M. P. and M. C. Jones. 1995. *Kernel Smoothing*, London: Chapman and Hall.

Watson, G. S. 1964. "Smooth regression analysis," *Sankhya Series A*, **26**:101-116.

Webb, Andrew. 2011. *Statistical Pattern Recognition, Third Edition*, Oxford: Oxford University Press.

Wegman, E. 1986. *Hyperdimensional Data Analysis Using Parallel Coordinates*, Technical Report No. 1, George Mason University Center for Computational Statistics.

Wegman, E. 1988. "Computational statistics: A new agenda for statistical theory and practice," *Journal of the Washington Academy of Sciences*, **78**:310-322.

Wegman, E. 1990. "Hyperdimensional data analysis using parallel coordinates," *Journal of the American Statistical Association*, **85**:664-675.

Wegman, E. and J. Shen. 1993. "Three-dimensional Andrews plots and the grand tour," *Proceedings of the 25th Symposium on the Interface*, pp. 284-288.

Wegman, E., D. Carr, and Q. Luo. 1993. "Visualizing multivariate data," in *Multivariate Analysis: Future Directions*, C. R. Rao, ed., The Netherlands: Elsevier Science Publishers, pp. 423-466.

Weiss, Neil. 1999. *Introductory Statistics*, New York: Addison Wesley Longman.

Wilcox, Rand R. 1997. *Introduction to Robust Estimation and Hypothesis Testing*, New York: Academic Press.

Wilk, M. and R. Gnanadesikan. 1968. "Probability plotting methods for the analysis of data," *Biometrika*, **55**:1-17.

Wilkinson, Leland. 1999. *The Grammar of Graphics*, New York: Springer-Verlag.

Wilkinson, G. N. and C. E. Rogers. 1973. "Symbolic description of factorial models for analysis of variance," *Journal of the Royal Statistical Society, Series C (Applied Statistics)*, **22**:392-399.

Wold, Herman. 1966. "Estimation of principal components and related models by iterative least squares," in *Multivariate Analysis*, P. R. Krishnaiah, ed., New York: Academic Press, pp. 391-420.

Wold, Herman. 1975. "Path models with latent variables: The NIPALS approach," in *Quantitative Sociology: International Perspectives on Mathematical and Statistical Model Building*, H. B. Blalock et al., ed., New York: Academic Press, pp. 307-357.

Wolfe, J. H. 1970. "Pattern clustering by multivariate mixture analysis," *Multivariate Behavioral Research*, **5**:329-350.

Young, Forrest W., Pedro M. Valero-Mora, and Michael Friendly. 2006. *Visual Statistics: Seeing Data with Interactive Graphics*, Hoboken, NJ: Wiley & Sons.

Young, G. and A. S. Householder. 1938. "Discussion of a set of points in terms of their mutual distances," *Psychometrika*, **3**:19-22.

Zambon, M., R. Lawrence, A. Bunn, and S. Powell. 2006. "Effect of alternative splitting rules on image processing using classification tree analysis," *Photogrammetric Engineering & Remote Sensing*, **72**:25-30.

Subject Index

A

Acceleration parameter, bootstrap, 287
Acceptance-rejection method
  continuous, 90
  discrete, 92
AdaBoost, 413, 434
adabstdemo, 415
Adaptive mixtures density estimation, 339
Additive model, 593
Adjacent values, 139
adjrand, 475
Agglomerative clustering, 444
Agglomerative clustering, model-based, 460
agmbclust, 461
Alternative hypothesis, 231
AMISE, 303, 307, 320
Andrews curves, 167, 173, 175
andrewsplot, 171
Anscombe residual, 513
Aperiodicity, 619
Arcing classifiers, 416
Arc-x4, 418
area, 670
ARESLab, 603
Arithmetic operators, 667
Arrays, 659
Autoregressive generating density, 627
Average linkage, 445
Averaged shifted histogram, 312
Axes, 672
axis, 679

B

Backfitting algorithm, 594

Backslash operator, 483
Bagging, 410
Bagplot, 179
Band matrix, 566
bar, 305, 670
bar3, 152, 670
Basic Fitting, 485, 538
Basis functions, 600
Bayes decision rule, 367, 370
Bayes decision theory, 361
Bayes' theorem, 19, 362, 367, 614
Bayesian inference, 614
Bayesian information criterion, 463
Bernoulli random variable, 24
beta, 40
Beta distribution, 40
  random number generation, 99
betafit, 352
betainc, 41
betapdf, 41, 101
betarnd, 100, 633
Between-class scatter matrix, 474
Bias, 66
BIC, 463
Bin smoother, 545
Bin width, 300
binocdf, 26
Binomial, 499
Binomial distribution, 24
  random number generation, 106
Binomialness plot, 132, 136
binopdf, 26, 634
binornd, 633
Bisquare method, 556
Blind source separation, 202
Boosting by sampling, 413
Boosting by weighting, 413
bootci, 265, 292
bootfun, 264
Bootstrap, 252, 277
  bias, 257
  bias corrected interval, 286
  confidence interval, 258
  jackknife-after-bootstrap, 289
  parametric bootstrap, 254
  percentile interval, 262, 286
  replications, 253

Printed in the United States
by Baker & Taylor Publisher Services